EVOLUTION OF THE HUMAN DI

HUMAN EVOLUTION SERIES

African Biogeography, Climate Change, and Human Evolution
edited by Timothy G. Bromage and Friedemann Schrenk

Meat-Eating and Human Evolution
edited by Craig B. Stanford and Henry T. Bunn

The Skull of Australopithecus afarensis
William H. Kimbel, Yoel Rak, and Donald C. Johanson

Early Modern Human Evolution in Central Europe: The People of Dolní Věstonice and Pavlov
edited by Erik Trinkaus and Jiří Svoboda

Evolution of the Human Diet: The Known, the Unknown, and the Unknowable
edited by Peter S. Ungar

EVOLUTION OF THE HUMAN DIET

The Known, the Unknown, and the Unknowable

EDITED BY
Peter S. Ungar

2007

OXFORD
UNIVERSITY PRESS

Oxford University Press, Inc., publishes works that further
Oxford University's objective of excellence
in research, scholarship, and education.

Oxford New York
Auckland Cape Town Dar es Salaam Hong Kong Karachi
Kuala Lumpur Madrid Melbourne Mexico City Nairobi
New Delhi Shanghai Taipei Toronto

With offices in
Argentina Austria Brazil Chile Czech Republic France Greece
Guatemala Hungary Italy Japan Poland Portugal Singapore
South Korea Switzerland Thailand Turkey Ukraine Vietnam

Published by Oxford University Press, Inc.
198 Madison Avenue, New York, New York 10016

www.oup.com

Oxford is a registered trademark of Oxford University Press

Library of Congress Cataloging-in-Publication Data
Evolution of the human diet : the known, the unknown, and the unknowable / edited by Peter S. Ungar.
p. cm. — (Human evolution series)
Includes bibliographical references.
ISBN-13 978-0-19-518346-7; 978-0-19-518347-4 (pbk.)
ISBN 0-19-518346-0; 0-19-518347-9 (pbk.)
1. Prehistoric peoples—Food. 2. Fossil hominids. 3. Human remains (Archaeology) 4. Dental anthropology.
5. Teeth, Fossil. 6. Diet—History. I. Ungar, Peter S. II. Series.
GN282.E84 2006 930.1—dc22 2005036120

9 8 7 6 5 4 3 2 1

Printed in the United States of America
on acid-free paper

To my mentors

Preface

Ralph Gomory, President of the Alfred P. Sloan Foundation, wrote in *Scientific American* (272:120), "we are all taught what is known, but we rarely learn about what is not known, and we almost never learn about the unknowable. That bias can lead to misconceptions about the world around us." The Sloan Foundation has a program called The Known, the Unknown and the Unknowable (KUU). This program funds conferences that bring together producers and consumers of knowledge to explore the limits of that knowledge. I thought this was a pretty neat idea and that the evolution of human diet might be a good candidate for a small workshop. The Sloan Foundation agreed, so thirteen of us gathered on the campus of the University of Arkansas at Fayetteville in August 2003. A report of that workshop can be found in *Evolutionary Anthropology* (13:45–46).

The idea for this volume stemmed from that KUU workshop. Many researchers with a wide range of academic backgrounds are interested in the evolution of hominin diets. Many of these workers come from disparate fields, belong to different academic organizations, and publish in different journals. All of these researchers nevertheless contribute (and face the same limits) to our knowledge of early hominin diets. This volume expands on the initial dialogue that began in Fayetteville in 2003. The chapters present what is known today about Plio-Pleistocene hominin diets from a diverse constellation of perspectives. Many also offer an assessment of the state of the science, the limits to our knowledge, and prospects and possible directions for future research.

This volume has at its core four main sections: (1) dietary reconstructions based on the hominin fossil themselves (tooth size, shape, structure, wear, and chemistry, as well as mandibular biomechanics); (2) the archaeological evidence of subsistence (stone tools and modified bones); (3) models of early hominin diets stemming from analogy with living humans and nonhuman primates, paleoecology, and energetics; and (4) nutritional analyses and their implications for evolutionary medicine. Introductory chapters present historical perspectives and backgrounds on the hominins, and a summation chapter appears at the end.

The works that follow demonstrate that knowledge is expanding into new frontiers as new approaches and theories produce new insights, and more sophisticated models help us refine our questions. We nudge the boundaries of what is known and knowable as we begin to colligate the disparate lines of evidence.

Acknowledgments

I thank all of the contributors who so generously gave of themselves for this volume. All chapters were peer-reviewed, and I am grateful to each referee for insightful appraisals and constructive, helpful suggestions. I also thank Jessica Scott for her tireless assistance in assembling the individual chapters. The staff of the University of Arkansas, especially Sarah Taylor, Allison Hogge, Melissa Blouin, Carl Hitt, and Mike Miller, were very helpful and supportive during the Alfred P. Sloan–sponsored symposium that led to the idea for this volume. Finally, I am most grateful to my wife, Diane, and my daughters, Rachel and Maya, for their encouragement, support, and tolerance above and beyond the call of duty.

Peter S. Ungar

Contents

Contributors

Robert J. Blumenschine, Ph.D.
Center for Human Evolutionary Studies
Department of Anthropology
Rutgers University
New Brunswick, NJ 08901-1414, U.S.A.

Henry T. Bunn, Ph.D.
Department of Anthropology
University of Wisconsin
Madison, WI 53706, U.S.A.

Loren Cordain, Ph.D.
Department of Health and Exercise Science
Colorado State University
Fort Collins, CO 80523, U.S.A.

David J. Daegling, Ph.D.
Department of Anthropology
University of Florida
Gainesville, FL 32611, U.S.A.

S. Boyd Eaton, M.D.
Departments of Anthropology and Radiology
Emory University
Atlanta, GA 30322, U.S.A.

Frederick E. Grine, Ph.D.
Departments of Anthropology and Anatomical Sciences
Stony Brook University
Stony Brook, NY 11794-4364, U.S.A.

Amanda G. Henry
Center for the Advanced Study of Hominid Paleobiology and
Hominid Paleobiology Doctoral Program
George Washington University
Washington, DC 20052, U.S.A.

Joanna E. Lambert, Ph.D.
Department of Anthropology
University of Wisconsin
Madison, WI 53706, U.S.A.

Julia Lee-Thorp, Ph.D.
Department of Archaeological Sciences
University of Bradford
Bradford BD7 1DP, U.K.

William R. Leonard, Ph.D.
Department of Anthropology
Northwestern University
Evanston, IL 60208-1330, U.S.A.

Peter W. Lucas, Ph.D.
Department of Anthropology
George Washington University
Washington, DC 20052, U.S.A.

Charles R. Peters, Ph.D.
Department of Anthropology and Institute of Ecology
Baldwin Hall
University of Georgia
Athens, GA 30606, U.S.A.

Briana L. Pobiner
Departments of Anthropology
Rutgers University
New Brunswick, NJ 08901-1414, U.S.A.
and Smithsonian Institution
Washington, DC 20013-7012, U.S.A.

Amy L. Rector
Institute of Human Origins
School of Human Evolution and Social Change
Arizona State University
Tempe, AZ 85281, U.S.A.

Kaye E. Reed, Ph.D.
Institute of Human Origins
School of Human Evolution and Social Change
Arizona State University
Tempe, AZ 85281, U.S.A.

Marcia L. Robertson, Ph.D.
Department of Anthropology
Northwestern University
Evanston, IL 60208-1330, U.S.A.

Darryl de Ruiter, Ph.D.
Department of Anthropology
Texas A&M University
College Station, TX 77843, U.S.A.

Margaret J. Schoeninger, Ph.D.
Department of Anthropology
University of California at San Diego
La Jolla, CA 92092-0532, U.S.A.

Jeanne Sept, Ph.D.
Department of Anthropology
Indiana University
Bloomington IN 47405, U.S.A.

John J. Shea, Ph.D.
Department of Anthropology
Stony Brook University
Stony Brook, NY 11794-4364, U.S.A.

J. Josh Snodgrass, Ph.D.
Department of Anthropology
University of Oregon
Eugene, OR 97403

Matt Sponheimer, Ph.D.
Department of Anthropology
University of Colorado at Boulder
Boulder, CO 80309, U.S.A.

Mark F. Teaford, Ph.D.
Center for Functional Anatomy and Evolution
Johns Hopkins University School of Medicine
Baltimore, MD 21205, U.S.A.

Peter S. Ungar, Ph.D.
Department of Anthropology
University of Arkansas
Fayetteville, AR 72701, U.S.A.

Alan Walker, Ph.D.
Departments of Anthropology and Biology
Pennsylvania State University
University Park, PA 16802-5301 U.S.A.

Bernard Wood, Ph.D.
Center for the Advanced Study of Hominid Paleobiology
Department of Anthropology
George Washington University
Washington, DC 20052, U.S.A.

Richard Wrangham, Ph.D.
Department of Anthropology
Harvard University
Cambridge, MA 02138, U.S.A.

PART I

INTRODUCTION

1

Early Hominin Diets

Overview and Historical Perspectives

ALAN WALKER

There are known knowns. There are things we know we know. We also know there are known unknowns. That is to say, we know there are some things we do not know. But there are also unknown unknowns, the ones we don't know we don't know.
—Donald Rumsfeld, United States Secretary of Defense, February 12, 2002

Being a paleontologist rather than a historian, I decided to introduce this field and give some background about what had been accomplished over the last quarter century. This should show whether analytical techniques that were established by then or had been suggested then have improved or have been superseded, and it would also show what new techniques have been developed. I will also have an opportunity to suggest new avenues of research. I chose twenty-five years as that much time has elapsed since I presented at a Royal Society meeting in London an overview of the methods then available for deciphering the dietary habits of early hominins (Walker, 1981).

We are interested in the evolution of hominin diets for several reasons. One reason is our fundamental concerns over our own present-day eating habits and the consequences of societal choices, such as unhealthy obesity in some cultures and equally unhealthy starvation in others. Another is that humans have invented many ways of feeding in extremely varied environments, and these adaptations, which are different in important ways from our closest biological relatives, must have historical roots of varying depths. The third reason—why most paleoanthropologists are interested in this question—is that a species' trophic level and feeding adaptations impose constraints on variables such as body size, locomotion, life history strategies, geographic range, habitat choice, and social behavior.

The Knowns and the Unknowns

Of course, the level of resolution we can expect to recover from the past varies according to the questions asked and the methods used. As a paleontologist, I am accustomed to low-resolution answers about feeding habits. I cannot expect to be able to reconstruct a list of food items by their species' names, as a primatologist can for instance, nor tell much about the variation in diets that we can observe in living species. Archaeologists can expect far better resolution by finding food refuse or even documenting foods available, as depicted on ancient murals, for instance.

Twenty-five years ago, I listed the following methods that could be applied to the fossil record of extinct hominins. I shall briefly examine what progress has been made in each:

1. Interspecific comparisons of tooth morphology.
2. Biomechanical reconstructions.
3. Inspection of tooth microwear.
4. Carbon isotope analysis.
5. Trace element analysis.
6. Application of ecological "rules."
7. Analysis of "food refuse" from archeological sites.
8. Diagnosis of cases of metabolic diseases caused by diet.

Functional Tooth Morphology

Although functional studies of tooth morphology are among the oldest, considerable progress has been made. This mostly has been brought about in two ways—by more quantification (Ungar, chapter 4, this volume) and by more theoretical considerations of the material properties of various foods (Lucas, chapter 3). Lucas (2004) gives a fuller account of the last approach.

Biomechanics of Mastication

Progress in understanding the biomechanics of mastication in anthropoid primates has been made through many physiological studies, although the work of Hylander and his colleagues in particular must be highlighted (see chapter 6). Strain gage and cine x-ray studies on living and freshly dead jaws and faces, together with finite element modeling, has improved our knowledge of the forces applied during chewing and biting and how the skull handles those functions. But as Daegling and Grine (chapter 6) stress, we are still far from having a complete understanding of the bony responses to masticatory forces and how those are associated with different diets. It follows that trying to make inferences from extinct anthropoid skulls is even more demanding.

Tooth Microwear

Progress in tooth microwear studies has been patchy. We have shown clearly that extremes of diet can often be easily distinguished: leaf eating versus hard-fruit eating, bone crunching versus just meat eating, grazing versus browsing, for instance. But quantification of more subtle differences has proved difficult. This is due to the purely historical accident of using the wrong high-resolution microscope (the scanning electron microscope, or SEM). This machine can produce seductively three-dimensional-looking images. But this is not sufficient for accurately counting wear features, which often have unclear boundaries and are superimposed on other features. Grine, Ungar, and Teaford (2002) have shown that intra- and interobserver errors in counting SEM features reduce statistical power substantially. Substantial advances have been made in which white-light confocal microscopy and completely automated software that measures wear surface texture with no human intervention are used (Ungar et al., 2003; Scott et al., 2005). This development will enable us to distinguish subtly different wear surfaces of both extant and extinct species. Teaford (chapter 7) summarizes the state of this field.

Isotope Studies

Twenty-five years ago, only carbon isotopes were being looked at for their dietary signal, but now both nitrogen and oxygen isotopes can give us information about possible diets, climate, and local vegetation (Sponheimer et al., chapter 8; Schoeninger, chapter 9). In my original article, I cautioned that changes due to fossilization might prove difficult to overcome, but happily this has not been a limitation in many cases, although the work involved is not trivial (see especially chapter 9).

Trace Element Analysis

Trace element analysis has been used to study the South African hominins (see, e.g., Sillen et al., 1995). This work, like that involving stable isotopes, needs substantial baseline research on modern ecosystems to understand the results, but it seems that a strong dietary signal can be recovered from fossils. Lee-Thorp (2002) has suggested that the relationships between strontium-calcium ratios and oxygen isotopes might prove very interesting and powerful for elucidating aspects of diet in extinct species. She also points out that developments in miniaturization of analytical sampling techniques will help resolve issues of seasonality. The ability to study smaller samples might persuade curators to allow more fossils to be sampled.

Ecological "Rules"

The first "application of ecological rules" I considered in 1980 was the Jarman-Bell principle in which large-bodied animals are unable to subsist on energetically rich, but rare, foods. But since then several groups of researchers have gone much further. Some have tried to reconstruct entire paleoecosystems (see Reed and Rector, chapter 14; Sept, chapter 15). These reconstructions depend, of course, on many disciplines, including actualistic studies of species distribution in similar habitats (see Peters, chapter 13). Others have modeled hominin behavior based on the ecology of living apes (see Lambert, chapter 17). Altogether such reconstructions provide constraining ecological bounds within which early hominins must have operated. Others have looked at energy expenditure in various ways. Here the comparative costs of basal metabolism, reproduction, locomotion, maintenance, and other activities are examined for clues to caloric demands. Leonard and colleagues (chapter 18) look at two energy costs, those for locomotion and those for growing and maintaining a large brain. Shipman and Walker (1989) have indicated that large herbivorous mammals spend much more time feeding and moving to feed than carnivorous ones. Because most agree that early members of the genus *Homo* became much more carnivorous than any other extant hominoid, that change would not only supply more energy-rich food but could also mean a reduction in locomotion costs.

Archeological Site Refuse

Archaeological site refuse includes the remains of plants and animals at sites where early hominin activities took place. Identifying these sites has been a controversial matter that has been clarified by taphonomic studies exemplified by, for example, Brain (1993). Scanning electron microscopy studies have revealed differences between carnivore tooth marks and stone tool cut marks, which helps identify bone collection agents at sites. In this volume, the contributions of Blumenschine and Pobiner (chapter 10) and of Bunn (chapter 11) deal with the zooarchaeological evidence for early hominin diets. I did not consider stone tool wear analysis in 1980. If Keeley's experimental work had been published at the time of the Royal Society meeting in 1980, I was not aware of it, although I ought to have known of slightly earlier attempts (Keeley and Newcomer, 1977). Keeley and Toth (1981) published microwear studies on stone tools from East Africa, but I am, frankly, not surprised at the lack of progress in this field. Unlike teeth that are made of the same mineral, stone tools are made of a variety of rocks with different mineral compositions, so the lack of fine-grained siliceous rocks such as flint in the early Paleolithic makes microwear studies very difficult (N. Toth, personal communication). I also have yet to be convinced by claims that phytoliths found on the surface of teeth or stone tools (e.g., Ciochon, Piperno, and Thompson, 1990; Dominguez-Rodrigo et al., 2001) necessarily have any functional relationship with the tooth or tool.

Metabolic Disease

Cases of metabolic diseases caused by diet are rare and often controversial. I included the possibility in my original methods list because of the case of KNM-ER 1808, an early African *Homo erectus* that is a possible case of hypervitaminosis A (Walker, Zimmerman, and Leakey, 1982). Because such cases are rare, there has been little progress in this field.

Tooth Structure

The study of the detailed structure of teeth and its functional significance was in its infancy in 1980, but considerable progress has been made in understanding mammalian enamel and its resistance to wear and crack formation. This has involved detailed knowledge of the prism patterns in three dimensions. But progress on hominins has been relatively limited, in part because of small samples available for destructive study and in part because there are not very major differences between taxa (Teaford, chapter 5).

New Ways of Determining Diet in Extinct Species

Coprolites

One method that I did not consider in 1980 was the study of coprolites. Boaz (1977) made SEM studies of coprolites from the Plio-Pleistocene of Omo, Ethiopia, and made claims about hominin diet, but assigning coprolites to species is difficult, if not impossible, unless there are diagnostic criteria such as those of crocodiles deposited in water-lain sediments. As it happens, hyenas usually produce feces of sufficient hardness to escape the dung beetles' notice and to fossilize, and it is likely that Boaz was studying their coprolites. In any case, this seems not to be a fruitful endeavor.

Parasite Relationships

Work by Hoberg and colleagues (2001) on the phylogenetic relationships of mammalian tapeworms has revealed that humans did not get their worms from domesticated stock. The study showed that the three human tapeworms had sister taxa to those parasitizing African carnivores—hunting dogs, hyenas, and lions. Early hominins were infected by eating the same animals as carnivores. The cladogram of Hoberg and colleagues is not well resolved, so given the large errors of molecular clock dates, their estimate of when this infection first happened is not very soundly based. The estimate of the divergence between the two main types of parasite is between 780,000 years and 1.71 Ma—during the time of *Homo erectus*. Further work on tightening this estimate as well as looking at the relationships among other gut parasites is clearly warranted.

Microbial Ecology

Another growing area of research is the microbial ecology of guts. Gut microbes act in digestion in various ways, for instance, by producing enzymes that break down otherwise indigestible vegetable fibers. Through coevolution with our own microbes our guts have many times more enzymes than they would on their own (Eckburg et al., 2005). But it has recently been demonstrated that some microbes can manipulate their hosts' genes, so dramatic changes in diet can cause dramatic changes in gut flora (Ley et al., 2005) and vice versa, so comparative studies of the gut floras of living apes and of humans with different diets could be especially illuminating.

Comparative Genomics

Comparative genomics and genetics of digestion will soon become an enlightening research field. Adult lactase persistence segregates as a dominant trait in families, and the prevalence of persistence is correlated with a long history of milk consumption after weaning (Weiss, 2005). Populations that have practiced this have only had domesticated stock for a few thousand years, so it follows that the genes that remain activated for lactase persistence after weaning have reached fixation in a very short time. This may well be true of other genes involved in digestion. For instance, humans have five copies of an amylase gene, three of which are expressed in saliva. Amylase is needed to break down amylin, the otherwise indigestible envelope of starch grains. Samuelson and colleagues (1990) in their discussion of amylase gene evolution show that a triplication of the salivary gene in the human lineage occurred after the last common ancestor of chimps and humans and suggest that this triplication took place around 1 million years ago. But this might also be a case of a relatively recent event contingent on the domestication of cereals or tubers and the ingestion of greater amounts of starch. Now that a draft chimpanzee genome has been completed (The Chimpanzee Sequencing and Analysis Consortium, 2005) we can expect an examination of the status of all genes related to digestion in both genera. When the macaque genome is completed shortly, we will also have an outgroup other than the mouse with which to test phylogenetic hypotheses of gene evolution. Care must be taken with conclusions from genomic studies because single genes can have several functions. For instance, the amylase gene is also involved in bacterial resistance (H. A. Larson, personal communication).

Cooking food, especially starchy food, as an innovation has, perhaps more than any other in human history, enabled people to extend their ranges into habitats that were impossible to live in before (see Lucas, chapter 3). Wrangham and colleagues (1999) proposed that cooking was being used as long ago as the first *Homo erectus* (about 2 million years ago) on the basis of hominin anatomy and a theoretical model of their social system. Now all human societies cook, and it is difficult to imagine people not doing it (see Wrangham, chapter 16). But there had to have been a transition from non-cooking to cooking and the timing of that transition is what we want to know. Wrangham and colleagues propose that early hominin females were cooking tubers regularly. A quick back-of-the-envelope calculation shows that if there were only 200 female hominins making a fire a week in the Koobi Fora area for a period of 2 millions years (the approximate accumulation time of the sediments there), the number of fires would have been in the billions. Yet there are only a very few signs of fires in the Koobi Fora sediments and whether these were natural or hominin made has been strongly debated. So for the moment, when cooking originated is the same question as when humans first used fire, and that question has also been strongly debated.

The Pull of the Present and Consilience of Evidence

Both paleontologists and historians know of the danger of the pull of the present (presentism), whereby the past can be distorted by our knowledge and immersion in present-day circumstances. It is extremely important not to extend modern human behaviors far into the past without good evidence. Several ideas and cases in this book suggest that the pull of the present is particularly acute. The use of modern hunter-gatherers as analogs is one such case, where these people are now mostly found in extreme environments. They mostly have not been able to extend into more favorable ecosystems occupied by other people, and they cope with ecological harshness by sophisticated cultural adaptations. Another is cooking, a cultural behavior of all humans, and yet another modern human physiology. We are beginning to understand how rapidly our gut flora can evolve in response to diet and how our own genes respond to dietary shifts such as animal domestication. We should be cautious, then, in assuming that our genome is "adapted for a Paleolithic lifestyle" (Cordain, chapter 19; Eaton, chapter 20), for not only was the Paleolithic lifestyle most probably very variable in time and place, but we can easily underestimate the power and speed of natural selection, especially in numerous and rapidly reproducing gut microbes.

As well as what can be known about the past, there is also the question of how certain we are about it. Clearly, the more lines of evidence that converge on a single answer the better, because different sorts of evidence have different sources of error. We can be reassured when all evidence points to the same conclusion, but if there is not consilience, then it can be difficult to decide why. A recent case concerning the trophic adaptations of an extinct primate exemplifies this. Godfrey, Semprebon, Schwartz, and colleagues (2005) attempted to find the dietary adaptations of *Hadropithecus stenognathus* by pulling together evidence from its locomotion, carbon isotopic signal, tooth wear, enamel thickness, and environmental context. These indicated that this species was a predominantly terrestrial species, with an isotopic signal typical of eating C4 grasses, teeth that did not have thick decussated enamel but rather mimicked those of the grazing primate *Theropithecus gelada*, and living in an environment with few grazers. But these authors found that the tooth wear signal did not match those of the grazing mammals in their comparative sample. They concluded that *Hadropithecus* was not a specialized grazer as had been previously hypothesized. But the only piece of evidence not in favor with this hypothesis was the tooth wear results. The credibility of their tooth wear methods has been lessened because the authors could not distinguish between dentin exposures that would have occurred regardless of the foods eaten and features caused by tooth-food interactions (Godfrey, Semprebon, Jungers, et al., 2005; Sembrebon et al., 2005). The only evidence not in concordance is the tooth wear study, and clearly that needs to be taken with a great deal of skepticism.

The Unknowable

And finally to Rumsfeld's unknown unknowns: Some mammals engage in feeding practices that would be difficult to imagine if they had not been observed. An example of this is grizzly bears of Yellowstone National Park feeding on cutworm moth aggregations. These very large mammals seek out and consume vast numbers of moths that collect on high alpine talus slopes (White, Kendal, and Picton, 1998). If early hominins had seasonal feeding of such an unusual kind that would be difficult to imagine, like the grizzlies and moths, we would not even know how to discover it. Fortunately, such things are rare.

References

Boaz, N.T., 1977. *Paleoecology of Early Hominidae in Africa*. Kroeber Anthropological Society Papers. University of California, Berkeley, 50, 37–62.

Brain, C.K. (Ed.), 1993. Swartkrans: A cave's chronicle of early man. Transvaal Museum Monographs 8, Pretoria.

The Chimpanzee Sequencing and Analysis Consortium, 2005. Initial sequence of the chimpanzee genome and comparison with the human genome. *Nature* 437, 69–87.

Ciochon, R.L., Piperno, D.R., and Thompson, R.G., 1990. Opal phytoliths found on the teeth of the extinct ape *Gigantopithecus blacki*: Implications for paleodietary studies. *Proc. Natl. Acad. Sci. USA* 87, 8120–8124.

Dominguez-Rodrigo, M., Serrallonga, J., Juan-Tresserras, J., Alcala, L., and Luque, L., 2001. Woodworking activities by early humans: A plant residue analysis on Acheulian stone tools from Peninj (Tanzania). *J. Hum. Evol.* 40, 289–299.

Eckburg, P.B., Bik, E.M., Bernstein, C.N., Purdom, E., Dethlefsen, L., Sargent, M., Gill, S.R., Nelson, K.E., and Relman, D.A., 2005. Diversity of the human intestinal microbial flora. *Science* 308, 1635–1638.

Godfrey, L.R., Semprebon, G.M., Jungers, W.L., Sutherland, M.R., Simons, E.L., and Solounias, N., 2005. Erratum. Dental use wear in extinct lemurs: evidence of diet and niche differentiation. *J. Hum. Evol.* 49, 662–663.

Godfrey, L.R., Semprebon, G.M., Schwartz, G.T., Burney, D.A., Jungers, W.L., Flanagan, E.K., Cuozzo, F.P., and King, S.J., 2005. New insights into old lemurs: The trophic adaptations of the Archaeolemuridae. *Int. J. Primatol.* 26, 825–854.

Grine, F.E., Ungar, P.S., and Teaford, M.F., 2002. Error rates in dental microwear quantification using scanning electron microscopy. *Scanning* 24, 144–153.

Hoberg, E.P., Alkire, N.L., de Queiroz, A., and Jones, A., 2001. Out of Africa: Origins of the Taenia tapeworms in humans. *Proc. R. Soc. Lond.* B 268, 781–787.

Keeley, L.H., and Newcomer, M.H., 1977. Microwear analysis of experimental flint tools: A test case. *J. Archaeol. Sci.* 4, 29–62.

Keeley, L.H., and Toth, N., 1981. Microwear polishes on early stone tools from Koobi Fora, Kenya. *Nature* 293, 464–465.

Lee-Thorp, J., 2002. Hominid dietary niches from proxy chemical indicators in fossils: the Swartkrans example. In: Ungar, P.S., and Teaford, M.F., (Eds.), *Human Diet.* Bergin & Garvey, Westport, CT, pp. 123–141.

Ley, R.E., Bäckhed, F., Turnbaugh, P., Lozupone, C.A., Knight, R.D., and Gordon, J.I., 2005. Obesity alters gut microbial ecology. *Proc. Natl. Acad. Sci. USA* 102, 11070–11075.

Lucas, P.W., 2004. *Dental Functional Morphology*. Cambridge University Press, Cambridge.

Samuelson, L.C., Wiebauer, K., Snow, C.M., and Meisler, M.H., 1990. Retroviral and pseudogene insertion sites reveal the lineage of human salivary and pancreatic amylase genes from a single gene during primate evolution. *Mol. Cell. Biol.* 10, 2513–2520.

Scott, R.S., Ungar, P.S., Bergstrom, T.S., Brown, C.A., Grine, F.E., and Teaford, M.F., 2005. Dental microwear texture analysis reflects diets of living primates and fossil hominins. *Nature* 436, 693–695.

Sembrebon, G.M., Godfrey, L.R., Solounias, N., Sutherland, M.R., and Jungers, W.L., 2005. Erratum. Can low-magnification stereomicroscopy reveal diet? *J. Hum. Evol.* 49, 662–663.

Shipman, P., Walker, A., 1989. The costs of becoming a predator. *J. Hum. Evol.*18, 373–392.

Sillen, A., Hall, G., Richardson, S., and Armstrong, R., 1995. Strontium calcium ratios (Sr/Ca) and strontium isotope ratios (^{87}Sr/^{86}Sr) of *Australopithecus robustus* and *Homo* sp. from Swartkrans. *J. Hum. Evol.* 28, 277–285.

Ungar, P.S., Brown, C.A, Bergstrom, T.S., and Walker, A., 2003. Quantification of dental microwear by tandem scanning confocal microscopy and scale-sensitive fractal analysis. *Scanning* 25, 185–193.

Walker, A., 1981. Dietary hypotheses and human evolution. *Philos. Trans. R. Soc.* 292, 57–64.

Walker, A., Zimmerman, M.R., and Leakey, R.E.F., 1982. A possible case of hypervitaminosis A in *Homo erectus. Nature* 296, 248–250.

Weiss, K.M., 2005. The reluctant calf. *Evol. Anthropol.* 14, 127–131.

White, D., Kendall, K.C., and Picton, H.D., 1998. Grizzly bear feeding activity at alpine army cutworm moth aggregation sites in northwest Montana. *Can. J. Zool.* 76, 221–227.

Wrangham R.W., Jones J.H., Laden G., Pilbeam D., and Conklin-Brittain N., 1999. The raw and the stolen: Cooking and the ecology of human origins. *Curr. Anthropol. 40*, 567–594.

2

Whose Diet?

An Introduction to the Hominin Fossil Record

AMANDA G. HENRY
BERNARD WOOD

This chapter introduces the hominin[1] fossil record by using two alternative hominin taxonomies, a "splitting" one and a "lumping" one, and reviews the taxa within each of six crude grades. It uses tables to provide summaries of each taxon, including the regions of the skeleton represented within each hypodigm. Finally, we focus on the fossil evidence from the relatively well-dated East African hominin fossil record that spans the time interval between 2.5 and 1.5 Myr. We indicate the quality and quantity of the fossil evidence available within each taxon for the anatomical regions (i.e., the dentition, jaws, and face) relevant to the reconstruction of diet.

Taxonomy

Within paleoanthropology, most taxon-based reconstructions of diet are made at the level of the species rather than the individual. The way individual fossils are distributed among hominin taxa will obviously affect the parameters of those taxa, and therefore they potentially influence the inferences that are drawn about the diets of the taxa. It is thus important that species-level inferences about an important functional adaptation such as diet are based on a sound taxonomy.

Eldredge (1993), building on the ideas of Ghiselin (1974) and Hull (1976, 1978), suggested that species should be regarded as "individuals" with their own "history." A species' history begins with a speciation event. Its "middle" phase lasts as long as the species persists, and its history ends when it either becomes extinct or participates in another speciation event. Within the hominin fossil record, the same species may be sampled once or several times during its history. Thus, when paleoanthropologists try to interpret the fossil record, they must decide whether they are looking at several samples belonging to the same taxon or samples of several different taxa. When making these judgments, they should strive to neither grossly underestimate nor extravagantly overestimate the actual number of species represented in the hominin fossil record.

One factor complicating these judgments is time. We observe variation within living species by using skeletal collections in museums, but these collections reflect merely a "snapshot" in the evolutionary history of these organisms. In contrast, the hypodigms of fossil hominin taxa are almost always spread over many tens of thousands of years and often several hundred thousands of years. Researchers who use comparative samples from museums as yardsticks for assessing the taxonomic significance of variation in their fossil samples must try to account for differences in the time depths of the extant and fossil samples. But exactly how this is best done is a matter of debate. Another factor paleoanthropologists must be aware of is that they have to work with a fossil record that is mostly confined to remains of the hard tissues (bones and teeth). We know from living animals that many uncontested species are difficult to distinguish using bones and teeth (e.g., *Cercopithecus* species). Thus, there are logical reasons to suspect that a hard tissue–bound fossil record is always likely to underestimate rather than overestimate the number of species.

Taxonomies are hypotheses. Researchers who place more emphasis on cladogenesis and on morphological discontinuities tend to recognize a larger number of taxa. Researchers who emphasize anagenesis and morphological continuity or who use the concept of allotaxa and thus allow a single species to manifest substantial regional geographic variation (e.g., Jolly, 2001; Antón, 2003) tend to recognize fewer taxa. We present two taxonomic schemes, one speciose (fig. 2.1), and the other less speciose (fig. 2.2). Although some researchers might contest the details of each taxonomy, these two schemes allow us to address the influence of different taxonomic hypotheses on the way we interpret the evolution of hominin diet.

Some comment and explanation of the figures are in order. First, the ages of the earliest (called the *first appearance datum*, or FAD) and latest (called the *last appearance datum*, or LAD) fossil evidence of a hominin taxon almost certainly underestimate the temporal span of that taxon, and new finds from new and existing sites may extend the range of a taxon backward and forward in time. Nonetheless, the existing FADs and LADs provide an approximate temporal sequence for the hominin taxa. Second, we have included boxes with question marks in them to remind researchers that in the early phase of the hominin fossil record we are almost certainly working with an incomplete data set. Researchers have only recently begun intensive exploration of sediments in the pre−4 Myr phase of hominin evolution, and this has been restricted in its geographical scope. For both of these reasons, it is likely we have an incomplete record of early hominin taxonomic diversity before 4 Myr. Researchers need to bear this in mind when formulating and testing hypotheses about any aspect of hominin evolution, including the evolution of diet. Third, we made a deliberate decision not to specify ancestor/descendant relationships in these diagrams. This reflects our view that, within the constraints of existing knowledge, only two relatively well-supported subclades exist within the hominin clade, one for *Paranthropus* taxa and the other for post–*Homo ergaster*

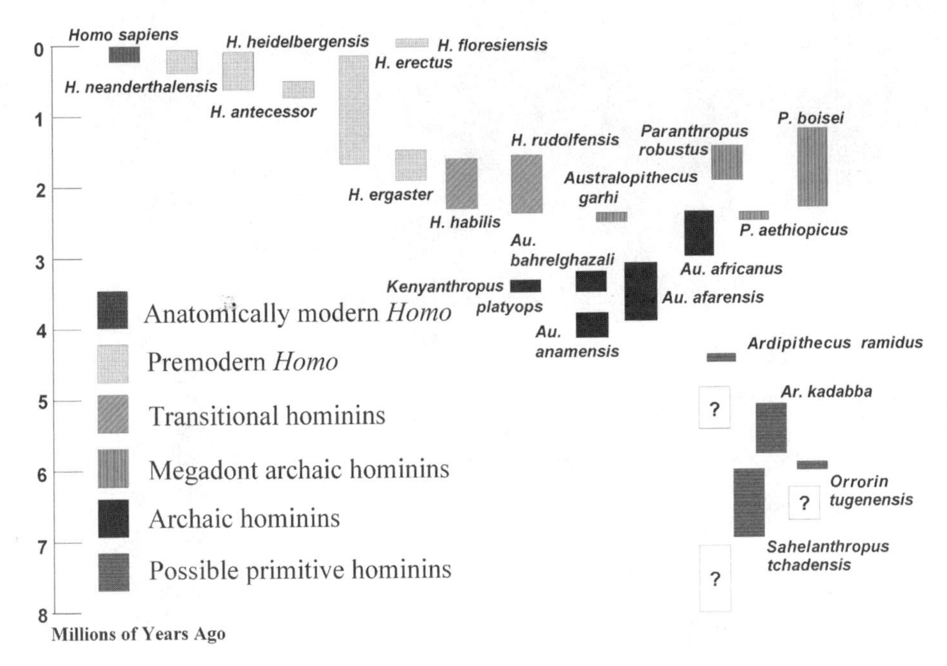

Figure 2.1 Each hominin taxon is assigned to a grade, as indicated by the shading. First appearance datum and last appearance datum are conservative estimates of the span of time each species existed. Boxes with question marks are meant to indicate the possibility that there are as-yet-unknown possible primitive hominins. See text for more details.

taxa assigned to the *Homo* clade. Without well-supported subclades, it is probably unwise to begin to try to identify specific taxa as ancestors or descendants of other taxa.

For each taxonomic scheme, we present basic information about its component taxa and indicate the anatomical regions represented in the hypodigm for each taxon (table 2.1). Each hominin taxon is placed in one of six informal grade groupings: (1) possible primitive hominins, (2) archaic hominins, (3) megadont archaic hominins, (4) transitional hominins, (5) premodern *Homo*, and (6) anatomically modern *Homo*.

Unless homoplasy (shared morphology not derived from the most recent common ancestor) is much more common than even we anticipate, there is little doubt that the taxa in the premodern *Homo* grade are more closely related to modern humans than to chimpanzees. All the taxa in the premodern *Homo* grade share medium- or large-sized brains; they have modest-sized canines, jaws, and chewing teeth; and they are all obligate bipeds. The closer we get to the split between hominins and panins, the more difficult it is to find features that we can be sure fossil hominins possessed and fossil panins did not. At the early stages in hominin evolution it may be either the lack of derived panin features or relatively subtle derived differences in the size and shape of the canines or in the detailed morphology of the limbs that distinguish primitive hominins from either primitive panins or the most

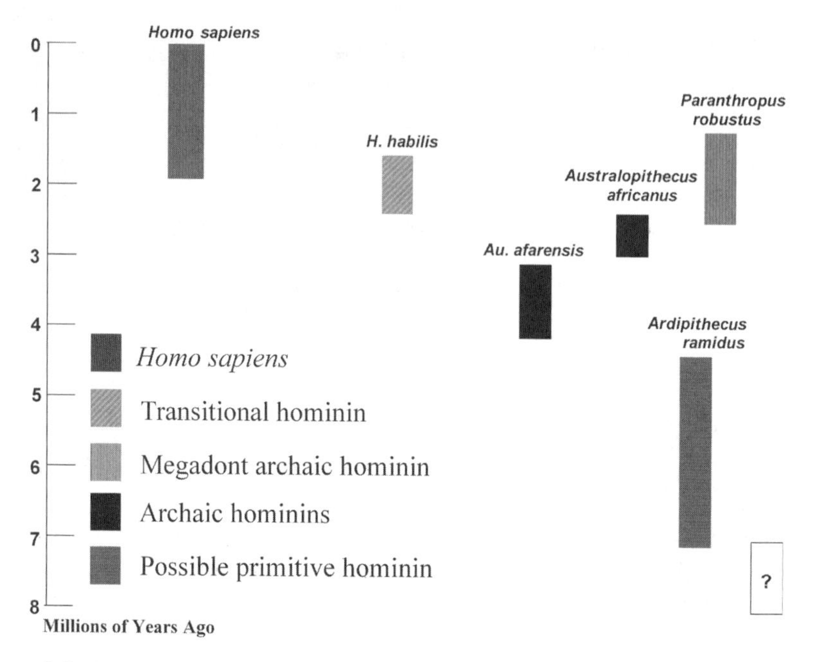

Figure 2.2 A nonspeciose taxonomy shrinks the number of species to six, each encompassing substantial temporal and geographic variation. See text for more details.

recent common ancestor of hominins and panins. It is also possible, and indeed probable, that clades other than those with extant representatives existed in the period between 8 and 4 Myr.

Some taxon names are used in different senses in the splitting and lumping taxonomies. When we refer to the hypodigm of a taxon in the splitting taxonomy, we use the taxon name followed by *sensu stricto* (e.g., *Australopithecus afarensis sensu stricto* or its abbreviation *Au. afarensis s. s.*). When we refer to the hypodigm in the lumping taxonomy (i.e., the hypodigm is larger and more inclusive), the Linnaean binomial is followed by *sensu lato* (e.g., *Au. afarensis sensu lato* or *Au. afarensis s. l.*).

Further details about most of the taxa and a more extensive bibliography can be found in Wood and Richmond (2000), Hartwig (2002), and Constantino and Wood (2004).

Possible Primitive Hominins

Researchers have put forward four species belonging to three genera as potential early hominins. In the lumping taxonomy, they are all combined in a single species. One of the main problems faced by researchers who are trying to determine whether a taxon is a primitive hominin is the relatively small amount of information we have about each of the four candidate taxa. Currently, the published information for one taxon consists of a cranium, lower jaws, and teeth but no useful limb bones. The

Table 2.1 "Splitter" and "Lumper" Taxonomies

Informal Group	"Splitter" Taxonomy	Age (Myr)	Type Specimen	Crania	Dentition	Axial	Upper Limb	Lower Limb
"Splitter" Hominin Taxonomy and Skeletal Representation								
Possible primitive hominins	*Sahelanthropus tchadensis*	7.0–6.0	TM 266-01-060-1	X	X			
	Orrorin tugenensis	6.0	BAR 1000′00		X		X	X
	Ardipithecus ramidus s. s.	5.7–4.5	ARA-VP-6/1	X	X		X	ff
	Ar. kadabba	5.8–5.2	ALA-VP-2/10		X		X	X
Archaic hominins	*Australopithecus anamensis*	4.2–3.9	KNM-KP 29281	ff	X	ff	X	X
	Au. afarensis s. s.	4.0–3.0	LH 4	X	X	X	X	X
	Kenyanthropus platyops	3.5–3.3	KNM-WT 40000	X	X			
	Au. bahrelghazali	3.5–3.0	KT 12/H1		X			
	Au. africanus	3.0–2.4	Taung 1	X	X	X	X	X
Megadont archaic hominins	*Au. garhi*	2.5	BOU-VP-12/130	X	X		?	?
	Paranthropus aethiopicus	2.5–2.3	Omo 18.18	X	X			
	P. boisei s. s.	2.3–1.3	OH 5	X	X		?	?
	P. robustus	2.0–1.5	TM 1517	X	X		X	X
Transitional hominins	*Homo habilis s. s.*	2.4–1.6	OH 7	X	X	X	X	X
	H. rudolfensis	2.4–1.6	KNM-ER 1470	X	X			?
Premodern *Homo*	*H. ergaster*	1.9–1.5	KNM-ER 992	X	X	X	X	X
	H. erectus s. s.	1.8–0.2	Trinil 2	X	X		X	X
	H. floresiensis	0.074–0.012	LB1	X	X	ff	ff	X
	H. antecessor	0.7–0.5	ATD6-5	X	X			
	H. heidelbergensis	0.6–0.1	Mauer 1	X	X		ff	X
	H. neanderthalensis	0.2–0.03	Neanderthal 1	X	X	X	X	X
Modern *Homo*	*H. sapiens s. s.*	0.16–present	None designated	X	X	X	X	X

(*continued*)

Table 2.1 *(Continued)*

Informal Group	"Lumper" Taxonomy	Age (Myr)	Taxa Included from Splitting Taxonomy
"Lumper" Hominin Taxonomy			
Probable primitive hominin	*Ar. ramidus s. l.*	7.0–4.5	*Ar. ramidus s. s., Ar. kadabba, S. tchadensis, O. tugenensis*
Archaic hominins	*Au. afarensis s. l.*	4.5–3.0	*Au. afarensis s. s., Au. anamensis, Au. bahrelghazali, K. platyops*
	Au. africanus	3.0–2.4	*Au. africanus*
Megadont archaic hominins	*P. robustus s. l.*	2.5–1.3	*P. boisei s. s., P. aethiopicus, Au. garhi, P. robustus s. s.*
Transitional hominin	*H. habilis s. l.*	2.4–1.6	*H. habilis s. s., H. rudolfensis*
Premodern and modern *Homo*	*H. sapiens s. l.*	1.9–pres	*H. erectus s. s., H. ergaster, H. floresiensis, H. antecessor, H. heidelbergensis, H. neanderthalensis, H. sapiens s. s.*

Note: This table summarizes the taxa in a more speciose, or "splitter," and a less speciose, or "lumper," taxonomy. *s. s.* = sensu stricto; *s. l.* = sensu lato. Skeletal representation key: X = present; ff = fragmentary specimens; ? = taxonomic affiliation of fossil specimen(s) uncertain.

hypodigm of another taxon consists of some teeth and proximal femora. For the third taxon, the fossil evidence is presently limited to teeth and some small hand and foot bones, and the published evidence about the fourth taxon consists of parts of several lower jaws and teeth, fragmentary postcranial bones, and two partial cranial bases.

The oldest of the four taxa is *Sahelanthropus tchadensis*, the remains of which were found at a 7- to 6-Myr-old site called Toros-Menalla in Chad, West Central Africa. Geological and paleontological evidence suggests that *S. tchadensis* lived in a complex habitat of lakes, grassy woodland, and rivers bordered by forests (Vignaud et al., 2002). The braincase of the *S. tchadensis* cranium is chimp-sized, but the upper part of its face has browridges like those seen in hominins less than half its geological age. The mandible is thicker than the jaws of living chimps, and the canines are worn down only at the tip and not on the sides as they are in chimpanzees. Researchers involved with *S. tchadensis* suggest that the browridges, the relatively orthognathic face and robust mandible, the canines that wear down only at the tip, the "intermediate" thickness of the tooth enamel (Brunet et al., 2005), and the placement of the foramen magnum (Zollikofer et al., 2005, but see Ahern, 2005) are sufficient evidence to be sure that *S. tchadensis* is a primitive hominin and not the common ancestor of chimpanzees and humans, or a member of the panin lineage, or a member of another extinct hominin clade. More evidence is needed in order to tell if their suggestion is a sound one.

The second oldest potential primitive hominin species is *Orrorin tugenensis*, the name given to circa 6-Myr-old fossils found in sediments in the Tugen Hills of northern Kenya. The evidence for *O. tugenensis* is still frustratingly fragmentary. The case for *O. tugenensis* being a hominin is based on two lines of evidence, one cranial, the other postcranial (Senut et al., 2001). Researchers claim that *O. tugenensis* has thick enamel unlike that of panins and like that found in later unambiguous members of the hominin clade. The postcranial case for including *O. tugenensis* in the tribe Hominini is based on the internal morphology of the femoral neck. In primates that are habitual climbers, the outer, or cortical, bone is equally thick all round the neck of the femur, but in habitual bipeds, the thickening is greatest at the top and bottom of the neck. Pickford and Senut (2004) claim that the cortical bone of the neck of the *O. tugenensis* femora is also preferentially thickened on the top and bottom of the neck. Critics of the view that these fossils belong to an early hominin make three points. First, they suggest that the morphology of the *O. tugenensis* femur is not much different from that of primates that spend some time moving around in trees. Second, they stress that it has not been demonstrated that within higher primates thick enamel is confined to the hominin clade. Third, as Galik and colleagues (2004) admit, much of the morphology of the teeth of *O. tugenensis* is "apelike."

The two other collections of fossils that might be from a primitive early hominin are both included in the same genus, *Ardipithecus*. The older fossil collection, dated to 5.7–5.2 Myr, is assigned to *Ardipithecus kadabba* and comes from the Middle Awash region of Ethiopia. Little of the morphology of the fossils in this collection resembles that of the archaic hominins we discuss next. The case for regarding *Ar. kadabba* as a hominin is not a strong one.

The second collection of *Ardipithecus* fossils dates from circa 4.5 Myr and comes from the Middle Awash and Gona regions of Ethiopia. It is assigned to a separate species, *Ardipithecus ramidus,* because its discoverers think that its canines are less apelike than those of *Ar. kadabba*. Several features link *Ar. ramidus* with hominins, the strongest evidence being the position of the foramen magnum. In *Ar. ramidus* this opening is further forward than in chimpanzees, though not as far forward as in modern humans. In spite of changes in the teeth and base of the skull in *Ar. ramidus* that link it with archaic hominins (discussed next), in overall appearance *Ar. ramidus* would have been much more like a chimpanzee than like a modern human.

Some researchers argue that too little is known about these taxa to allocate them to four separate species distributed across three genera. These researchers instead propose that for the time being all this material should be allocated to a single genus and possibly a single species (Haile-Selassie, Suwa, and White, 2004). If all these remains did belong to a single species, the species name with priority would be *Ar. ramidus sensu lato*.

Archaic Hominins

All of these species are more recent than any of the taxa previously discussed, and they share more of their morphology with modern humans than they do with chimpanzees. However, they do not show the changes in jaw and tooth size and in body size and shape that characterize hominin species we include within our own genus *Homo*.

Fossils from sites in Kenya and Ethiopia that date from 4.2 to 3.9 Myr have provided evidence of *Australopithecus anamensis*, a hominin that might the immediate ancestor of *Australopithecus afarensis*. The canines of *Au. anamensis* are more chimplike than those of *Au. afarensis*; yet the chewing teeth are very different from those of chimps.

The hypodigm of *Au. afarensis* consists of more than four hundred fossils, including a skull, several well-preserved crania, many lower jaws, and limb bones, that are spread unevenly over the period between circa 4 and 3 Myr. The incisor teeth of *Au. afarensis* are smaller than those of chimpanzees, but the postcanine teeth are larger. This suggests that its diet included more hard-to-chew items than does the diet of chimps. The shape and size of the pelvis and lower limb remains suggest that *Au. afarensis* was capable of walking bipedally but probably only for short distances. The 3.6-Myr-old trails of hominin footprints excavated at Laetoli, Tanzania, provide graphic evidence that a contemporary early hominin, presumably *Au. afarensis* (but some researchers dissent from this association), was capable of walking bipedally.

Three and a half million years old hominin fossils collected in 1995 in Chad not far from the site where *S. tchadensis* was found have been assigned to *Australopithecus bahrelghazali*. However, many researchers claim, probably correctly, that these remains belong to a geographical variant of *Au. afarensis*.

The last East African archaic hominin we consider, the 3.5- to 3.3-Myr-old *Kenyanthropus platyops*, was assigned to both a new species and a new genus (Leakey et al., 2001). The best-preserved specimen is a cranium, but it is deformed by many matrix-filled cracks that permeate the face and rest of the cranium. This deformation has led White (2003) and others to question the validity of this taxon. Despite the cracking there are features of the face that do not match the face of *Au.*

afarensis, the archaic hominin best known in this time period in East Africa. Meave Leakey's team is convinced their find is distinct from *Au. afarensis*, and they point to similarities between it and *Homo rudolfensis*.

At least one archaic hominin taxon is known from the cave sites of southern Africa. Our current understanding of the circa 3- to 2.4-Myr-old *Australopithecus africanus* is that its physique was more primitive than that of *Au. afarensis,* but its chewing teeth were larger and its skull not as apelike. Its average endocranial volume is a little larger than that of *Au. afarensis*, but the ranges of endocranial volumes in both species overlap considerably. The postcranial skeleton suggests that, although *Au. africanus* could walk bipedally, it was also capable of climbing in trees. A remarkably complete hominin skeleton, Stw 573 has been found deep in the Sterkfontein cave. Its discoverers claim that it is considerably older (ca. 4 Myr) than the main hypodigm of *Au. africanus*, but other researchers claim it is at the younger end of the temporal range of *Au. africanus*. It is too early to tell whether it belongs to *Au. africanus* or to a more primitive australopith taxon. Hominins resembling *Au. africanus* recovered from even deeper within the Sterkfontein cave system, in the Jacovec Cavern, may be at least 4 Myr old.

Megadont Archaic Hominins

The first megadont archaic hominin to be discovered and recognized is the 2- to 1.5-Myr-old *Paranthropus robustus* from southern African cave sites. Its chewing teeth are larger than those of *Au. africanus*, its face is broader, and its brain is slightly bigger. Some researchers think that the locomotion of *P. robustus* may have differed from that of *Au. africanus*, but there is not enough evidence to be sure of this (Robinson, 1972).

The first East African archaic megadont hominin taxon to be recognized was *Paranthropus boisei*. The features that make *P. boisei* so distinctive are found in the cranium, mandible, and the dentition. It is the only presently known hominin to combine a massive, wide, flat face with very large chewing teeth and small incisors and canines. Despite these large jaws and chewing teeth, its brain (around 450 cc) is similar in size to the brains of *Au. afarensis*. The earliest evidence of *Paranthropus* in East Africa is a variant that has a more projecting face, larger incisor alveolae, and a more apelike cranial base. Some researchers assign these pre–2.3-Myr fossils to a separate species, *Paranthropus aethiopicus*.

The latest East African megadont archaic hominin taxon to be recognized is the 2.5-Myr-old *Australopithecus garhi*. Several limb bones found in the same strata as the cranial remains assigned to *Au. garhi* show evidence of bipedal locomotion, but as they are not associated with the crania, we cannot be sure they belong to the same taxon. The chewing teeth of *Au. garhi* are as large as those of *P. boisei*. No stone tools have been found with the *Au. garhi* fossils, but animal bones found close by show telltale signs that flesh had been removed using a sharp-edged tool.

There are many resemblances between *P. aethiopicus*, *P. boisei*, and *P. robustus*. Some researchers, including the authors, interpret these as shared derived features that justify supporting the working hypothesis that these three taxa form a subclade within the hominin clade. Researchers who are "lumpers" go further and would subsume the two East African *Paranthropus* species hypodigms within a single taxon,

P. robustus sensu lato. Not enough published data are available for *Au. garhi* to tell whether it too should be subsumed into *P. robustus sensu lato*.

Transitional Hominins

It is not possible to allocate two fossil hominin taxa, *Homo habilis* and *Homo rudolfensis*, with confidence to either the archaic hominin or premodern *Homo* grades: we refer to them as "transitional hominins." They lack the more derived features (e.g., facial shape and postcanine tooth morphology) of *P. boisei*, a synchronic archaic hominin, but also lack some important derived features of *Homo* (e.g., more gracile mandibles; smaller, less complex postcanine tooth crowns and roots; and more modern-humanlike skeletal proportions).

Premodern *Homo*

All the fossil hominin taxa we have considered thus far are relatively small (ca. 60–120 lb) compared with most modern humans. Limb proportions are only known for a few individuals belonging to archaic and transitional hominin taxa, but in all cases where there is enough information to make even a rough estimate of limb length, they have relatively shorter legs than modern humans. A little less than 2 million years ago we begin to see in some of the fossils recovered from Koobi Fora in Kenya the first evidence of creatures that are more like modern humans than any archaic or transitional hominin. The formal name for this fossil evidence is *Homo ergaster*. Not all researchers use a separate species name for this material. Instead, they refer to it as belonging to "early African *Homo erectus*."

Homo ergaster is the earliest hominin with a body whose size and shape is more like that of modern humans than any of the archaic or transitional hominin taxa. In relation to the size of its body, the teeth and jaws of *H. ergaster* are smaller than those of the archaic and transitional hominins. This means *H. ergaster* either had a different diet than that of the archaic and transitional hominins or it was eating the same sorts of foods but was processing them outside instead of inside the mouth.

Currently, the earliest good fossil evidence of hominins beyond Africa comes from the site of Dmanisi in the Caucasus. There are no absolute dates for the sediments from the site, but the radioisotope age of the lava beneath the sediments and the biochronological age suggested by the fossil animals found with the hominins suggest a date of around 1.7–1.8 Myr. The hominins found there have yet to be studied in detail, but they appear to belong to a relatively primitive small-bodied and small-brained *H. ergaster*–like creature (Gabunia et al., 2000; Vekua et al., 2002).

By between 1.5 and 1 million years ago, evidence of *Homo erectus* (as opposed to *H. ergaster*) is found in Africa, China, and Indonesia. The crania of *H. erectus* are all low, with the greatest width toward the base of the cranium. There is a substantial and more or less continuous bony ridge, or torus, above the orbits; a depression, or sulcus, behind it; and a pronounced blunt ridge, or keel, of bone runs in the midline from the front to the back of the brain case: this is called a sagittal torus. At the back of the cranium, the sharply angulated occipital region has a well-defined sulcus above it. The walls of the brain case are thicker than in archaic hominins. All of these features are more obvious in the larger specimens. The volume of the cranial cavity of *H. erectus* varies from about 730 cm^3 for OH 12 (650 cm^3 if D2282 from Dmanisi is included) to about 1250 cm^3 for the Ngandong 6 (Solo V) calotte.

If the antiquity for the child's cranium for Modjokerto/Perning and the very recent date for the Ngandong remains are confirmed, then, even if *H. ergaster* from East Africa is excluded from the *H. erectus sensu stricto* hypodigm, the sets of dates from the two Indonesian sites suggest the temporal range of *H. erectus* was from circa 1.9 Myr to circa 50 Kyr. The diminutive *Homo floresiensis* may well be a dwarfed form of *H. erectus*.

By 600 Kyr we begin to see at sites in Africa, such as Bodo in Ethiopia and Kabwe in Zambia, evidence of hominins attributed to *Homo heidelbergensis* because they lack the distinctive features associated with *H. erectus*. These crania also have a braincase whose volume averages 1200 cm^3, as opposed to the means of less than 800 cm^3 and about 1000 cm^3, respectively, for *H. ergaster* and *H. erectus*. There is also a further reduction in the size of the jaws and chewing teeth. The postcranial bones lack some of the specialized features of the *H. erectus* skeleton, such as their flat shafts (front-to-back in the femur and side-to-side in the tibia), but even so the limb bones of *H. heidelbergensis* are substantially thicker and stronger, and the joint surfaces are larger than those of modern humans.

The best-known species in the "premodern *Homo*" category is *Homo neanderthalensis*. The earliest evidence of hominins that show signs of Neanderthal specializations comes from a 450- to 400-Kyr-old site in Spain called the Sima de los Huesos at Atapuerca. The distinctive features of *H. neanderthalensis* include a face that projects forward in the midline, a large nasal opening, a rounded top and back of the cranium, a cranial cavity that is on average larger than that of modern humans, and distinctive limb bones with thick shafts and large joint surfaces. The fossil evidence of Neanderthals is confined to Europe, the Near East, and western Asia.

Modern *Homo*

The earliest fossil evidence of hominins that are difficult to distinguish from modern humans (*Homo sapiens*) comes from East Africa. Compared with premodern *Homo*, these specimens have smaller faces, mandibles, and teeth but larger, more rounded, braincases. The earliest evidence was recovered some years ago from Kibish in the Omo Region in southern Ethiopia. On rather weak biochronological evidence, the Omo I cranium had been dated to circa 120 Kyr, but a recent attempt to date the Omo I cranium using isotope dating has suggested a substantially older date close to 200 Kyr (McDougall, Brown, and Fleagle, 2005). A collection of fossils from Herto, another Ethiopian site, also suggests that modern-humanlike fossil hominins were present in Africa around 150–160 Kyr.

Diet and Taxonomy

To illustrate how taxonomic hypotheses can influence the interpretation of diet, we have traced the changes in several dental and mandibular variables over the period from 2.5 to 1.5 Myr. Several nonhominin mammal lineages are known to show

Table 2.2 Preservation and Taxonomic Status of East African Hominin Mandibular Fossils between 1.5 and 2.5 Myr

Homo erectus s. s.	*H. habilis*	*Paranthropus boisei s. s.*	*Homo* sp. indet.
1 G/RT	3 G/RT	5 G/RT	1 M/LRT
1 M/RT	8 M/RT	23 M/RT	2 P/LRT
1 M/LRT	1 P/RT	11 P/RT	*? Homo*
	1 P/LRT	9 M/LRT	6 M/LRT
H. ergaster		3 P/LRT	1 P/LRT
1 G/RT	*H. rudolfensis*		
3 M/RT	3 M/RT	*P. aethiopicus*	
1 P/RT	1 P/RT	14 M/RT	
1 P/LRT		2 P/RT	

Note: Relevant fossils assigned to taxon scored for preservation. G = good preservation; M = moderate preservation; P = poor preservation; RT = reliable taxonomic status; and LRT = less reliable taxonomic status.

significant extinction and speciation events during this time interval, and some researchers link a climate shift in Africa with this turnover in species (e.g., Vrba, 1995).

Even in a less speciose taxonomy, it is clear that at least two hominin species, *P. boisei s. l.* and *H. erectus s. l.*, coexisted during this time in East Africa. In the more speciose taxonomy, the number of species in this time interval jumps to six: *P. aethiopicus*, *P. boisei s. s.*, *H. habilis*, *H. ergaster*, *H. erectus s. s.*, and *H. rudolfensis*. It is possible that the species (or two groups of species, *P. aethiopicus* and *P. boisei* vs. *H. habilis*, *H. ergaster*, *H. erectus*, and *H. rudolfensis*) differentiated by adopting different dietary strategies. We focus on mandibular and mandibular dental data because these elements are relatively well represented in the hominin fossil record, and because they are likely to give a strong signal about diet. We focused on East Africa because the regular volcanic activity gives more constrained dates for the fossils than one can get from the biochronology that has to be used in the southern African cave sites.

The mandibles and isolated teeth for which there are good published descriptions are in varying stages of preservation and are sometimes reliably assigned to a particular taxon and sometimes not. In table 2.2 we summarize the preservation and taxonomic status of the mandibular fossils used in the analysis. Specimens assigned to *Homo* but not to any particular species are included in *Homo* sp. indet. Specimens less reliably assigned to *Homo* are indicated by ? *Homo*. Note that we did not take into consideration other sources of variation, such as sex, and we included poorly preserved specimens in our analysis.

We chose to focus on three of the more commonly measured variables that relate to diet to highlight the implications of the two taxonomies. We show in table 2.3 parameters of the breadth, height, and robustness (defined as breadth/height × 100) of the mandibular corpus for each of the taxa in the two taxonomic schemes, including the number of fossils, the mean, minimum and maximum values, and the standard deviation for each variable.

The point to emphasize about these data is that, unsurprisingly, the number of fossils assigned to each taxon in the more speciose taxonomy is small. Some of the

Table 2.3 Mandibular Corpus Variables and Parameters by Taxon

Taxon	Breadth	Height	Robusticity	Taxon	Breadth	Height	Robusticity
***Homo habilis* RT**	$N=5$	$N=4$	$N=4$	***H. ergaster* RT**	$N=2$	$N=2$	$N=2$
Mean	19.0	28.2	64.1	Mean	20.0	31.5	63.5
Max	23.0	30.3	68.3	Max	20.0	32.0	64.5
Min	17.0	26.5	58.6	Min	20.0	31.0	62.5
SD	2.6	1.8	4.2	SD	0.0	0.7	1.4
***H. habilis* LRT**	$N=1$	$N=1$	$N=1$	***H. ergaster* LRT**	$N=1$	$N=1$	$N=1$
Value	21.0	30.0	70.0	Value	19.0	27.0	70.4
***H. rudolfensis* RT**	$N=6$	$N=4$	$N=4$	***Homo* sp. indet.**	$N=2$	$N=3$	$N=2$
Mean	22.2	35.3	63.7	Mean	18.8	31.7	60.6
Max	27.0	38.0	71.1	Max	19.5	33.0	62.1
Min	20.0	31.0	58.8	Min	18.0	29.0	59.1
SD	2.8	3.4	5.4	SD	1.1	2.3	2.1
***H. erectus s. s.* RT**	$N=2$	$N=2$	$N=2$	***? Homo***		$N=1$	
Mean	20.0	23.8	84.1	Value		31.0	
Max	20.5	24.4	88.4	***H. erectus s. l.***	$N=20$	$N=19$	$N=17$
Min	19.5	23.2	79.9	Mean	20.2	30.5	66.4
SD	0.7	0.8	6.0	Max	27.0	38.0	88.4
				Min	17.0	23.2	58.6
***H. erectus s. s.* LRT**	$N=1$	$N=1$	$N=1$	SD	2.3	3.9	8.0
Value	19.0	31.5	60.3				
***Paranthropus boisei s. s.* RT**	$N=31$	$N=32$	$N=31$	***P. boisei s. s.* LRT**	$N=1$	$N=1$	$N=1$
Mean	29.0	42.3	68.7	Value	18.0	30.0	60.0
Max	37.0	50.0	80.4				
Min	24.0	34.0	57.5				
SD	3.4	4.1	5.2				
***P. aethiopicus* RT**	$N=3$	$N=2$	$N=2$	***P. boisei s. l.***	$N=35$	$N=35$	$N=34$
Mean	24.6	35.3	67.4	Mean	28.3	41.5	68.3
Max	26.2	35.5	74.3	Max	37.0	50.0	80.4
Min	21.5	35.0	60.6	Min	18.0	30.0	57.5
SD	2.7	0.4	9.7	SD	4.0	4.7	5.4

Note: Abbreviations are the same as in table 2.2. The number of fossils within each taxon for which corpus height, breadth, and robusticity could be measured is indicated, and the parameters for each of these measurements (where applicable) are provided. *s. s.* = sensu stricto; *s. l.* = sensu lato.

taxa have only one mandible and sometimes only one measurement on that one mandible. In the less speciose taxonomy, that is, when we combine *P. boisei s. s.* and *P. aethiopicus* into *P. boisei s. l.* and all of the *Homo* species into *H. erectus s. l.*, all of the variables of interest are represented in both taxa. Second, the standard deviation for all of the variables for *P. boisei s. s.* is similar to that for *H. erectus s. l.*, even though *P. boisei s. s.* contains only one species, and *H. erectus s. l.* may subsume up to four species. This suggests that either *P. boisei s. s.* is a "good" species with a higher level of intraspecific variation than species within the *Homo* clade (perhaps due to sexual dimorphism or to nontrending changes through time [Silverman, Richmond,

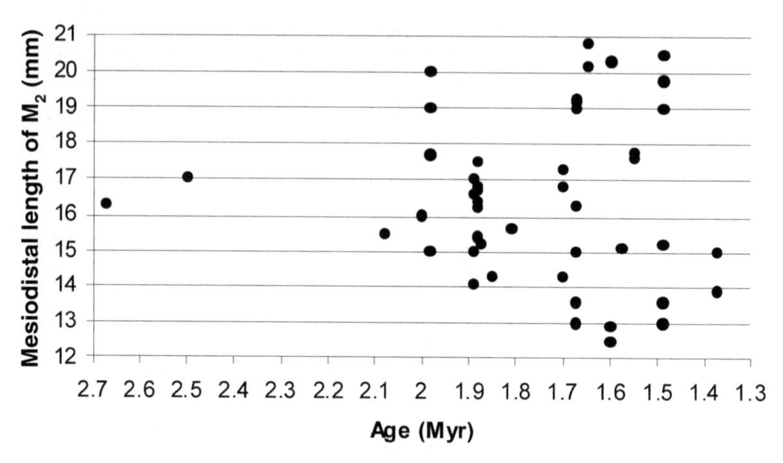

Figure 2.3 Second mandibular molar mesiodistal size variation through time. Each point on this graph represents the measured mesiodistal length of a particular fossil. Often, fossils were assigned a range of dates, so to simplify this analysis, the average date for each fossil was used. Some of the specimens have average dates that fall outside the 2.5 to 1.5 Myr range, but they were included in this analysis because at least part of their range of dates falls within this time period.

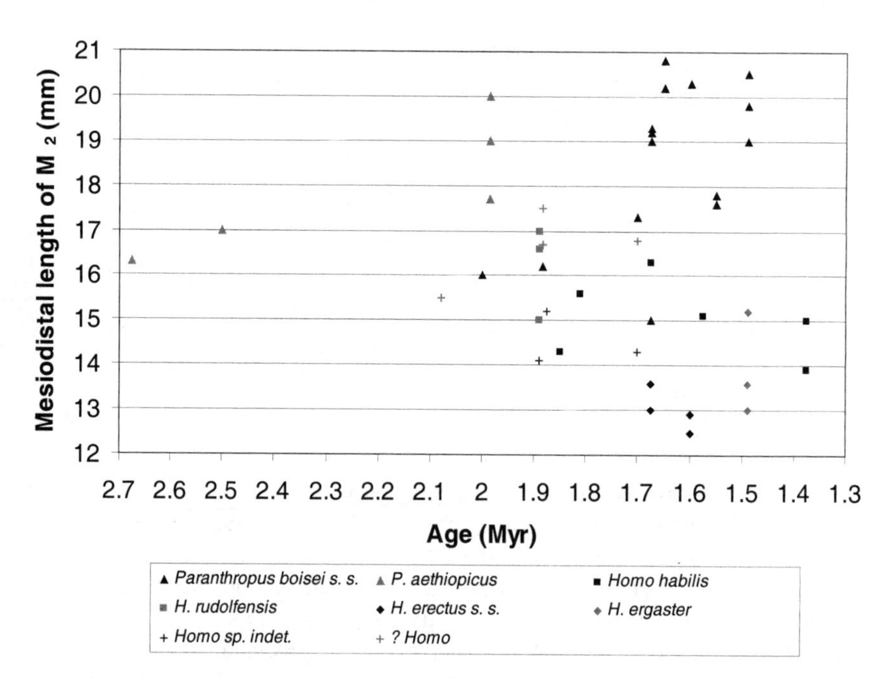

Figure 2.4 Second mandibular molar mesiodistal size variation by taxon. Fossils for which the taxonomic status is questionable are identified by a question mark. See text for more details.

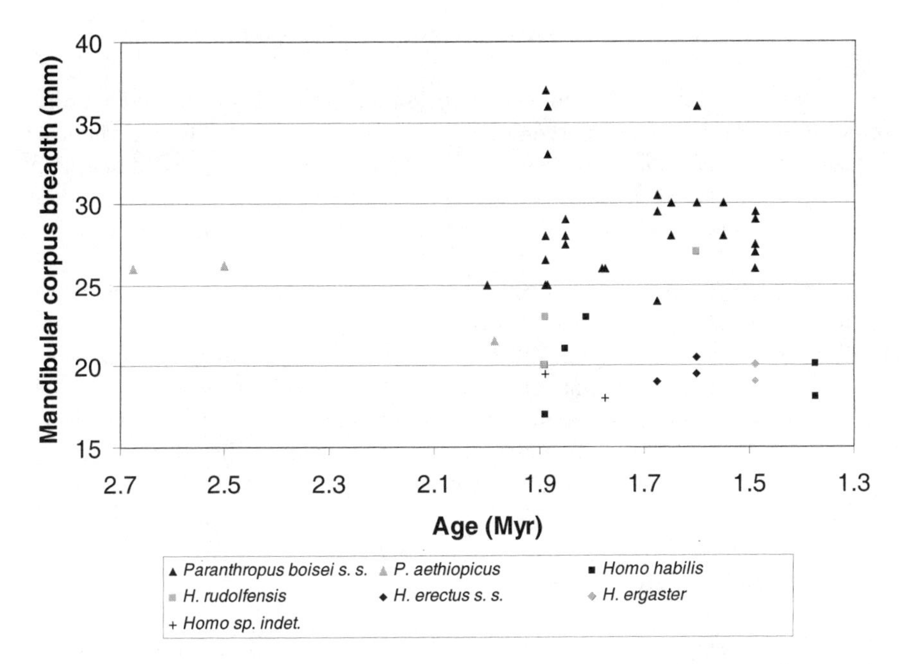

Figure 2.5 Mandibular corpus breadth variation through time and by taxon. See text for more details.

and Wood, 2001]) or that *H. erectus s. l.* is a "good" species, and the fossil groups included in *H. erectus s. l.* should not be considered separate species.

It is instructive to compare the "taxon-free" and taxonomic representations of how the variables change over time. As can be seen in figure 2.3, the taxon-free plot of second molar mesiodistal length appears to show an increase in variation over time. However, figure 2.4 shows that the increase in variation can be explained by the emergence of *P. boisei* and *H. erectus s. l.* Furthermore, trends within each of these lineages are also evident. For example, *H. erectus s. s.* and *H. ergaster* postcanine teeth tend to be shorter mesiodistally than those assigned to *H. habilis* or *H. rudolfensis* or the several specimens attributed to *Homo* sp. indet. This pattern suggests that, at least in this variable, there is an appreciable difference among the species subsumed within *H. erectus s. l.* Therefore, a taxonomy that does not distinguish between *H. habilis* and *H. erectus s. s.* or *s. l.* would be less able to explain this variation. This distinction makes it likely that the several fossils assigned to *Homo* sp. indet. should be included in the *H. habilis* or *H. rudolfensis* hypodigm, as they all fall above the values for *H. erectus s. s.* and *H. ergaster*. In figure 2.5 corpus breadth at M1 by taxon is plotted over time. As with second molar mesiodistal length, there is a distinction between the *Homo* and *Paranthropus* lineages, but the values for the species within the *Homo* lineage overlap and show no separation, suggesting that a "lumping" taxonomy would adequately explain the variation seen in these data. In summary, inferences about the diet of taxa are inevitably affected by taxonomic decisions made about individual specimens.

The "Known," "Unknown," and "Unknowable"

We are conscious that the taxonomy of early hominins will always be "unknowable" in the sense that taxonomies are always hypotheses. Thus, even if researchers can use several lines of evidence to recover reliable information about the diet of a single fossil hominin specimen, that specimen and therefore those data still have to be attributed to a taxon to enable those data to inform us about the paleobiology of early hominins.

It is evident that the more derived a hominin taxon, the easier it is to assign specimens to that taxon. This means that we are more likely to assemble sound hypodigms of *Paranthropus* species than species that belong to the *Homo* clade, and are therefore able to know more about their diet. This will disappoint those whose interest in the evolution of diet is confined to our direct ancestors. However, it will please those of us who are as interested, if not more interested, in the diets of taxa in extinct hominin subclades such as *Paranthropus*.

Conclusion

As seen in the examples above, the choice of taxonomy directly affects how one can analyze and interpret the fossil data available. Researchers must be aware of the limitations of the fossil record (e.g., small sample sizes, fossils with less reliable taxonomic status, and poorly preserved fossils). The reconstruction of diet and the quantification of what we can know about any particular species' diet hinges on how we choose to define that species and what specimens are included in its hypodigm. A sound taxonomy is the first and crucial step in any study of hominin diet.

Acknowledgments We thank Peter Ungar for inviting us to contribute to this volume. B.W. is supported by the Henry Luce Foundation, the National Science Foundation (NSF), and the George Washington University Vice-President for Academic Affairs. A.G.H. is supported by an NSF-IGERT grant and GWU.

Note

1. "Hominin" is the vernacular version of the tribe Hominini. This is the Linnaean category many researchers are now using for the clade that includes modern humans and all the extinct species of higher primates judged to be more closely related to modern humans than to any other living taxon. The equivalent terms for chimpanzees are "panin" and "Panini."

References

Ahern, J.C.M., 2005. Foramen magnum position variation in *Pan troglodytes*, Plio-Pleistocene hominids, and recent *Homo sapiens*: Implications for recognizing the earliest hominids. *Am. J. Phys. Anthropol.* 127, 267–276.

Antón, S.C., 2003. Natural history of *Homo erectus*. *Yearb. Phys. Anthropol.* 46, 126–170.
Brunet, M., Guy, F., Pilbeam, D., Lieberman, D.E., Likius, A., Mackaye, H., Ponce de León, M.S., Zollikofer, C.P.E., and Vignaud, P., 2005. New material of the earliest hominid from the Upper Miocene of Chad. *Nature* 434, 752–755.
Constantino, P., and Wood, B.A., 2004. *Paranthropus* paleobiology. In: Zona Arqueológica (Eds.), *Miscelánea en homenaje a Emiliano Aguirre*. Vol. 3: Paleoanthropología. Museo Arqueologico Regional, Madrid, pp. 137–150.
Eldredge, N., 1993. What, if anything, is a species? In: Kimbel, W.H., and Martin, L.B. (Eds.), *Species, Species Concepts, and Primate Evolution*. Plenum Press, New York, pp. 3–20.
Gabunia, L., Vekua, A., Lordkipanidze, D., Swisher III, C.C., Ferring, R., Justus, A., Niordaze, M., Tvalchrelidze, M., Antón, S.C., Bosinski, G., Joris, O., de Lumley, M.-A., Majsoradze, G., and Movskhelishvili, A., 2000. Earliest Pleistocene hominid cranial remains from Dmanisi, Republic of Georgia: Taxonomy, geological setting, and age. *Science* 288, 1019–1025.
Galik, K., Senut, B., Pickford, M., Gommery, D., Treil, J., Kuperavage, A.J., and Eckhardt, R.B., 2004. External and internal morphology of the BAR 1002'00 *Orrorin tugenensis* femur. *Science* 305, 1450–1453.
Ghiselin, M. T., 1974. A radical solution to the species problem. *Syst. Zool.* 23, 536–544.
Haile-Selassie, Y., Suwa, G., and White, T.D., 2004. Late Miocene teeth from Middle Awash, Ethiopia, and early hominid dental evolution. *Science* 303, 1503–1505.
Hartwig, W.C. (Ed.), 2002. *The Primate Fossil Record.* Cambridge University Press, Cambridge.
Hull, D.L., 1976. Are species really individuals? *Syst. Zool.* 25, 174–191.
Hull, D.L., 1978. A matter of individuality. *Philos. Sci.* 45, 335–360.
Jolly, C.J., 2001. A proper study for mankind: Analogies from the Papionin monkeys and their implications for human evolution. *Yearbook Phys. Anthropol.* 44, 177–204.
Leakey, M.G., Spoor, F., Brown, F.H., Gathogo, P.N., Kiarie, C., Leakey, L.N., and McDougall, I., 2001. New hominin genus from eastern Africa shows diverse middle Pliocene lineages. *Nature* 410, 433–440.
McDougall, I., Brown F.H., and Fleagle, J.G., 2005. Stratigraphic placement and age of modern humans from Kibish, Ethiopia. *Nature* 433, 733–736.
Pickford, M., and Senut, B., 2001. The geological and faunal context of Late Miocene remains from Lukeino, Kenya. *C. R. Acad. Sci., Paris, Earth Planet. Sci. Lett.* 332, 145–152.
Robinson, J.T., 1972. *Early Hominid Posture and Locomotion.* University of Chicago Press, Chicago.
Senut, B., Pickford, M., Gommery, D., Mein, P., Cheboi, K., and Coppens, Y., 2001. First hominid from the Miocene (Lukeino Formation, Kenya). *C. R. Acad. Sci., Paris* 332, 137–144.
Silverman, N., Richmond, B., and Wood, B., 2001. Testing the taxonomic integrity of *Paranthropus boisei sensu stricto*. *Am. J. Phys. Anthropol.* 112, 167–178.
Vekua, A., Lordkipanidze, D., Rightmire, G.P., Agusti, J., Ferring, R., Maisuradze, G., Mouskhelishvili, A., Nioradze, M., Ponde de Leon, M., Tappen, M., Tvalchreldze, M., and Zollikofer, C., 2002. A new skull of early *Homo* from Dmanisi, Georgia. *Science* 297, 85–89.
Vignaud, P., Duringer, P., Mackaye, H.T., Likius, A., Blondel, C., Boisserie, J.-R., de Bonis, L., Eisenmann, V., Etienne, M.-E., Geraads, D., Guy, F., Lehmann, T., Lihoreau, F., Lopez-Martinez, N., Mourer-Chauvire, C., Otero, O., Rage, J.-C., Schuster, M., Viriot, L., Zazzo, A., and Brunet, M., 2002. Geology and paleontology of the Upper Miocene Toros-Menalla hominid locality, Chad. *Nature* 418, 152–155.
Vrba, E., 1995. The fossil record of African antelopes (Mammalia, Bovidae) in relation to human evolution and paleoclimate. In: Vrba, E.S., Denton, G.H., Partridge, T.C., and Burckle, L.H. (Eds.), *Paleoclimate and Evolution, with Emphasis on Human Origins.* Yale University Press, New Haven, CT, pp. 385–424.
White, T., 2003. Early hominids—diversity or distortion? *Science* 299, 1994–1997.

Wood, B., and Richmond, B., 2000. Human evolution: Taxonomy and paleobiology. *J. Anat.* 196, 19–60.

Zollikofer, C.P.E., Ponce de León, M.S., Lieberman, D.E., Guy, F., Pilbeam, D., Likius, A., Mackaye, H., Vignaud, P., and Brunet, M., 2005. Virtual cranial reconstruction of *Sahelanthropus tchadensis*. *Nature* 434, 755–759.

PART II

THE HOMININ FOSSIL RECORD

3

The Evolution of the Hominin Diet from a Dental Functional Perspective

PETER W. LUCAS

No other part of the modern human body suffers from disruptions as our mouths; they are so frequently diseased that an entire health profession, dentistry, specializes in them. Its various parts do not seem to fit together properly. The teeth, which develop in the jaws, often have insufficient space to erupt into its cavity. This causes rotation, twisting, tilting, impaction, and sometimes even a reversal of development, where even if some tooth germs form, they fail to mineralize and are resorbed. All this has been documented very thoroughly in dental journals along with a battery of techniques for coping with these problems. What have not been thoroughly examined are the root causes or the functional consequences of these disturbances, basically because the dentition tends to be treated as semiredundant. The dental profession adopts treatment protocols that are protective of dental state and of structures known to be vulnerable, notably the temporomandibular joint, but has done relatively little to investigate dental function in a broad context. Yet it is not just dentists who feel this way. Physical anthropologists have long regarded the size reduction of the dentition in the genus *Homo* (Brace, Rosenberg, and Hunt, 1987; Calcagno and Gibson, 1991) as indicative of its gradual loss of function due to increased tool use (Brace, 1963, 1964). Recently, Wrangham and colleagues (1999) have suggested that cooking, rather than tool use, explains a trend toward the reduction of tooth size as far back as 1.9 million years ago.

Teeth act on solid foods, and it is the physical properties of these foods that form the major selective pressure on them (Lucas, 2004). These physical properties can be sensed as food texture (Bourne, 2002). Cooking greatly affects texture, and because individuals can both detect and select (within the limits of what particular modes of cooking can do) desired textures, this allows very rapid evolutionary changes in the dentition in response. Foods can be cooked in many ways, and the scientific understanding of what happens to the food structure when it is cooked is still far from perfect (Barham, 2001). Nevertheless, study of the evolution of cooking seems to offer good prospects, not only for understanding tooth form during human evolution but also as a model for dental evolution in mammals in a broader sense.

Factors Affecting Tooth Size: The Physiological Viewpoint

The major importance of tooth size that can be demonstrated physiologically is that the larger teeth are, the more likely they can strike their target effectively. The incisors serve to get food into the mouth, a process called *ingestion*. Thus, broader incisors remove more of the outer peel of a fruit with each action than narrower ones. Analogously, larger canines in male anthropoids are more likely to penetrate the skin of male conspecifics. There are exceptions to this though. Many primates use their incisors mainly as frictional devices, for example, to strip leaves from their supports by pulling small branches through their mouths (Ungar, 1994; Yamashita, 2003), rather than use their bladed surface specifically to induce fractures. Provided that Amonton's laws of friction apply, then effectiveness in stripping should be unrelated to the size of the incisal edges. However, when teeth are intended either for fracture or for (canine) displays that indicate the propensity for this, then size is important in two senses: larger teeth are most likely to produce that fracture and, because they are used repeatedly, are likely to withstand being worn for longer.

The posterior teeth, the premolars and molars, act usually to reduce the particle size of foods, a process called *mastication*. Fragmented particles are then mixed with saliva before being swallowed. Larger posterior teeth increase the chance of hitting food particles that are being masticated (many studies cited by Lucas, 2004). Measurement of this probability is called the *selection function* in comminution processes (Epstein, 1947), contrasting with the *breakage function*, which is the degree of fragmentation that follows when a particle is hit (Gardner and Austin, 1962). Measurements of the selection and breakage functions in human mastication have been proved capable of replicating the particle size distributions that humans produce in chewing (Baragar, van der Glas, and van der Bilt, 1996).

Lucas and colleagues (1985, 1986) interpreted tooth size trends within hominins based on an analysis using comminution functions. They suggested that ingestion of small volumes of small chemically protected objects (i.e., ones that were immune to digestion unless broken by the teeth), particularly ones that were abrasive, would be liable to select for larger posterior, but smaller anterior, teeth. The most likely food items constituting such a dietary intake would be either leaves or seeds. These items often go together in primate diets, particularly because the fermentation chambers of the digestive systems of certain primates work effectively both on the cell walls that impede digestive access to leaf mesophyll and in detoxifying the defensive mechanisms of seeds. From another direction of thought, Lucas (1989) suggested that the amount of work done by each class of tooth, contrasting the incisors anteriorly with the postcanines posteriorly, would be reflected in their relative sizes. This argument focused on the consumption of fruit tissues, which are usually the most abundant in primate diets, and also on the treatment of seeds within them. Primates that tend to ingest whole fruits, swallowing both seeds and flesh without much processing in the mouth, would tend to have the smallest dentitions. Those primates attempting always to separate seeds from the surrounding flesh, (when fruits are large with the front teeth, but when fruits are small with the postcanines, (the latter only usually being possible in cheek-pouch-possessing Old World monkeys) would have very large teeth.

All these suggestions are based on applying a "probability" approach to the analysis of tooth size. It is unclear how that could help interpret the evolutionary response of tooth size to cooking because the latter changes the internal properties of foods, such as food toughness. Changes to the external properties of foods, such as particle size, are much more likely to be due to processing with tools.

A simple amendment to previous models may help to produce the right approach. I hypothesize that tool use in human evolution has its major influence on the size of anterior teeth by reducing input particle size. Cooking influences the size of the postcanine teeth by modifying foods such that their fracture is made easier. This may come about by reducing the toughness or the stress-strain gradient of foods. However, there is another important and common effect of cooking: it may also make foods more "notch sensitive." Most solid materials are notch sensitive, a term implying that structural defects (defects made experimentally are in the form of notches, thus giving the property its name) weaken objects made from these materials out of proportion to the size of those defects. All such materials, whether relatively homogeneous, have tight structural connections. Many foods are not like this; they have regions within them that impose barriers to the transfer of the stored (strain) energy that dental loads impart. Such regions are exemplified by the ground substance between the fibers of animal connective tissue, or the perimysium, between bundles of muscle fibers that effectively disconnect the structure. Gordon (1978) considers many types of biological tissues designed to take tensile loads like this. The advantage is that the failure of any of the fibrous components of the structure, which normally carry the tensile stresses, are isolated and have minimal effect on the stress capacity of other fibers. Thus, the sprain of a joint (ligamentous damage), the rupture of a tendon, or the ripping of muscle fibers rarely results in subdivision of the entire ligament, tendon, or muscle. Without such notch insensitivity, failure would tend to be catastrophic. What cooking may often do is to better connect such tissues. Collagen, for example, is gelatinized above usual cooking temperatures and connective tissue seems to be bound together more tightly. Thus, a potential food like meat, which normally obstructs free-running fracture via notch insensitivity, becomes much easier to consume with the relatively blunt human dentition when cooked (Lucas, 2004). However, all this depends greatly on a definition of what "cooked," as a textural term, really means, not just in terms of its treatment (boiled in water vs. heated in air) but also with respect to "cooked" as opposed to "undercooked" or "overcooked" because this really pinpoints the direction of selection on the teeth.

Factors Affecting Tooth Size: The Morphological Viewpoint

Most of the morphological literature on tooth size in recent years has dealt with teeth from an allometric perspective (Ungar, 1998). It is appropriate to think of the subject in this way because a trend to enlargement is pervasive in mammalian evolution, and, in fact, the first effort in the field was directed at hominins (Pilbeam and Gould, 1974). These authors put forward an argument, amplified by Gould (1975), that to service the body's energy requirements, the surface areas of the posterior teeth of a mammal would be scaled to that mammal's basal metabolic rate. Such rates are related in most animals to the body mass (*M*) raised to the power ¾, that is, to $M^{0.75}$ (Kleiber, 1961). Thus, larger mammals, requiring posterior tooth areas proportional to $M^{0.75}$, would have relatively larger cheek teeth and thus need similarly enlarged jaws to accommodate them. However, Fortelius (1985) disputed the accuracy of this prediction and presented an enhancement of it. Noting that larger mammals chew more slowly than smaller ones, he presented theory and evidence that chewing rates decline with increasing body weight as $M^{-0.25}$. All else being equal, larger mammals would need to ingest much larger intakes than Pilbeam and Gould envisaged, proportional to $M^{0.75}/M^{-0.25} = M^{1.0}$, to survive. However, Fortelius argued that this does not mean that larger mammals need even larger teeth than Pilbeam and Gould (1974) suggested because the latter authors forgot that the surfaces of teeth contact the surfaces of food particles, not their volumes. So given the same food shape, Fortelius (1985) said that anterior and posterior teeth would scale as $M^{0.67}$. In other words, teeth remain in proportion to body size whatever the body size of a mammal.

Sadly, this explanation does not help in understanding why tool use or cooking might have affected tooth size in human evolution. There is apparently nothing mechanical about the above arguments. Both Pilbeam and Gould (1974) and Fortelius (1985) assume similarity of stress levels, that is, foods of any given composition, whatever their particle size, fracture at similar stresses (that elements of the food mouthful fail at the same force per unit area). If both large and small mammals were eating the same-sized food objects, just ingesting them in different quantities, then these general arguments may be right. But what if larger mammals cannot find dense enough patches of these food objects to ingest larger quantities of them? They may instead have to search for larger objects. Then, we need to take into account that solid objects of a given material do not break at characteristic stresses. If prepared in the same shapes and subjected to the same loading regime, larger objects fail at lower stresses. The property that controls fracture is toughness, which is defined as the energy generating unit area of crack. The scaling arguments related to crack generation are rigorous and proven (Gurney and Hunt, 1967), and investigations in many types of industrial processes have validated it, particularly when the behavior of large and small structures is compared (Atkins and Mai, 1985), for example, in the comminution industry (Kendall, 2001).

Adding a mechanical extension to Fortelius's argument based on the energetics of fracture, Lucas (2004) proposed that in mammals eating elastic "Hookean" foods (those in which stress is proportional to strain nearly until the onset of fracture), posterior tooth areas would scale only as $M^{0.5}$, meaning that larger mammals would be

Table 3.1 Features of the Teeth and Mouth Predicted by Either Geometric, Metabolic, or Fracture Scaling Patterns Discussed in the Text

Feature	Geometric Scaling	Metabolic Scaling[a]	Fracture Scaling
Incisal width	$M^{0.33}$	Not predicted	$M^{0.33}$
Postcanine tooth area	$M^{0.66}$	$M^{0.75}$	$M^{0.50}$
Jaw length	$M^{0.33}$	Not predicted, but probably $M^{0.375}$	$M^{0.33}$
Volume of oral cavity	$M^{1.0}$	Not predicted	$M^{1.0}$
Mouth slit (treated as an area)	$M^{0.66}$	Not predicted, but probably $M^{0.75}$	$M^{0.66}$
Muscle cross-sections[b]	$M^{0.66}$	Not predicted	$M^{0.5}$
Muscle volume (or weight)	$M^{1.0}$		$M^{0.83}$

Note: Predictions are given as exponents of body mass, *M*. Geometric scaling follows standard mechanical explanations, whereby solid objects of a given composition fail as a given fixed fracture stress.
[a]Pilbeam and Gould (1974).
[b]For muscles either active in jaw opening or closing muscles that permit opening by passively stretching.

likely to have relatively smaller posterior teeth to do the same job as those of smaller mammals. Jaw lengths must be scaled to food particle sizes because they have to admit the food intake and anterior teeth should be scaled similarly because their job, particularly in higher primates, is to regulate the bite size. Thus, anterior and posterior tooth sizes have different scaling patterns. Table 3.1 shows these in relation to predictions from Pilbeam and Gould (1974) and as would be expected if any given food fails at a fixed stress.

Fracture scaling reverses the predictions of Pilbeam and Gould (1974) to some extent. Larger mammals are predicted to have smaller posterior teeth. However, they would not have smaller jaws because, if they did, they would need to gape wider to admit those particles. If jaw size were proportional to tooth size, then larger animals would face the sorts of problems with muscle efficiencies discussed long ago by Herring and Herring (1974).

The prediction from the simplest food behaviors is clear: larger mammals would have anterior teeth that "fit" the jaw, but there would be free space in the postcanine region because of the somewhat undersized dentition further back. Dwarfing mammalian lineages, of which there are several well-documented examples (Fortelius, 1985), including one that now belongs to the genus *Homo* (Brown et al., 2004), would however be in a difficult situation. As they get absolutely smaller, the posterior teeth would get relatively larger and might actually run out of available jaw space. In this situation, tooth size reduction or even tooth loss might become inevitable.

Dietary Disturbances Produced by Tool Use and Cooking

Tool use by modern humans is often directed at reducing particle size. A comminution industry, sucking much of the power that industry generates, is defined by it (Lowrison, 1974). Suppose that tool use in paleolithic times had a similar comminuting motivation and that the target of this comminution was essentially food. If so, very little technology would be needed to drastically reduce the food particle sizes of many foods, so selecting for reduction in incisal dimensions, as has long been suggested for "small-object feeders" (Jolly, 1970). The gape of the human mouth is controlled by the need to ingest food particles because, with reduced canines, there is little to gain by threatening social competitors. If canines were still involved in threat gestures, then no reduction would be possible. However, without this factor, jaw lengths and incisal sizes can be thought of as tied together in this argument and would reduce together if ingested food particle size were reduced by tool use.

Cooking is a universal practice in modern human groups and may be almost as old as stone tool use (Wrangham et al., 1999). Lucas (2004) suggested a fracture similarity criterion for explaining cheek tooth size that predicts that it varies with the material properties of foods in the diet. At the very simplest level, the linear dimensions of the cheek teeth vary as the cube root of change in food toughness. A general result of cooking is often reduction in food toughness. If so, then tooth size would decline in accord.

If so, it does not take much effort to see that tool use could reduce food particle size (to which jaw length and anterior tooth size are proportional) much more rapidly than cooking could change the ratio of cooked-to-raw food toughness. During human evolution, it is quite feasible to suggest that this led to a degree of jaw crowding sufficient, say, to deny space to third molars as regular members of the tooth row.

The Known, Unknown, and Unknowable in the Evolution of Diet in the Genus *Homo*

Overall, little is known about dental-dietary relationships in mammals, but I contend that diet can be "read" from the dentition of many of them if enough theory is developed to understand the mechanical interactions involved in dental function. Hominin fossils provide the necessary knowledge of tooth and jaw sizes, and the more fossils found, the more intriguing the evidence gets. Australopithecines as a whole tend to have relatively large postcanine teeth, suggesting, among other things, that they were not cooking food (Teaford and Ungar, 2000). Robust australopithecines such as *Australopithecus robustus* and *Australopithecus boisei* had relatively small incisors: whether tool use was involved, they must have been consuming small objects (Jolly, 1970). However, it is within the genus *Homo* that much of the theory developed here could be applied. The published material of *Homo floresiensis* from Flores, Indonesia, shows that dental crowding occurred in a species of *Homo* other than humans (Brown et al., 2004). Such crowding does not appear so in Neanderthals, which show a variably large tooth-free area behind the third molar (Franciscus and Trinkaus, 1995). This suggests differences in diet. Cooking is much more critical for understanding tooth size trends in humans than tool use. Wrangham et al. (1999) contend that starchy underground storage organs were the initial objectives of cooking. There are physiological correlates that support the importance of starches in the human diet. The multiple copies of amylase in the human genome, with three copies expressing themselves in the saliva, are unparalleled in primates (Samuelson et al., 1990). A small survey of starchy foods in modern human diets found that they seem to exhibit a variety of behaviors dependent on cooking mode: some change their toughness when cooked, while others do not (Lucas, 2004). Generally, the form of the stress-strain behavior changes with cooking, and the directions of cracking change as well (Lillford, 2000). All this complicates the application of any theory, but the chances of developing hypotheses that can relate cooking to tooth size remain good.

References

Atkins, A.G., and Mai, Y.-W., 1985. *Elastic and Plastic Fracture.* Ellis Horwood, Chichester.

Baragar, F.A., van der Glas, H.W., and van der Bilt, A., 1996. An analytic probability density for particle-size in human mastication. *J. Theor. Biol.* 181, 169–178.

Barham, P., 2001. *The Science of Cooking*. Springer, New York.

Brace, C.L., 1963. Structural reduction in evolution. *Am. Nat.* 97, 39–49.

Brace, C.L., 1964. The probable mutation effect. *Am. Nat.* 98, 453–455.

Brace, C.L., Rosenberg, K., and Hunt, K.D., 1987. Gradual change in human tooth size in the late Pleistocene and post-Pleistocene. *Evolution* 41, 705–720.

Bourne, M.C., 2002. *Food Texture and Viscosity: Concept and Measurement.* 2nd ed. Academic Press, London.

Brown, P. Sutikna, T., Morwood, M.J., Soejono, R.P., Jatmiko, Wayhu Saptomo, E., and Rokus Awe Due, 2004. A new small-bodied hominin from the Late Pleistocene of Flores, Indonesia. *Nature* 431, 1055–1061.

Calcagno, J.M., and Gibson, K.R., 1991. Selective compromise: Evolutionary trends and mechanisms in hominid tooth size. In: Kelley, M.A., and Larsen, C.S. (Eds.), *Advances in Dental Anthropology*. Wiley, New York, pp. 59–76.

Epstein, B., 1947. The mathematical description of certain breakage functions leading to the logarithmico-normal distribution. *J. Franklin Inst.* 244, 471–477.

Fortelius, M., 1985. Ungulate cheek teeth: Developmental, functional and evolutionary interrelationships. *Acta. Zool. Fenn.* 180, 1–76.

Franciscus, R.G., and Trinkaus, E., 1995. Determinants of retromolar space presence in Pleistocene *Homo* mandibles. *J. Hum. Evol.* 28, 577–595.

Gardner, R.P., and Austin, L.G., 1962. A chemical engineering treatment of batch grinding. In: Rumpf, H. (Ed.), *Zerkleinern Symposion.* Verlag Chemie, Düsseldorf, pp. 217–248.

Gordon, J.E., 1978. *Structures*. Plenum Press, New York.

Gould, S.J., 1975. On the scaling of tooth size in mammals. *Am. Zool.* 15, 351–362.

Gurney, C., and Hunt, J., 1967. Quasi-static crack propagation. *Proc. R. Soc. Lond.* A 299, 508–524.

Herring, S.W., and Herring, S.E., 1974. The superficial masseter and gape in mammals. *Am. Nat.* 108, 561–575.

Jolly, C.J., 1970. The seed-eaters: A new model of hominid differentiation based on a baboon analogy. *Man,* n.s., 5, 1–26.

Kendall, K., 2001. *Molecular Adhesion*. Kluwer/Plenum Press, New York.

Kleiber, M. 1961. *The Fire of Life: An Introduction to Animal Energetics.* Wiley Press, New York.

Lillford, P.J., 2000. The materials science of eating and food breakdown. *MRS Bull.* December, 38–43.

Lowrison, G.C., 1974. *Crushing and Grinding.* Butterworths, London.

Lucas, P.W., 1989. A new theory relating seed processing by primates to their relative tooth sizes. In: Schmitt, L.H., Freedman, L., and Bruce, N.W. (Eds.), *The Growing Scope of Human Biology*. Centre for Human Biology, University of Western Australia, Perth, pp. 37–49.

Lucas, P.W., 2004. *Dental Functional Morphology*. Cambridge University Press, Cambridge.

Lucas, P.W., Corlett, R.T., and Luke, D.A., 1985. Plio-Pleistocene hominids: An approach combining masticatory and ecological analysis. *J. Hum. Evol.* 14, 187–202.

Lucas, P.W., Corlett, R.T., and Luke, D.A., 1986. New approach to postcanine tooth size applied to Plio-Pleistocene hominids. In: Else, J.G., and Lee, P.C. (Eds.), *Primate Evolution*. Cambridge University Press, Cambridge, pp. 191–201.

Pilbeam, D.R., and Gould, S.J., 1974. Size and scaling in human evolution. *Science* 186, 892–901.

Samuelson, L.C., Wiebauer, K., Snow, C.M., and Meisler, M.H., 1990. Retroviral and pseudogene insertion sites reveal the lineage of human salivary and pancreatic amylase genes from a single gene during primate evolution. *Mol. Cell Biol.* 10, 2513–2520.

Teaford, M.F., and Ungar, P.S., 2000. Diet and the evolution of the earliest human ancestors. *Proc. Natl. Acad. Sci.* 97, 13506–13511.

Ungar, P.S., 1994. Patterns of ingestive behavior and anterior tooth use differences in sympatric anthropoid primates. *Am. J. Phys. Anthropol.* 95, 197–219.

Ungar, P.S., 1998. Dental allometry, morphology, and wear as evidence for diet in fossil primates. *Evol. Anthropol.* 6, 205–217.

Wrangham, R.W., Jones, J.H., Laden, G., Pilbeam, D., and Conklin-Brittain, N.L., 1999. The raw and the stolen: Cooking and the ecology of human origins. *Curr. Anthropol.* 40, 567–594.

Yamashita, N., 2003. Food procurement and tooth use in two sympatric lemur species. *Am. J. Phys. Anthropol.* 121, 125–133.

4

Dental Functional Morphology

The Known, the Unknown, and the Unknowable

PETER S. UNGAR

Begun (2004) recently wrote, "We will never be able to describe the behavior of fossils as if they were extant organisms" (497). This begs the question what *can* be known? In this chapter, I review the "knowns," "unknowns," and "unknowables" about reconstructing the diets of early hominins and other fossil primates through the study of their occlusal morphologies. The basic process involves the comparative method. If we can identify form-function relationships between teeth and diet in living primates, we can infer diet from the teeth of fossil forms.

The Known

Researchers have recognized for a long time that tooth form reflects function (e.g., Owen, 1840–1845). Gregory (1922) was among the earliest to study dental functional morphology in detail, suggesting that primate molar shape evolved to improve mechanical efficiency for chewing. Subsequent workers (Simpson, 1933; Crompton and Sita-Lumsden, 1970; Hiiemae and Kay, 1972) built on this idea and viewed teeth as guides for masticatory movements. Kay and Hiiemae (1974), for example, associated specific dental morphologies with specific chewing behaviors, noting that insectivorous primates have reciprocally concave blades well suited to shearing tough insect chitin between the leading edges of crown crests, whereas frugivore molars possess cusp tips oriented more parallel to the occlusal plane for crushing and grinding three-dimensional fruit flesh and seeds (fig. 4.1). These basic associations between form and function were subsequently confirmed for various primate taxa (Rosenberger and Kinzey, 1976; Kinzey, 1978; Seligsohn and Szalay, 1978).

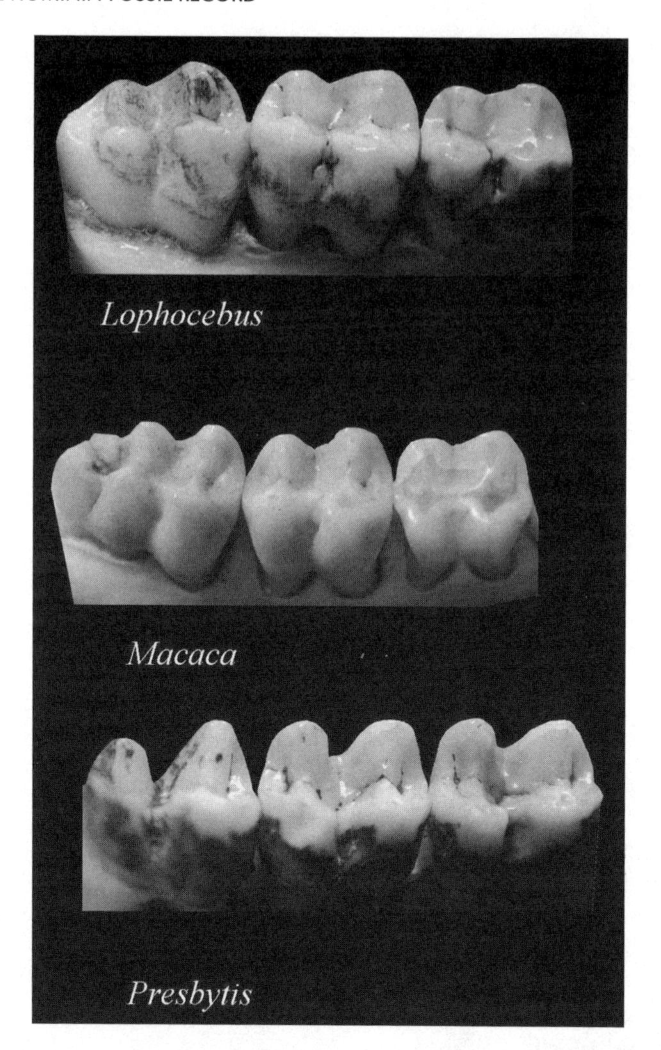

Figure 4.1 Shearing crest development in Old World monkeys with different diets. *Lophocebus albigena* (*top*) consumes harder, brittler foods than *Macaca fascicularis* (*middle*), and *Presbytis rubicunda* (*bottom*) eats more tough, elastic leaves than the other primates.

Quantifying Functional Aspects of Tooth Form

Kay (1978, 1984) and Kay and Covert (1984) then recognized the need for a quantitative approach to characterizing tooth shape and so developed the shearing quotient (SQ). This technique involves measurement of the lengths of mesiodistal crests on unworn molars of several closely related species with similar diets (fig. 4.2). A least squares regression line is fit to summed crest length and mesiodistal occlusal surface length in logarithmic space. Shearing quotients are computed as residuals or deviations from the regression line. Comparative studies have shown that folivores and

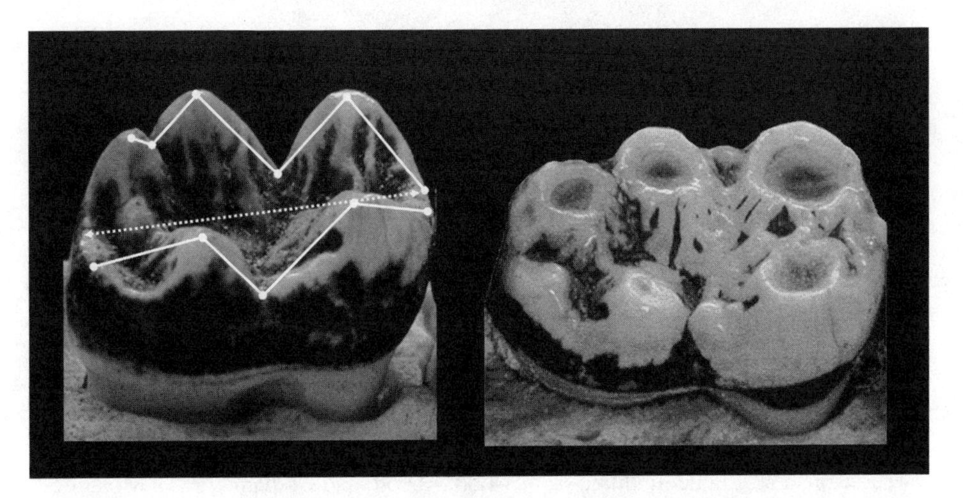

Figure 4.2 Measurement of shearing crest lengths (*solid*) and mesiodistal occlusal table length (*dashed*) on a gorilla molar. Note that tooth wear makes accurate measurement of crest lengths for the specimen on the right difficult.

insectivores have higher SQs than do closely related frugivores and that, among frugivores, soft-fruit flesh feeders have higher SQs than hard-object feeders. This approach has been shown to track diets for all higher-level modern primate taxa (Kay, 1977; Kay, 1984; Anthony and Kay, 1993; Strait, 1993a; Meldrum and Kay, 1997).

Researchers commonly apply SQ studies to the primate fossil record. Kay and colleagues, for example, have reconstructed the diets of numerous anthropoids from the Oligocene and Miocene (Kay, 1977; Kay and Simons, 1980; Anthony and Kay, 1993; Ungar and Kay, 1995; Fleagle, Kay, and Anthony, 1997; Kay and Ungar, 1997; Kirk and Simons, 2001; Ungar, Teaford, and Kay, 2004), and similar techniques have been used to study early primates from the Eocene (Strait, 1993a; 1993b; Williams and Covert, 1994).

Work on early hominin SQs though, has been limited to *Australopithecus africanus* and *Paranthropus robustus* (Ungar, Teaford, and Grine, 1999). Results suggest that the australopiths both had relatively flat and blunt molar teeth compared with extant frugivorous hominoids but that *A. africanus* had higher SQ values on average than did *P. robustus*. This is consistent with Grine's (1981) observation that "gracile" australopiths had more occlusal relief than did "robust" forms and suggests that while *A. africanus* molars may have been better able to shear than could those of *P. robustus*, neither could have had a very tough diet (Kay, 1985; Teaford, Ungar, and Grine, 2002).

Tooth Wear and Diet

Although these studies clearly demonstrate relationships between primate tooth shape and diet, most such work has been limited to unworn teeth. Researchers are now starting to realize that this presents two problems. First, most teeth, whether they are in museum collections or are still in the mouth, are in fact worn. Molars begin to wear as soon as they enter the occlusal battery, and samples suitable for SQ analyses are often abysmally small, especially for fossil taxa. There are, for example, less than ten unworn M_2s (the teeth most often used in functional studies) are in the entire published assemblage of early hominins from South Africa!

The other problem is that reliance on unworn teeth leads to an incomplete picture of the form-function relationship. Wear is a normal phenomenon and natural selection should also act on worn teeth, favoring morphologies that wear in a manner that keep them mechanically efficient for fracturing foods (Kay, 1981; Teaford, 1983; Kay, 1985; Ungar and Williamson, 2000). We are potentially missing a lot of information if we exclude worn teeth in our analyses.

It is, however, not easy to measure occlusal form on worn teeth. Traditional dental morphometrics depend on measuring distances between landmarks that are quickly obliterated by wear (see, e.g., Teaford, 1981). Smith (1999) attempted to control for wear using a technique modified from Wood, Abbott, and Graham (1983). Molar occlusal views were captured on video, and individual cusp areas were identified on a computer screen by a mouse-driven cursor. This allowed calculation of relative two-dimensional projected (planimetric) areas of cusps on unworn to moderately worn teeth (as long as cusp boundaries were identifiable). Smith's results group chimpanzees with gibbons rather than with gorillas, suggesting that there might be a functional signal in this.

This approach is still limited because specimens have to be sufficiently unworn to distinguish individual cusp boundaries. These boundaries can become obliterated rapidly in thin-enameled molars such as those of chimpanzees and gorillas. More important, planimetric area studies provide no information about the third dimension. Masticatory movements are three dimensional, and occlusal relief, crest orientation, and facet inclination all play an important role in the mechanics of food fracture (e.g., Simpson, 1933; Crompton and Sita-Lumsden, 1970; Hiiemae and Kay, 1972; Thexton and Hiiemae, 1997; Lucas, 2004). Surface relief is critical to the angle of approach between mandibular and maxillary teeth as facets come into occlusion during mastication. This, in turn, determines the biomechanical efficiency with which items of given mechanical properties are fractured (e.g., whether foods are sheared or crushed). Thus, the ability to collect data on tooth shape in three-dimensional space is vital to studies of dental functional morphology.

Dental Topographic Analysis

What is needed is a way to study functional morphology of worn teeth in three dimensions. This is where dental topographic analysis comes in. Dental topographic analysis is a landmark-free, three-dimensional method for characterizing functional aspects of occlusal morphology. First, a point cloud representing an occlusal surface is generated using a three-dimensional scanner (e.g., a confocal microscope, laser scanner, or mechanical piezo scanner). Then elevation data are sampled at fixed intervals along the mesodistal and buccolingual axes of the tooth and imported into geographic information systems (GIS) software as an ASCII file of x-, y-, z-coordinates. Finally, surfaces are interpolated to generate a "digital elevation model" of the tooth, where cusps become mountains, fissures are valleys, and so forth. Geographic information systems software allows measurement of surface slope, aspect, area, and other attributes. Researchers have used many different combinations of scanning devices and GIS software packages for dental topographic analysis (Zuccotti et al., 1998; Jernvall and Selänne, 1999; Ungar and Williamson, 2000; Evans et al., 2001; Ungar and M'Kirera, 2003).

Because this approach uses the entire occlusal surface for measurement, it does not depend on specific landmarks that change or disappear as the tooth wears. Thus, it works equally well on worn and unworn teeth. It even allows us to examine changes in occlusal morphology on specific teeth with wear. Dennis and colleagues (2004), for example, studied changes in molar morphology of *Alouatta palliata* individuals sampled at fixed intervals over the course of ten years at Hacienda La Pacifica in Costa Rica. These authors found that occlusal morphology of individual monkeys changed in consistent, predictable ways with wear. The implication of this for fossil work is that different specimens at different stages of occlusal wear can be used to put together a species-specific wear sequence with sufficient samples. This is a necessary prerequisite if we are to include variably worn teeth in functional studies.

The process by which dental topographic analysis can be used to characterize functional aspects of occlusal form can be illustrated by example (fig. 4.3). Here I combine data for M_2s of *Gorilla gorilla gorilla* ($n = 47$), *Pan troglodytes troglodytes* ($n = 54$), and *Pongo pygmaeus pygmaeus* ($n = 51$) (M'Kirera and Ungar, 2003; Ungar and M'Kirera, 2003; Ungar and Taylor, 2005). These taxa are of relevance here because they have similar cusp patterns to early hominins and also allow for a measure of phylogenetic control. These species also differ from one another to a modest degree, both in molar morphology and in diet, making them a good baseline series for comparison with early hominins.

All three of these great apes prefer soft, succulent fruits when available. The central African chimpanzee and western lowland gorillas take many of the same foods where these apes are sympatric, but they also differ in their diets, especially at times of fruit scarcity. At such times, gorillas fall back more on tough, fibrous foods than do chimpanzees (Tutin et al., 1991; Remis, 1997). Average annual food-type proportions reported for *Pan troglodytes troglodytes* include about 70–80 percent fruit flesh, as compared with 45–55 percent fruit flesh for *Gorilla gorilla gorilla* (Williamson et al., 1990; Kuroda, 1992; Nishihara, 1992; Tutin et al., 1997). It is more difficult to compare these taxa with the Bornean orangutan because of differing resource availabilities and methods of data collection. Nevertheless, average annual fruit to leaf proportions for *Pongo pygmaeus pygmaeus* are intermediate between those described for *P. troglodytes troglodytes* and *G. gorilla gorilla*, with an average fruit percentage of about 55–65 percent reported for the orangutans (Rodman, 1977; MacKinnon, 1977).

Data on average surface slope are presented in table 4.1 and illustrated in figure 4.4. Occlusal surfaces were scanned using a laser scanner as point clouds with lateral and vertical resolutions of 25.4 μm (Ungar and Williamson, 2000; M'Kirera and Ungar, 2003; Ungar and M'Kirera, 2003; Dennis et al., 2004). Resulting data files were opened as tables in ArcView 3.2 (ESRI Corp.) GIS software, and digital elevation models were cropped to exclude areas below the lowest point of the occlusal

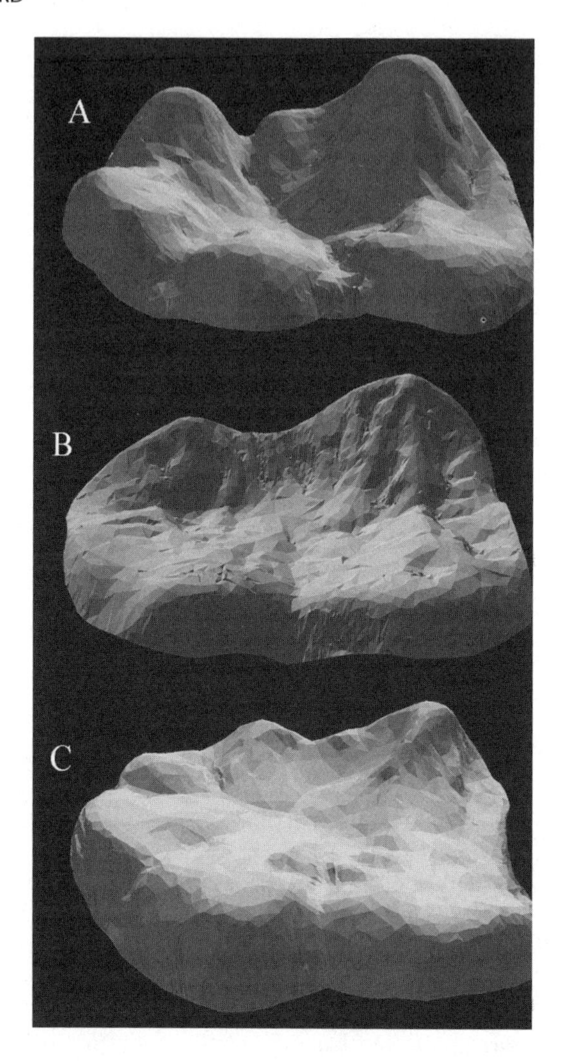

Figure 4.3 Triangulated irregular network models of M_2s of (*A*) *Gorilla gorilla gorilla*, (*B*) *Pongo pygmaeus pygmaeus*, and (*C*) *Pan troglodytes troglodytes*. All specimens at wear stage 2.

basin. Average slope between adjacent points (surface slope) was then recorded for each specimen.

The species overlapped in three wear stages (as defined in Ungar, 2004). As expected, more worn molar surfaces tended to show less occlusal relief and shallower slopes for each of the taxa. At any given stage of wear, however, gorillas had the steepest slopes followed by orangutans. Chimpanzees had the shallowest molar cusps (fig. 4.4).

This example suggests several things. First, tooth shape changes with wear. As teeth wear down, they become flatter. This is not rocket science! Second, cusp slope values mirror leaf-to-fruit ratios reported in the literature for these taxa, suggesting that we are picking up functional aspects of form. Finally, and perhaps most important, the differences among the species are of about same magnitude at each of the recorded wear stages. This suggests that differences between species remain consistent

Table 4.1 Occlusal Slope Comparisons

A. *Summary Statistics*

Wear Stage	*Gorilla* Mean	*Gorilla* SD	*Gorilla* n	*Pongo* Mean	*Pongo* SD	*Pongo* n	*Pan* Mean	*Pan* SD	*Pan* n
1	—	—	—	37.83	2.389	5	—	—	—
2	37.75	5.036	7	36.38	2.447	24	32.88	5.859	5
3	36.29	2.665	10	32.86	2.274	16	30.15	5.771	28
4	32.13	5.069	14	30.32	2.170	6	26.48	4.680	18
5	27.53	4.290	13	—	—	—	25.69	8.607	2
6	32.14	6.445	3	—	—	—	29.31	—	1

B. *Two Factor ANOVA: Occlusal Slope Data (ranked)*

Source	SS	df	MS	*F*-Ratio	*P*
Species	32,065	2	16,032	11.027	0.000
Wear stage	47,936	2	23,968	16.485	0.000
Interaction	34,844		871	0.599	0.664
Error	53,065	119	1,286		

C. *Bonferroni's Multiple Comparison Tests*

	Gorilla	*Pongo*	*Pan*
Gorilla	0.0		
Pongo	–49.366*	0.0	
Pan	–45.960*	–26.105*	0.0

*$p < 0.05$

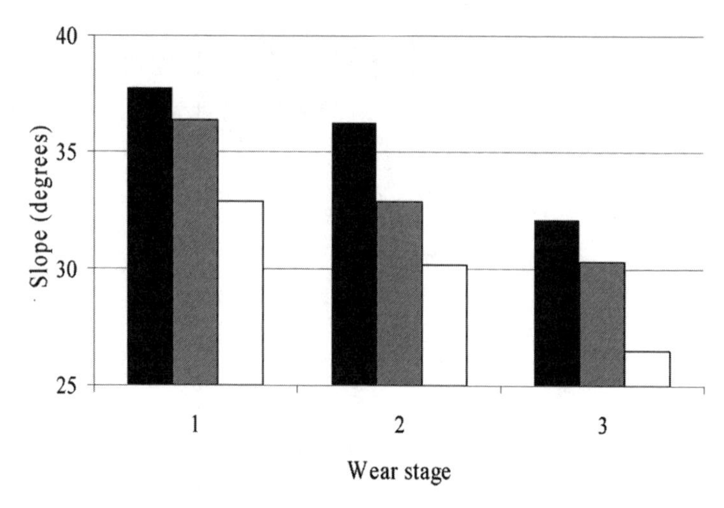

Figure 4.4. Average specimen surface slope at each of three wear stages (see text). Black bars = *Gorilla gorilla gorilla;* grey bars = *Pongo pygmaeus pygmaeus*; white bars = *Pan troglodytes troglodytes.*

through the wear sequence, so we can compare chimps, gorillas, and orangutans at any given wear stage and get the same results. This is consistent with two-way ANOVA results (table 4.1) showing significant differences between species and between wear stages but not for the interaction between the two factors.

The implications of these results are that species need not be represented by unworn teeth as long as there is a baseline of comparative data for specimens with similar degrees of wear. This will allow us to reconstruct the diets of a whole new assortment of fossil taxa that could not be analyzed in the past for lack of available methods.

Dental topographic analysis on early hominin teeth is just beginning, but results are encouraging. Comparisons of *Australopithecus afarensis* and early *Homo* specimens show significant differences between these taxa (Ungar, 2004). Results indicate that at any given stage of wear, *A. afarensis* shows flatter molar teeth with less occlusal relief than do early *Homo* specimens. This suggests that early *Homo* would have been capable of more efficiently fracturing tough foods than would their australopith predecessors. Indeed, differences between these two taxa are of about the same magnitude as those separating chimpanzees and gorillas. This may imply similar degrees of difference in their diets. If so, perhaps both ate many of the same sorts of foods. Early members of our own genus may have occasionally consumed more tough foods, especially at times of resource stress, whereas the australopiths "fell back" on more hard, brittle foods.

Fallback Foods and Dental Functional Morphology

This brings up another important point. Apes have a penchant for succulent, sugar-rich foods—a legacy of the ancestral catarrhine dietary adaptation (Milton, 1993; Ross, 2000; Ungar, 2005). The same was likely true of early hominins. Differences in diet among living catarrhines often hinge on a seasonal shift to fallback foods taken when preferred resources are less available (Rogers et al., 1992; Lambert et al., 2004). Fecal studies for gorillas and chimpanzees at Lopé in the Gabon, for example, show 60 percent to 80 percent plant species overlap (Williamson et al., 1990; Tutin and Fernandez, 1993). These apes both eat fruits much of the year but diverge at "crunch times" when preferred fruits are scarce. At such times, gorillas fall back more on leaves and other fibrous plant parts. The same is true for sympatric mountain gorillas and chimpanzees at the Bwindi Impenetrable National Park in Uganda (Stanford and Nkurunungi, 2003).

Preferred resources may be easy to digest, may offer a low cost-benefit ratio, and may not result in selective pressures that would tax dental functional morphology. However, less desirable but seasonally critical fallback foods might require some morphological specialization (Robinson and Wilson, 1998). Kinzey (1978) recognized decades ago that dental specializations need not be restricted to preferred foods. Although *Callicebus moloch* and *C. torquatus* are both primarily frugivorous, the former have longer shearing crests for slicing leaves and the latter have larger talonid basins for crushing insect chitin. It is not surprising then that differences in occlusal morphology between living great apes relate more to fallback foods than to preferences per se. This effectively shifts the null hypothesis from resource preferences to fallback food exploitation when interpreting differences in occlusal morphology between early hominins.

Function and Phylogeny

Another important issue for dental functional morphologists is the parsing of function and phylogeny signals. This conundrum has plagued generations of graduate students facing their comprehensive examinations! While Hartman's (1988) analysis of hominoid molars suggests that tooth form differences among apes more reflect diet than evolutionary propinquity, phylogenetic inertia clearly plays an important role in how adaptations manifest themselves (Kay and Ungar, 1997). As Kay and Covert (1984) showed, for example, while SQs track diet within higher-level anthropoid taxa, Old World monkeys have relatively longer shearing crests than apes, and apes have longer shearing crests than New World monkeys independent of diet. These differences relate to phylogeny, which determines morphological "starting points." Thus, it is particularly important to take care when choosing extant baseline taxa for the interpretation of morphology of an extinct species. This is especially true for basal taxa or those with uncertain phyletic affinities.

In some cases, it may be possible to separate effects of function and phylogeny on occlusal morphology when considering higher-level radiations of primates. For example, early Miocene catarrhines tend to have less well developed shearing crests than do extant hominoids—though their ranges of SQ values are similar (Kay and Ungar, 1997). Dental microwear analysis, a nongenetic signal for diet, can be used to "anchor" fossil group ranges and confirm that living hominoids are "upshifted" relative to their early Miocene predecessors (Ungar, Teaford, and Kay, 2004). Bivariate comparisons of microwear and SQ results for given taxa help us unravel the effects of function and phylogeny in this case (Ungar, 1998).

The Unknown

The more we learn about dental functional morphology, the more we realize how little we actually know. Recent studies have raised more questions than they have answered and have pointed to new avenues for future research. The "unknowns" come in three categories: (1) functional aspects of tooth shape, (2) mechanical properties of foods, and (3) the mechanics of how tooth shape accomplishes food fracture.

Functional Aspects of Tooth Shape

Shearing crest lengths and overall occlusal slope give some measure of functional aspects of tooth form, but more precision is almost certainly possible. Occlusal slope provides important information about angles of approach between opposing teeth, but other aspects of morphology, such as surface angularity (see Ungar and M'Kirera, 2003), likely also reflect function. Blade theory dictates that jaggedness, or serratedness, of a cutting surface affects the directions of forces acting on a food item and so can have dramatic effects on fracture efficiency (Frazzetta, 1988). Primate occlusal surface jaggedness is determined in part by tooth wear as enamel gives way to softer dentin, and those dentin exposures develop steeply angled walls. This phenomenon may be comparable, though to a lesser extent, to that seen in herbivorous ungulates that have complex infoldings and lophs adapted to form sharp edges with dentin exposure for shearing and grinding tough foods (Rensberger, 1973; Fortelius, 1985). Perhaps this explains the thin enamel found in many primate folivores (Kay, 1981).

We still have much to learn, though, about how tooth wear affects shape. Because preliminary work suggests that individuals within primate species wear their teeth down in similar ways (Dennis et al., 2004), we can posit that natural selection favors teeth that wear in a specific manner that keep them mechanically efficient for fracturing foods. Wear should sculpt tooth surfaces in predictable ways in part because of differences in hardness between enamel and dentin. If so, an understanding of relationships between enamel crown shape and the shape of the underlying dentin cap should lead to a better understanding of occlusal form changes with wear. New technologies, such as computerized x-ray microtomography, hold the promise of allowing us to work out these relationships (e.g., Bjørndal et al., 1999).

Dental Biomechanics and the Mechanical Properties of Foods

We also have much to learn about how teeth achieve food fracture. Dental functional morphology has recently begun to take more of a biomechanical perspective, focusing on relationships between tooth shape and the strength, toughness, and deformability of foods (Strait, 1993a; Lucas and Teaford, 1994; Spears and Crompton, 1996; Yamashita, 1998; Lucas et al., 2004). Models of the mastication process continue to become more sophisticated, but a great deal of work remains to be done to gain a complete understanding of the many factors involved (Lucas, 2004).

Such studies depend on a solid understanding of the material properties of foods eaten by primates. While much progress is being made in efforts to document fracture properties of foods that are commercially produced, distributed, and processed (see Agrawal et al., 1997; Lucas et al., 2002), the same cannot be said of most wild foods. We assume that underground storage organs are hard and brittle and that edible game tissues are elastic and tough, but the fracture properties of these and other wild foods remain undocumented. Before we can really look closely at tooth form-function relationships for nonhuman primates, we need a better handle on the salient properties of their foods.

The Unknowable

As Gomory (1995) has noted, "We are all taught what is known, but we rarely learn about what is not known, and we almost never learn about the unknowable." It is difficult to assess knowability because its limits are determined by our frame of reference. Few in medieval Europe would have thought it possible to predict when and where a tsunami will next likely hit or when and where a volcano will erupt. Obsessing on the "unknowable." then, runs the risk of setting the bar too low. Nevertheless, discussion can be useful for keeping us grounded as we explore future research directions. Limits to knowability in the study of the evolution of human diet also take on some significance as nutritionists and physicians look to evolutionary medicine as they prepare menus to combat chronic degenerative diseases (e.g., Eaton and Konner, 1985; Eaton, Eaton, and Cordain, 2002).

Begun (2004) recently commented on the limits of functional morphology, noting that reconstructions of *Dryopithecus* and *Choloepus* label both taxa "frugivorous arboreal quadrupeds." He wrote, "While it is frustrating to be unable to describe a fossil hominoid's behavior with sufficient detail to be able to distinguish it from an edentate, that is probably as good as it gets" (497). This perception echo's Lauder's (1995) conclusion that, while we can predict basic ecological categories from functional morphology, we have much greater difficulty when trying to infer more detail.

Why the difficulty? Limits to our abilities to infer diet from teeth come on several levels, from those specific to dental form-function relationships to those inherent to all historical sciences. On the most basic level, we cannot infer more from occlusal morphology than food fracture properties. Differences in tooth form relate to differences in mechanical properties of the foods they are adapted to comminute. Occlusal morphology tells us nothing about nutrients or other attributes that are independent of food fracture characteristics. Tooth shape will never allow us to infer dietary proportions of fiber, carbohydrates, and protein, let alone ratios of omega 6 to omega 3 fatty acids.

The complex nature of food fracture and dental biomechanics also contributes to the limits of our ability to "predict" fossil hominin diets from occlusal morphology. Our ability to understand any natural phenomenon depends on the number of factors that affect that phenomenon (e.g., Gomory, 1995). As we identify more and more variables involved in the relationships between tooth form and function, the more complex, cumbersome and difficult to predict our models become (see Lucas, 2004). The more we learn about dental functional morphology and food fracture, the more unknowable the details seem to become. Perhaps this explains why we have been unable to get beyond basic ecological categories.

The inference of diet from dental functional morphology is also subject to the fundamental limits of the comparative method. Kay and Cartmill (1977) laid out the following requirements for use of the comparative method to infer a given function (F) from a given trait (T) found in a fossil species: (1) T must be present in some extant species; (2) for extant species with T, T must have F; (3) T must not have become fixed in a lineage before it assumed F; (4) T must have a functional relationship with F. The limitations of this approach are many. First, the comparative method requires classic inductive inference. No matter how many times a given tooth form is associated with a specific diet, we cannot prove that this association has held or will hold in all cases. This "problem of induction" is endemic to science and has plagued philosophers of science such as Bacon, Hume, Mills, von Wright, and others for more than a quarter of a millennium.

Although most of these philosophers were concerned with whether future observations would bear out general principles induced from past events, historical scientists attempt to infer past events from general principles induced from observations of the present. An additional problem with the comparative method then relates to a dependence on extant taxa and the kinds of inferences that can be drawn from them (e.g., Leroi et al., 1994; Doughty, 1996). The notion that the present limits our ability to understand the past has been a source of concern and debate for many paleoanthropologists. One big issue has been the belief that uniformitarian assumptions preclude us from discovering and understanding evolutionary novelties. Does reliance on modern analogs make it impossible to interpret patterns in the past that no longer exist? Kelley (1993) has argued that the misuse of uniformitarianism guarantees "a world of the past that is an exact replica of the present." If we assume that the past was substantively different than today, though, how could we falsify hypotheses? In the end, we have no choice but to use the present to interpret the past, or we risk leaving paleobiological interpretation in "indecipherable chaos" (Martin, 1991).

One solution is to focus on underlying processes rather than their outcomes. As Teaford, Walker, and Mugaisi (1993) note, "It is these processes, not the resultant morphology and behavior, that remain relatively constant in the eyes of uniformitarianism" (389). From this perspective, one can certainly document evolutionary novelties using the comparative method (e.g., Solounias, Teaford, and Walker, 1988). Knowing that the range of shearing quotients in late Miocene European apes was greater than that those of living hominoids, for example, allows us to infer that the ranges of ape diets were greater in the past than today (Ungar and Kay, 1995).

A related problem is parsing function from phylogeny. In some cases, such as the late Miocene radiation of hominoids, we may be able to control for phylogeny, in others we may not. While we know, for example, that living cercopithecoids have longer shearing crests and larger incisors than platyrrhines independent of diet, we cannot know the effects of phylogenetic heritage on basal taxa or those with uncertain phyletic affinities, at least unless we have other evidence of their behaviors (Kay and Ungar, 1997; Ungar, 1998; Ungar, Teaford, and Kay, 2004; see Teaford, chapters 5 and 7, and Sponheimer, Lee-Thorp, and de Ruiter, chapter 8).

Conclusions

Can we hope to get a complete picture of food fracture properties in early hominins based on their occlusal morphology? Of course not. Does this mean we should stop trying? Of course not. We certainly know more today about hominin diets than we ever have before, and new inferences show little sign of abatement. We know, for example, that early *Homo* was better able to process tough, elastic foods than could their *Australopithecus* predecessors. We also need to recognize that we have access to many different lines of evidence, as presented in other chapters of this book. Taken together, the various approaches can allow us to develop an increasingly detailed picture of the diets of our ancestors. Still, we should acknowledge that there are many things we do not know and will probably never know.

Acknowledgments I am grateful to several colleagues for discussions of dental topographic analysis and/or early hominin paleoecology over the years, including John Dennis, Ken Glander, Fred Grine, Rich Kay, Fred Limp, Peter Lucas, Francis M'Kirera, Gildas Merceron, Alejandro Pérez-Pérez, Mike Plavcan, Jerry Rose, Rob Scott, Sarah Taylor, Mark Teaford, Alan Walker, Malcolm Williamson, John Wilson, and Lucy Zuccotti. Access to the ape collections used in this chapter was granted by curators at the Cleveland Museum of Natural History and the State Collection of Anthropology and Palaeoanatomy in Munich. Dental replicas were prepared with the help of Mark Teaford, and much of the data on the extant primates were originally generated by Francis M'Kirera and Sarah Taylor. Research described in this chapter was funded in part by the L.S.B. Leakey Foundation, and the U.S. National Science Foundation.

References

Agrawal, K.R., Lucas, P.W., Prinz, J.F., and Bruce, I.C., 1997. Mechanical properties of foods responsible for resisting food breakdown in the human mouth. *Arch. Oral Biol.* 42, 1–9.

Anthony, M.R.L., and Kay, R.F., 1993. Tooth form and diet in ateline and alouattine primates: Reflections on the comparative method. *Am. J. Sci.* 293A, 356–382.

Begun, D.R., 2004. The three "Cs" of behavior reconstruction in fossil primates. *J. Hum. Evol.* 46, 497–503.

Bjørndal, L., Carlsen, O., Thuesen, G., Darvann, T., and Kreiborg, S., 1999. External and internal macrotomography in 3D-reconstructructed maxillary molars using computerized X-ray microtomography. *Int. Endodont. J.* 32, 3–9.

Crompton, A.W., and Sita-Lumsden, A.G., 1970. Functional significance of therian molar pattern. *Nature* 227, 197–199.

Dennis, J.C., Ungar, P.S., Teaford, M.F., and Glander, K.E., 2004. Dental topography and molar wear in *Alouatta palliata* from Costa Rica. *Am. J. Phys.* Anthropol. 125, 152–161.

Doughty, P., 1996. Statistical analysis of natural experiments in evolutionary biology: Comments on recent criticisms of the use of comparative methods to study adaptation. *Am. Nat.* 148, 943–956.

Eaton, S.B., Eaton S.B., III, and Cordain, L., 2002. Evolution, diet and health. In: Ungar, P.S. and Teaford, M.F. (Eds.), *Human Diet: Its Origin and Evolution.* Bergen & Garvey, Westport, CT, pp. 7–18.

Eaton, S.B., and Konner, M., 1985. Paleolithic nutrition: A consideration of its nature and current implications. *New Eng. J. Med.* 312, 283–289.

Evans, A.R., Harper, I.S., and Sanson, G.D., 2001. Confocal imaging, visualization and 3-D surface measurement of small mammalian teeth. *J. Microsc.* 204, 108–118.

Fleagle, J.G., Kay, R.F., and Anthony, M.R.L., 1997. Fossil New World monkeys. In: Kay, R.F., Madden, R.H., Cifelli, R.L., and Flynn, J.J. (Eds.), *Vertebrate Paleontology in the Neotropics*. Smithsoninan Institution Press, Washington DC, pp. 473–495.

Fortelius, M., 1985. Ungulate cheek teeth: Developmental, functional and evolutionary interrelations. *Acta Zool. Fenn.* 180, 1–76.

Frazzetta, T.H., 1988. The mechanics of cutting and the form of shark teeth (Chondrichthyes, Elasmobranchii). *Zoomorphology* 108, 93–107.

Gomory, R.E., 1995. The known, the unknown and the unknowable. *Sci. Am.* 272, 120.

Gregory, W.K., 1922. *The Origin and Evolution of Human Dentition*. Williams and Wilkins, Baltimore.

Grine, F.E., 1981. Trophic differences between "gracile" and "robust" australopithecines: A scanning electron microcope analysis of occlusal events. *S. Afr. J. Sci.* 77, 203–230.

Hartman, S.E., 1988. A cladistic analysis of hominoid molars. *J. Hum. Evol.* 17, 489–502.

Hiiemae, K., and Kay, R.F., 1972. Trends in evolution of primate mastication. *Nature* 240, 486–487.

Jernvall, J., and Selänne, L., 1999. Laser confocal microscopy and geographic information systems in the study of dental morphology. *Palaeontol. Electron.* 2, 18 p.

Kay, R.F., 1977. Diets of early Miocene African hominoids. *Nature* 268, 628–630.

Kay, R.F., 1978. Molar structure and diet in extant Cercopithecidae. In: Butler, P.M., and Joysey, K.A. (Ed.), *Development, Function, and Evolution of Teeth.* Academic Press, New York, pp. 309–339.

Kay, R.F., 1981. The nut-crackers: A new theory of the adaptations of the Ramapithecinae. *Am. J. Phys. Anthropol.* 55, 141–151.

Kay, R.F., 1984. On the use of anatomical features to infer foraging behavior in extinct primates. In: Rodman, P.S., and Cant, J.G.H. (Eds.), *Adaptations for Foraging in Nonhuman Primates: Contributions to an Organismal Biology of Prosimians, Monkeys and Apes*. Columbia University Press, New York, pp. 21–53.

Kay, R.F., 1985. Dental evidence for the diet of *Australopithecus*. *Annu. Rev. Anthropol.* 14, 315–341.

Kay, R.F., and Cartmill, M., 1977. Cranial morphology and adaptations of *Palaechthon nacimienti* and Other Paromomyidae (Plesiadapoidea, Primates), with a description of a new genus and species. *J. Hum. Evol.* 6, 19–53.

Kay, R.F., and Covert, H.H., 1984. Anatomy and behavior of extinct primates. In: Chivers, D. J., Wood, B. A., and Bilsborough, A. (Eds.), *Food Acquisition and Processing in Primates*. Plenum Press, New York, pp. 467–508.

Kay, R.F., and Hiiemae, K.M., 1974. Jaw movement and tooth use in recent and fossil primates. *Am. J. Phys. Anthropol.* 40, 227–256.

Kay, R.F., Simons, E.L., 1980. The ecology of Oligocene African Anthropoidea. *Int. J. Primatol.* 1, 21–37.

Kay, R.F., and Ungar, P.S., 1997. Dental evidence for diet in some Miocene catarrhines with comments on the effects of phylogeny on the interpretation of adaptation. In: Begun, D.R., Ward, C., and Rose, M. (Eds.), *Function, Phylogeny and Fossils: Miocene Hominoids and Great Ape and Human Origins.* Plenum Press, New York, pp. 131–151.

Kelley, J., 1993. Taxonomic implications of sexual dimorphism in *Lufengpithecus*. In: Kimbel, W.H., and Martin, L.B. (Eds.), *Species, Species Concepts, and Primate Evolution.* Plenum Press, New York, pp. 429–458.

Kinzey, W.G., 1978. Feeding behavior and molar features in two species of titi monkey. In: Chivers, D.J., and Herbert, J. (Eds.), *Recent Advances in Primatology*. Vol. 1: *Behavior*. Academic Press, New York, pp. 373–385.

Kirk, E.C., and Simons, E.L., 2001. Diets of fossil primates from the Fayum Depression of Egypt: A quantitative analysis of molar shearing. *J. Hum. Evol.* 40, 203–229.

Kuroda, S., 1992. Ecological interspecies relationships between gorillas and chimpanzees in the Ndoki-Nouabale Reserve, northern Congo. In: Itoigawa, N., Sugiyama, Y., Sackett,

G.P., and Thompson, R.K.R. (Eds.), *Topics in Primatology*. Vol. 2: *Behavior, Ecology and Conservation*. University of Tokyo Press, Tokyo, pp. 385–394.

Lambert, J.E., Chapman, C.A., Wrangham, R.W., and Conklin-Brittain, N.L., 2004. The hardness of cercopithecine foods: implications for the critical function of enamel thickness in exploiting fallback foods. *Am. J. Phys. Anthropol.* 125, 363–368.

Lauder, G.V., 1995. On the inference of function from structure. In: Thomason, J. (Ed.), *Functional Morphology in Vertebrate Paleontology.* Cambridge University Press, Cambridge, pp. 1–18.

Leroi, A.M., Rose, M.R., and Lauder, G.V., 1994. What does the comparative method reveal about adaptation. *Am. Nat.* 143, 381–402.

Lucas, P.W., 2004. *Dental Functional Morphology: How Teeth Work.* Cambridge University Press, New York.

Lucas, P.W., Prinz, J.F., Agrawal, K.R., and Bruce, I.C., 2002. Food physics and oral physiology. *Food Qual. Pref.* 13, 203–213.

Lucas, P.W., Prinz, J.F., Agrawal, K.R., Bruce, I.C., 2004. Food texture and its effect on ingestion, mastication and swallowing. *J. Text. Stud.* 35, 159–170.

Lucas, P.W., and Teaford, M.F., 1994. Functional morphology of colobine teeth. In: Davies, A.G., and Oates, J.F. (Eds.), *Colobine Monkeys: Their Ecology, Behaviour and Evolution*. Cambridge University Press, Cambridge, pp. 173–203.

M'Kirera, F., and Ungar, P.S., 2003. Occlusal relief changes with molar wear in *Pan troglodytes troglodytes* and *Gorilla gorilla gorilla*. *Am. J. Primatol.* 60, 31–41.

MacKinnon, J., 1977. A comparative ecology of Asian Apes. *Primates* 18, 747–772.

Martin, L., 1991. Paleoanthropology: Teeth, sex and species. *Nature* 352, 111–112.

Meldrum, D.J., and Kay, R.F., 1997. *Nuciruptor rubricae*, a new pitheciin seed predator from the Miocene of Colombia. *Am. J. Phys. Anthropol.* 102, 407–427.

Milton, K., 1993. Diet and primate evolution. *Sci. Am.* 269, 86–93.

Nishihara, T., 1992. A preliminary report on the feeding habits of western lowland gorillas (*Gorilla gorilla gorilla*) in the Ndoki Forest, Northern Congo. In: Itoigawa, N., Sugiyama, Y., Sackett, G.P., and Thompson, R.K.R. (Eds.), *Topics in Primatology*. Vol. 2: *Behavior, Ecology and Conservation*. University of Tokyo Press, Tokyo, pp. 225–240.

Owen, R., 1840. *Odontography*. Hippolyte Bailliere, London.

Remis, M.J., 1997. Western lowland gorillas (*Gorilla gorilla gorilla*) as seasonal frugivores: Use of variable resources. *Am. J. Primatol.* 43, 87–109.

Rensberger, J.M., 1973. Occlusion model for mastication and dental wear in herbivorous mammals. *J. Paleontol.* 47, 515–528.

Robinson, B.W., and Wilson, D.S., 1998. Optimal foraging, specialization, and a solution to Liem's paradox. *Am. Nat.* 151, 223–235.

Rodman, P. S., 1977. Feeding behaviour of orangutans of the Kutai Nature Reserve, East Kalimantan. In: Clutton-Brock, T.H. (Ed.), *Primate Ecology: Studies of Feeding and Ranging Behaviour in Lemurs, Monkeys and Apes*. Academic Press, London, pp. 383–413.

Rogers, M.E., Maisels, F., Williamson, E.A., Tutin, C.E., and Fernandez, M., 1992. Nutritional aspects of gorilla food choice in the Lopé Reserve, Gabon. In: Itoigawa, N., Sugiyama, Y., Sackett, G.P., and Thompson, R.K.R. (Eds.), *Topics in Primatology*. Vol. 2: *Behavior, Ecology and Conservation*. University of Tokyo, Tokyo, pp. 267–281.

Rosenberger, A.L., and Kinzey, W.G., 1976. Functional patterns of molar occlusion in platyrrhine primates. *Am. J. Phys. Anthropol.* 45, 281–297.

Ross, C.F., 2000. Into the light: The origin of Anthropoidea. *Annu. Rev. Anthropol.* 29, 147–194.

Seligsohn, D., and Szalay, F.S., 1978. Relationship between natural selection and dental morphology: Tooth function and diet in *Lepilemur* and *Hapalemur*. In: Butler, P.M., and Joysey, K.A. (Eds.), *Development, Function and Evolution of Teeth*. Academic Press, New York, pp. 289–307.

Simpson, G.G., 1933. Paleobiology of Jurassic mammals. *Paleobiology* 5, 127–158.

Smith, E., 1999. A functional analysis of molar morphometrics in living and fossil hominoids. PhD diss., University of Toronto.

Solounias, N., Teaford, M., and Walker, A., 1988. Interpreting the diet of extinct ruminants: The case of a non-browsing giraffid. *Paleobiology* 14, 287–300.

Spears, I.R., and Crompton, R.H., 1996. The mechanical significance of the occlusal geometry of great ape molars in food breakdown. J. *Hum. Evol.* 31, 517–535.

Stanford, C.B., and Nkurunungi, J.B., 2003. Do wild chimpanzees and mountain gorillas compete for food ? *Am. J. Phys. Anthropol.* 198–199.

Strait, S.G., 1993a. Differences in occlusal morphology and molar size in frugivores and faunivores. *J. Hum. Evol.* 25, 471–484.

Strait, S.G., 1993b. Molar morphology and food texture among small bodied insectivorous mammals. *J. Mammal.* 74, 391–402.

Teaford, M.F., 1981. Molar wear patterns in *Macaca fascicularis, Presbytis cristatus*, and *Presbytis rubicunda*: A photogrammetric analysis.PhD diss., University of Illinois.

Teaford, M.F., 1983. The morphology and wear of the lingual notch in macaques and langurs. *Am. J. Phys. Anthropol.* 60, 7–14.

Teaford, M.F., Ungar, P.S., and Grine, F.E., 2002. Paleontological evidence for the diets of African Plio-Pleistocene hominins with special reference to early *Homo*. In: Ungar, P.S., and Teaford, M.F (Eds.), *Human Diet: Its Origin and Evolution.* Bergin & Garvey, Westport, CT, pp. 143–166.

Teaford, M.F., Walker, A., and Mugaisi, G.S., 1993. Species discrimination in *Proconsul* from Rusinga and Mfangano Islands, Kenya. In: Kimbel, W.H., and Martin, L.B. (Eds.), *Species, Species Concepts, and Primate Evolution*. Plenum Press, New York, pp. 373–392.

Thexton, A., and Hiiemae, K.M., 1997. The effect of food consistency upon jaw movement in the macaque: A cineradiographic study. *J. Dent. Res.* 76, 552–560.

Tutin, C.E.G., and Fernandez, M., 1993. Composition of the diet of chimpanzees and comparisons with that of sympatric lowland gorillas in the Lope Reserve, Gabon. *Am. J. Primatol.* 30, 195–211.

Tutin, C.E.G., Fernandez, M., Rogers, M.E., Williamson, E.A., and Mcgrew, W.C., 1991. Foraging profiles of sympatric lowland gorillas and chimpanzees in the Lope Reserve, Gabon. *Philos. Trans. R. Soc. Lond.* B 334, 179–186.

Tutin, C.E.G., Ham, R.M., White, L.J.T., and Harrison, M.J.S., 1997. The primate community of the Lope Reserve, Gabon: Diets, responses to fruit scarcity, and effects on biomass. *Am. J. Primatol.* 42, 1–24.

Ungar, P., 1998. Dental allometry, morphology, and wear as evidence for diet in fossil primates. *Evol. Anthropol.* 6, 205–217.

Ungar, P.S., 2004. Dental topography and diets of *Australopithecus afarensis* and early *Homo. J. Hum. Evol.* 46, 605–622.

Ungar, P.S., 2005. Dental evidence for the diets of fossil primates from Rudabánya, northeastern Hungary with comments on extant primate analogs and “noncompetitive” sympatry. *Palaeontogr. Ital.* 90, 97–111.

Ungar, P.S., and Kay, R.F., 1995. The dietary adaptations of European Miocene catarrhines. *Proc. Natl. Acad. Sci.* 92, 5479–5481.

Ungar, P.S., and M’Kirera, F., 2003. A solution to the worn tooth conundrum in primate functional anatomy. *Proc. Natl. Acad. Sci.* 100, 3874–3877.

Ungar, P.S., and Taylor, S.R., 2005. Dental topographic analysis: Tooth wear and function. *Am. J. Phys. Anthropol.* Suppl. 40, 210.

Ungar, P.S., Teaford, M.F., Grine, F.E., 1999. A preliminary molar study of occlusal relief in *Australopithecus africanus* and *Paranthropus robustus*. *Am. J. Phys. Anthropol.* Suppl. 28, 269.

Ungar, P.S., Teaford, M.F., and Kay, R.F., 2004. Molar microwear and shearing crest development in Miocene catarrhines. *Anthropology* 42, 21–35.

Ungar, P.S., and Williamson, M., 2000. Exploring the effects of tooth wear on functional morphology: A preliminary study using dental topographic analysis. *Palaeontol. Electron.* 3, 18 p.

Williams, B.A., and Covert, H.H., 1994. New early Eocene anaptomorphine primate (Omomyidae) from the Washakie Basin, Wyoming, with comments on the phylogeny and paleobiology of anaptomorphines. *Am. J. Phys. Anthropol.* 93, 323–340.

Williamson, E.A., Tutin, C.E.G., Rogers, M.E., and Fernandez, M., 1990. Composition of the diet of lowland gorillas at Lope in Gabon. *Am. J. Primatol.* 21, 265–277.

Wood, B.A., Abbott, S.A., and Graham, S.H., 1983. Analysis of the dental morphology of Plio-Pleistocene hominids. 2. Mandibular molars: Study of cusp areas, fissure pattern and cross-sectional shape of the crown. *J. Anat.* 137, 287–314.

Yamashita, N., 1998. Functional dental correlates of food properties in five Malagasy lemur species. *Am. J. Phys. Anthropol.* 106, 169–188.

Zuccotti, L.F., Williamson, M.D., Limp, W.F., and Ungar, P.S., 1998. Technical note: Modeling primate occlusal topography using geographic information systems technology. *Am. J. Phys. Anthropol.* 107, 137–142.

5

What Do We Know and Not Know about Diet and Enamel Structure?

MARK F. TEAFORD

Teeth are the most common elements in the human fossil record—largely because they're the most resilient structures in the body. For the most part, they are made of inorganic materials, and they tend to remain intact well after death, unless they are subject to postmortem abrasion or erosion. Thus, it's no surprise that they've provided many clues about the evolution of human diet.

However, when you consider the range of "diets" currently sold through the mass media, and when you look at the intricate complexities of nonhuman primate diets in the wild, it becomes all too obvious that "diet" is an incredibly complicated topic. So if we're ever going to have any hope of documenting the origins and evolution of human diet, we will have to use every available piece of evidence—each with its strengths and weaknesses. The focus of this chapter will be on dental enamel, specifically, its thickness and microstructure.

What Do We Know about Dental Enamel?

Teeth are composed of a number of materials—most notably enamel and dentin. The focus here is on enamel because it forms the outer coating of a tooth—at least until portions of it are worn away. Thus, it is the most likely component to come into contact with food during an animal's lifetime. It is one of those rare materials in the body that *is not* remodeled during life or replaced if it's lost (that is, unless you're an aye-aye with ever-growing incisors). Once your teeth are in, what you see is what you get! So enamel needs to survive as long as possible. If it cannot withstand the processing of certain foods, and if those foods are crucial for survival, then evolutionary changes in its structure will ultimately be dictated, in part, by the demands of food processing. Its structure is not simple, as it is a complex composite, made of crystals spun together in a complicated fashion forming what are known as *prisms* (Boyde, 1964; Carlson, 1990). From this complexity, however, insight can result, and there are two main areas of interest when it comes to dietary inferences: (1) variations in enamel thickness and (2) variations in enamel microstructure.

Enamel Thickness

One way enamel can survive the demands of wear and mastication is to start with a very thick cap of it on the tooth. Not surprisingly, when investigators began looking at tooth structure (e.g., Butler, 1956; Robinson, 1956), they discovered that some species seemed to have thick enamel, while others seemed to have thin enamel. Initial surveys of enamel thickness in primates reiterated this point and noted some intriguing differences between species (Gantt, 1977; Molnar and Gantt, 1977; Kay, 1981). For instance, primates characterized as hard-object feeders (e.g., *Cebus apella*) tended to have thicker molar enamel than did soft-fruit eaters (e.g., *Ateles belzebuth*). However, humans also seemed to have relatively thick molar enamel, raising the possibility that we, or our ancestors, were also somehow adapted for eating hard objects.

These early studies were based on linear measurements of enamel thickness, wherever and however they could be obtained, and therein lay the dilemma. To really document enamel thickness, we have to somehow "see" inside the tooth. Thus, Gantt used prepared sections of teeth, while Kay used measures of enamel thickness taken at the margins of dentin exposures. Both approaches had their limitations. The former required, at the very least, damaging the tooth. Even then, it only yielded measures of enamel thickness along the plane of section. By contrast, measuring enamel thickness at the margins of dentin exposures yielded a larger sample of individuals, but an even more limited sample of locations across the tooth, not to mention the fact that the enamel thickness in each case was ultimately revealed by an unknown amount of wear, so however much enamel had been previously worn away remained unknown. Neither approach gave an accurate measure of enamel thickness across the entire tooth crown (Schwartz, 2000a, 2000b), and, of course, both approaches needed to take into account differences in body size. Martin (1983, 1985) subsequently developed a more sophisticated measure of relative enamel thickness, incorporating the length of the enamel-dentin junction and the area of the enamel cap. But it still only yielded a summary measure of thickness, which may mask functionally important differences across the tooth crown (Macho and Thackeray, 1992; Macho and Berner, 1993; Macho, 1994; Schwartz, 2000b). Given the complexity of crown shape and development, it is not surprising then that there have been continuing discussions about the proper methods of analysis (e.g., Conroy, 1991; Grine, 1991, 2002, 2005; Macho and Thackeray, 1992; Macho and Berner, 1993; Spoor, Zonneveld, and Macho, 1993; Macho, 1994; Dumont, 1995; Schwartz et al., 1998; Schwartz, 2000a, 2000b; Grine, Stevens, and Jungers, 2001; Kono, Suwa, and Tanijiri, 2002; Kono, 2004; Suwa and Kono, 2005). In essence, most researchers are still torn between using a very precise method of measurement (histological analyses of sectioned crowns), which is inherently destructive and thus of limited usefulness on fossils, or various noninvasive, radiographic techniques, which have some methodological limitations (Grine 1991, 2005; Spoor, Zonneveld, and Macho, 1993; Schwartz et al., 1998; Grine, Stevens, and Jungers, 2001; Suwa and Kono, 2005) but which may also be the standard for future analyses (see Shimizu, 2002; Kono, 2004; and Suwa and Kono, 2005, for recent refinements of this approach). The advent of high-resolution laser scanning has provided another option: scanning the tooth crown and scanning the enamel-dentin junction, then electronically comparing the two over the entire surface of the enamel-dentin junction (Kono, Suwa, and Tanijiri, 2002). Unfortunately, this method requires the physical removal of the crown from the enamel-dentin junction, making analyses of rare museum material impossible. The net effect of these methodological difficulties is that we still have little more than a varied assortment of analyses of molar enamel thickness in modern and fossil primates. As noted by Grine (2005) and Kono (2004), few studies of modern primates have used samples large enough for statistical analysis, and analyses of fossils have been limited in scope and number.

So where does that leave us? As might be expected, the best data are from modern humans, where we have an ample supply of teeth. There, investigators have shown that not only do distal molars tend to have thicker enamel than more proximal ones (Macho and Berner 1993, 1994; Spears and Macho, 1995, 1998; Macho and Spears, 1999; Schwartz, 2000a, 2000b; Gantt et al., 2001; Grine, 2005), but there are also differences in enamel thickness between the buccal and lingual cusps of molar crowns, perhaps reflecting differences in the functional demands placed on them (Shillingburg and Grace, 1973; Molnar and Gantt, 1977; Grine and Martin, 1988; Macho and Berner, 1993, 1994; Spears and Macho, 1995; Schwartz, 2000a, 2000b; Gantt et al., 2001). From this perspective, thicker enamel would then help the tooth withstand higher occlusal forces thought to occur on either more distal molars, or the so-called functional cusps of molars. However, more recent analyses (Kono, Suwa, Tanijiri, 2002; Grine, 2005; Suwa and Kono, 2005) have shown that the correlation between intercusp differences in enamel thickness and dental function is not that simple. Still, patterns of variation in enamel thickness are making researchers think about how teeth are actually used and the relative magnitude of forces applied along the tooth row.

As for nonhuman primates, detailed analyses are lagging behind those of humans. After the pioneering work of Molnar and Gantt (1977), Martin (1983, 1985) put interspecific comparisons of primates on firmer methodological footing. Shellis and colleagues (1998), Schwartz (2000b), and Martin, Olejniczak, and Maas (2003) then used larger samples to make more robust statistical comparisons between species. If one were to use a summary "measure" to characterize molar enamel thickness in modern hominoids, it would probably be something as follows: *Gorilla* and *Pan* have relatively thin enamel, *Pongo* has medium-thick enamel, depending on which teeth are examined, and *Homo* has thick enamel (Shellis et al., 1998; Schwartz, 2000b). These variations are probably due to differences in the *duration* of crown formation rather than the rate of enamel production (Beynon, Dean, and Reid, 1991). However, as in humans, these summaries may also mask important functional differences across the tooth crown. For instance, while *Pongo* may have slightly thinner enamel than *Homo*, the distribution of enamel across their molar crowns is also different with *Pongo* having relatively thick enamel in areas associ-

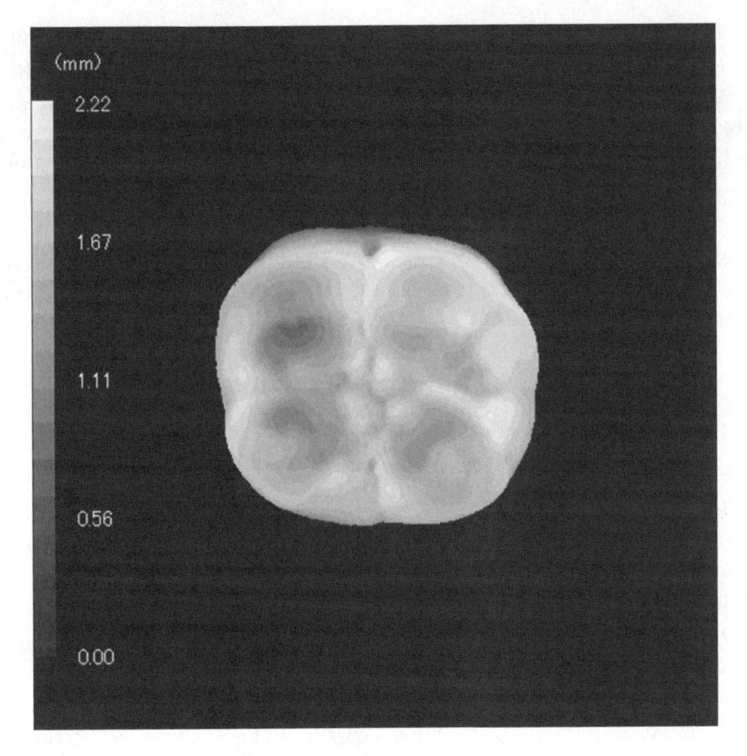

Figure 5.1 Grayscale rendering of enamel thickness across the crown of a human permanent molar. Modified from Kono et al. (2002) and kindly provided by Reiko Kono.

ated with crushing (Schwartz, 2000b; Kono, 2004) and *Homo* having thicker enamel across the entire crown (Kono, 2004). Taking analyses one step further in cercopithecoids, Shimizu (2002) and Kono (2004) have noted correlations between patterns of molar enamel thickness and the location and morphology of dentin exposures which can be of major functional significance as tooth wear progresses (see below). In light of these studies, perhaps it would be better to focus on enamel *distribution* in the future (e.g., Kono, 2004; fig. 5.1), rather than merely enamel thickness.

Of course, studies of enamel thickness or distribution in human ancestors have been forced to use whatever information is obtainable however it might be obtained. Occasionally, fossil teeth have been sectioned (e.g., Martin, 1983, 1985; Grine and Martin, 1988). However, more often than not, measures of enamel thickness have been taken at fortuitous breaks in fossil teeth, along the margins of dentin exposures or via various radiographic techniques. Obviously, the latter hold great promise for future work as improvements in resolution and three-dimensional analyses are made. For the moment though, a number of basic patterns have emerged. Initially, it was felt that the earliest Miocene apes had relatively thin enamel, and thus, this might be a primitive trait for apes and humans (Martin, 1983, 1985; Andrews and

Martin, 1991). More detailed analyses have subsequently shown that Miocene apes show a range of molar enamel thickness, from relatively thin in *Proconsul major* and *P. africanus* to relatively thick in *P. nyanzae* and various sivapthecids (Beynon, Dean, et al., 1998; Smith, Martin, and Leakey, 2003). Of course, prehistoric species are not immune to the effects of body size, thus estimates of relative enamel thickness have proven problematic for taxa with fragmentary or incomplete remains. This is especially true for some of the earliest hominins, where glimpses of absolute measures suggest thicker enamel in some taxa (e.g., *Sahelanthropus*) and thinner enamel in others (e.g., *Ardipithecus*). But without clearer indications of body size, the estimates are little more than tempting morsels for discussion and debate. Still, despite methodological differences between studies and the use of small sample sizes (Martin, 1985; Beynon and Wood, 1986; Grine and Martin, 1988; Beynon, Dean, and Reid, 1991; Macho and Thackeray, 1992; Spoor, Zonneveld, and Macho, 1993; Macho, 1994), there is a general consensus that the early hominins had relatively thick enamel compared with modern humans and most other living primates. Moreover, analyses suggest that *Paranthropus* species had thicker enamel than other early hominins, including early *Homo* (Beynon and Wood, 1986; Grine and Martin, 1988; Macho and Thackeray, 1992; Schwartz et al., 1998). Finally, while it may be difficult to separate early *Homo* from some australopiths based on isolated measures of enamel thickness (Tobias, 1991; Ramirez-Rozzi, 1998), *Homo erectus* appears to have the thinnest enamel of those Plio-Pleistocene hominins analyzed by Beynon and Wood (1986). This raises the possibility that, with the arrival of *H. erectus*, and probably a heavier emphasis on tool use, enamel thickness may have started a gradual decline that continues to this day. Of course, this raises questions about the selective mechanisms behind such a trend.

So why have thicker, or thinner, molar enamel? Two adaptive explanations are usually given in the literature: (1) thick enamel may have evolved to prolong the functional life of teeth exposed to an abrasive diet (Macho and Spears, 1999), or (2) it evolved to strengthen teeth given high occlusal forces generated by a diet that includes hard foods (Kay, 1981; Dumont, 1995). The first explanation is akin to that used to explain hyposodonty in grazing ungulates (Janis and Fortelius, 1988). However, most researchers have recently viewed the second explanation as more acceptable for primates because primate hard-object specialists have been shown to have thicker tooth enamel than closely related taxa that habitually eat softer foods (Kay, 1981; Dumont, 1995). Thin tooth enamel, however, may be better suited to a tough food diet, as dentin exposure yields sharp edges on an occlusal surface that can improve shearing and slicing efficiency (Delson, 1973; Lumsden and Osborn, 1977; Kay, 1981, 1985; Benefit, 1987; Dean, Jones, and Pilley, 1992; Lucas and Teaford, 1994; Spears and Crompton, 1996; Ungar and Williamson, 2000; Shimizu, 2002; Ungar and M'Kirera, 2003; Lucas, 2004). Thus thin enamel may itself be of evolutionary significance under certain dietary regimens (by aiding in the maintenance of dental function with wear) rather than simply being what is left in the absence of thick enamel.

Interestingly, functional generalizations such as these, while useful to a point, are now being modified by information from two other sources, as researchers realize that intratooth variations in enamel thickness are only partially explained by adaptations to the functional demands of mastication (Kono, Suwa, and Tanijiri, 2002;

Kono, 2004; Grine, 2005; Suwa and Kono, 2005). The first source of new information concerns the physical properties of primate foods and how primates actually process those foods. In essence, investigators have grossly oversimplified primate foods by lumping them into categories like fruits versus leaves, hard versus soft, and so forth. Recent work has shown that primate foods are far more physically complex than workers had previously imagined (Kinzey and Norconk, 1990, 1993; Lucas and Pereira, 1990; Lucas and Teaford, 1994; Lucas et al., 1995, 2000; Strait, 1997; Wrangham, Conklin-Brittain, and Hunt, 1998; Yamashita, 1998; Lambert et al., 2004; Teaford et al., 2006). This is of particular importance for interpreting the functional significance of molar enamel thickness, where investigators have often equated "seed eating" with "hard-object feeding" (e.g., Godfrey et al., 2004). As it turns out, many seeds eaten by primates are *not* hard but instead are pliant and tough (Lucas and Teaford, 1994), and even those seeds that *are* encased in hard pericarp are often opened with the front teeth, leaving the molars to process seeds that are again surprisingly soft (Kinzey and Norconk, 1990). The potential impact on interpretations of the functional significance of variations in molar enamel thickness is huge, as some primate seed eaters may have relatively thin enamel, and some primate hard-object feeders may still be using their molars to process relatively soft foods. Clearly, we need a better understanding of primate feeding and the properties of primate foods. However, we also need a better understanding of dental microstructure, as enamel thickness, by itself, only gives a limited view of dental function.

Enamel Microstructure

Because enamel is indeed a complex composite, the intricate details of its development and morphology have provided a wealth of research possibilities, ranging from permanent markers of developmental history, to structural anisotropy of functional significance. For instance, close examination of enamel prisms has revealed so-called cross-striations laid down in a circadian fashion (Schour and Hoffman, 1939; Massler and Schour, 1946; Boyde, 1964; Fitzgerald, 1998). This, coupled with longer-term surface markers known as *perikymata*, has allowed researchers to estimate the amount of time necessary for crown completion in modern hominoids (Dean and Wood, 1981; Kuykendall, 1996; Beynon, Clayton, et al., 1998; Reid et al., 1998; Shellis, 1998; Dean, 2000). It has also given insights into tooth formation time and age at death in fossils (Bromage and Dean, 1985; Beynon and Dean, 1988; Dean et al., 1993), thereby suggesting that most of the early hominins had an apelike pattern of dental development.

Before histological studies of the timing of dental development, researchers felt that the shape of prisms in prepared tooth sections (prism packing patterns) could be used in phylogenetic studies, as certain patterns might be characteristic of certain taxonomic groups (Shellis and Poole, 1977; Gantt, 1979, 1983). However, subsequent work showed that results were often dependent on methods of specimen preparation (Boyde, Jones, and Reynolds, 1978; Vrba and Grine, 1978). Detailed analyses of enamel at controlled depths subsequently suggested that hominoids might exhibit an unusual preponderance of "type 3" enamel (Boyde and Martin, 1982; Martin, Boyde, and Grine, 1988). However, more work is still necessary to document the range of possibilities within and between large samples of teeth.

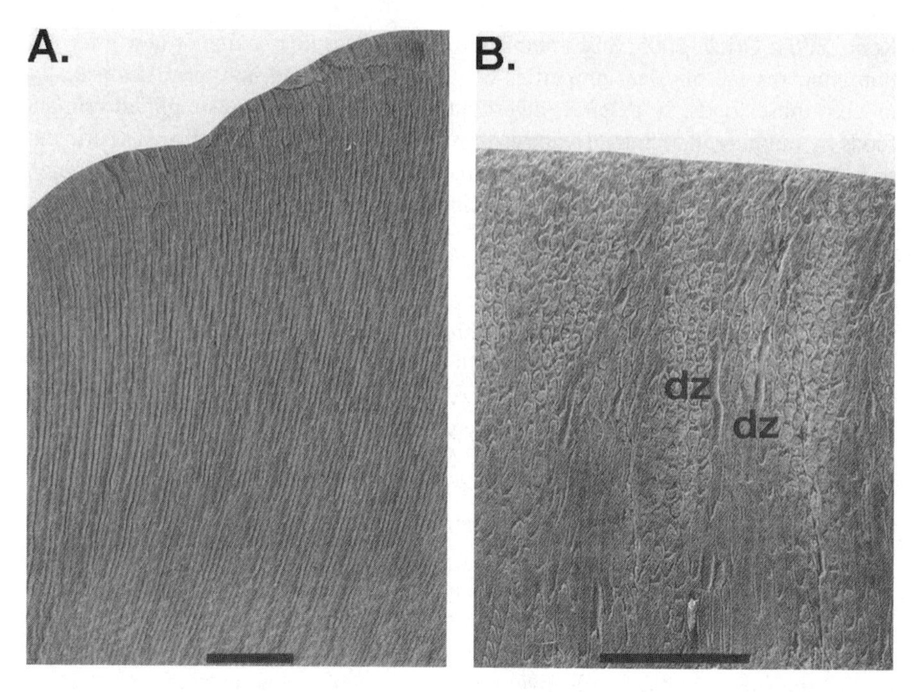

Figure 5.2 Contrast between radial enamel (*A*) and decussating enamel (*B*) in Oligocene primates from the Fayum of Egypt (from Teaford, et al., 1996; "DZ" = decussating zones of prisms). Original micrographs prepared by Mary Maas.

More germane to this discussion, what are the *functional* implications of these variations in enamel microstructure? Work with a wide range of mammals has shown that the building blocks of enamel (i.e., crystallites and prisms) can be arranged in innumerable ways (Koenigswald and Clemens, 1992; Koenigswald and Sander, 1997). As a result, the functional implications of variations in enamel microstructure are complex, to say the least (Koenigswald, 1980; Rensberger and Koenigswald, 1980; Fortelius, 1985; Boyde and Fortelius, 1986; Pfretzschner, 1986; Koenigswald, Rensberger, and Pfretzschner, 1987; Maas, 1991, 1993, 1994; Rensberger, 1993, 1997, 2000; Koenigswald and Sander, 1997; Maas and Dumont, 1999). For instance, differences in the orientation of the hydroxyapatite crystallites may affect the resistance of a tooth to abrasion and relatively small-scale tissue loss. By contrast, differences in the orientation of prisms, or groups of prisms, may affect the resistance of a tooth to fracture and thus larger-scale tissue loss (Rasmussen et al., 1976; Boyde, 1989). Thus some animals have enamel that is deposited in a simple radial pattern, with all enamel prisms aligned parallel to each other (fig. 5.2). Because all the prisms are aligned in a uniform orientation, these teeth are subject to breakage if a hard object, or an opposing tooth, happens to strike the tooth surface with enough force at precisely the wrong angle. In light of this possibility, it is not surprising that only smaller animals (i.e., those that generate lower bite force) tend to have radial enamel (Koenigswald, Rensberger, and Pfretzschner, 1987). As it turns out, all of our ancestors are large enough to generate high forces in chewing,

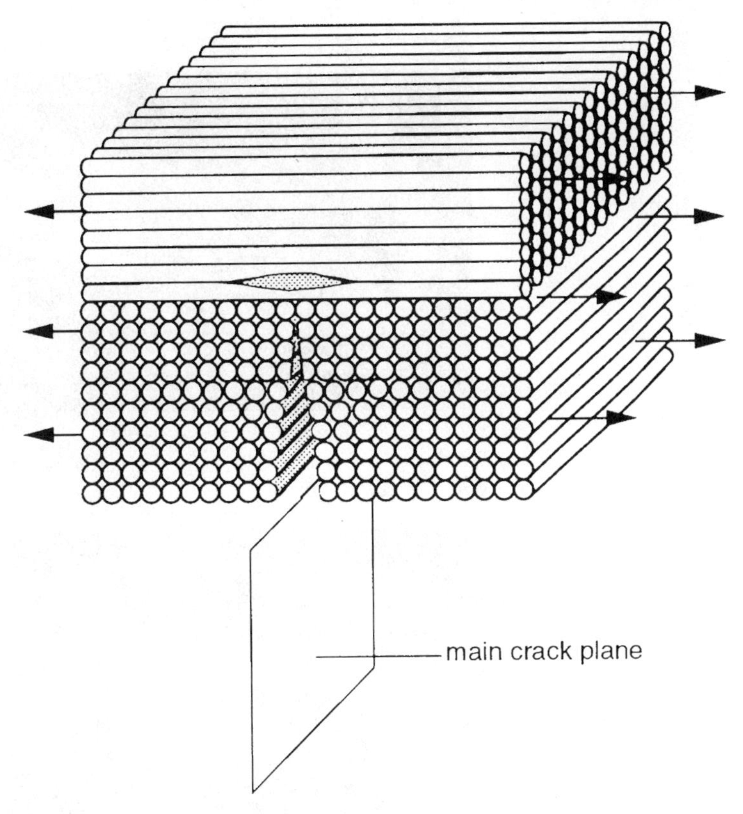

Figure 5.3 Inhibition of crack propagation by prism decussation (from Rensberger, 2000, in *Development, Function and Evolution of Teeth*, Teaford, Smith and Ferguson, eds., with permission of author and Cambridge University Press).

and thus all of them have enamel in which the prisms are interwoven in a decussating pattern that resists such breakage by serving as an admirable crack-stopping mechanism (Martin, Boyde, and Grine, 1988; Koenigswald and Clemens, 1992; Rensberger, 1997, 2000; Maas and Dumont, 1999; fig. 5.3). However, this arrangement also leaves decussated enamel susceptible to tensile stresses acting across prisms (Rasmussen et al., 1976).

At first glance, this would seem to tell us little new information about the origins and evolution of human diet. And how does it relate to variations in enamel thickness? To better understand the functional significance of variations in enamel thickness and microstructure, we need to look at both traits together (e.g., Teaford, Maas, and Simons, 1996). Martin, Olejniczak, and Maas (2003) demonstrated an admirable approach to the problem in their study of pitheciin molar enamel (fig. 5.4). The Neotropical primates *Pithecia*, *Chiropotes*, and *Cacajao* are all known to eat seeds with hard pericarps, especially in times of marked resource stress (Kinzey and Norconk, 1990). On the basis of traditional assumptions, one would expect them all to exhibit thick molar enamel. However, they do not, probably for two main reasons.

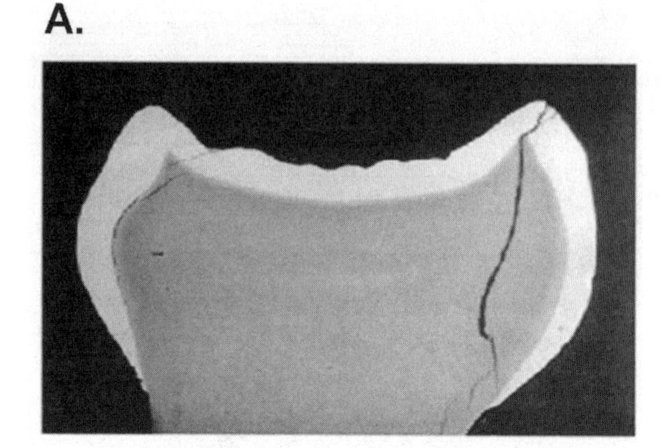

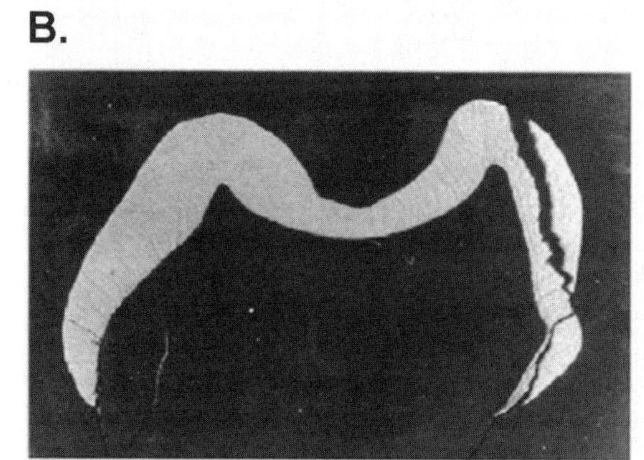

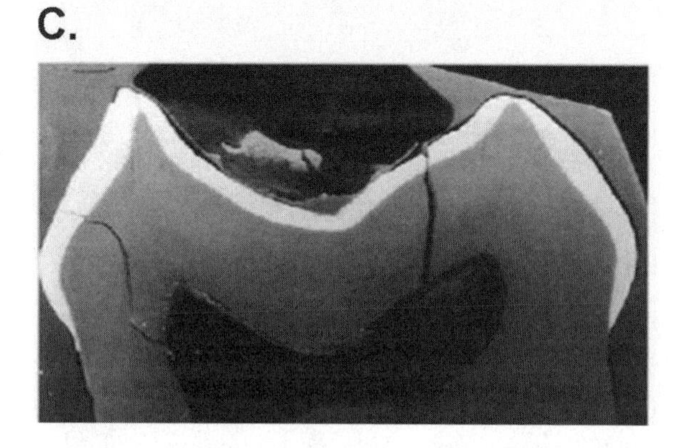

Figure 5.4 Representative images of enamel thickness in *Chiropotes* (*A*), *Cebus* (*B*), and *Ateles* (*C*). Reprinted from *Journal of Human Evolution*, vol. 45, Martin, Olejniczak, and Maas, p. 355, 2003, with permission of the authors and Elsevier.

First, the hard pericarp is broken primarily with their specialized anterior dentition. Second, their thin molar enamel is extremely decussated, thus providing protection against cracking. From this perspective, thick, decussated enamel would seem to be the best combination for resisting all types of wear. Of course, that's assuming most tooth wear is abrasive wear. As noted by Shellis and colleagues (1998), thick enamel may be an adaptive response to a number of factors (e.g., acid erosion), and even adhesive wear may be an important factor in the processing of certain foods (Puech, Prone, and Albertini, 1981; Puech, 1986; Teaford and Runestad, 1992). So, it may indeed be that the early hominins had relatively thick, decussated enamel, which left them equipped to deal with a variety of foods (Teaford and Ungar, 2000); and it may also be that members of our genus had progressively thinner enamel, allowed (in part) by the advantages of tool use (Teaford, Ungar, and Grine, 2002). However, we must not forget the limitations of our data and that our evolutionary explanations are still little more than "just so" stories until we have a better understanding of the complex interplay between dental structure and dental function, and any selective mechanisms that might be in operation.

Unfortunately, for analyses of enamel microstructure, the limitations are much the same as those for analyses of enamel thickness, for example, visualization of prism decussation usually requires some degree of damage to the tooth (in the form of acid etching and/or sectioning), and noninvasive techniques still do not have the ideal resolution for such analyses. Also, the quantification of the form and degree of prism decussation has proven difficult (Rensberger, 1993; Martin, Olejniczak, and Maas, 2003). So, despite the potential insights from variations in enamel microstructure across the tooth crown (Martin, Olejniczak, and Maas, 2003), interpretations are still limited by analyses of small samples from a very limited range of perspectives.

Finite Element Modeling

One way around the methodological difficulties of analyzing variations in enamel microstructure and distribution is to use any available data to create finite element models, to better understand how teeth behave under specific loads. By their very nature, such studies have the capability of shedding new light on internal, microscopic responses to stress and strain because traditional biomechanical tests, of necessity, focus on the response of the material, as a whole, rather than small areas within the material (Popowics, Rensberger, and Herring, 2004). The one "catch" in the whole enterprise is that any finite element model (FEM) is only as good as its underlying data. Thus, while initial efforts (e.g., Spears et al., 1993; Spears, 1997; Spears and Macho, 1998) gave insights into the locations of maximum stress points on tooth crowns under specific loading regimes, the models were, in essence, simple forays into a world of untold complexity. Recent work (Macho et al., 2005; Shimizu, Macho, and Spears, 2005) has brought FEM analysis down to prism resolution, as indicated by the number of elements per tooth (increased from previous levels of approximately eight hundred to over two hundred fifty thousand). However, certain underlying assumptions must still be made to make computations feasible. Thus, for instance, prisms are kept to a specific decussating path throughout the enamel cap, different patterns of crystallite orientation within prisms are held constant throughout the tooth, the shape of the prisms is held constant, the chemical composition of the enamel is held constant throughout the cap, physical properties are standardized within each element, and there are no differences in any of these measures between species. Clearly, as we learn more about the properties and composition of enamel (e.g., Cuy et al., 2002; Teaford et al., 2003; see "What We Don't Know about Enamel Properties" below), these assumptions will be viewed as gross oversimplifications that need to be modified. In the process, however, the models will become even more informative. As is, they have already forced investigators to look at possible correlations between (1) the distribution of stress over the tooth crown and the arrangement of enamel prisms and (2) prism orientations and rates of tooth wear (Shimizu, Macho, and Spears, 2005). They have also been used in an attempt to gain a new perspective on the dietary adaptations of one of the earliest human relatives, *Australopithecus anamensis* (Macho et al., 2005), although interpretations are somewhat tainted by misunderstandings of the distinction between the "phases" of chewing and puncture crushing. Still, the work is making people think about whether patterns of prism orientation can be correlated with primary movements in mastication or with the processing of foods of different properties. Thus, while FEM analysis will always be an exploratory tool, limited by its underlying data, it can continue to help move interpretations forward. If the most up-to-date data, on dental microstructure and physical properties, are routinely entered into FEM analyses, then they will, in turn, nudge researchers to collect even finer-resolution data, which will then yield more useful FEM analyses. The end result will be more insights into the origin and evolution of human diet.

What Do We *Not* Know about Dental Enamel?

Limitations in Our Data on Enamel Thickness and Enamel Microstructure

Until we are capable of providing high-resolution, three-dimensional surface characterizations of enamel and dentin surfaces in a noninvasive, nondestructive manner (e.g., something analogous to Ungar's topographic analyses of teeth; Ungar and Williamson, 2000; Ungar and M'Kirera, 2003), our sample sizes of modern nonhuman primates will always be small, and our samples of fossils will be even smaller. Will refinements of computed topography, or some other method of medical imaging, allow us to safely measure the inside of teeth with impunity? Only further work will tell. For now, we desperately need statistical analyses (e.g., of traditional measurements of enamel thickness and decussation) of a wide range of modern and fossil primates and analyses at various points across the tooth crown. More important, as simplistic as this might sound, we need analyses of other teeth. If different primates are processing foods at different points along the tooth row, then we need to know the structure of those teeth as well. At the present time, there are clinical studies of enamel thickness for some other teeth (e.g., Harris and Hicks, 1988), but nothing similar for other primates. We also need documentation of how enamel changes with wear. If surface layers of enamel are often prismless (Martin, Boyde, and Grine, 1988; Maas and Dumont, 1999), how long do they last during the life of a tooth? Do they leave the tooth susceptible to certain types of wear, and by their presence or absence, give us additional clues about diet? How does prism decussation vary within and between species? If it varies significantly between species, and even within individual teeth (Martin, Olejniczak, and Maas, 2003), what sorts of dietary clues are we missing? From a different perspective, what about dentin? We tend to think of it as a passive support mechanism for the tooth; yet, its shape and survival may be of crucial importance to the survival of the tooth as a whole. Of course, all of these questions assume that dental structure is only an indirect indicator of diet. In other words, if a certain variation in dental structure somehow allows an individual to survive and reproduce better, that variant may become more common in the population. But as we come to know more about the intricacies of dental development (Weiss, Stock, and Zhao, 1998; Salazar-Ciudad, Jernvall, and Newman, 2003), could there be epigenetic effects that we are overlooking? Could deciduous tooth use be influencing permanent tooth structure? If so, what would be the driving force of such influence, foods most frequently processed, or foods of critical importance to survival at certain periods in life (Kay, 1975; Kinzey, 1978)?

What We Don't Know about Enamel Properties

What we know about enamel structure also masks an incredible lack of knowledge about the physical properties of teeth. For instance, most people think of enamel as a homogeneous substance, like steel or water, with uniform properties. Thus, statements like "the hardness of enamel is . . ." are extremely common in the literature. Yet, it is built of prisms and crystals imperfectly joined together with points of weakness and even microscopic porosities between them (Shellis 1984; Shellis and Dibdin, 2000). This structural complexity might be expected to cause its mechanical properties to vary as its constituents change (Shimizu, Macho, and Spears, 2005). As a result, we cannot expect enamel to respond uniformly when loaded by chewing forces.

As is so often the case in science, our simplistic notions of enamel properties stem largely from methodological limitations. Quite simply, enamel is incredibly difficult to test, and the methods that have been used have, until recently, been fairly crude. The net result is that we have little more than summary characterizations of properties that could be of crucial importance to dental function (e.g., Craig and Peyton, 1958; Waters, 1980).

The hardness of enamel is defined as the pressure necessary to cause its permanent or plastic deformation. It has been the most thoroughly studied of enamel's properties, with enamel being routinely identified as the hardest tissue in the body (Waters, 1980). Most techniques used to measure hardness involve pressing a hard object, like diamond, into an enamel surface until permanent (plastic) deformation occurs. Measurement of the resultant indents ultimately yields an estimate of enamel hardness. Unfortunately, the indents are usually far larger than enamel prisms, and most have been created at an extremely limited number of locations on the tooth. Of course, since these approaches are dependent on permanent deformation, they also do not allow for the direct measurement of another major property of enamel, its elasticity.

Most people think of enamel as a brittle substance. So why should elasticity be important? Appreciation of that fact once again hinges on an understanding of the composite nature of enamel. Its complex array of prisms and crystals sits on a dentine core with different properties. As a result, when loaded, enamel does not respond entirely rigidly. It can bend, ever so slightly, and that flexibility is crucial for understanding the distribution of stresses within the tooth. Unfortunately, the movements involved are so slight as to be difficult to detect.

While we are beginning to get more refined measures of these properties of enamel, we still have a long way to go before we can really understand them. A relatively new technique, with the potential to add new insights to this problem, is nanoindentation technology. As with many other techniques, it involves pressing a hard object into the tooth surface, but there the similarities end because the indenter can be positioned at a resolution of one micron, the applied loads are minute (a few μN's), and displacements of less than 1 nm can be measured, allowing for the computation of both hardness and elasticity at an extremely fine level of resolution. This means that measures of hardness are generally computed for areas of about 1–2 enamel prism diameters, and measures of elasticity probably encompass the diameter of 8–10 prisms (i.e., enough resolution to shed light on correlations between prism orientation and physical properties; contra Shimizu, Macho, and Spears, 2005).

Results of analyses for a small sample of human teeth (Cuy et al., 2002) suggest that the hardness and elasticity of enamel vary within and between teeth. Enamel is harder and stiffer on the surface, with each measure decreasing by greater than 50 percent as one moves to the enamel-dentin junction. These variations mirror differences in mineralization that normally occur throughout the crown (Angmar, Carlström, and Glas, 1963; Staines, Robinson, and Hood, 1981; Theuns et al., 1983; Svalbe et al., 1984; Spears, 1997), and they do not seem to vary with differences in enamel prism orientation. There are also variations in hardness and stiffness between cusps and between teeth, which seem to match biomechanical theoretical expectations (Mansour and Reynick, 1975; Koolstra et al. 1988; Khera et al., 1990; Macho and Berner, 1994; Spears and Macho, 1998; Macho and Spears, 1999; Schwartz, 2000a, 2000b). In essence, the hardest and stiffest regions are those normally subjected to greater force in chewing. Interestingly, preliminary results suggest that the permanent molars of *Alouatta palliata* do not show the same patterns of intratooth variability, as intermediate levels of hardness and elasticity are maintained throughout the depth of the tooth crown (Teaford et al., 2003; fig. 5.5). Because *Alouatta* has relatively thin enamel (Shellis et al., 1998), fast tooth wear (Teaford and Glander, 1991), and medium body size, this raises interesting functional questions, which in turn raise a host of other questions.

In essence, we have a glimpse of how enamel properties might vary within and between some teeth, but we need far more work on a variety of teeth at various stages of wear. And we have no idea how the physical properties of enamel vary between species. If enamel properties vary within teeth, and if those patterns of variation differ between species, then not all enamel is alike. Thus, simple measures of enamel thickness may be nothing more than crude estimates of functional potential; that is, we may need to consider enamel *quality* in addition to enamel *quantity*. This raises innumerable possibilities for future research. For instance, do the mechanical properties of enamel and dentin always vary in concert? If the stresses in chewing, or puncture crushing, vary between teeth, will the mechanical properties vary as well? How might different types of enamel (or dentin) vary in their mechanical properties? If enamel properties are tied to degree of mineralization, could some species with thin enamel be ingesting foods that are, in essence, increasing the hardness of their enamel through the topical application of minerals, or are interspecific differences in the degree of prism decussation enough to compensate for the poorer

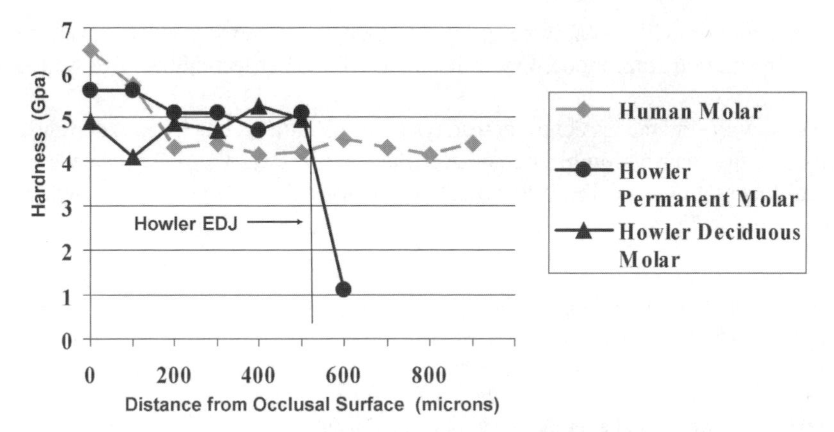

Figure 5.5 Differences in hardness between molars of *Homo sapiens* and *Alouatta palliata* (since human enamel is much thicker than howler enamel, the values for the human molar continue on off the chart; from Teaford et al., 2003).

wear resistance of thin enamel? Could other species with thick enamel be ingesting acidic foods that habitually soften the surface enamel? Another unknown effect is that of tooth wear—as a tooth wears down, what happens to its physical properties? For instance, if the interior of a tooth is softer than its exterior, does that mean that tooth wear exposes softer and softer enamel to "the elements"? What about fossils? Could we ever compare the enamel properties of different fossils? If the properties *do* vary in conjunction with variations in mineralization, and yet the process of fossilization replaces what little organic material there is in teeth with more minerals, then the physical properties of fossil enamel might reflect the mode of fossilization rather than diet. Similarly, diagenesis is a complex process that does not focus solely on the replacement of organic material. Still, if food items could directly affect enamel properties, then enamel properties might, in turn, give us further direct evidence of diet, perhaps on a paleontological site-by-site basis. Of course, because current methods of testing the properties of dental materials generally involve plastic deformation, albeit on a microscopic scale, the chances of such techniques being used on rare fossils are very slim.

What Is Unknowable about Dental Enamel?

With those limitations in mind, is there anything we can never know about enamel and its properties? The key lies in the resolution of detail obtainable in noninvasive techniques. If imaging techniques continue to be refined, then mapping the distribution of enamel, and even the location and orientation of prisms, across any tooth may be possible. Will the resolution of detail ever be sufficient to document the three-dimensional position and orientation of crystallites? Only time will tell.

If we are dependent on techniques that damage the tooth, then we will never know all we can about fossils, and we will know far less about modern teeth than we would like. So if the only way to test the hardness of enamel involves plastic deformation, that information will probably never be obtainable for some fossils. However, it will yield information from modern teeth invaluable for other techniques, like finite element analysis.

Thus, while enamel structure nowadays gives us indirect clues about diet through traits like the distribution of enamel over the tooth crown, the presence/absence of prism decussation, and the orientation of prisms, there is much more we do not know about it. In essence, we have only begun to tap its potential.

Acknowledgments I would like to thank (1) Peter Ungar for inviting me to participate in the symposium The Evolution of Human Diet: The Known, the Unknown, and the Unknowable and (2) The Sloan Foundation and the University of Arkansas for making that session possible. I would also like to thank all of the symposium participants for their insightful discussions, before, during, and after the sessions. Special thanks go to Peter Ungar and Fred Grine for their constant feedback and continuing collaborative efforts in the study of the evolution of human diet. Fred, Alan Walker, and an anonymous reviewer added significant improvements to this manuscript. Reiko Kono also deserves special thanks for modifying the figure reproduced as figure 5.1 here. As with any review chapter, this one has benefited from the help of innumerable colleagues and curators from around the world, and it also would not have been possible without the support of the National Science Foundation, whose help is gratefully acknowledged.

References

Andrews, P., and Martin, L., 1991. Hominoid dietary evolution. *Philos. Trans. R. Soc. Lond.* 334, 199–209.

Angmar, B., Carlstrőm, D., and Glas, J.-E., 1963. Studies on the ultrastructure of dental enamel. IV. The mineralization of normal human enamel. *J. Ultrastruct. Res.* 8, 12–23.

Benefit, B.R., 1987. The molar morphology, natural history and phylogenetic position of the Middle Miocene monkey *Victoriapithecus*. PhD diss., New York University.

Beynon, A.D., Clayton, C.B., Ramirez-Rozzi, F.V., and Reid, D.J., 1998. Radiographic and histological methodologies in estimating the chronology of crown development in modern humans and great apes: A review, with some applications for studies on juvenile hominids. *J. Hum. Evol.* 35, 351–370.

Beynon, A.D., and Dean, M.C., 1988. Distinct dental development patterns in early fossil hominids. *Nature* 335, 509–514.

Beynon, A.D., Dean, M.C., Leakey, M.G., Reid, D.J., and Walker, A., 1998. Comparative dental development and microstructure of *Proconsul* teeth from Rusinga Island, Kenya. *J. Hum. Evol.* 35, 163–209.

Beynon, A.D., Dean, M.C., and Reid, D.J., 1991. Histological study on the chronology of the developing dentition in gorilla and orangutan. *Am. J. Phys. Anthropol.* 86, 189–203.

Beynon, A.D., and Wood, B.A., 1986. Variation in enamel thickness and structure in East African hominids. *Am. J. Phys. Anthropol.* 70, 177–193.

Boyde, A., 1964. The structure and development of mammalian enamel. PhD diss., University of London.

Boyde, A., 1989. Enamel. In: Oksche, A., and Vollrath, L. (Eds.), *Handbook of Microscopic Anatomy*. Vol. V/6: Teeth. Springer, Berlin, pp. 309–473.

Boyde, A., and Fortelius, M., 1986. Development, structure and function of rhinoceros enamel. *Zool. J. Linn. Soc.* 87, 181–214.

Boyde, A., Jones, S.J., and Reynolds, P.S., 1978. Quantitative and qualitative studies of enamel etching with acid and EDTA. *Scanning Electron Microsc.* 2, 991–1002.

Boyde, A., and Martin, L.B., 1982. Enamel microstructure determination in hominoid and cercopithecoid primates. *Anat. Embryol.* 165, 193–212.

Bromage, T.G., and Dean, M.C., 1985. Re-evaluation of age at death of immature fossil hominids. *Nature* 317, 525–527.

Butler, P.M., 1956. The ontogeny of molar pattern. *Biol. Rev.* 31, 30–70.

Carlson, S.J., 1990. Vertebrate dental structures. In: Carter J.G. (Ed.), *Skeletal Biomineralization: Patterns Processes and Evolutionary Trends*. Vol. 1. Van Nostrand Reinhold, New York, pp. 531–556.

Conroy, G.C., 1991. Enamel thickness in South African australopithecines: Noninvasive determination by computed tomography. *Palaeontol. Afr.* 28, 53–59.

Craig, R.G., and Payton, F.A., 1958. The microhardness of enamel and dentin. *J. Dent. Res.* 37, 661–668.

Cuy, J.L, Mann, A.B., Livi, K.J., Teaford, M.F., and Weihs T.P., 2002. Nanoindentation mapping of the mechanical properties of molar enamel. *Arch. Oral Biol.* 47, 281–291.

Dean, M.C., 2000. Progress in understanding hominid dental development. *J. Anat.* 197, 77–101.

Dean, M.C., Beynon, A.D., Thackeray, J.F., and Macho, G.A., 1993. Histological reconstruction of dental development and age at death of a juvenile *Paranthropus robustus* specimen, SK 63, from Swartkrans, South Africa. *Am. J. Phys. Anthropol.* 91, 401–419.

Dean, M.C., Jones, M.E., and Pilley, J.R., 1992. The natural history of tooth wear, continuous eruption and periodontal disease in wild shot great apes. J. Hum. Evol. 22, 23–39.

Dean, M.C., and Wood, B.A., 1981. Developing pongid dentition and its use for ageing individual crania in comparative cross-sectional growth studies. *Folia Primatol.* 36, 111–127.

Delson, E., 1973. Fossil colobine monkeys of the circum-Mediterranean region and the evolutionary history of the Cercopithecidae (Primates, Mammalia). PhD diss., Columbia University.

Dumont, E.R., 1995. Enamel thickness and dietary adaptation among extant primates and chiropterans. *J. Mammal.* 76, 1127–1136.

Fitzgerald, C.M., 1998. Do enamel microstructures have regular time dependency? Conclusions from the literature and a large-scale study. *J. Hum. Evol.* 35, 371–386.

Fortelius, M., 1985. Ungulate cheek teeth: Developmental, functional, and evolutionary interrelationships. *Acta Zool. Fenn.* 180, 1–76.

Gantt, D.G., 1977. Enamel of primate teeth: Its thickness and structure with reference to functional and phyletic implications. PhD diss., Washington University.

Gantt, D.G., 1979. A method of interpreting enamel prism patterns. *Scanning Electron Microsc.* 2, 975–981.

Gantt, D.G., 1983. The enamel of Neogene hominoids: Structural and phyletic implications. In: Ciochon, R.L., and Corruccini, R.S. (Eds.), *New Interpretations of Ape and Human Ancestry*. Plenum Press, New York, pp. 249–298.

Gantt, D.G., Harris, E.F., Rafter, J.A., and Rahn, J.K., 2001. Distribution of enamel thickness on human deciduous molars. In: Brook, A. (Ed.), *Dental Morphology 2001*. Sheffield Academic Press, Sheffield, pp. 167–190.

Godfrey, L.R., Semprebon, G.M., Jungers, W.L., Sutherland, M.R., Simons, E.L., and Solonouias, N., 2004. Dental use wear in extinct lemurs: Evidence of diet and niche differentiation. *J. Hum. Evol.* 47, 145–169.

Grine, F.E. 1991. Computed tomography and the measurement of enamel thickness in extant hominoids: implications for its palaeontological application. *Palaeontol. Afr.* 28, 61–69.

Grine, F.E., 2002. Scaling of tooth enamel thickness, and molar crown size reduction in modern humans. *S. Afr. J. Sci.* 98, 503–509.

Grine, F.E., 2005. Enamel thickness of deciduous and permanent molars in modern *Homo sapiens*. *Am. J. Phys. Anthropol.* 126, 14–31.

Grine, F.E., and Martin, B., 1988. Enamel thickness and development in *Australopithecus* and *Paranthropus*. In: Grine, F.E. (Ed.), *Evolutionary History of the "Robust" Australopithecines*. Aldine de Gruyter, New York, pp. 3–42.

Grine, F.E., Stevens, N.J., and Jungers, W.L., 2001. Evaluation of dental radiograph accuracy in the measurement of enamel thickness. *Arch. Oral Biol.* 46, 1117–1125.

Harris, E.F., and Hicks, J.D., 1988. Enamel thickness in maxillary human incisors: A radiographic assessment. *Arch. Oral Biol.* 43, 825–831.

Janis, C.M., and Fortelius, M., 1988. On the means whereby mammals achieve increased functional durability of their dentitions, with special reference to limiting factors. Biol. Rev. 63, 197–230.

Kay, R.F., 1975. The functional adaptations of primate molar teeth. *Am. J. Phys. Anthropol.* 42, 195–215.

Kay, R.F., 1981. The nut-crackers—a new theory of the adaptations of the Ramapithecinae. *Am. J. Phys. Anthropol.* 55, 141–151.

Kay, R.F., 1985. Dental evidence for the diet of *Australopithecus*. *Annu. Rev. Anthropol.* 14, 315–341.

Khera, S.C., Carpenter, C.W., Vetter, J.D., and Staley, R.N., 1990. Anatomy of cusps of posterior teeth and their fracture potential. *J. Prosth. Dent.* 64, 139–147.

Kinzey, W.G., 1978. Feeding behavior and molar features in two species of titi monkey. In: Chivers, D.J., and Herbert, J. (Eds.), *Recent Advances in Primatology*. Vol. 1: *Behavior*. Academic Press, New York, pp. 373–385.

Kinzey, W.G., and Norconk, M.A., 1990. Hardness as a basis of fruit choice in two sympatric primates. *Am. J. Phys. Anthropol.* 81, 5–15.

Kinzey, W.G., and Norconk, M.A., 1993. Physical and chemical properties of fruit and seeds eaten by *Pithecia* and *Chiropotes* in Surinam and Venezuela. *Int. J. Primatol.* 14, 207–226.

Koenigswald, W.v., 1980. Schmelstruktur und Morphologie in den Molaren der Arvicolidae (Rodentia). *Abh. Senck. Natur. Gesell.* 539, 1–129.

Koenigswald, W.v., and Clemens, W.A., 1992. Levels of complexity in the microstructure of mammalian enamel and their application in studies of systematics. *Scanning Microsc.* 6, 195–218.

Koenigswald, W.v., Rensberger, J.M., and Pfretzschner, H.U., 1987. Changes in the tooth enamel of early Paleocene mammals allowing increased diet diversity. *Nature* 328, 150–152.

Koenigswald, W.v., and Sander, P.M. (Eds.), 1997. *Tooth Enamel Microstructure*. A. A. Balkema, Rotterdam.

Kono, R.T., 2004. Molar enamel thickness and distribution patterns in extant great apes and humans: new insights based on a 3-dimensional whole crown perspective. *Anthropol. Sci.* 112, 121–146.

Kono, R.T., Suwa G., and Tanijiri, T., 2002. A three-dimensional analysis of enamel distribution patterns in human permanent first molars. *Arch. Oral Biol.* 47, 867–875.

Koolstra, J.H., van Eijden, T.M.J.G., Weihs, W.A., and Naeije, M., 1998. A three-dimensional mathematical model of the human masticatory system predicting maximum bite forces. *J. Biomech.* 21, 563–576.

Kuykendall, K.L., 1996. Dental development in chimpanzees (*Pan troglodytes*): The timing of tooth calcification stages. *Am. J. Phys. Anthropol.* 99, 135–157.

Lambert, J.E., Chapman, C.A., Wrangham, R.W., and Conklin-Brittain, N.L., 2004. The hardness of cercopithecine foods: implications for the critical function of enamel thickness in exploiting fallback foods. *Am. J. Phys. Anthropol.* 125, 363–368.

Lucas, P.W., 2004. *Dental Functional Morphology: How Teeth Work.* Cambridge University Press, New York.

Lucas, P.W., Darvell, B.W., Lee, P.K.D., Yuen, T.D.B., and Choong, M.F., 1995. The toughness of plant cell walls. *Philos. Trans. R. Soc. Lond* B 348, 363–372.

Lucas, P.W., and Pereira, B.P., 1990. Estimation of the fracture toughness of leaves. *Funct. Ecol.* 4, 235–236.

Lucas, P.W., and Teaford, M.F., 1994. Functional morphology of colobine teeth. In: Davies, A.G., and Oates, J.F. (Eds.), *Colobine Monkeys: Their Ecology, Behaviour and Evolution*. Cambridge University Press, New York, pp. 173–204.

Lucas, P.W., Turner, I.M., Dominy, N.J., and Yamashita, N., 2000. Mechanical defences to herbivory. *Ann. Bot.* 86, 913–920.

Lumsden, A.G., and Osborn, J.W., 1977. The evolution of chewing: A dentist's view of palaeontology. *J. Dent.* 5, 269–287.

Maas, M.C., 1991. Enamel structure and microwear: an experimental study of the response of enamel to shearing force. *Am. J. Phys. Anthropol.* 85, 31–49.

Maas, M.C., 1993. Enamel microstructure and molar wear in the greater galago, *Otolemur crassicaudatus* (Mammalia, Primates). *Am. J. Phys. Anthropol.* 92, 217–233.

Maas, M.C., 1994. Enamel microstructure in Lemuridae (Mammalia, Primates): Assessment of variability. *Am. J. Phys. Anthropol.* 95, 221–241.

Maas, M.C., and Dumont, E.R., 1999. Built to last: The structure, function, and evolution of primate dental enamel. *Evol. Anthropol.* 8, 133–152.

Macho, G.A., 1994. Variation in enamel thickness and cusp area within human maxillary molars and its bearing on scaling techniques used for studies of enamel thickness between species. *Arch. Oral Biol.* 39, 783–792.

Macho, G.A., and Berner. M.E., 1993. Enamel thickness of human maxillary molars reconsidered. *Am. J. Phys. Anthropol.* 92, 189–200.

Macho, G.A., and Berner, M.E., 1994. Enamel thickness and the helicoidal occlusal plane. Am. J. Phys. Anthropol. 94, 327–337.

Macho, G.A., Shimizu, D., Jiang, Y, and Spears, I.R., 2005. *Australopithecus anamensis*: A finite-element approach to studying the functional adaptations of extinct hominins. *Anat. Rec.* A 283, 310–318.

Macho, G.A., and Spears, I.R., 1999. Effects of loading on the biomechanical behavior of molars of *Homo*, *Pan*, and *Pongo*. *Am. J. Phys. Anthropol.* 109, 211–227.

Macho, G.A., and Thackeray, J.F., 1992. Computer tomography and enamel thickness of maxillary molars of Plio-Pleistocene hominids from Sterkfontein, Swartkrans and Kromdraai (South Africa): An exploratory study. Am. J. Phys. Anthropol. 89, 133–143.

Mansour, R.M., and Reynick R.J., 1975. *In vivo* occlusal forces and moments. I. Forces measured in terminal hinge position and associated movements. *J. Dent. Res.* 54, 114–120.

Martin, L.B., 1983. The relationships of the Later Miocene Hominoidea. PhD diss., University of London.

Martin, L.B., 1985. Significance of enamel thickness in hominoid evolution. *Nature* 314, 260–263.

Martin, L.B., Boyde, A., and Grine F.E., 1988. Enamel structure in primates—a review of scanning electron microscope studies. *Scanning Microsc.* 2, 1503–1526.

Martin, L.B., Olejniczak, A.J., and Maas M.C., 2003. Enamel thickness and structure in pitheciin primates, with comments on dietary adaptations of the Middle Miocene hominoid *Kenyapithecus*. *J. Hum. Evol.* 45, 351–367.

Massler, M., and Schour, I., 1946. The appositional life span of the enamel and dentine-forming cells. *J. Dent. Res.* 25, 145–156.

Molnar, S., Gantt, D.G., 1977. Functional implications of primate enamel thickness. *Am. J. Phys. Anthropol.* 46, 447–454.

Pfretzschner, H.U., 1986. Structural reinforcement and crack propagation in enamel. In: Russell, D.E., Santoro, J.P., and Sigoneau-Russell, D. (Eds.), *Teeth Revisited: Proceedings of the 7th International Symposium on Dental Morphology.* Museum of Natural History, Paris, pp. 133–143.

Popowics, T.E., Rensberger, J.M., and Herring, S.W., 2004. Enamel microstructure and microstrain in the fracture of human and pig molar cusps. *Arch. Oral Biol.* 49, 595–605.

Puech, P.-F., 1986. Dental microwear features as an indicator for plant food in early hominids: A preliminary study of enamel. *Hum. Evol.* 1, 507–515.

Puech, P.-F., Prone, A., and Albertini, H., 1981. Reproduction expérimentale des processus d'altération de la surface dentaire par friction non abrasive et non adhésive: application à l'étude de alimentation de L'Homme fossile. *C. R. Acad. Sci., Paris* D 293, 729–734.

Ramirez-Rozzi, F.V., 1998. Can enamel microstructure be used to establish the presence of different species of Plio-Pleistocene hominids from Omo, Ethiopia? *J. Hum. Evol.* 35, 543–576.

Rasmussen, S.T., Patchin, R.E., Scott, D.B., and Heuer, A.H., 1976. Fracture properties of human enamel and dentine. *J. Dent. Res.* 55, 154–164.

Reid, D.J., Schwartz, G.T., Dean, C., and Chandrasekera, M.S., 1998. A histological reconstruction of dental development in the common chimpanzee, *Pan troglodytes. J. Hum. Evol.* 35, 427–448.

Rensberger, J.M., 1993. Adaptation of enamel microstructure to differences in stress intensity in the Eocene perissodactyl *Hyracotherium.* In: Kobayashi, I., Mutvei, H., and Sahni, A. (Eds.), *Structure, Formation, and Evolution of Fossil Hard Tissues.* Tokai University Press, Tokyo, pp. 131–145.

Rensberger, J.M., 1997. Mechanical adaptation in enamel. In: Koenigswald, W.v., and Sanders, P.M. (Eds.), *Tooth Enamel Microstructure.* A. A. Balkema, Rotterdam, pp. 237–258.

Rensberger, J.M., 2000. Pathways to functional differentiation in mammalian enamel. In: Teaford, M.F. Smith, M.M., and Ferguson, M.W.J. (Eds.), *Development, Function and Evolution of Teeth.* Cambridge University Press, Cambridge, pp. 252–268.

Rensberger, J.M., and Koenigswald, W.v., 1980. Functional and phyletic interpretations of enamel microstructure in rhinoceroses. *Paleobiology* 6, 477–495.

Robinson, J.G., 1956. The dentition of the australopithecines. *Trans. Mus. Mem.* 9, 1–179.

Salazar-Ciudad, I, Jernvall, J., and Newman, S.A., 2003. Mechanisms of pattern formation in development and evolution. *Development* 130, 2027–2037.

Schour, I., and Hoffman, M.M., 1939. Studies in tooth development. II. The rate of apposition of enamel and dentine in man and other animals. *J. Dent. Res.* 18, 91–102.

Schwartz, G.T., 2000a. Enamel thickness and the helicoidal wear plane in modern human mandibular molars. *Arch. Oral Biol.* 45, 401–409.

Schwartz, G.T., 2000b. Taxonomic and functional aspects of the patterning of enamel thickness distribution in extant large-bodied hominoids. *Am. J. Phys. Anthropol.* 111, 221–244.

Schwartz, G.T., Thackeray, J.F., Reid, C., and van Reenen, J.F., 1998. Enamel thickness and the topography of the enamel-dentine junction in South African Plio-Pleistocene hominids with special reference to the Carabelli Trait. *J. Hum. Evol.* 35, 523–542.

Shellis, R.P., 1984. Relationship between human enamel structure and the formation of caries-like lesions *in vitro. Arch. Oral Biol.* 29, 975–981.

Shellis, R.P., 1998. Utilisation of periodic markings in enamel to obtain information about tooth growth. *J. Hum. Evol.* 35, 427–448.

Shellis, R.P., Beynon, A.D., Reid, D.J., and Hiiemae, K.M., 1998. Variations in molar enamel thickness among primates. *J. Hum Evol.* 35, 507–522.

Shellis, R.P., Dibdin, and G.H., 2000. Enamel microporosity and its functional implications. In: Teaford, M.F., Smith, M.M., and Ferguson, M.W.J. (Eds.), *Development, Function and Evolution of Teeth.* Cambridge University Press, Cambridge, pp. 242–251.

Shellis, R.P., and Poole D.F.G., 1977. The calcified dental tissues of primates. In: Lavelle C.L.B., Shellis, R.P., and Poole, D.F.G. (Eds.), *Evolutionary Changes to the Primate Skull and Dentition.* Thomas, Springfield, IL, pp. 197–279.

Shillingburg, H., Jr., and Grace, C., 1973. Thickness of enamel and dentin. *J. S. Calif. Dent. Assoc.* 41, 33–52.

Shimizu, D., 2002. Functional implications of enamel thickness in the lower molars of red colobus (*Procolobus badius*) and Japanese macaque (*Macaca fuscata*). *J. Hum. Evol.* 43, 605–620.

Shimizu, D., Macho, G.A., and Spears, I., 2005. Effect of prism orientation and loading direction on contact stresses in prismatic enamel of primates: Implications for interpreting wear patterns. Am. J. Phys. Anthrop. 126, 427–434.

Smith, T.M., Martin, L.B., and Leakey, M.G., 2003. Enamel thickness, microstructure and development in *Afropithecus turkanensis. J. Hum. Evol.* 44, 283–306.

Spears, I.R., 1997. A three-dimensional finite element model of prismatic enamel: a reappraisal of the data on the Young's modulus of enamel. *J. Dent. Res.* 76, 1690–1697.

Spears, I.R., and Crompton, R.H., 1996. The mechanical significance of the occlusal geometry of great ape molars in food breakdown. *J. Hum. Evol.* 31, 517–535.

Spears, I.R., and Macho, G.A., 1995. The helicoidal occlusal plane—a functional and biomechanical appraisal of molars. In: Radlanski, R.J., and Renz, H. (Eds.), Proceedings of the 10th International Symposium on Dental Morphology. "M" Marketing Services, Berlin, pp. 391–397.

Spears, I.R., and Macho, G.A., 1998. Biomechanical behavior of modern human molars: implications for interpreting the fossil record. *Am. J. Phys. Anthropol.* 106, 467–482.

Spears, I.R., van Noort, R., Crompton, R.H., Cardew, G.E., and Howard, I.C., 1993. The effects of enamel anisotropy on the distribution of stress in a tooth. *J. Dent. Res.* 72, 1526–1531.

Spoor, C.F., Zonneveld, F.W., and Macho, G.A., 1993. Linear measurements of cortical bone and dental enamel by computed tomography: applications and problems. *Am. J. Phys. Anthropol.* 91, 469–484.

Staines, M., Robinson, W.H., and Hood, J.A.A., 1981. Spherical indentation of tooth enamel. *J. Mat. Sci.* 16, 2551–2556.

Strait, S.G., 1997. Tooth use and the physical properties of food. *Evol. Anthropol.* 5, 199–211.

Suwa, G., and Kono, R.T., 2005. A micro-CT based study of linear enamel thickness in the mesial cusp section of human molars: Reevaluation of methodology and assessment of within-tooth, serial, and individual variation. *Anthropol. Sci.*,.113, 273–289.

Svalbe, I.D., Chaudhri, M.A., Traxel, K., Ender, C., and Mandel, A., 1984. Microprobe profiling of fluorine and other trace elements to large depths in teeth. *Nucl. Instr. Meth. Phys. Res.* B 3, 648–650.

Teaford, M.F., and Glander, K.E., 1991. Dental microwear in live, wild-trapped *Alouatta palliata* from Costa Rica. *Am. J. Phys. Anthropol.* 85, 313–319.

Teaford, M.F., Lucas, P.W., Ungar, P.S., and Glander, K.E., 2006. Mechanical defenses in leaves eaten by Costa Rican *Alouatta palliata*. Am. J. Phys. Anthropol. 129, 99–104.

Teaford, M.F., Maas, M.C., and Simons, E.L., 1996. Dental microwear and microstructure in early Oligocene Fayum primates: Implications for diet. *Am. J. Phys. Anthropol.* 101, 527–543.

Teaford, M.F., and Runestad, J.A., 1992. Dental microwear and diet in Venezuelan primates. *Am. J. Phys. Anthropol.* 88, 347–364.

Teaford, M.F., and Ungar, P.S., 2000. Diet and the evolution of the earliest human ancestors. *Proc. Natl. Acad. Sci.* 97, 13506–13511.

Teaford, M.F., Ungar, P.S., and Grine, F.E., 2002. Fossil evidence for the evolution of human diet. In: Ungar, P.S., and Teaford, M.F. (Eds.), *Human Diet: Its Origins and Evolution.* London, Bergen & Garvey, Westport, CT, pp. 143–166.

Teaford, M.F., Weiner, M., Darnell, L., and Weihs, T.P., 2003. Mechanical properties of molar enamel in *Homo sapiens* and *Alouatta palliata*. *Am. J. Phys. Anthropol.*, suppl., 36, 206–207.

Theuns, H.M., van Dijk, J.W.E., Jongebloed, W.L., and Groeneveld, A., 1983. The mineral content of human enamel studied by polarizing microscopy, microradiography and scanning electron microscopy. *Arch. Oral Biol.* 28, 797–803.

Tobias, P.V., 1991. *Olduvai Gorge*. Vol. 4: *The Skulls, Endocasts and Teeth of* Homo habilis. Cambridge University Press, Cambridge.

Ungar, P.S., and M'Kirera, F., 2003. A solution to the worn tooth conundrum in primate functional anatomy. *Proc. Natl. Acad. Sci.* 100, 3874–3877.

Ungar, P.S., and Williamson, M., 2000. Exploring the effects of tooth wear on functional morphology: A preliminary study using dental topographic analysis. *Palaeontol. Electron.* 3, 18.

Vrba, E.S., and Grine, F.E., 1978. Australopithecine enamel prism patterns. *Science* 202, 890–892.

Waters, N.E., 1980. Some mechanical and physical properties of teeth. In: Vincent J.F.V., and Curry J.D. (Eds.), *The Mechanical Properties of Biological Materials.* Cambridge University Press, Cambridge, pp. 99–135.

Weiss, K.M., Stock, D.W., and Zhao, Z., 1998. Dynamic interactions and the evolutionary genetics of dental patterning. *Crit. Rev. Oral Biol. Med.* 9, 369–98.

Wrangham, RW, Conklin-Brittain, NL, and Hunt, K.D., 1998. Dietary response of chimpanzees and cercopithecines to seasonal variation in fruit abundance. II. Nutrients. *Int. J. Primatol.* 19, 971–987.

Yamashita, N., 1998. Functional dental correlates of food properties in five Malagasy lemur species. *Am. J. Phys. Anthropol.* 106, 169–188.

6

Mandibular Biomechanics and the Paleontological Evidence for the Evolution of Human Diet

DAVID J. DAEGLING
FREDERICK E. GRINE

Many Darwinian biologists accept that organisms and their constituent parts exhibit optimal design (given certain architectural, developmental, and phylogenetic constraints) so as to maximize reproductive survival. Strictly adaptationist approaches to explaining design perfection in biological has been rightly criticized as Panglossian storytelling (Gould and Lewontin, 1979), but there is ample evidence that the configurations of at least some morphological attributes are explicable in reference to the function(s) that they serve. However, exactly how closely the structure of a trait can be related to its apparent biological function remains a topic of discovery. There is an abundant literature indicating that mechanical (as well as physiological) environments may profoundly influence skeletal structure (e.g., Frost, 1964; Cowin, 1986; Martin and Burr, 1989; Fung, 1993; Martin, Burr, and Sharkey, 1998; Currey, 2002). But while it is evident that mechanical loads influence the skeletal system, it is not clear to what extent the mechanical environment dictates a particular morphological configuration.

This problem is especially pertinent to the relationship between an organism's diet and the structure of its masticatory apparatus. To what extent does a species' diet influence mandibular morphology? As a corollary, if diet does indeed influence jaw form, is it possible to examine variation in mandibular morphology to infer anything about the dietary proclivities of the species to which it belongs? Most of the research that has been undertaken to elucidate the function-form relationship has focused on the long bones of the postcranial skeleton because of their functional linkage to locomotor forces, and because they are more commonly fractured and of greater clinical concern. Nevertheless, considerable effort has been expended in attempts to better understand the structure of the primate mandible and its relationship to the mechanical demands that may be placed on it by different dietary regimens (e.g., Biegert, 1956; Lindblom, 1960; Smith, 1978; Hylander, 1979b, 1979c, 1985a, 1985b, 1991; Bouvier and Hylander, 1981; Smith, 1983; Smith, Petersen, and Gipe, 1983; Demes, Preuschoft, and Wolff, 1984; Hiiemae, 1984; Bouvier, 1986; Weijs, 1989; Daegling, 1990, 1992, 2002a; Chen, 1995; Spencer, 1995, 1999; Wall, 1995; McGraw and Daegling, 1999; Daegling and Hylander, 2000; Taylor, 2002; Daegling and Hotzman, 2003). At the same time, certain aspects of mandibular form have been examined from the standpoint of what they might tell us about the dietary habits of extinct members of the human family tree (Wolpoff, 1975; DuBrul, 1977; Ward and Molnar, 1980; Hylander, 1988; Grine et al., 1989; Daegling and Grine, 1991; Spencer and Demes, 1993; Chen and Grine, 1997; Wood and Aiello, 1998; Daegling and Hylander, 2000). This interest in mandibular form and function is, of course, understandable, because diet is a fundamental component of an organism's biology, and the masticatory system is directly related to food processing.

In this chapter, we review and evaluate the various lines of evidence pertaining to mandibular biomechanics that have been brought (or that might profitably be brought) to bear in the reconstruction of the dietary habits of various hominin species from the Pliocene and early Pleistocene. We then examine how some of these biomechanical principles can be applied to the fossil record using the South African australopiths as a case study.

A fully comprehensive model of mandibular biomechanics entails the incorporation of a bewildering array of distinct, but not disparate, variables that are all components of the jaw as it functions in food ingestion and mastication. Such features include the absolute and relative size of the jaw, the morphology of the temporomandibular joint, the height of the temporomandibular joint above the occlusal plane, the distance between the temporomandibular joint and the tooth row, the size and configuration of the masticatory muscles, the structure and geometry of the mandibular corpus (including the symphysis), the distribution and mechanical properties of cortical bone within the corpus, the presence of tooth roots and the periodontal ligaments that anchor them in bony alveoli, and the networks of trabecular bone struts in the interior of the corpus. The structural complexity of the mandible makes it much more difficult to model than the long bones of the skeleton, which are neither recurved nor interrupted by irregularly spaced and shaped cavities that are plugged with dentine. This complexity, coupled with the rather catholic and eclectic diets of many primate taxa, complicates the task of identifying specific aspects of jaw morphology that have been influenced by variations in chewing stresses that may be related to dietary differences.

Indeed, developmental and/or allometric constraints, as well as features unrelated to the generation or distribution of chewing stresses, may have sufficient impact on the biomechanics of the masticatory system to further obscure the relationship between diet and mandibular morphology. Thus, while allometric patterns are no less subject to selection than morphology (Gould, 1966), the robusticity of the mandibular corpus may be related more directly to allometric than to dietary factors (Ravosa, 2000). Indeed, some aspects of mandibular morphology, such as symphyseal depth, may be artifacts of somatic scaling that have secondary effects on masticatory biomechanics (Smith, 1993). Similarly, selection for features such as canine size and gape may impact the biomechanics of the masticatory system by deepening the anterior corpus and lowering the height of the mandibular condyle, respectively (Lucas, 1981; Smith, 1984).

Biomechanical Modeling of the Mandible

Biomechanical models describe the ways in which the functional components of the masticatory system interact. By necessity, such descriptions rely on simplifications that typically take the form of a theoretical analogy or construct (e.g., modeling a bone as a hollow beam). Because biomechanical studies tend to produce simplified descriptions of intricate anatomical systems, they have been criticized as reductionist (Smith, 1982). This problem of oversimplification, however, may be alleviated to some degree by employing experimental data (e.g., bone strain or cineradiography) to evaluate the mechanical effects of morphological variation that can be generalized to a comparative setting. These data will influence the type of model that is chosen to describe the mechanical attributes of various components of the system.

Lever Mechanics: The Mandible as a Rigid Body

Although the mandible has been modeled as an anatomical link that converts muscular force directly into bite force without any load applied to the temporomandibular joint (TMJ) (Gingerich, 1971; Roberts and Tattersall, 1974; Taylor, 1986), most of the arguments presented in its favor have been demonstrated to be incorrect (Hylander, 1975; Picq, Plavcan, and Hylander, 1987). More important, in vivo experimental studies have demonstrated that the TMJ is compressively loaded during mastication (Hylander, 1979a, 1985b; Hylander and Bays, 1979; Brehnan et al., 1981; Hohl and Tucek, 1982; Boyd et al., 1982, 1990). In this regard, the TMJ is generally loaded more on the balancing (nonbiting) side than on the working (biting) side, such that the fulcrum is formed by the balancing-side mandibular condyle. Under certain conditions, tensile forces occur at the working-side joint (Hylander, 1985b).

The functional morphology and possible linkage of TMJ form to diet have been the subject of a number of studies (e.g., Biegert, 1956; Lindblom, 1960; Smith, 1978; Noble, 1979; Smith, Petersen, and Gipe, 1983; Mack, 1984; Bouvier, 1986a, 1986b; Kantomaa, 1989; Boyd et al., 1990; Wall, 1995; Williams et al., 2002, Vinyard et al., 2003). With one or two possible exceptions, however, no clear relationship between dietary variables, such as food object size or consistency, and morphological variables of the joint has emerged from these analyses. One exception appears to be the degree of sagittal curvature of the mandibular condyle, where relatively flatter condyles are generally (but not uniformly) found in primates that chew more resistant foods (Smith, Petersen, and Gipe, 1983; Wall, 1995). Those studies with a more ecological focus (i.e., examining feeding behavior rather than food items exclusively; Williams et al., 2002; Vinyard et al., 2003) have successfully identified morphological variables that covary with behavioral aspects of food processing. For example, Vinyard and colleagues (2003) discovered that condylar position is associated with tree-gouging specialization in primates, while condylar morphology suggests, counterintuitively, that these behaviors do not produce unusually large forces in the skull.

The dominant framework that currently underlies functional interpretations of primate mandibular form holds that it functions as a third-class lever, with the TMJ acting as its fulcrum (Hylander, 1975, 1985; Greaves, 1978; Smith, 1978; Walker, 1978; Demes, Creel, and Preuschoft, 1986; Picq, Plavcan, and Hylander, 1987; Daegling, 1990; Spencer and Demes, 1993; Chen, 1995; Spencer, 1995, 1999; Kieser, 1999; Thompson, Biknevicius, and German, 2003). As such, the lever can be modeled in two dimensions, where the jaw adductor muscles (temporalis, masseter, and medial pterygoid) apply forces to the mandible that are resisted by forces at the TMJ (joint reaction force) and the bite point (bite force). The component vectors for these three muscles can be combined to produce a single-resultant muscle force vector. In the two-dimensional model, bilateral muscle and joint reaction forces are also combined into single vectors and projected in the sagittal plane (Throckmorton and Throckmorton, 1985; Baragar and Osborn, 1987; Wood, 1994). This model is further simplified by projecting the three forces—joint reaction force, bite force, and muscle resultant force—onto a common reference line (usually the occlusal plane) and considering only those force components that act perpendicular to that plane. Under this model, the mandible acts as a rigid body, and the muscle and bite forces exert some rotatory action (referred to as a *moment*, or *torque*) relative to the fulcrum of the TMJ. The height of the TMJ above the occlusal plane is a potentially confounding variable in this model. Although this may affect the magnitudes of the bite and joint reaction forces through differences in the moment arm lengths of the muscles (Spencer, 1995; Wall, 1995), joint position is also clearly related to functional requirements related to gape (Herring and Herring, 1974; Carlson, 1977; Smith, 1984; Ravosa, 1987).

While two-dimensional models of the masticatory system are instructive, they preclude consideration of the differential activities of the adductor musculature on the working and balancing sides of the jaw and of the differential loading of these two joints. By contrast, three-dimensional models combine separate two-dimensional analyses of forces in both the sagittal and coronal planes to provide a more comprehensive assessment of jaw loading (Smith, 1978). In this way, forces in the sagittal plane are constructed as in the two-dimensional model, but separate working-side and balancing-side joint reaction forces can be realized in frontal projection, and it is possible to calculate a resultant vector for components of all six jaw adductor muscles that occupies both planes. Greaves (1978) has provided a method by which the spatial relationships of all forces can be visualized simultaneously, where the mandible functions as a constrained third-class lever with three points of resistance, providing a "triangle of support." To maintain static equilibrium, the reaction forces at the three corners of the triangle must sum to the magnitude of the muscle resultant force. According to this model, as the bite point moves posteriorly, the side of the triangle that connects the bite point and the balancing-side joint more closely approximates the midline muscle resultant. Thus, more posterior bite points cause the balancing-side joint to resist a greater proportion of the load than the working-side joint. At the point where the midline muscle resultant intersects the side of the triangle, all of the muscle resultant force will be resisted by the bite point and balancing-side joint alone. Continued posterior movement of the bite point places the midline muscle resultant beyond the triangle of support so that the mandible will tend to rotate around the line between the bite point and balancing-side condyle. This produces tensile stress on the working-side joint. This unstable situation can be averted by shifting the muscle resultant force from the midline along a transverse line by differential recruitment of the working- and balancing-side adductors (Greaves, 1978). Reduction in the activity of the balancing-side adductors will serve to shift the muscle resultant force back into the triangle of support but at a cost to total muscle force magnitude.

Greaves's (1978) model results in two regions of possible bite-point distribution (Spencer, 1995, 1999). Bite points situated more anteriorly (Region I) produce a triangle of support that envelops a midline muscle resultant associated with full adductor recruitment. However, any bite point located within the posterior region (Region II) produces a triangle of support that contains the muscle resultant only if it is repositioned laterally through a reduction in balancing-side muscle force.

The transition from Region I to Region II in modern humans occurs just distal to the first permanent molar because the muscle resultant is located approximately coincident with the coronoid processes of the mandibular ramus. This results in humans displaying a distalward gradient of reduced occlusal masticatory force from M_1 to M_3 (Spencer, 1995, 1998; Kieser, Gebbie, and Ksiezycka, 1996; Kieser, 1999; Grine et al., 2005). This contradicts models that have postulated increased distal bite forces in humans (e.g., Molnar and Ward, 1977; Osborn and Baragar, 1985; Janis and Fortelius, 1988; Koolstra et al., 1988; Osborn, 1996), but it is consistent with observations that molar crown size as well as tooth root complexity and surface area decrease from M_1 to M_3 in modern humans (Müller, 1959; Nikolai, 1985; Kieser, 1990; Kupczik, Spoor, and Dean, 2003).

The Mandible as a Deformable Body

If the assumption of a rigid body is removed, the mandible can also be modeled as a bent beam, a sheared block, and a twisted member. The major forces involved in mastication will generate bending moments in both corpora of the mandible; on the balancing side, there will generally be tension in the alveolar region and compression along the base (Hylander, 1977, 1979a, 1979b, 1981; Hylander and Crompton, 1986; Hylander, Johnson, and Crompton, 1987). Moreover, in association with the vertical components of muscular, biting, and reaction forces, the working-side corpus is also directly sheared (Hylander, 1979a, 1979b, 1981). Experimental data indicate that the working- and balancing-side mandibular corpora are not only bent but also twisted about their long axes during the power stroke of mastication and in unilateral molar biting (Hylander, 1979a, 1981, Hylander and Crompton, 1986; Hylander, Johnson, and Crompton, 1987). This is due to the fact that the resultants of occlusal and muscular forces do not pass through the shear center of corpus sections.

Of the various loading regimes, twisting (i.e., torsion) of the postcanine corpus about its long axis is probably the most important for causing large stress concentrations (Hylander, 1988; but see Dechow and Hylander, 2000). As noted by Hylander (1988), torsion results from the principal adductor forces of the temporalis and masseter acting on the mandible lateral to the corpus, thus everting its lower border and inverting its alveolar process. While the various loading regimes act on the corpus more or less simultaneously, their relative contribution to stress and strain magnitudes has been assessed by considering principal strain orientations observed in vivo, principally in *Macaca fascicularis* (Hylander, 1979a, 1979b, 1981).

The same forces that produce bending, twisting, and shear in the postcanine region induce specific loading regimes in the anterior corpus as well. Dorsoventral shear results from the vertical components of working-side and balancing-side resultant forces, twisting of the postcanine corpora induces coronal bending of the anterior corpus, and lateral components of biting and muscular forces produce lateral transverse bending ("wishboning") of the symphyseal region (Hylander, 1984). This wishboning engenders high strains on the lingual surface of the symphysis. Interestingly, while twisting of the symphysis about its transverse axis is theoretically possible, the in vivo data indicate that this load does not occur in macaques (Hylander, 1984).

Can Diet Be Linked to Masticatory Loads?

Functional morphological studies of the mandible often assume a relationship between a species' diet and its jaw morphology. Are the mechanical properties of the mandible adaptively influenced in some manner by the material properties of the-foods that are ingested and masticated? Is such a relationship detectable by the sorts of comparative approaches that are currently employed by researchers? Some studies of the mandibles of living primates with different dietary habits suggest that such differences may have predictable effects on jaw morphology (Hylander, 1979a, 1979b; Bouvier and Hylander, 1981; Bouvier, 1986a, 1986b; Daegling, 1992; Ravosa, 1991, 1996a). However, the functional link between the material properties of food items and primate jaw mechanics may be more subtle than is commonly assumed (Daegling and McGraw, 2001). Moreover, the cross-sectional geometry of the corpus may reflect a stronger phylogenetic than functional signal (Daegling, 2002a).

If a relationship between diet and jaw morphology exists, and if it can be detected in some manner, the ultimate question, of course, is whether this relationship can be extrapolated to make inferences about diet from mandibular morphology in the hominin fossil record. To address these questions, it is necessary to identify variables that are relevant to the biomechanical behavior of the mandible, and to determine those that are amenable to comparative analysis. This latter caveat is important from a practical standpoint: incorporation of all the relevant variables for inference of mechanical behavior is prohibitively complex in a comparative context. Consequently, simplifying assumptions about bone structure (e.g., homogeneity and isotropy) are used not because they are correct but because they can be applied to large samples.

Biomechanics of Bone

An area of biomechanics known as strength of materials is concerned with the deformation of biological bodies (in this case, bone) that are acted on by mechanical forces. Application of this approach has been particularly fruitful for understanding the functional morphology of bone, both structurally and materially. Bone is a richly innervated and vascularized composite (biphasic) material comprising inorganic (crystalline mineral) and organic (cells and extracellular matrix) components. As such, a strong, albeit brittle material is embedded within a weaker, but more flexible compartment. This results in the combined substance being stronger per unit weight than either component alone (Albright, 1987). A structural analogy might be fiberglass or steel-reinforced concrete.

The most important properties of bone from the standpoint of mechanical function are its strength, stiffness, and toughness (Nordin and Frankel, 1989; Currey, 2002). *Stiffness* refers to the degree of elasticity or the relationship between the applied load and the deformation experienced in the tissue. Over an initial range of deformation, bone behaves elastically; that is, when a load is removed, the deformed material returns to its original shape. The elastic limit, or "yield point," of the bone refers to the point after which further increase in loading will result in permanent structural deformation. *Strength* refers to the load and concomitant deformation that can be sustained, and the amount of energy that can be stored before the bone experiences ultimate failure (i.e., fracture) is known as *toughness*. Strength and stiffness are intimately related to the central concepts of stress and strain. Stress (σ) is a unit load, or internal resistance that develops within a bone in response to an externally applied force; it is reckoned as force per unit area. Strain (ε) is the deformation that develops in a bone in response to an externally applied force. Specifically, strain refers to the change in length experienced by a loaded object as a proportion of its original length ($\Delta l/l$, a dimensionless quantity). The relationship of stress and strain for a given material is conveniently summarized as the elastic modulus E, simply defined as their ratio (σ/ε). As such, strain and stress are related through the material properties (elastic modulus) of the bone. The relationship between stress and strain is essentially the same as between force (load) and deformation with regard to the stiffness and strength of a whole bone. Comparative studies typically assume modulus to be invariant within elements as well as between species.

Influence of Microscopic and Macroscopic Structural Factors on the Biomechanical Behavior of Bone

Several factors alluded to earlier will have an effect on bone behavior because they affect its material properties. We can envisage at least five such factors that are worthy of consideration in any attempt to appreciate the biomechanical behavior of the mandible.

The first factor might be the mechanical properties of the fibrous periosteum that surrounds the bone and serves as the site of muscular and ligamentous attachment (Popowics, Zhu, and Herring, 2002). The transfer of forces from this layer to the underlying osteogenic (cambium) layer of the periosteum and to the bone itself is not well understood, although the biomechanical properties of the periosteum may influence intramembranous bone deposition (Taylor, 1992; Covell and Herring, 1995). Although some aspects of periosteal biology in the mandibles of rats, rabbits, and pigs have been investigated, no such comparative data exist for primates.

Another factor that might profitably be considered is the differential mechanical properties of primary versus osteonal (Haversian) compact bone (Reilly and Burstein,

1974, 1975; Martin and Burr, 1989; Martin, Burr, and Sharkey et al., 1998; Schwartz-Dabney and Dechow, 2003). Loading and deformation of bone strongly influence the modeling and remodeling of its cortical structure (Lanyon, 1991; Skedros et al., 1996; Martin, Burr, and Sharkey, 1998). As primary woven bone is replaced with osteons, its degree of anisotropy increases and its bending fatigue strength is reduced (Martin, Burr, and Sharkey, 1998). This is presumably related to the presence of less highly calcified bone matrix in more recently formed secondary osteons, the increased porosity of Haversian bone, and the increase in anisotropy that results from the alteration of collagen fiber orientation and the introduction of cement (reversal line) interfaces that surround the osteons. Whereas bone tissue strength appears to be diminished through osteonal remodeling, the presence of osteons may enhance bone durability and toughness on a structural level through the repair of fatigue microdamage (Martin, Burr, and Sharkey, 1998). Mechanical strain is positively correlated with increased Haversian remodeling (Lanyon and Rubin, 1984; Lanyon, 1991; Weinbaum, Cowin, and Zeng, 1994; Bouvier and Hylander, 1996), presumably through the influence of microdamage on bone cells such as osteocytes (Bentolila et al., 1997). This is the likely basis for the well-known relationship between individual age and secondary osteonal remodeling (Stout, 1992; Robling and Stout, 2000), although factors such as nutritional status and hormonal levels are also influential.

Bouvier and Hylander (1981) demonstrated significantly higher incidences of secondary osteons in the mandibular cortices of rhesus macaques (*Macaca mulatta*) that were fed hard diets in comparison to those fed softer foods. There are limited, unpublished data pertaining to the incidence of osteonal bone in the mandibles of platyrrhine monkeys that eat soft foods (e.g., *Alouatta seniculus*) and those that eat hard foods (e.g., *Cebus apella*), which suggest that the latter exhibit more secondary osteons per mm^2 (Wall and Grine, n.d.). A thin section of cortical bone from the base of the mandibular corpus of a single specimen of the Plio-Pleistocene hominin, *Paranthropus robustus* has been examined for evidence of Haversian remodeling (fig. 6.1). This particular fossil reveals a strikingly low incidence of secondary osteons and is more similar to those recorded for *Alouatta* and *Ateles* than for *Cebus* and *Chiropotes*.

At present, the apparently low incidence of remodeling in this *P. robustus* jaw is difficult to interpret. We are uncertain about the extent of individual variation in Haversian remodeling of the mandible in different primate species (especially large-bodied catarrhines) or the effect that age and sex might have on this parameter. Thus, the apparently low incidence of remodeling in this *P. robustus* jaw may be interpreted in several ways. Does this finding indicate that masticatory stresses were somehow reduced in these forms, or is it that bone modeling was such that fatigue damage was mitigated to the point that remodeling was reduced? The questions are not mutually exclusive. Whatever interpretation is offered, the underlying assumption is that peak strain magnitude, in conjunction with the number of daily loading cycles, has a direct role in governing the remodeling response. It is, however, prudent to recognize that there are other candidate variables likely to be more directly related to remodeling activity (Rubin et al., 1991, 2001).

A third factor that may affect the biomechanical behavior of a bone is the presence of cancellous, or trabecular, bone lying deep to the cortex. Trabecular structure

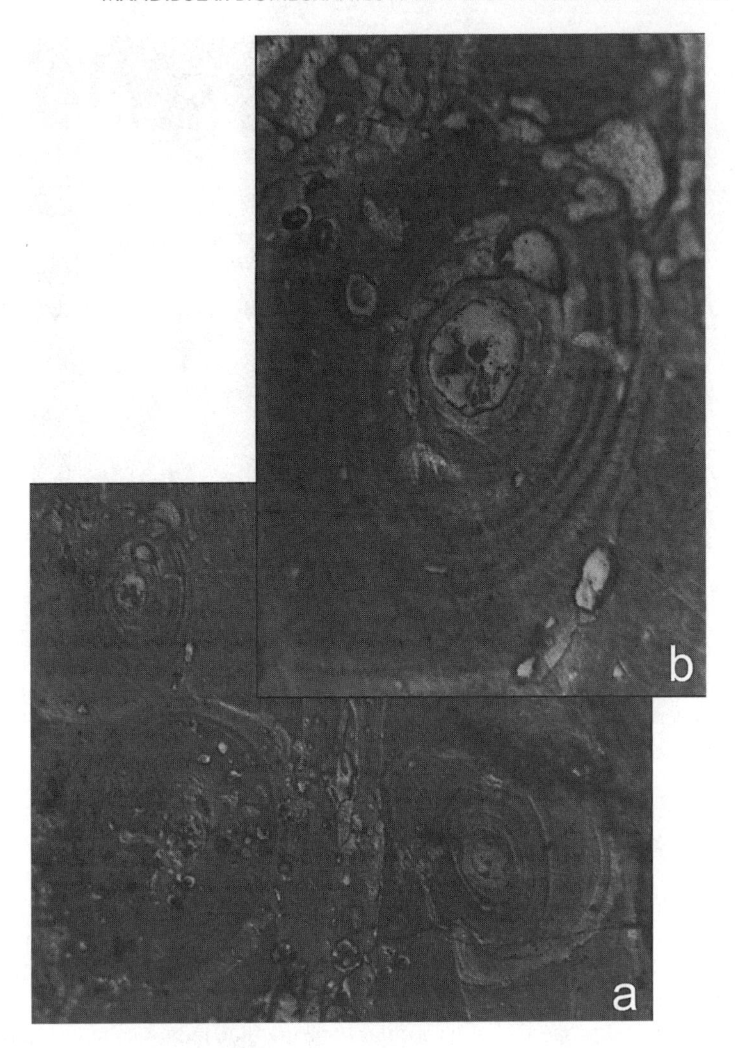

Figure 6.1 Reflected light micrographs of Haversian bone in the inferior aspect of the corpus of an adult *Paranthropus robustus* individual from the site of Swartkrans, South Africa. (*a*) Micrograph showing three osteons surrounding central Haversian canals in upper left, lower left and lower right corners. (*b*) Enlargement of the upper left osteon in (*a*) showing concentric lamellae. The irregular "bubbles" are inclusions of an unspecified nature. Osteonal bone represents only about 6 percent of cortical area in this specimen.

has a significant relationship with bone strength (Rockoff, Sweet, and Bleustein, 1969; Martin, Burr, and Sharkey, 1998; Ito et al., 2002), although the mechanical contribution of trabecular bone to structural rigidity in long bone shafts has been disputed (Burr and Piotrowski, 1982; Ruff, 1983). The mechanical properties of trabecular bone differ from those of cortical bone at both the microscopic (material) and macroscopic (structural) levels (Gibson, 1985; Keaveny and Hayes, 1993; Martin, Burr, and Sharkey, 1998; Currey, 2002). Relative to cortical bone, cancellous bone has a lower mineral content and tissue density. In addition, the osteons in

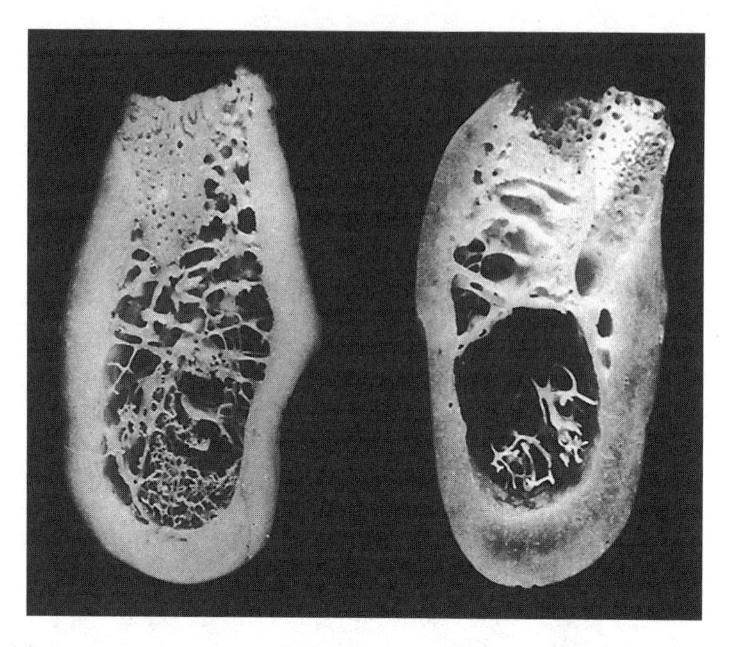

Figure 6.2 Sections of two human mandibles taken at the level of the M_1 showing the variation in trabecular networks. This variation complicates the task of developing a single biomechanical model of the corpus that will be appropriate even to all individuals of a single species (from Daegling, 1989, p. 102).

trabeculae are shaped differently than in compact bone; in the former, the cement lines intersect the surface, resulting in the bony struts being less resistant to damage (Martin, Burr, and Sharkey, 1998). At the macroscopic level, cancellous bone probably has reduced capacity for elastic energy storage than cortical bone, assuming its yield strain is lower (Martin, Burr, and Sharkey, 1998), and its yield strength is reduced, given its greater compliance (Rho, Ashman, and Turner, 1993).

The mechanical properties of trabecular bone are dictated by its apparent density (i.e., dry mass/volume), which varies considerably between and even within skeletal elements (Goldstein et al., 1983; Morgan, Bayraktar, and Keaveny, 2003). Trabecular architecture also dictates potent nonlinear effects by adding a layer of structural anisotropy to the intrinsic anisotropy of the tissue (Judex, Whiting, and Zernicke, 1999). While some workers have questioned whether the trabeculae are arranged in relation to the principal stresses that are generated during loading (e.g., Cowin, 1997), a number of studies suggest that cancellous struts are generally organized this way (Lanyon, 1974; Fiala and Heřt, 1993; Heřt, 1994; Biewener et al., 1996).

The impact of cancellous bone on the biomechanical behavior of the mandibular corpus has yet to be investigated. With regard to modern humans and apes, Daegling (1989, 1990) has shown that these bony struts are quite variable in their distribution within the corpus of a single individual, and at homologous locations between individuals (fig. 6.2). This variation will have a significant effect on the choice of biomechanical model to represent the behavior of the mandibular corpus. In those cases

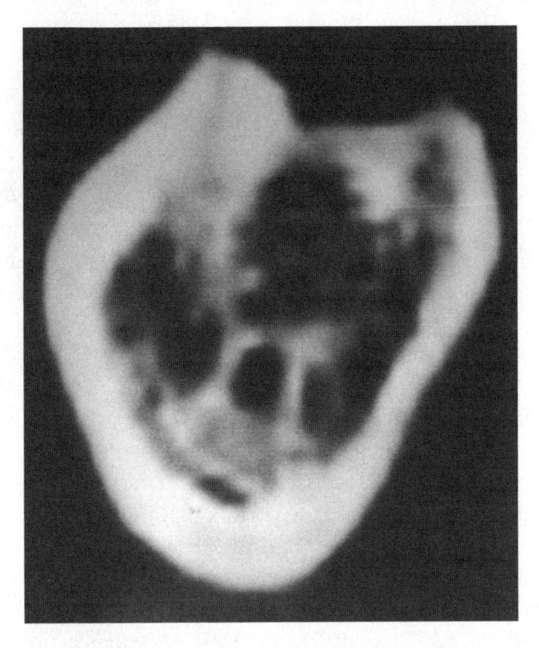

Figure 6.3 Computed-tomography (CT) section of *Paranthropus robustus* mandible SKW 5013 showing the extent of trabecular networks in the subendosteal space below the M_2. The lateral aspect of the corpus is to the left. Grine et al. (1989) have shown that such struts are not artifacts of fossilization, but represent mineral deposits surrounding preexisting trabeculae (from Daegling, 1989, p. 103).

where cancellous bone fills the space deep to the cortical bone (e.g., fig. 6.2, left), the most appropriate model might be that of a reinforced porous beam, while in cases where there is comparatively little trabeculation (e.g., fig. 6.2, right), the preferred model might be a hollow beam.

Similar variation exists in the amount and distribution of trabeculation in the corpus of South African australopith mandibles (Daegling, 1989; Grine et al., 1989; Daegling and Grine, 1991). While such variation (fig. 6.3) will affect biomechanical behavior, it has yet to be incorporated into analytical models. Some of these cancellous bony struts may play a significant role in the dissipation of bite forces transmitted from the teeth, given their position in connecting the lamina dura to remotely located endosteal borders.

Finally, and of particular relevance to the mandible, the presence of tooth roots and the periodontal ligaments that attach them to the bone in the alveolar crypts should be considered in biomechanical models. These irregularly shaped and sized roots, the ligaments, and the thin shell of alveolar bone to which they attach add a further layer of structural heterogeneity to the jaw. It has been argued that the intrusion of these structures into the corpus provides justification for modeling the mandible as a beam with an open section (Hylander, 1979c; Smith, 1983). Shear stress induced by torsion increases significantly in a bone with an open section defect (Frankel and Burstein, 1970). In closed sections (e.g., of a long-bone diaphysis), all the shear stress acts to resist the torque. In an open section, however, only the shear stress at the periphery resists the applied torque; stresses at the discontinuity change direction, and those along the internal aspect flow in the direction of the applied torques. As a result, in an open section, the amount of bone resisting the load is greatly decreased, such that the degree of load stress to bone failure was reduced by as much as 90 percent in human tibiae with a continuously elongate open section defect (Frankel and Burstein, 1970). Although single or intermittent holes cause less of a reduction in torsional rigidity than a continuously open section, such "stress raisers" significantly decrease bone strength by as much as 60 percent (Brooks, Burstein, and Frankel, 1970; Burstein et al., 1972; Hipp, Edgerton, and Hayes, 1990).

In vitro experiments and finite element models suggest that the mandibular corpus behaves as a closed section in torsion (Daegling et al., 1992; Daegling and Hylander, 1996; Daegling and Hylander, 1998; Chen and Chen, 1998). Nevertheless, while it is certain that tooth roots and the periodontal ligament have an effect on load distributions in the mandible (Melcher and Walker, 1976; Wills, Picton, and Davies, 1978; Mandel, Dalgaard, and Viidik, 1986; Asundi and Kishen, 2000; Imbeni et al., 2003), it remains unclear exactly how well these structures render the intermittent alveolar holes mechanically invisible. At the very least, the accuracy of biomechanical models would seem to benefit from the incorporation of dental and periodontal elements (Smith, 1983; Daegling, 1989; Chen and Chen, 1998).

A final factor to be considered is material anisotropy. It has long been recognized that bone is not equally stiff in all directions; yet, the assumption of isotropy is universal in comparative studies. Dechow's work (Dechow and Hylander, 2000) has established that primate mandibular bone is orthotropic, with the stiffest material axis aligned with the corpus long axis in the postcanine region. Variation in material properties is also regionally specific (Rapoff et al., 2003; Schwartz-Dabney and Dechow, 2003), meaning that a full accounting of material properties throughout the jaw will require monumental time and resources to be applied in a comparative context. The interpretive cost of assuming that mandibular bone is isotropic is dependent on the question under study; however, it is clear that estimates of strength and stiffness based on that assumption are inaccurate to some degree.

Influence of Size, Shape, and Geometry on the Biomechanical Behavior of Bone

The cross-sectional geometry of a bone significantly affect its biomechanical behavior. Three of the most important factors that affect the mandibular corpus under shear, bending, and torsion are (1) its cross-sectional area, (2) the distribution of bone around the neutral axis of bending (which in a simply loaded homogeneous structure passes through the centroid of a section), and (3) the distribution of bone around the twisting axis of neutrality (also passing through the centroid of idealized structures). In both bending and torsion, stress magnitudes increase with their distance from the neutral axis and twisting axis of neutrality, respectively. In bending, tensile stresses act on one side of the neutral axis, and compressive stresses act on the other; in both instances the normal stresses act perpendicular to the plane of section. These stresses increase linearly with their distance from the neutral axis. Under torsional loading, tensile and compressive stresses act on a plane that is diagonal (at 45 degrees) to the long axis of the bone, and this results in maximal shear stresses acting along a plane that is perpendicular to the long axis (Nordin and Frankel, 1989). Because the body of the primate mandible is typically deeper than it is broad, its cross-sectional shape approximates an ellipse more closely than a circle. Consequently, torsion of the corpus results in greater stress concentrations along its medial and lateral margins than over its superior and inferior borders because of the bone tissue deep to the medial and lateral margins being distributed closer to the neutral twisting axis (Frankel and Burstein, 1970; Hylander, 1979b, 1981).

Resistance to bending is quantified by variables referred to as second moments of area, or area moments of inertia. The second moment of area essentially measures how material is distributed in a cross-section with respect to a specified neutral axis about which bending occurs (or about which one wishes to calculate theoretical stresses). These variables incorporate information about the amount and distribution of material in a section by summing the product of elemental areas and their squared distances from the neutral axis, and are thus expressed in units of length to the fourth power. As such, its value will increase as the distance between a given area of bone and the neutral axis of zero normal stress increases. Second moments of area (denoted I) can be calculated about any axis (such that the area moment about a defined x-x axis is I_{xx}). In any cross-section, there is a principal axis that passes through the centroid about which the second moment of area will be maximized (I_{max}), and one perpendicular to it about which the second moment of area will be minimized (I_{min}). In a circular cross-section, these two principal axes have identical values and are not uniquely defined. In any section that deviates from axial symmetry, there are uniquely defined principal axes and their associated area moments will differ ($I_{max} > I_{min}$.). Area moments calculated about other sets of axes will have values between these extremes. Because stresses increase linearly with the distance from the neutral axis, the absolute dimensions of a cross-section have important consequences for stress resistance, where larger cross-sections will provide more resistance. A hollow cross-section represents a more economical use of material than a solid cross-section because more material (bone tissue) is distributed further away from the neutral axis where the bending moments can be more effectively resisted.

It is obvious that the relationship of I_{max} to I_{min} will change depending on the geometry of the cross-section. A circular beam is ideal if the bending moments are unpredictable (e.g., variable loading from multiple behaviors), whereas an elliptical section might be expected to have a more stereotypical loading regime (i.e., in the mandible, vertical loading from the primarily vertical components of muscular and bite forces).

With regard to the mandible, a closer relationship of I_{min} to I_{max} (or of I_{yy} to I_{xx}) signifies enhanced resistance for horizontal bending in comparison to vertical bending. The former results from wishboning of the mandible during the power stroke of mastication, where persistent activity in the balancing-side deep masseter and the lateral component of bite force on the working side tends to draw the left and right corpora away from the midline (Hylander, 1984, 1988). A transversely broader postcanine mandibular corpus (or labiolingually thicker symphysis) represents an effective morphological strategy for countering this load.

While the ideal shape of a section for resisting bending moments depends on the variety and orientation of these loads, the ideal cross-sectional geometry for torsion in unambiguous. A cylindrical cross section (in which $I_{min}/I_{max} = 1.0$) is ideally suited to resist twisting moments because the peripheral shear stresses are equitably distributed around the outer fibers of the section.

A variable related to the second moment of area that is used to measure a cross section's ability to resist torsion is known as the polar second moment of area, or the polar moment of inertia (J). It is calculated about a point (the centroid) rather than an axis or plane and simply represents the summed values of I_{max} and I_{min} (or any two orthogonal area moments). Recalling Pythagoras, this variable can also be directly calculated by summing the squared radial distances of each element of area from the centroid. Because the polar second moment of area assumes axial symmetry (i.e., a circular cross section), it is largely a function of area rather than geometric shape. Employment of the variable assumes that stress magnitude is a simple function of distance from the centroid, but as stated above, in noncircular sections this is not the case. As a result, the suitability of J for the depiction of torsional resistance in noncircular sections is highly suspect (Burr, 1980; Daegling, 2002b).

The Influence of Variation in Bone Mass Distribution

Returning to the factors that most importantly affect a bone's biomechanical behavior in bending and torsion, recent studies have moved beyond the exclusive use of second moments of area and have turned to a consideration of variation in cortical bone thickness as it relates to load history and resistance (Daegling, 1992, 2002a; Chen and Grine, 1997; Chen and Chen, 1998). Patterns of cortical thinning within cross sections have a modest influence on calculation of second moments of area and subsequent predictions of maximum stresses in bending. With respect to torsion, however, patterns of cortical thinning in primate jaws have such a pronounced effect on stress distributions that the contribution of cross-sectional geometry to torsional resistance is obscured (Daegling and Hylander, 1998). Specifically, the lingual thinning of bone that is typical of sections in the molar region of the corpus results in high strains and stresses, such that reliance on measures based on area moments will fail to predict structural integrity under torsion. It is thought that this thinning of bone is due to the subtractive effects of superposed twisting and shearing loads in the lingual cortex of the posterior corpus (Demes, Preuschoft, and Wolff, 1984), and this possibility enjoys some support from in vitro strain data (Daegling and Hotzman, 2003). What has been established in vivo is that, for whatever reason, mastication creates higher strains on the lateral cortical bone of the posterior corpus (Dechow and Hylander, 2000). Collectively, these findings raise the interesting possibility that local differences in bone mass are reflective of similar variation in stress and strain gradients. Appropriate tests of this hypothesis have yet to be conducted.

Paleontological Inference of Diet from Jaw Form: The Case of the Early Hominins *Australopithecus africanus* and *Paranthropus robustus*

Mandibular proportions in the early hominins *Australopithecus* and *Paranthropus* are unique among primates and mammals in general (fig. 6.4). Attempts to invoke contemporary analogs for feeding behavior and diet in the two genera have resulted in wildly divergent interpretations of trophic adaptations (cf. Jolly, 1970; Szalay, 1972; Cachel, 1975; DuBrul, 1977), presumably due to unwarranted (and often unstated) assumptions about paleoecological variables. Of these studies, arguably DuBrul's (1977) more strict adherence to a comparative biomechanical approach yielded the most credible (i.e., testable) account of dietary divergence between the so-called gracile and robust forms. This conclusion echoed the "dietary hypothesis" that emerged from Robinson's (1954) study of the dentition, which proposed that the gracile *Australopithecus africanus* was omnivorous, while the robust *P. robustus* was a more dedicated herbivore.

The implication of divergent dietary adaptations for biomechanical comparisons is straightforward, if one assumes that bone tissue models and remodels itself to maintain stress levels within some optimal interval. Specifically, given that mastication of fibrous and tough foods engenders higher strains in the mandible (Hylander,

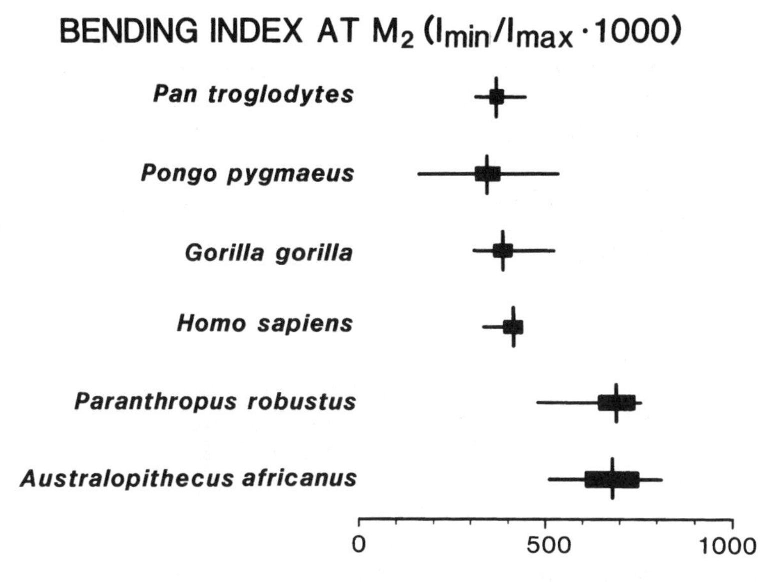

Figure 6.4 Ratio of minimum to maximum moments of inertia at M_2 sections in great apes, humans, and South African australopiths. The cross-sectional shape of the fossil hominins is unique among hominoids. The functional significance of this shape difference is unclear, although it is not a simple effect of postcanine megadontia. Sample sizes are $N = 10$ for extant species (with males and females equally represented), $N = 4$ for *Australopithecus*, and $N = 6$ for *Paranthropus* (from Daegling and Grine, 1991, p. 327).

1981, 1988), then a more demanding diet for *Paranthropus* would require a mandible that was structurally suited to accommodate greater masticatory forces while maintaining stress magnitudes comparable to *Australopithecus*. In the vernacular, *Paranthropus* should have a more robust jaw than *Australopithecus*.[1]

Whether *Paranthropus* and *Australopithecus* experienced similar stress magnitudes in their jaws is—from an empirical standpoint—impossible to determine. Even though moments of inertia can be calculated from cortical bone contours visualized in computed tomography (CT) images, and reasonable proxies for moment arms in bending can be obtained from partial and complete specimens, we have no way of reliably estimating resultant muscle vectors from fossil material. Indeed, this is currently beyond our capabilities in living animals. In effect, the only question we can address is whether stress magnitudes would be comparable given the same level of adductor recruitment in the two taxa.

Robinson's notion of *Australopithecus* and *Paranthropus* as having been functionally distinct can be evaluated from the perspective of masticatory biomechanics using living hominoids as a baseline of comparison. Scaling the appropriate second moments of area against moment arms acting at particular corpus sections, both fossil species are functionally distinct from the great apes in having relatively stronger mandibles (Daegling, 1990). With respect to one another, however, they are not distinguishable biomechanically. In other words, there is no basis for assuming that *P. robustus* jaws could accommodate larger bending forces than *A. africanus* mandibles and experience similar stress magnitudes at the same time (fig. 6.5).

Bending is but one of the major loading regimes acting on the mandibular corpus during mastication. Direct shear and torsion are also important sources of stress (Hylander, 1979b, 1981). Resistance to shear is essentially a function of cross-sectional area. Because the overall corpus dimensions of both *A. africanus* and *P. robustus* are typically larger than those of modern hominoids, and because the amount of bone per unit of subperiosteal area is comparable to living forms (Daegling 1990), it follows that the fossil hominins were better equipped to deal with large shearing forces. Evaluating torsional strength in a comparative setting, however, is more difficult. Hylander (1979b, 1988) has argued that the large dimensions of early hominin jaws in combination with their more cylindrical cross-sectional shape suggests a particular structural response to counter large twisting moments. As noted above, however, cross-sectional shape is probably less important for resisting torsion given the asymmetrical distribution of cortical bone within sections (Daegling and Hylander, 1998). An alternative measure of torsional strength that accounts for this asymmetry (a variable known as Bredt's formula in the engineering literature) suggests that *A. africanus* and *P. robustus* jaws were, in fact, absolutely stronger than those of extant hominoids (fig. 6.6). But this finding of absolute strength differences is not terribly informative because it does not correct for the different magnitudes of twisting moments that may (or may not) operate in the different species. Ideally, measures of torsional resistance or strength should be scaled against estimates of the moment arm in twisting (if not estimates of the moments themselves), and these cannot be reliably inferred from skeletal material because they depend not on jaw length (as is the case for bending loads), but on the position of the adductor resultant relative to the section of interest. Thus, we lack a robust test of the hypoth-

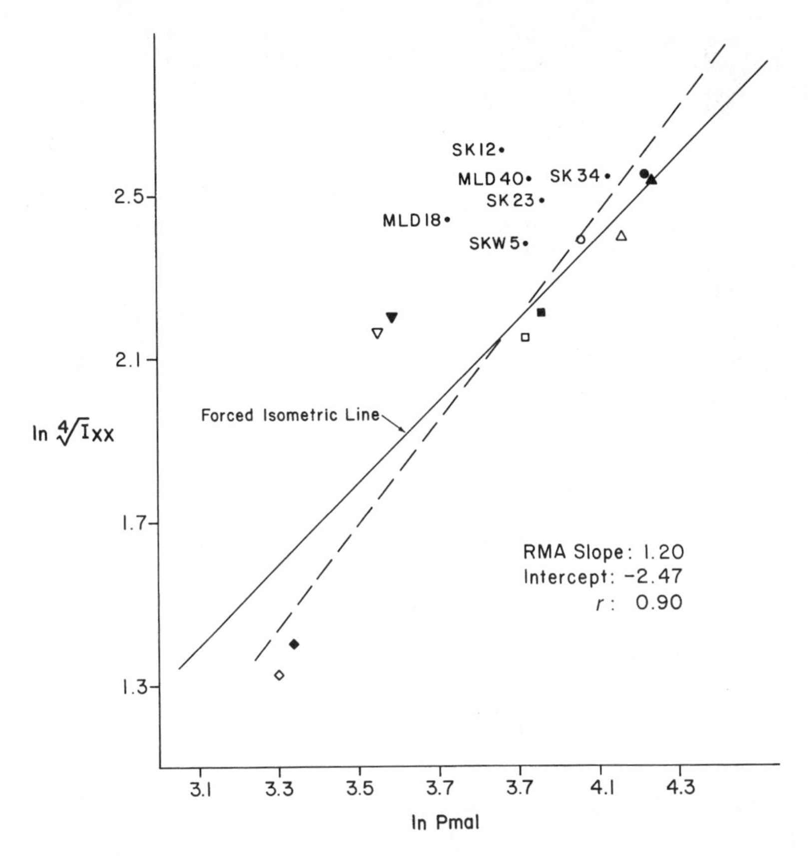

Figure 6.5 Regression of the moment of inertia that resists parasagittal bending on that load's associated moment arm. Empirical (*dashed*) and forced isometric lines are shown. Individual fossil hominin specimens are shown, with mean values for extant hominoids shown for males (*filled symbols*) and females (*open symbols*) separately in each taxon. The empirical and forced isometric lines were calculated using the mean values for *Hylobates* (*diamonds*), *Pan* (*squares*), *Pongo* (*circles*), *Gorilla* (*triangles*), and *Homo* (*inverted triangles*). The positioning of the fossils of both australopith taxa indicates enhanced resistance and strength with respect to parasagittal bending loads.

esis that these fossil jaws were subject to particularly high twisting loads, even if corpus morphology suggests that interpretation.

An interesting finding in regressions of mechanical rigidity and strength variables against moment arm surrogates is that modern human mandibles tend to fall above empirical best-fit and forced isometric lines (figs. 6.5 and 6.6), thus prompting the interpretation that, by an allometric criterion, modern humans have relatively strong jaws (Daegling, 1990, 1993). But the difference between modern human and early hominin jaws indicates that strength is achieved in fundamentally distinct ways. Modern human jaws are strong because they are short (i.e., corpus size and proportions are

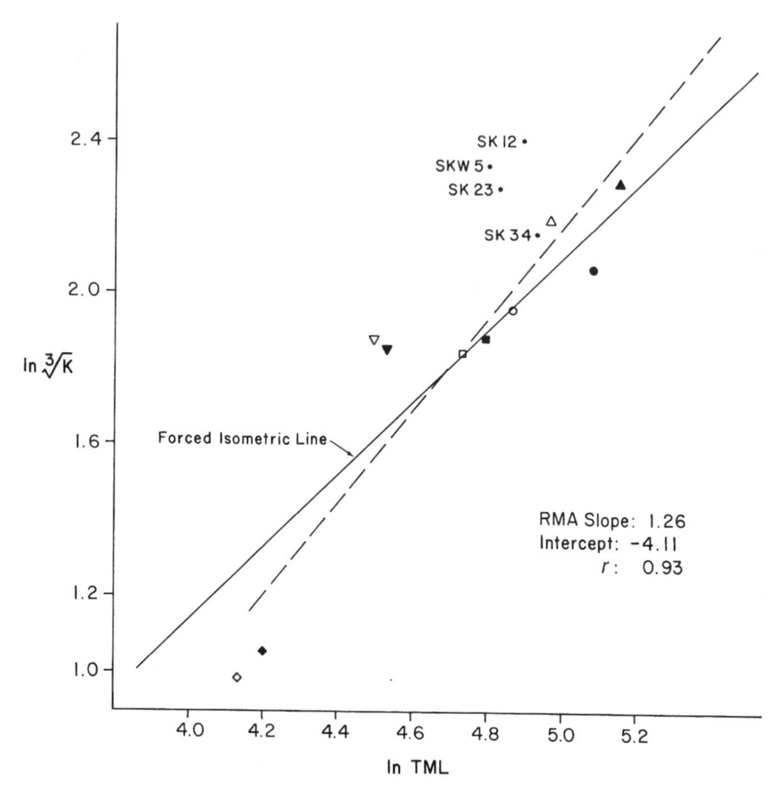

Figure 6.6 Regression of a measure of torsional strength (K) on total mandibular length (TML) in extant hominoids and a sample of *Paranthropus* mandibles (individual specimens plotted). The variable K is figured in mm^3 such that taking its third root makes it dimensionally equivalent to jaw length. Forced isometric and empirical (*dashed lines*) lines were calculated using mean values for males and females separately for the five extant hominoid taxa. The *Australopithecus* sample did not have specimens sufficiently intact to permit their inclusion in the plot. Jaw length does not provide an estimate of the moment arm in torsion, but is intended as a general size surrogate only. Symbols as in figure 6.5.

unremarkable for a hominoid but the moment arms of the associated masticatory loads are much smaller). This contrasts with the early hominin case, in which massive cross-sectional dimensions are incorporated in a mandible that has an essentially apelike configuration. Whether this means that stress levels are comparable in these cases is arguable: masticatory muscle hypertrophy is reasonably inferred from cranial form (especially in *Paranthropus*), whereas it is generally believed that the increasing gracility of *Homo* crania through time is linked to a reduction in adductor muscle mass. If both suppositions are true then the evolution of the Homininae may be marked by a temporal trend toward reduction in masticatory stresses. Under this interpretation, the mandibles of *Homo sapiens* are not functionally diminutive, they are actually overdesigned (i.e., needlessly strong) with respect to masticatory loads.

What do the biomechanical properties of *A. africanus* and *P. robustus* jaws tell us about diet in these early hominins? An important consideration is that postcanine megadontia is a feature of both taxa. It has been argued that functional requirements for supporting these large teeth account for the large corpus proportions (Wolpoff, 1975), but subsequent analysis suggests that mandibular dimensions are not determined by the size of the dentition (Daegling and Grine, 1991; Plavcan and Daegling, in press). Instead, the most parsimonious interpretation of early hominin jaws is that the large corpus dimensions were resisting elevated adductor muscle forces used to create high-magnitude bite forces. Because these bite forces were distributed across an enlarged postcanine dental battery, it is likely that the bite pressures across the occlusal surfaces were comparable to living hominoids (Demes and Creel, 1988).

The combination of elevated masticatory forces and postcanine megadontia could be interpreted as indicating selection for processing greater amounts of food, while not necessarily indicating that the food items themselves were qualitatively different from those processed by great apes today (e.g., see Walker, 1981). The concept of equivalent bite pressures is crucial to the interpretation because it underlies the interpretation that these early hominins were simply processing relatively more of the same types of foods that apes eat today. In other words, it is the quantity, rather than the quality, of the diet, that accounts for the morphology of the fossils.

Such an assumption is flawed, however. Even though the concept of equivalent bite pressures is sound in theory, it does not follow that early hominins were no more mechanically efficient than their modern counterparts among the Hominoidea. If ingestion of small, hard objects was part of the behavioral repertoire of early hominins (Grine, 1981), the application of bite force during the initial phases of food breakdown would have been localized to the object borders (assuming a three-dimensional geometry). In this case, the stress imparted to the object would be much higher than implied by bite pressures because the adductor force would be applied as a point load (or for multiple objects a series of point loads, with local stresses many multiples of a hypothetical bite pressure imparted over an entire occlusal area). Thus, postcanine megadontia does not necessarily imply that bite pressures applied to food items could not have been very large. Instead, megadontia could be seen as the result of selection for mitigating the effects of attrition imposed by a diet involving the repeated application of high, localized occlusal stresses. The thick enamel characteristic of *A. africanus* and *P. robustus* molars is consistent with this interpretation.

By a strictly biomechanical criterion, the mandibles of *P. robustus* and *A. africanus* are indistinguishable (given the comparative variables chosen). This functional similarity stands in contrast to the cranial and dental differences that distinguish the two taxa. These differences have traditionally, and sensibly, in our view, been related to distinctive trophic adaptations. The question naturally arises as to why the mandibles of these forms—elements so clearly functionally linked to mastication—do not appear to reflect the dietary differences to which the teeth and crania seem to point. We view the most profitable avenue of investigation to be the reassessment of how mechanical competence is measured and quantified. In this discussion of the fossils, we have implicitly assumed that the corpus has been modeled accurately as an isotropic, homogeneous structure and that the reference variables for scaling biomechanical properties are comparably measured across different taxa.

The masticatory model used—*Macaca fascicularis*—is the best available, but by employing it, we assume that the mechanics of mastication are essentially the same in all catarrhines. As we have emphasized throughout this chapter, all of the above caveats remain open questions, but they can also all be addressed by appropriate research designs.

Jaw Mechanics and Diet: The Known, the Unknown, and the Unknowable

The material properties of dietary items can influence mandibular morphology over the long term, whether that is reckoned over ontogenetic or evolutionary timescales (Smith and Savage, 1959; Turnbull, 1970; Bouvier and Hylander, 1981). The theoretical connection between diet and jaw form is straightforward. Tougher or stronger food items require more muscular force to create greater bite forces, and these loads create a more hazardous stress environment in the mandible without some sort of morphological compensation. Among primates, this compensation seems to take the form of increased bone mass rather than the alteration of material properties (although this remains an understudied area). In addition, certain primate clades exhibit more efficient arrangements of masticatory muscle position for producing large bite forces in association with relatively tough or obdurate foods (Ravosa, 1991, 1996a; Spencer, 1999).

Broad allometric surveys of jaw form among primates have been employed to explore the idea of a functional linkage of dietary and mandibular variation. The scaling of cross-sectional dimensions relative to body size (Smith, 1983), or moment arm proxies (i.e., mandibular length [Bouvier, 1986; Ravosa, 1991, 1996a, 1996b]), do not yield consistent patterns of postcanine mandibular scaling across higher taxa. This is somewhat surprising because a sound theoretical argument can be made that larger primates should have both absolutely and relatively stronger jaws. Given that dietary quality tends to decrease with body size in primates and among mammals in general (Ravosa, 1996a, 1996b, 1999), a finding of disproportionally large mandibles in bigger primates would be predicted because of the addition of more bulk fiber in the diets of larger species. In the laboratory setting, primates fed tougher dietary items show higher levels of bone strain (Hylander, 1979c), but patterns of mandibular scaling do not invariably show a positive allometric trend in cross-sectional dimensions that might be expected if larger primates are processing tougher foods. Thus, many of the connections between diet, allometry, and jaw morphology remain to be resolved because the plethora of comparative studies reveal a myriad of patterns that do not lend themselves to a single, unifying explanation. This caveat does not apply to the anterior corpus, where patterns of scaling, symphyseal fusion, and geometry are readily explained with reference to muscle recruitment patterns that have more tangible connections to diet (Hylander et al., 1998; Ravosa, 1999). This does not necessarily mean that the anterior corpus experiences more stereotypical loading patterns in anthropoids; instead, there is a clear architectural reason why symphyseal dimensions scale the way they do (Hylander, 1985a). That the morphology of the posterior corpus, which also is prone to large stresses and strains, is not so clearly tied to masticatory mechanics suggests that the variables chosen to represent mechanical performance in comparative studies need to be reassessed.

A related issue is whether—owing to factors of diet, size, or both—increased load magnitudes are themselves instrumental in producing stronger jaws or whether a greater number of loading cycles is responsible (Hylander, 1979b). The problem is that with fibrous foods, primates generate larger forces, but they also engage in more masticatory cycles to process them. Thus, the effects of load magnitude versus load frequency on mandibular morphology are not easily disentangled.

A more perplexing unknown involves those cases where there are clear dietary differences between primate species but few corresponding morphological differences between them (Daegling and McGraw, 2001). If (as we do here) one starts from the premise that bone as a tissue responds in consistent fashion to changes in the mechanical environment, there is no ready explanation about why some dietary contrasts are borne out by morphology and others are not. One can always invoke ontogenetic and phylogenetic constraints as clouding a dietary signal, but in doing so, the initial premise is abandoned in favor of a factor that is deemed to be present but is not specifically identified (but see Vinyard and Ravosa [1998] for an example of an ontogenetic constraint that may be mechanically mediated). These constraints are probably real, but it reveals something about our ignorance of how constraint operates when they are invoked only when a favored hypothesis is overturned.

There are also unknowns that go unrecognized, owing to misjudgments about data quality. Lab or museum-based studies properly invoke field data on diet and feeding to perform what Bock (1980) referred to as the synthetic method for unraveling biological adaptations. The approach is desirable and valid, but the field data will seldom be adequate unless the morphologist consulted with the field scientist in advance about what ecological variables need to be sampled. Without such prior consultation, the field data are essentially anecdotal. It is unlikely that a prior field study will have collected the kinds of data that permit a rigorous test of hypotheses emerging from lab-based research.

Stating what is unknowable is perilous in scientific investigation because it presumes that the limits of knowledge are absolute. Methodological and analytical innovations, however, are difficult to anticipate, and they lead to insights that are predictable only in retrospect. Still, the case study of the South African early hominins can be used to glean some understanding of the limits of our discrimination. Biomechanically speaking, *A. africanus* and *P. robustus* have mandibles that do not seem to be functionally distinct from one another. Cranial form, dental proportions, and patterns of dental microwear, however, clearly distinguish these early hominins. Their diets were probably different, but it does not follow that their jaws must therefore be morphologically distinct for them to have been mechanically competent for their respective functional environments. Diet per se does not affect jaw morphology, but the properties of the food items in combination with feeding behavior does. This distinction is important because it recognizes that mandibular form provides limited insight into diet, even with the understanding that there is a tangible functional linkage between them.

These difficulties are not restricted to the problem at hand. If we recognize biomechanical capabilities as adaptations (a risky undertaking, but implicit in most comparative studies), then there is an essential historical component to the emergence and maintenance of these capabilities (Lauder, Leroi, and Rose, 1993; Lauder et al., 1995; Leroi, Rose, and Lauder, 1994). Lauder (2003) regards the integration of biomechanics with evolutionary biology to be a persistent intellectual challenge, one that cannot be ignored if morphological studies are to reside within an evolutionary paradigm. Biomechanics without history ignores the evolutionary context, and retrofitting comparative research to that context usually results in historical narratives that take the form of "just so stories" with a few equations and *P* values thrown in to dress up the argument. This does not mean that pure biomechanics have nothing to offer the biologist. On the contrary, figuring out how something works has to occur before the question of how it evolved can be intelligently addressed. But an ahistorical mechanical perspective, even one that recognizes the reality of biological constraints, cannot help us distinguish adaptations from their fortuitous effects (Williams, 1966; Bock, 1980).

Conclusions

Our ability to infer diet from mandibular morphology in hominin evolution is compromised, not so much by the paucity of the fossil record as by our limited understanding of the relationship between the mechanical stress environment and bone form. The loading environment of the mandible is among the better understood of any skeletal element among primates; yet, the prediction of stress levels from morphological criteria is currently an imprecise endeavor. Recognizing patterns of variation in mandibular form is relatively unproblematic; the current challenge is to disentangle the interrelationship between bone modeling and remodeling, the mechanical stress environment, material property variation, and whole-bone structural properties. Once the processes of functional adaptation are better understood, the appropriate variables for comparative analysis will be more easily and confidently specified.

Acknowledgments We thank Peter Ungar for organizing and conducting an outstanding conference on the evolution of human diet. L. Betti provided needed assistance with some of the figures, and A. Rapoff and an anonymous reviewer provided helpful criticism on earlier drafts.

Note

1. The term "robust" is here used to mean large and (therefore) strong. With respect to hominin fossils, the term is sometimes used as a shape descriptor—the "robusticity index" of mandibular corpus breadth over height. In this latter sense, the term has imprecise biomechanical significance; for example, by this criterion, galagos have more robust jaws than anthropoid hard-object specialists, such as mangabeys and pithecines.

References

Albright, J.A., 1987. Bone: physical properties. In: Albright, J.A., and Brand, R.A. (Eds.), *The Scientific Basis of Orthopedics*, 2nd Ed. Appleton-Century-Crofts, New York, pp. 213–240.

Asundi, A., and Kishen, A., 2000. A strain gauge and photoelastic analysis of in vivo strain and in vitro stress distribution in human dental supporting structures. *Arch. Oral Biol.* 45, 543–550.

Baragar, F.A., and Osborn, J.W., 1987. Efficiency as a predictor of human jaw design in the sagittal plane. *J. Biomech.* 20, 447–457.

Bentolila, V., Hillam, R.A., Skerry, T.M., Boyce, T.M., Fyhrie, D.P., and Schaffler, M.B., 1997. Activation of intracortical remodeling in adult rat long bones by fatigue loading. *Trans. Orthopaed. Res. Soc.* 22, 578–589.

Biegert, J., 1956. Das Kiefergelenk der Primaten. Pap. Anthropol. Inst. Univ. Zurich 1956, 249–404.

Biewener, A.A., Fazzalari, N.L., Konieczynski, D.D., and Baudinette, R.V., 1996. Adaptive changes in trabecular architecture in relation to functional strain patterns and disuse. *Bone* 19, 1–8.

Bock, W.J., 1980. The definition and recognition of biological adaptation. *Am. Zool.* 20, 217–227.

Bouvier, M., 1986a. A biomechanical analysis of mandibular scaling in Old World monkeys. *Am. J. Phys. Anthropol.* 69, 473–482.

Bouvier, M., 1986b. Biomechnical scaling of mandibular dimensions in New World monkeys. *Int. J. Primatol.* 7, 551–567.

Bouvier, M., and Hylander, W.L., 1981. The effect of strain on cortical bone structure in macaques (*Macaca mulatta*). *J. Morphol.* 167, 1–12.

Bouvier, M., and Hylander, W.L., 1996. The mechanical or metabolic function of secondary osteonal bone in the monkey *Macaca fascicularis*. *Arch. Oral Biol.* 41, 941–950.

Boyd, R.L., Gibbs, C.H., Mahan, P.E., Richmond, A.F., and Laskin, J.L., 1990. Temporomandibular joint forces measured at the condyle of *Macaca arctoides*. *Am. J. Orthod. Dentofac. Orthoped.* 97, 472–479.

Boyd, R.L., Gibbs, C.H., Richmond, A.F., Laskin, J.L., and Brehnan, K., 1982. Temporomandibular joint forces in monkey measured with piezoelectric foil. *J. Dent. Res.* 61, 351.

Brehnan, K., Boyd, R.L., Laskin, J., Gibbs, and C.H. Mahan, P., 1981. Direct measurements of loads at the temporomandibular joint in *Macaca arctoides*. J. Dent. Res. 60, 1820–1824.

Brooks, D.B., Burstein, A.H., and Frankel, V.H., 1970. The biomechanics of torsional fractures: The stress concentration of a drill hole. *J. Bone Joint Surg.* 52A, 507–514.

Burr, D.B., 1980. The relationship among the physical, geometrical and mechanical properties of bone. *Yearb. Phys. Anthropol.* 23, 109–146.

Burr, D.B., and Piotrowski, G., 1982. How do trabeculae affect the calculation of structural properties of bone? *Am. J. Phys. Anthropol.* 57, 341–352.

Burstein, A.H., Currey, J.D., Frankel, V.H., and Reilly, D.T., 1972. The ultimate properties of bone tissue: The effects of yielding. *J. Biomech.* 5, 35–44.

Cachel, S., 1975. A new view of speciation in *Australopithecus*. In Tuttle, R. (Ed.), *Paleoanthropology, Morphology and Paleoecology*. Mouton, The Hague, pp. 183–201.

Carlson, D.S., 1977. Condylar translation and the function of the superficial masseter muscle in the rhesus monkey (*M. mulatta*). *Am. J. Phys. Anthropol.* 47, 53–63.

Chen, X., 1995. Biomechanics of the hominoid masticatory apparatus. PhD diss., Yale University.

Chen, X., and Chen, H., 1998. The influence of alveolar structures on the torsional strain field in a gorilla corporeal cross-section. *J. Hum. Evol.* 35, 611–633.

Chen, X., and Grine, F.E., 1997. The effect of cortical thickness on mandibular torsional strength of South African early hominids. *J. Hum. Evol.* 32, A6.

Covell, D.A., and Herring, S.W., 1995. Periosteal migration in the growing mandible: An animal model. *Am. J. Orthod. Dentofac. Orthoped.* 108, 22–29.

Cowin, S.C., 1986. *Bone Mechanics*. CRC Press, Boca Raton, FL.

Cowin, S.C., 1997. The false premise of Wolff 's law. *Forma* 12, 247–262.

Currey, J.D., 2002. *Bones: Structure and Mechanics*. Princeton University Press, Princeton, NJ.

Daegling, D.J., 1989. Biomechanics of cross-sectional size and shape in the hominoid mandibular corpus. *Am. J. Phys. Anthropol.* 80, 91–106.
Daegling, D.J., 1990. Geometry and biomechanics of hominoid mandibles. PhD diss., Stony Brook University.
Daegling, D.J., 1992. Mandibular morphology and diet in the genus *Cebus*. *Int. J. Primatol.* 13, 545–570.
Daegling, D.J., 1993. Functional morphology of the human chin. *Evol. Anthropol.* 1, 170–177.
Daegling, D.J., 2002a. Bone geometry in cercopithecoid mandibles. *Arch. Oral Biol.* 47, 315–325.
Daegling, D.J., 2002b. Estimation of torsional rigidity in primate long bones. *J. Hum. Evol.* 43, 229–239.
Daegling, D.J., and Grine, F.E., 1991. Compact bone distribution and biomechanics of early hominid mandibles. *Am. J. Phys. Anthropol.* 86, 321–339.
Daegling, D.J., and Hotzman, J.L., 2003. Functional significance of cortical bone distribution in anthropopid mandibles: An in vitro assessment of bone strain under combined loads. *Am. J. Phys. Anthrop.* 122, 38–50.
Daegling, D.J., and Hylander, W.L., 1996. Bone strain in the mandibular corpus: Experimental validation of theoretical models. *Am. J. Phys. Anthropol.* Suppl. 22, 92.
Daegling, D.J., and Hylander, W.L., 1998. Biomechanics of torsion in the human mandible. *Am. J. Phys. Anthropol.* 105, 73–87.
Daegling, D.J., and Hylander, W.L., 2000. Experimental observation, theoretical models, and biomechanical inference in the study of mandibular form. *Am. J. Phys. Anthropol.* 112, 541–551.
Daegling, D.J., and McGraw, W.S., 2001. Feeding, diet and jaw form in West African *Colobus* and *Procolobus*. *Int. J. Primatol.* 22, 1033–1055.
Daegling, D.J., Ravosa, M.J., Johnson, K.R., and Hylander, W.L., 1992. Influence of teeth, alveoli, and periodontal ligaments on torsional rigidity in human mandibles. *Am. J. Phys. Anthropol.* 89, 59–72.
Dechow P.C., and Hylander W.L., 2000. Elastic properties and masticatory bone stress in the macaque mandible. *Am. J. Phys. Anthropol.* 112, 553–574.
Demes, B., and Creel, N., 1988. Bite force, diet, and cranial morphology of fossil hominids. *J. Hum. Evol.* 17, 657–670.
Demes, B., Creel, N., and Preuschoft, H., 1986. Functional significance of allometric trends in the hominoid masticatory apparatus. In: Else, J.G., and Lee, P.C. (Eds.), *Primate Evolution*. Cambridge University Press, Cambridge, pp. 229–237.
Demes, B., Preuschoft, H., and Wolff, J.E.A., 1984. Stress-strength relationships in the mandibles of hominoids. In: Chivers, D.J., Wood, B.A., and Bilsborough, A. (Eds.), *Food Acquisition and Processing in Primates*. Plenum, New York, pp. 369–390.
DuBrul, E.L., 1977. Early hominid feeding mechanisms. *Am. J. Phys. Anthropol.* 47, 305–320.
Fiala, P., and Heřt, J., 1993. Principal types of functional architecture of cancellous bone in man. *Funct. Dev. Morph.* 3, 91–99.
Frankel, V.H., and Burstein, A.H., 1970. *Orthopedic Biomechanics*. Lea & Febiger, Philadelphia.
Frost, H.M., 1964. *The Laws of Bone Structure*. C. R. Thomas, Springfield, IL.
Fung, Y.C., 1993. *Biomechanics: Mechanical Properties of Living Tissues*. Springer, New York.
Gibson, L.J., 1985. The mechanical behavior of cancellous bone. *J. Biomech.* 18, 341–349.
Gingerich, P.D., 1971. Functional significance of mandibular translation in vertebrate jaw mechanics. *Postilla* 152, 1–10.
Goldstein, S.A., Wilson, D.L., Sonstegard, D.A., and Mathews, L.S., 1983. The mechanical properties of human tibial trabecular bone as a function of metaphyseal location. *J. Biomech.* 16, 965–969.
Gould, S.J., 1966. Allometry and size in ontogeny and phylogeny. *Biol. Rev.* 41, 587–640.
Gould, S.J., and Lewontin, R.C., 1979. The spandrels of San Marcos and the Panglossian paradigm, a critique of the adaptationist programme. *Proc. R. Soc. Lond.* B 205, 581–598.

Greaves, W.S., 1978. The jaw lever system in ungulates: A new model. *J. Zool. Lond.* 184, 271–285.

Grine, F.E., 1981. Trophic differences between "gracile" and "robust" australopithecines: A scanning electron microscope analysis of occlusal events. *S. Afr. J. Sci.* 77, 203–230.

Grine, F.E., Colflesh, D.E., Daegling, D.J., Krause, D.W., Dewey, M.M., Cameron, R.H., Brain, and C.K., 1989. Electron probe X-ray microanalysis of internal structures in a fossil hominid mandible and its implications for biomechanical modeling. *S. Afr. J. Sci.* 85, 509–514.

Grine, F.E., Spencer, M.A., Demes, B., Smith, H.F., Striat, D.S., and Constant, D.A., 2005. Molar enamel thickness in the chacma baboon, *Papio ursinus* (Kerr, 1792). *Am. J. Phys. Anthropol.* 128, 812–822.

Herring, S.W., and Herring, S.E., 1974. The superficial masseter and gape in mammals. *Am. Nat.* 108, 561–576.

Heřt, J., 1994. A new attempt at the interpretation of the functional architecture of the cancellous bone. *J. Biomech.* 27, 239–242.

Hiiemae, K.M., 1984. Functional aspects of primate jaw morphology. In: Chivers, D.J., Wood, B.A., and Bilsborough, A. (Eds.), *Food Acquisition and Processing in Primates.* Plenum, New York, pp. 257–281.

Hipp, J.A., Edgerton, K.N., and Hayes, W.C., 1990. Structural consequences of transcortical holes in long bones loaded in torsion. *J. Biomech.* 23, 1261–1268.

Hohl, T.H., and Tucek, W.H., 1982. Measurement of condylar loading forces by instrumental prosthesis in the baboon. *J. Maxillofac. Surg.* 10, 1–7.

Hylander, W.L., 1975. The human mandible: Lever or link? *Am. J. Phys. Anthropol.* 43, 227–242.

Hylander, W.L., 1977. In vivo bone strain in the mandible of *Galago crassicaudatus. Am. J. Phys. Anthropol.* 46, 309–326.

Hylander, W.L., 1979a. An experimental analysis of temporomandibular joint reaction force in macaques. *Am. J. Phys. Anthropol.* 51, 433–456.

Hylander, W.L., 1979b. The functional significance of primate mandibular form. *J. Morphol.* 160, 223–240.

Hylander, W.L., 1979c. Mandibular function in *Galago crassicaudatus* and *Macaca fascicularis*: An *in vivo* approach to stress analysis of the mandible. *J. Morphol.* 159, 253–296.

Hylander, W.L., 1981. Patterns of stress and strain in the macaque mandible. In: Carlson, D.S. (Ed.), *Craniofacial Biology*. Monograph 10, Craniofacial Growth Series, Center for Human Growth and Development, University of Michigan, Ann Arbor, pp. 1–37.

Hylander, W.L., 1984. Stress and strain in the mandibular symphysis of primates: A test of competing hypotheses. *Am. J. Phys. Anthropol.* 64, 1–46.

Hylander, W.L., 1985a. Mandibular function and biomechanical stress and scaling. Am. Zool. 25, 315–330.

Hylander, W.L., 1985b. Mandibular function and temporomandibular joint loading. In: Carlson, D.S., McNamara, J.A., and Ribbens, K.A. (Eds.), *Developmental Aspects of Temporomandibular Joint Disorders*. Monograph 16, Craniofacial Growth Series, Center for Human Growth and Development, University of Michigan, Ann Arbor, p. 19–35.

Hylander, W.L., 1988. Implications of in vivo experiments for interpreting the functional significance of "robust" australopithecine jaws. In: Grine, F.E. (Ed.), *Evolutionary History of the "Robust" Australopithecines*. Aldine de Gruyter, New York, pp. 55–83.

Hylander, W.L., 1991. Functional anatomy of the temporomandibular joint. In: Sarnat, B.G., and Laskin, D. (Eds.), *Temporomandibular Joint: A Biologic Basis for Clinical Practice*. Charles C. Thomas, Springfield, IL, pp. 60–92.

Hylander, W.L., and Bays, R., 1979. An in vivo strain gauge analysis of squamosal dentary joint reaction force during mastication and incision in *Macaca mulatta* and *Macaca fascicularis. Arch. Oral Biol.* 24, 689–697.

Hylander, W.L., and Crompton, A.W., 1986. Jaw movement and patterns of bone strain in the monkey *Macaca fascicularis. Arch. Oral Biol.* 31, 841–848.

Hylander, W.L., Johnson, K.R., and Crompton, A.W., 1987. Loading patterns and jaw movement during mastication in *Macaca fascicularis*: A bone strain, electromyographic and cineradiographic analysis. *Am. J. Phys. Anthropol.* 72, 287–314.

Hylander, W.L., Ravosa, M.J., Ross, C.F., and Johnson, K.R., 1998. Mandibular corpus strain in primates: Further evidence for a functional link between symphyseal fusion and jaw-adductor muscle force. *Am. J. Phys. Anthropol.* 107, 257–271.

Imbeni, V., Nalla, R.K., Bosi, C., Kinney, J.H., and Ritchie, R.O., 2003. *In vitro* fracture toughness of human dentin. *J. Biomed. Mat. Res.* 66A, 1–9.

Ito, M., Nishida, A., Koga, A., Ikeda, S., Shiraishi, A., Uetani, M., Hayashi, K., and Nakamura, T., 2002. Contribution of trabecular and cortical components to the mechanical properties of bone and their regulating parameters. *Bone* 31, 351–358.

Janis, C.M., and Fortelius, M., 1988. On the means whereby mammals achieve increased functional durability of their dentitions, with special reference to limiting factors. *Biol. Rev. Camb. Philos. Soc.* 63, 197–230.

Jolly, C.J., 1970. The seed-eaters: A new model of hominid differentiation based on a baboon analogy. *Man* 5, 5–28.

Judex, S., Whiting, W., and Zernicke, R., 1999. Bone biomechanics and fractures. In: Kumar, S. (Ed.), *Biomechanics in Ergonomics.* Taylor & Francis, Philadelphia, pp. 59–73.

Kantomaa, T., 1989. The relationship between mandibular configuration and the shape of the glenoid fossa in the human. *Eur. J. Orthod.* 11, 77–81.

Keaveny, T.M., and Hayes, W.C., 1993. A 20-year perspective on the mechanical properties of trabecular bone. *J. Biomech. Eng.* 115, 534–542.

Kieser, J., Gebbie, T., and Ksiezycka, K. 1996. A mathematical model for hypothetical force distribution between opposing jaws. *J. Dent. Assoc. S. Afr.* 51, 701–705.

Kieser, J.A., 1990. *Human Adult Odontometrics: The Study of Variation in Tooth Size.* Cambridge University Press, Cambridge.

Kieser, J.A., 1999. Biomechanics of masticatory force production. *J. Hum. Evol.* 36, 575–579.

Koolstra, J.H., van Eijden, T.M., Weijs, W.A., and Naeije, M., 1988. A three-dimensional mathematical model of the human masticatory system predicting maximum possible bite forces. *J. Biomech.* 21, 563–576.

Kupczik, K., Spoor, F., and Dean, M.C., 2003. Tooth root morphology and masticatory muscle force pattern in humans and non-human primates. *Am. J. Phys. Anthropol.* Suppl. 36, 134.

Lanyon, L.E., 1974. Experimental support for the trajectorial theory of bone structure. *J. Bone Joint Surg.* 56B, 160–166.

Lanyon, L.E., 1991. Biomechanical properties of bone and response of bone to mechanical stimuli: Functional strain as a controlling influence on bone modeling and remodeling behavior. In: Hall, B.K. (Ed.), *Bone Matrix and Bone Specific Products.* CRC Press, London, pp. 79–108.

Lanyon, L.E., and Rubin, C.T., 1984. Static vs. dynamic loads as an influence on bone remodeling. *J. Biomech.* 17, 897–905.

Lauder, G.V., 2003. The intellectual challenge of biomechanics and evolution. In: Bles, V.L., Gasc, J.-P., and Casinos, A. (Eds.), *Vertebrate Biomechanics and Evolution.* BIOS Scientific, Oxford, pp. 319–325.

Lauder, G.V., Huey, R.B., Monson, R.K., and Jensen, R.J., 1995. Systematics and the study of organismal form and function. *BioScience* 45, 696–704.

Lauder, G.V., Leroi, A.M., and Rose, M.R., 1993. Adaptations and history. *Trends Ecol. Evol.* 8, 294–297.

Leroi, A.M., Rose, M.R., and Lauder, G.V., 1994. What does the comparative method reveal about adaptation? *Am. Nat.* 143, 381–402.

Lindblom, G., 1960. On the anatomy and function of the temporomandibular joint. *Acta Odont. Scand.* Suppl. 28, 1–287.

Lucas, P.W., 1981. An analysis of canine size and jaw shape in some Old and New World non-human primates. *J. Zool. Lond.* 195, 437–448.

Mack, P.J., 1984. A functional explanation for the morphology of the temporomandibular joint of man. *J. Dent.* 12, 225–230.

Mandel, U., Dalgaard, P., and Viidik, A.,1986. A biomechanical study of the human periodontal ligament. *J. Biomech.* 19, 637–645.

Martin, R.B., and Burr, D.B., 1989. *Structure, Function, and Adaptation of Compact Bone.* Raven Press, New York.

Martin, R.B., Burr, D.B., and Sharkey, N.A., 1998. *Skeletal Tissue Mechanics.* Springer, New York.

McGraw, W.S., and Daegling, D.J., 1999. Jaws and diet among colobines: A biomechanical enigma. *Am. J. Phys. Anthropol.* Suppl. 28, 196.

Melcher, A.H., and Walker, T.W., 1976. The periodontal ligament in attachment and as a shock absorber. In: Poole, D.G.F., and Stack, M.V. (Eds.), *The Eruption and Occlusion of Teeth.* Butterworths, London, pp. 183–192.

Molnar, S., and Ward, S.C., 1977. On the hominid masticatory complex: biomechanical and evolutionary perspectives. *J. Hum. Evol.* 6, 557–568.

Morgan, E.F., Bayraktar, H.H., and Keaveny, T.M., 2003. Trabecular bone modulus-density relationships depend on anatomic site. *J. Biomech.* 36, 897–904.

Müller, J.J., 1959. Die Wurtzeloberfläche menschlicher Zähne und ihre Bedeutung für Ersatzkonstruktionen im Belastungsbereich des normalen und pathologischen Paradontiums. *Schwiz. Mschr. Zahnheilik.* 69, 193–199.

Nikolai, R.J., 1985. *Bioengineering Analysis of Orthodontic Mechanics.* Lea & Febiger, Philadelphia.

Noble, H.W., 1979. Comparative functional morphology of the temporomandibular joint. In: Zarb, G.A., and Carlsson, G.E. (Eds.), *Temporomandibular Joint Function and Dysfunction.* C. V. Mosby, St. Louis, pp. 1–41.

Nordin, M., and Frankel, V.H., 1989. Biomechanics of bone. In: Nordin, M., and Frankel, V.H. (Eds.), *Basic Biomechanics of the Musculoskeletal System.* 2nd ed. Lea & Febiger, Philadelphia, pp. 3–29.

Osborn, J.W., 1996. Features of human jaw design which maximize the bite force. *J. Biomech.* 29, 589–595.

Osborn, J.W., and Baragar, F.A., 1985. Predicted action of human muscle activity during clenching derived from a computer assisted model: Symmetrical vertical bite forces. *J. Biomech.* 18, 599–612.

Picq, P.G., Plavcan, J.M., and Hylander, W.L., 1987. Nonlever action of the mandible: The return of the hydra. *Am. J. Phys. Anthropol.* 74, 305–307.

Plavcan, J.M., and Daegling, D.J., in press. Interspecific and Intraspecific Relationships Between Tooth Size and Jaw Size in Primates. *J. Hum. Evol.*

Popowics, T.E., Zhu, Z., and Herring, S.W., 2002. Mechanical properties of the periosteum in the pig, *Sus scrofa. Arch. Oral Biol.* 47, 733–741.

Rapoff, A.J., Rinaldi, R.G., Johnson, W.M., Venkataraman, S., and Daegling, D.J., 2003. Heterogeneous anisotropic elastic properties in a *Macaca fascicularis* mandible. *Am. J Phys. Anthropol.* Suppl. 36, 174–175.

Ravosa, M.J., 1987. Functional aspects of maxillo-mandibular form in Old World monkeys. *Am. J. Phys. Anthropol.* 72, 245.

Ravosa, M.J., 1991. Structural allometry of the mandibular corpus and symphysis in prosimian primates. *J. Hum. Evol.* 20, 3–20.

Ravosa, M.J., 1996a. Jaw morphology and function in living and fossil Old World monkeys. *Int. J. Primatol.* 17, 909–932.

Ravosa, M.J., 1996b. Mandibular form and function in North American and European Adapidae and Omomyidae. *J. Morphol.* 229, 171–190.

Ravosa, M.J., 1999. Anthropoid origins and the modern symphysis. *Folia Primatol.* 70, 65–78.

Ravosa, M.J., 2000. Size and scaling in the mandible of living and extinct apes. *Folia Primatol.* 71, 305–322.

Reilly, D.T., and Burstein, A.H., 1974. The mechanical properties of cortical bone. *J. Bone Joint Surg.* 56A, 1001–1021.

Reilly, D.T., and Burstein, A.H., 1975. The elastic and ultimate properties of compact bone tissue. *J. Biomech.* 8, 393–405.

Rho, J.Y., Ashman, R.B., and Turner, C.H., 1993. Young's modulus of trabecular and cortical bone material: Ultrasonic and microtensile measurements. *J. Biomech.* 26, 111–119.

Roberts, D., and Tattersall, I., 1974. Skull form and the mechanics of mandibular elevation in mammals. *Am. Mus. Novit.* 2536, 1–9.

Robinson, J.T., 1954. Prehominid dentition and hominid evolution. *Evolution* 8, 324–334.

Robling, A.G., and Stout, S.D., 2000. Histomorphometry of human cortical bone: applications to age estimation. In: Katzenberg, M.A., and Saunders, S.R. (Eds.), *Biological Anthropology of the Human Skeleton.* Wiley-Liss, New York, pp. 187–213.

Rockoff, S.D., Sweet, E., and Bleustein, J., 1969. The relative contribution of trabecular and cortical bone to the strength of the human lumbar vertebrae. *Calcif. Tissue Res.* 3, 163–175.

Rubin, C.T., McLeod, K.J., Gross, T.S., and Donahue, H.J. 1991. Physical stimuli as potent determinants of bone morphology. In: Carlson, D.S., and Goldstein, S.A. (Eds.), *Bone Biodynamics in Orthodontic and Orthopedic Treatment.* Vol. 27. Craniofacial Growth Series. Center for Human Growth and Development, University of Michigan, Ann Arbor, pp. 75–91.

Rubin C.T., Turner, A.S., Bain, S., Mallinckrodt, C., and McLeod, K., 2001. Low mechanical signals strengthen long bones. *Nature* 412, 603–604.

Ruff, C.B., 1983. The contribution of cancellous bone to long bone strength and rigidity. *Am. J. Phys. Anthropol.* 61, 141–143.

Schwartz-Dabney, and C.L., Dechow, P.C., 2003. Variations in cortical material properties throughout the human dentate mandible. *Am. J. Phys. Anthropol.* 120, 252–277.

Skedros, J.G., Mason, M.W., Nelson M.C., and Bloebaum R.D., 1996. Evidence of structural and material adaptation to specific strain features in cortical bone. *Anat. Rec.* 246, 47–63.

Smith, J.M., and Savage R.J.G., 1959. The mechanics of mammalian jaws. *School Sci. Rev.* 40, 289–301.

Smith, R.J., 1978. Mandibular biomechanics and temporomandibular joint function in primates. *Am. J. Phys. Anthropol.* 49, 341–350.

Smith, R.J., 1982. On the mechanical reduction of functional morphology. *J. Theor. Biol.* 96, 99–106.

Smith, R.J., 1983. The mandibular corpus of female primates: Taxonomic, dietary and allometric correlates of interspecific variations in size and shape. *Am. J. Phys. Anthropol.* 61, 315–330.

Smith, R.J., 1984. Comparative functional morphology of maximum mandibular opening (gape) in primates. In: Chivers, D.J, Wood, B.A., and Bilsborough, A. (Eds.), *Food Acquisition and Processing in Primates.* Plenum, New York, pp. 231–255.

Smith, R.J., 1993. Categories of allometry: Body size versus biomechanics. *J. Hum. Evol.* 24, 173–182.

Smith, R.J., Petersen, C.E., and Gipe, D.P., 1983. Size and shape of the mandibular condyle in primates. *J. Morphol.* 177, 59–68.

Spencer, M.A., 1995. Masticatory system configuration and diet in anthropoid primates. PhD diss., Stony Brook University.

Spencer, M.A., 1998. Force production in the primate masticatory system: Electromyographic tests of biomechanical hypotheses. *J. Hum. Evol.* 34, 25–54.

Spencer, M.A., 1999. Constraints on masticatory system evolution in anthropoid primates. *Am. J. Phys. Anthropol.* 108, 483–506.

Spencer, M.A., and Demes, B., 1993. Biomechanical analysis of masticatory system configuration in Neandertals and Inuits. *Am. J. Phys. Anthropol.* 91, 1–20.

Stout, S.D., 1992. Methods of determining age at death using bone microstructure. In: Saunders, S.R., and Katzenberg, M.A. (Eds.), *Skeletal Biology of Past Peoples: Research Methods.* Wiley-Liss, New York, pp. 21–35.

Szalay, F.S., 1972. Hunting-scavenging protohominids: A model for hominid origins. *Man* 10, 420–429.

Taylor, A.B., 2002. Masticatory form and function in the African apes. *Am. J. Phys. Anthropol.* 117,133–156.

Taylor, J.F., 1992. The periosteum and bone growth. In: Hall, B.K. (Ed.), Bone. CRC Press, New York, pp. 21–52.

Taylor, R.M.S., 1986. Nonlever action of the mandible. *Am. J. Phys. Anthropol.*, 70, 417–421.

Thompson, E.N., Biknevicius, A.R., and German, R.Z., 2003. Ontogeny and feeding function in the gray short-tailed opossum *Monodelphis domestica*: Empirical support for the constrained model of jaw biomechanics. *J. Exp. Biol.* 206, 923–932.

Throckmorton, G.S., and Throckmorton, L.S., 1985. Quantitative calculations of temporomandibularn joint reaction forces. I. The importance of the magnitude of jaw muscle forces. *J. Biomech.* 18, 445–452.

Turnbull, W.D., 1970. Mammalian masticatory apparatus. Field Mus. Nat. Hist. Fieldiana Geol. 18, 149–356.

Vinyard, C.J., and Ravosa, M.J., 1998. Ontogeny, function, and scaling of the mandibular symphysis in papionin primates. *J. Morphol.* 235, 157–175.

Vinyard, C.J., Wall, C.E., Williams, S.H., and Hylander, W.L., 2003. Comparative functional analysis of skull morphology of tree-gouging primates. *Am. J. Phys. Anthropol.* 120, 153–170.

Walker, A.C., 1978. Functional anatomy of oral tissues: mastication and deglutition. In: Shaw, J., Sweeney, E., Cappuccino, C., and Miller, S. (Eds.), *Textbook of Oral Biology*. W. B. Saunders, London, pp. 227–296.

Walker, A.C., 1981. Diet and teeth: Dietary hypotheses and human evolution. *Philos. Trans. R. Soc.* B 292, 57–64.

Wall, C.E., 1995. Form and function of the temporomandibular joint in anthropoid primates. PhD diss., Stony Brook University, Stony Brook, NY.

Ward, S.C., and Molnar, S., 1980. Experimental stress analysis of topographic diversity in early hominid gnathic morphology. *Am. J. Phys. Anthropol.* 53, 383–395.

Weijs, W.A., 1989. The functional significance of morphological variation of the human mandible and masticatory muscles. *Acta Morph. Neerl.-Scand.* 27, 149–162.

Weinbaum, S., Cowin, S.C., and Zeng, Y., 1994. A model for the excitation of osteocytes by mechanical loading-induced bone fluid shear stress. *J. Biomech.* 27, 339–360.

Williams, G.C., 1966. *Adaptation and Natural Selection*. Princeton University Press, Princeton, NJ.

Williams, S.H., Wall, C.E., Vinyard C.J., and Hylander, W.L., 2002 A biomechanical analysis of skull form in gum-harvesting galagids. *Folia Primatol.* 73, 197–209.

Wills, D.J., Picton, D.C.A., and Davies, W.I.R., 1978. The intrusion of the tooth for different loading rates. *J. Biomech.* 11, 429–434.

Wolpoff, M.H., 1975. Some aspects of human mandibular evolution. In: McNamara, J.A. (Ed.), *Determinants of Mandibular Form and Growth*. Center for Human Growth and Development, Ann Arbor, pp. 1–64.

Wood, B.A., and Aiello, L.C., 1998. Taxonomic and functional implications of mandibular scaling in early hominins. *Am. J. Phys. Anthropol.*, 105, 523–538.

Wood, C., 1994. The correspondence between diet and masticatory morphology in a range of extant primates. *Z. Morph. Anthropol.* 80, 19–50.

7

What Do We Know and Not Know about Dental Microwear and Diet?

MARK F. TEAFORD

At first glance, in thinking about what we do and do not know about the evolution of human diet, it might seem impossible to truly *know* anything about it—that is until someone invents a time machine! However, paleontologists are, in many ways, like forensic scientists who travel through time. They must use any available clues to help decipher what went on eons ago. Only by considering the total range of evidence can we appreciate what is known, unknown, and unknowable about this complex topic.

Unfortunately, much of that evidence is not *direct* evidence, in the sense of something visible on the bone or tooth, caused directly by something that happened during the individual's lifetime. For instance, the relative size of certain bones may or may not be indicative of what an animal actually did, as the animal may have, for example, relatively long hindlimbs simply because its ancestors did. So, when looking through the evidence, paleontologists are constantly forced to evaluate their data, to see what they can, and cannot, say about the hypotheses being tested.

The most common elements in the human fossil record are teeth, largely because they are the most resilient structures in the body. For the most part, they are made of inorganic materials, and they tend to remain intact well after death. Thus, it is perhaps no surprise they have provided many clues about the evolution of human diet. For instance, analyses of tooth shape have shown that species adapted to eat tough, elastic foods generally have longer molar shearing crests than do species adapted to eat hard and brittle foods (Kay, 1975; Kay and Hylander, 1978; Lucas, 1979, 2004). However, most of these studies have focused on analyses of unworn teeth (see Ungar, 2004, and chapter 4 for a revolutionary new perspective on this topic). Yet, like death and taxes, tooth wear is one of life's inevitabilities. As soon as a tooth reaches occlusion, it begins to wear down. In some cases (e.g., guinea pigs), wear even begins in utero (Ainamo, 1971; Teaford and Walker, 1983). Its first steps are imperceptible to the naked eye—microscopic scratches and pits nicking the surface. But those microscopic effects add up, leading to the formation of wear facets on the teeth and eventually dentin exposure, as the overlying enamel is worn away. So, while the shape of unworn teeth can tell us a great deal about what a tooth is capable of processing, tooth wear can give us insights into how a tooth was actually used. This chapter will focus on the evidence provided by microscopic wear patterns on the chewing surfaces of teeth: what is often referred to as dental microwear analyses. This is different from most other analyses of fossils because it is direct evidence of past behavior, ultimately based on microscopic wear caused by food, or abrasives on food, during an animal's lifetime. As a result, this technique has the potential to yield information about prehistoric diets at a unique level of resolution.

What Do We Know about Dental Microwear Analyses?

Postmortem Wear

The first question that springs to mind in contemplating dental microwear analyses of *fossils* is, if a tooth has been lying in the ground for thousands or millions of years, how do we know that the wear on it was really caused during the animal's lifetime? Actually, it's surprisingly easy (Teaford, 1988b) because the wear patterns caused during chewing are laid down in regular patterns at specific locations on teeth (see fig. 7.1). By contrast, when a tooth is buried in the ground it is subjected to wear at innumerable, unusual locations and angles (see fig. 7.2; Puech et al., 1985; Teaford, 1988b; King, Andrews, and Boz, 1999). This so-called postmortem wear is certainly a problem when analyzing fossils for two reasons. First, at low magnifica-

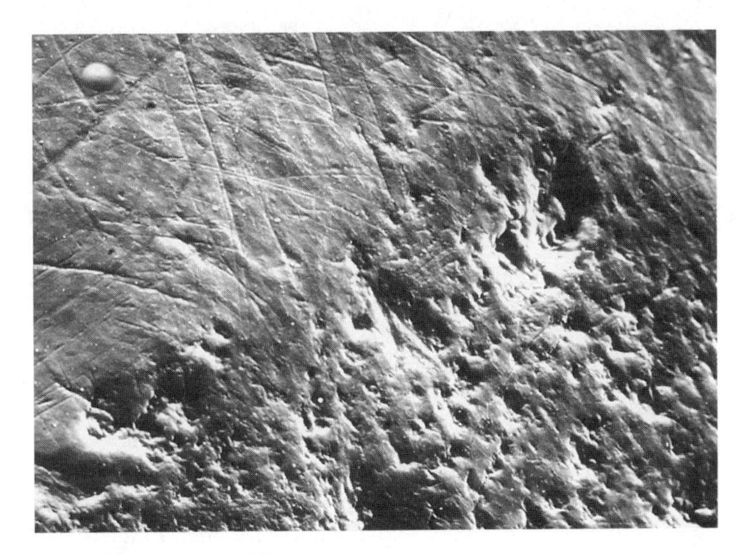

Figure 7.1 SEM of occlusal and nonocclusal surfaces of a molar of *Cebus apella* (from Teaford, 1988b).

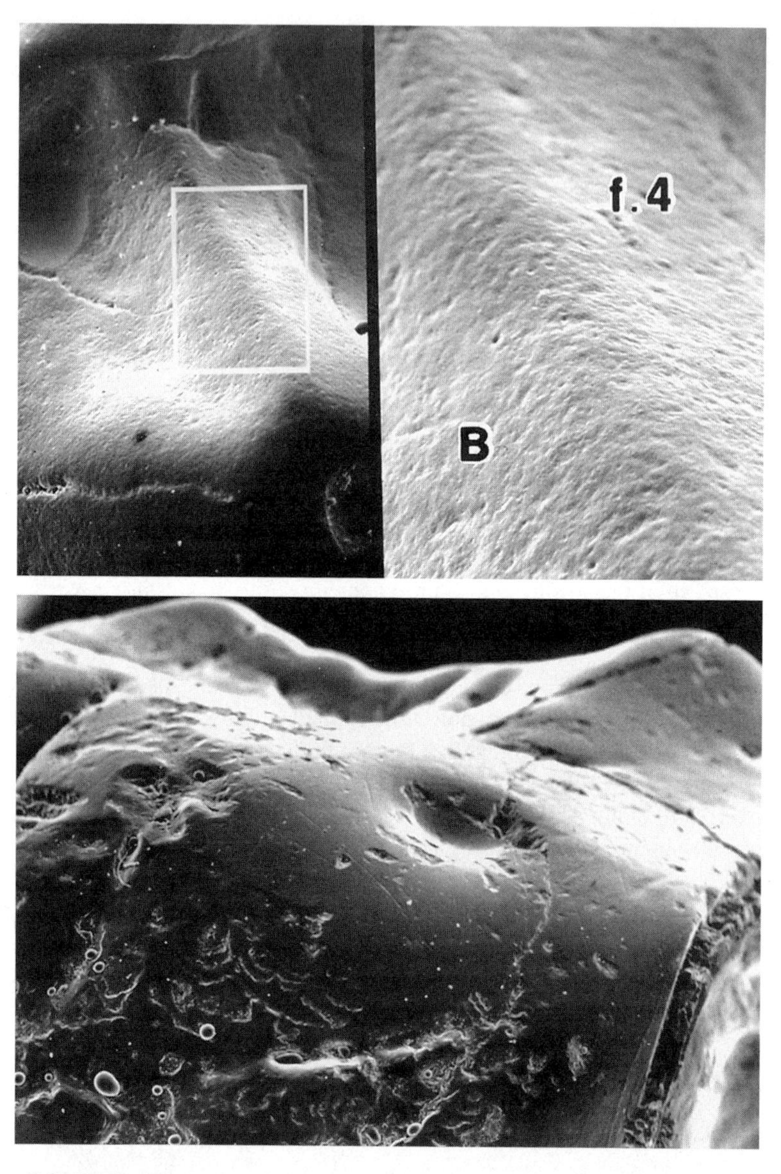

Figure 7.2 SEM micrographs of postmortem wear. Top image = postmortem abrasive wear on 50 million year old *Cantius* molar; note similar microwear patterns on facet 4 (*f.4*) and buccal side of tooth (*B*); from Teaford, 1988b). Bottom image = postmortem chemical wear of eartly hominin molar (LH4); note extension of pitting from side of tooth up onto occlusal surface.

tions, only the most obvious postmortem problems are visible. Thus, even if a specimen is being cleaned under a light microscope, the investigator cannot be absolutely sure it is clean until it is placed under something like a scanning electron microscope (SEM). For instance, in a recent study of early *Homo* from Africa (Ungar et al., 2006), only 28 percent of the specimens cleaned under a light microscope ultimately

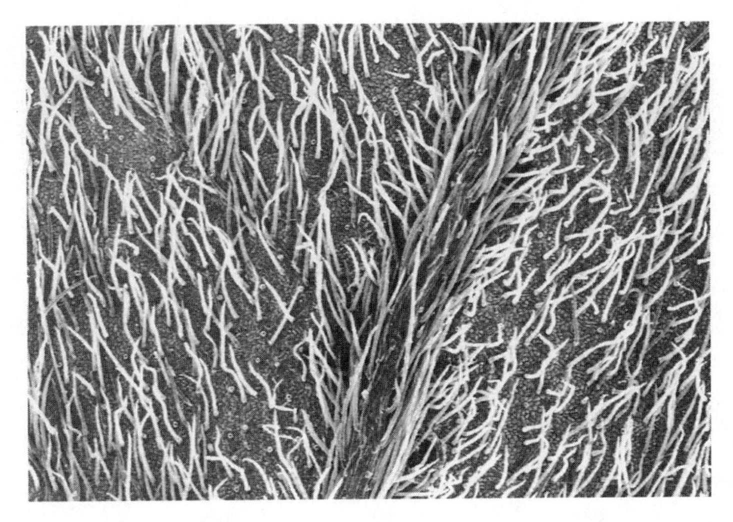

Figure 7.3 SEM of long silica "trichomes" on leaf routinely eaten by *Alouatta palliata* in Costa Rica (from Teaford et al., 2006).

proved usable for SEM analyses. Thus, the second problem is that, because postmortem wear *can* be recognized, many specimens may have to be eliminated from analyses. Obviously, the degree of postmortem wear can be a function of many factors, such as the length of time a specimen has been exposed to the elements; the presence of destructive acids in the postdepositional environment, whether the tooth was excavated, how it was prepared in cleaning, the types of preservative applied to its surface, and so on. As a result, the proportion of fossil specimens useful for dental microwear analyses may vary dramatically from site to site (e.g., less than 25% at Koobi Fora or more than 60% at Olduvai).

Analyses of Museum Material

Most of what we know about dental microwear stems from what might be called the *direct* effects of food on teeth. Of course, we should note at the outset that many foods are not hard enough to scratch teeth (Lucas, 1991). In fact, very few foods in our diets can scratch our teeth because our foods are so clean, cooked, and processed. Still, without such processing, many foods (e.g., certain leaves) include abrasives that can cause striations on teeth (Lucas and Teaford, 1995; Gugel et al., 2001; Teaford et al., 2006; fig. 7.3). Others include acids, which can etch the teeth (fig. 7.4; Puech, 1984b; Puech, Cianfarani, and Albertini, 1986; Teaford, 1988a, 1994; Ungar, 1994; King, Andrews, and Boz, 1999b).

Analyses of mammalian teeth from museum collections have served as one source of information, by demonstrating correlations between certain diets and certain microwear patterns. These correlations depend on which teeth are analyzed because the incisors and canines are used mainly in *ingesting* food, while the premolars and molars are used primarily in *chewing* food once it has been ingested. Analyses of incisor microwear have yielded two basic conclusions. First, animals

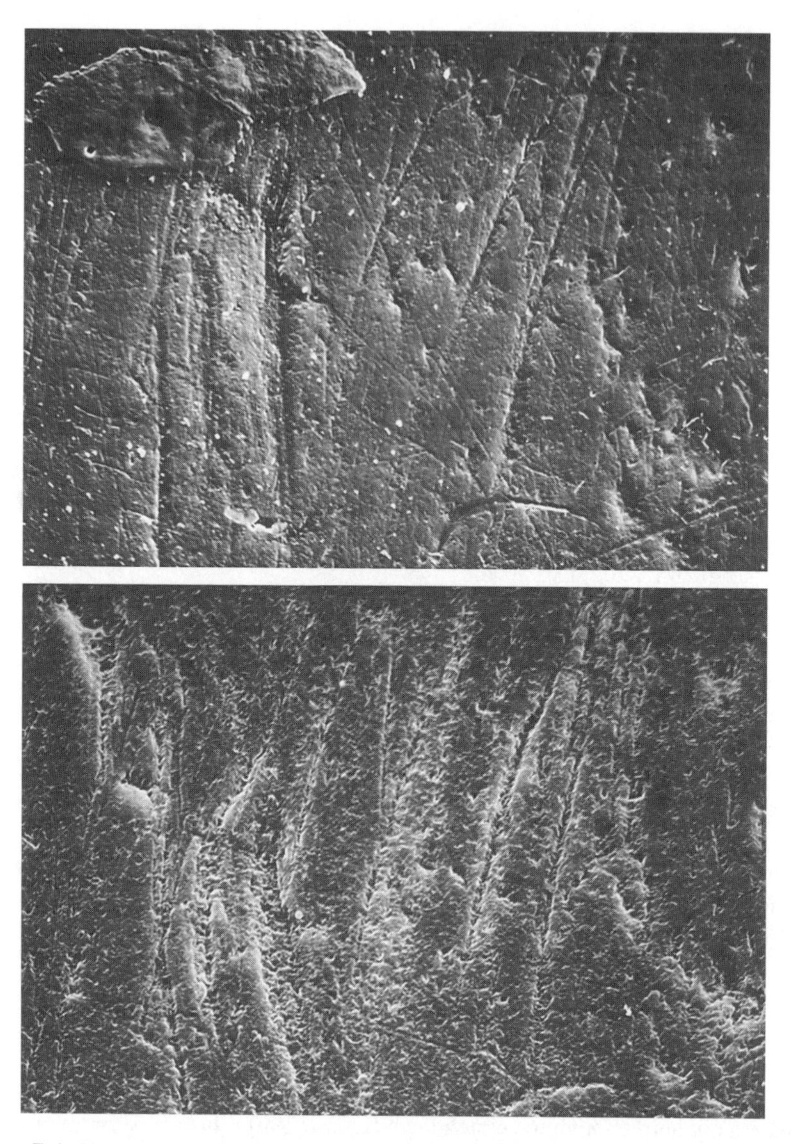

Figure 7.4 Chemical wear of a molar of *Alouatta palliata*. Top = baseline micrograph. Bottom = follow-up micrograph after two to three second exposure to a 30% solution of phosphoric acid (note the removal of smaller microwear features; from Teaford, 1994).

that use their incisors very heavily in the ingestion of food show higher densities of incisal microwear features (Ryan, 1981; Kelley, 1986, 1990; Ungar, 1990, 1994, 1996b). Second, the orientation of striations on the incisors reflects the direction of preferred movement of food (or other items) across the incisors (Walker, 1976; Rose, Walker, and Jacobs, 1981; Ryan, 1981; Ungar, 1994). Thus, for example, the orangutan, which generally uses its incisors a great deal in preparing food, shows more scratches on its incisors than does the gibbon, and those scratches often run in a more mesiodistal direction, reflecting a tendency to pull branches mesiodistally between the front teeth. Analyses of incisor microwear have also yielded an interesting insight that may be more generally applicable; that is, that the size of abrasives may be reflected in the size of microscopic scratches on the teeth, and that this, in turn, may be indicative of feeding height in the canopy, as phytoliths in leaves are generally larger than the abrasive particles in clay-based soils. (Ungar, 1990, 1994).

Further back in the mouth, analyses of molar microwear have demonstrated a few more points. Some of the earliest analyses focused on correlations between orientations of jaw movement and scratches on mammalian molars (Butler, 1952; Mills, 1955, 1963), and this work has continued on a variety of species (Gordon, 1984a; Rensberger, 1986; Young and Robson, 1987; Young, Brennan, and Marshall, 1990; Morel, Albuisson, and Woda, 1991). Early workers also issued cautions to future researchers about the need to take into account potential methodological difficulties (Gordon, 1988; Teaford, 1988a; Pastor, 1993) and biomechanical differences along the tooth row (Gordon, 1982, 1984b, 1988), cautions that are still coming back to haunt us (see "What Is Unknown about Dental Microwear Analyses" in this chapter).

More recent work has demonstrated that grazers tend to show more molar microwear than do browsers (Solounias and Moelleken 1992, 1994; Solounias and Semprebon, 2002), and primates that eat hard objects usually show large pits on their molars, while leaf eaters tend to have relatively more scratches than pits on their molar enamel (Teaford and Walker, 1984; Teaford, 1988a; fig. 7.5). Those "hard objects" can evidently include hard nuts but also smaller items such as

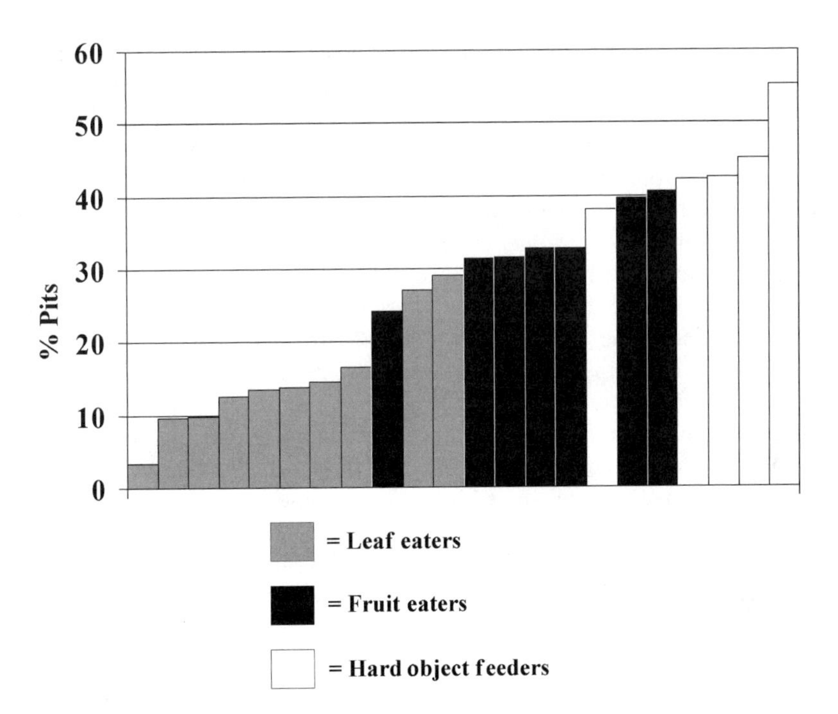

Figure 7.5 Histogram of the incidence of pitting on molars in primates with different diets. Data from Teaford (1988).

insect exoskeletons (Strait, 1993; Silcox and Teaford, 2002). Microwear is also occasionally found on the buccal or lingual ("nonocclusal") surfaces of molars, which may give additional indications of the abrasiveness of the diet, the size of food items, or even the degree of terrestriality (Puech, 1977; Lalueza Fox, 1992; Lalueza Fox and Pérez-Pérez, 1993; Ungar and Teaford, 1996). Even more interesting, museum analyses of molar microwear have yielded glimpses of subtler differences associated with differences in diet. These have ranged from differences between closely related genera (Solounias and Hayek, 1993; Teaford, 1993; Daegling and Grine, 1999; Oliveira, 2001; Solounias and Semprebon, 2002) to differences between subspecies (e.g., *Gorilla gorilla berengei* vs. *G.g. gorilla*; King, Aiello, and Andrews., 1999a) to differences between populations within the same species (Teaford and Robinson, 1989; Merceron, Viriot, and Blondel, 2004). Obviously, such analyses are only as good as the dates and locations of collection for the museum samples, and the published dietary information for those species. As a result, these studies are relatively rare.

Analyses of Live Primates

Unfortunately, double-checking correlations between dental microwear and diet in live animals (in the lab or in the wild) is not easy either. In fact, keeping animals in a laboratory setting is extremely difficult and expensive. Moreover, because the animals have to be anesthetized to make copies of their teeth, the exact timing and type of anesthesia is often a matter of discussion and debate, as most veterinarians prefer to stick with tried-and-true methods (e.g., the use of ketamine administered after eight to twelve hours of fasting), which leave the animals rigidly hard to work with, salivating excessively, and with thick organic films built up on their teeth. As a result, it is perhaps not surprising that the only successful laboratory study to date is one from the 1980s. Teaford and Oyen (1989a, 1989b) raised a group of vervet monkeys on hard and soft diets to check for the effects of food properties on craniofacial growth. As the hard diet consisted of monkey chow and apples, and the soft diet water-softened monkey chow and pureed applesauce, you might expect the effects on the teeth to be relatively similar because both diets had the same basic ingredients and were very abrasive. However, there were two surprising differences. First, the incisors of the soft-food animals were more heavily worn than those of the hard-food animals because the former were routinely rubbing handfuls of food across their incisors, whereas the latter were hardly using their incisors at all. Second, in the molar region, animals on the soft diet showed smaller pits on the occlusal surfaces, perhaps due to repeated tooth-tooth contacts in chewing. The laboratory study also reaffirmed what had been noted in museum studies: that molar facets used for shearing or crushing showed different microwear patterns. Finally, the laboratory study also showed that the turnover in dental microwear could be quite rapid in animals with an abrasive diet, as all of the microwear features in an area sampled by an SEM micrograph would change in one to two weeks, depending on whether the animal was raised on the hard or soft diet.

A more feasible option for studies of live primates might involve the use of human volunteers fed specific food items. However, regulations concerning the use of human subjects make such work difficult, if external funding is to be sought. A pilot study by Noble and Teaford (1995) using American foods normally thought to be hard or abrasive did reaffirm that few foods in our diet (e.g., popcorn kernels) scratch enamel. From a different perspective, rates of microscopic wear (Teaford and Tylenda, 1991) have also been used to gain insights into dental clinical problems, for instance, monitoring the incidence of tooth-grinding in patients with various symptoms of temporomandibular joint disease (Raphael, Marbach, and Teaford, 2003).

Of course, laboratory studies of living animals are limited in how they can change diets, and most laboratory diets are not nearly as diverse as diets in the wild, where seasonal, geographic, and annual differences in diet have the potential to have a huge effect on dental microwear patterns. Thus, work with animals in the wild is a potential gold mine of information, as demonstrated by the pioneering study of Walker, Hoeck, and Perez (1978) on hyraxes, where skulls were collected directly from the same area in which behavioral observations were recorded. Unfortunately, while studies of living primates in the wild have been attempted a number of times, they have usually met with little success. Primates often live in forested habitats where they are hard to see and even harder to catch. Even in open habitats (e.g., baboons in the East African savanna), results can still be disappointing, largely because of the aforementioned difficulties in cleaning and drying the teeth. Thus far, only two studies have consistently yielded high-quality copies of primate teeth in the wild. The first is the ongoing study at La Pacifica in the Guanacaste region of Costa Rica (Teaford and Glander, 1991, 1996; Ungar et al., 1995; Dennis et al., 2004). There, howling monkeys (*Alouatta palliata*) are regularly observed, captured, and released in a dry tropical forest setting. That work has certainly verified some of the standard correlations from museum analyses (e.g., leaf eating and scratches on teeth). It has also given us glimpses of other complicating factors. For instance, the amount of molar microwear may vary from season to season and between riverine and nonriverine microhabitats (Teaford and Glander, 1996). The studies at La Pacifica have also shown that tooth wear generally proceeds at a rapid pace in the wild—in fact, at about eight to ten times the pace of that in U.S. dental patients (Teaford and Glander, 1991). This has led to the idea of the "Last Supper" phenomenon (Grine, 1986); that is, in some situations, dental microwear may only record the effects of the most recently eaten foods on the teeth, although some investigators feel that microwear on the sides of the teeth may show far slower turnover (Pérez-Pérez, Lalueza, and Turbón, 1994).

Recently, a second long-term study has begun to yield high-resolution casts of primate teeth in the wild (Nystrom, Phillips-Conroy, and Jolly, 2004). The study populations, from the anubis-hamadryas hybrid zone of Awash National Park, Ethiopia, have been the focus of multidisciplinary work for more than thirty years (e.g., Nagel, 1973; Phillips-Conroy, 1978; Sugawara, 1979; Phillips-Conroy and Jolly, 1986; Phillips-Conroy, Jolly, and Brett, 1991; Szmulewicz et al., 1999; Phillips-Conroy, Bergman, and Jolly, 2000; Dirks et al., 2002) and have yielded fascinating insights into the behavioral, ecological, and anatomical ramifications of species hybridization in the wild. The precise timing of dental microwear analyses in this case (before the heavy onset of new leaves and grasses in this seasonal environment), allowed Nystrom, Phillips-Conroy, and Jolly (2004) to implicate "small-caliber environmental grit" as the main cause of the observed microwear patterns, which included no significant differences between the sexes, age groups, or different troops.

In Vitro Laboratory Studies

If studies of living primates are so difficult, why not do experimental studies of dental enamel abraded by different foods? Early studies showed that substances like acids could have a profound effect on enamel surfaces (Mannerberg, 1960; Boyde, 1964). Still, experimental work has proceeded in fits and spurts. Some studies have demonstrated that the orientation of scratches on a tooth's surface can indeed reflect the orientation of tooth-food-tooth movements (e.g., Ryan, 1979; Teaford and Walker, 1983; Gordon, 1984a; Walker, 1984; Teaford and Byrd, 1989). Other studies have shown that certain agents, such as wind-blown sand or various acids, can leave characteristic microwear patterns on teeth (Puech and Prone, 1979; Puech, Prone, and Kraatz, 1980; Puech, Prone, Albertini, 1981; Gordon, 1984c; Puech et al., 1985; Puech, 1986a; King, Andrews, and Boz, 1999b). However, there have been surprisingly few controlled studies of the wear patterns caused by different types of foods.

Peters (1982) used standard physical property-testing equipment while examining the effects of a range of African foods on dental microwear, ultimately showing that few foods could actually scratch enamel; instead extraneous abrasives were one of the prime culprits (see also Puech, Cianfarani, and Albertini, 1986). Only with more detailed analyses did subsequent work (e.g., Gugel, Grupe, and Kunzelmann, 2001) demonstrate the effects of specific foods on microwear patterns (e.g., "cereal-specific" microwear related to phytolith content in certain grains; fig. 7.6).

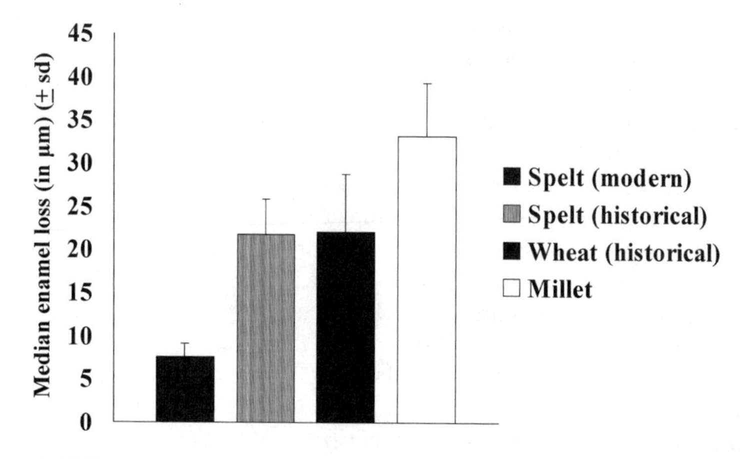

Figure 7.6 Differences in the amount of enamel lost through in vitro abrasion of enamel by different cereal grains. Data from Gugel et al. (2001).

Indirect Causes of Dental Microwear

Laboratory studies have also raised the possibility that some microwear patterns might be caused solely by what might best be termed the *indirect* effects of food on dental microwear. For instance, certain cooking procedures may introduce abrasives into foods, causing a high incidence of microscopic scratches on teeth, scratches not caused by the foods themselves but by the methods with which they were prepared (Pastor, 1992, 1993; Teaford and Lytle, 1996 ; fig. 7.7). Similarly, animals may also

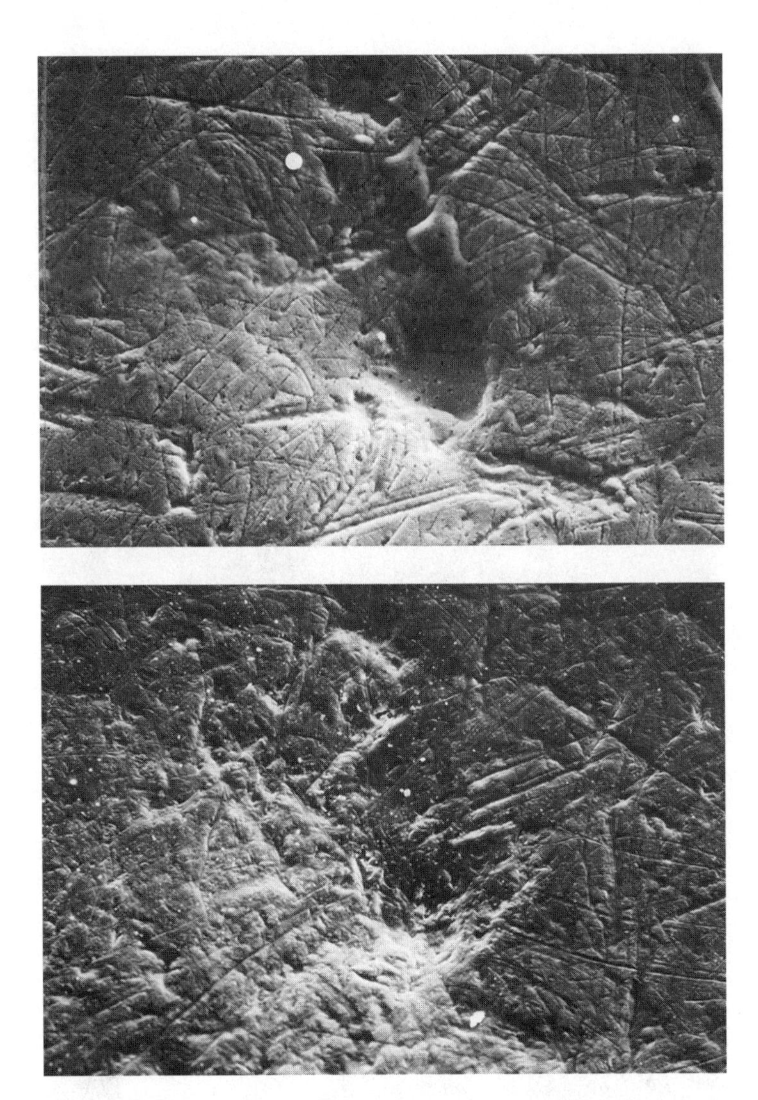

Figure 7.7 Changes in dental microwear caused by abrasives introduced by food processing. Top = baseline micrograph. Bottom = follow-up micrograph of same surface, after seven days of exposure to a new diet incorporating stone-ground maize in each meal (from Teaford and Lytle, 1996).

eat soft foods and still show many scratches on their teeth, if the food is coated with abrasives (e.g., earthworms coated with dirt or leaves and ivy coated with sand; Silcox and Teaford, 2002; Merceron, Viriot, and Blondel, 2004a).

If an animal has a soft but tough diet, can tooth-on-tooth wear yield characteristic microwear patterns as enamel edges penetrate the food and grind past each other? Analyses of museum collections of modern animals suggest this may be the case because animals that traditionally feed on tough bean pods (or fruits with a tough pericarp) show a high incidence of small pits on their teeth, pits probably caused by the adhesive wear of enamel on enamel (fig. 7.8; Puech, Prone, and Al-

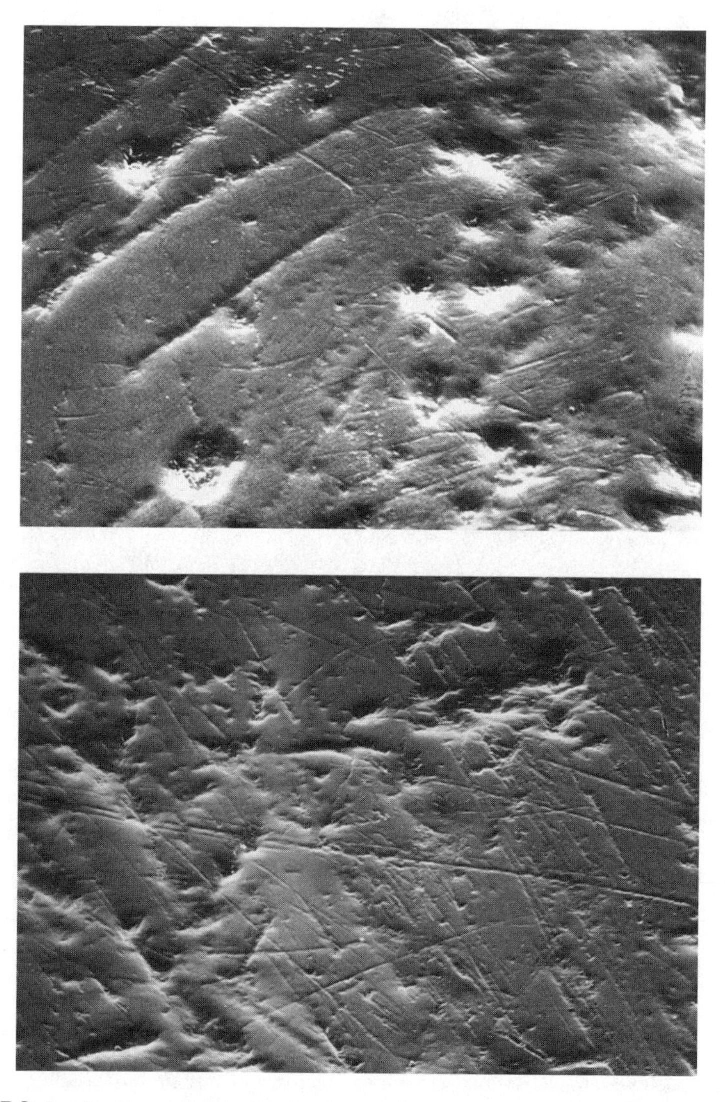

Figure 7.8 Small pits on the teeth of primates known to eat tough, pliant foods. Top = *Hapalemur*. Bottom = *Propithecus*.

bertini, 1981; Puech, 1984a, 1986a; Walker, 1984; Puech, Cianfarani, and Albertini, 1986; Radlanski and Jäger, 1989; Teaford and Runestad 1992; Rafferty, Teaford, and Jungers, 2002).

Analyses of Archeological and Paleontological Samples

When dental microwear analyses are aimed at the past, they often raise more questions than they answer, largely because they give new and different glimpses of the intricacies of previous behavior. Investigators have used inferences of both direct and indirect causes of microwear patterns to aid in their interpretations, and most analyses have generally been either of two types: (1) comparisons of populations suspected to have had diet differences (based on associated archaeological remains, functional morphology, historical records, etc.) or (2) samples bracketing major diet transitions (e.g., the change from gathering-hunting to agriculture). Of course, interpretations of results are dependent on our knowledge of present-day correlations between diet and dental microwear. Thus, while we often have significant differences in dental microwear between teeth from different sites or time periods, the exact meaning of those differences may be subject to discussion and debate until better data are available for modern species.

As might be expected, given the variable diets of modern humans and the effects of cooking and methods of food preparation on dental microwear, analyses of human archaeological material have yielded a wide range of results. For the anterior dentition, the distinction between abrasive foods and abrasives *on* foods has again been raised, this time in interpreting results for samples of Eskimo incisors. Historical records indicate that some Eskimo populations routinely ate large amounts of meat (Hrdlička, 1945), but because meat does not normally contain abrasives, any incisal scratches or chipping were probably caused by abrasives on the meat or by abrasives adhering to nonfood items, such as hides, normally prepared by the incisors (Merbs, 1968; Ryan and Johanson, 1989; Ungar and Spencer, 1999). Also, while the number of scratches on Eskimo incisors may be low, the width of those scratches is high, perhaps indicating high-incisal bite forces (Ungar and Spencer, 1999). Unfortunately, scratch orientation is probably not correlated with handedness in the use of stuff-and-cut behaviors in populations such as the Aleut (Bax and Ungar, 1999).

Going further back in time, along the southeastern coast of the United States, analyses of Native American populations do not show significant differences in the size or incidence of incisor microwear in prehistoric versus historic populations. However, the incisors of the prehistoric populations do show more variable scratch orientations, suggesting they were used for a greater variety of tasks than in the historic populations (Teaford et al., 2001).

As far as molars are concerned, early qualitative studies of Native American populations noted correlations between the amount of microwear and the abrasiveness of the diet and the presence of large pits and the consumption of hard objects (e.g., Shkurkin et al., 1975; Rose, 1984; Rose and Marks, 1985; Harmon and Rose, 1988). Subsequent work has extended analyses to other populations (e.g., predynastic and dynastic Egypt; Puech, Serratrice, and Leek, 1983) and refined analyses through quantification and examination of a greater variety of samples. Some of the most intriguing work has involved analyses of teeth where the actual abrasives causing the wear can be collected on-site, either imbedded in the teeth themselves (Lalueza Fox, Pérez-Pérez, and Juan, 1994) or from coprolites at the site (Danielson and Reinhard, 1998).

The transition from hunting-gathering to agriculture has left a complex signal in the microwear record, depending on which populations are examined, in which habitats, and so on. In central North America, there seems to be a transition in preagricultural populations to a slightly harder diet before the actual switch to agriculture (Schmidt, 2001), with subsequent agricultural populations having a softer, less variable diet (Bullington, 1991). Along the southeastern coast of the United States, however, microwear evidence for that dietary transition is complicated by significant local differences in microhabitat, most notably between coastal and inland sites, with the former evidently including significant amounts of large-grained abrasives in foods (Teaford, 1991, 2002; Teaford et al., 2001). In other regions of the world (e.g., the Indian subcontinent), interpretations are further complicated by changes in food-processing techniques, as stone grinding tools introduced significant abrasives into what would have otherwise been a fairly nonabrasive diet (Pastor, 1992; Pastor and Johnston, 1992).

Of course, once the change to agriculture was made, human diets did not simply stay stagnant. Some became more homogeneous, as evidenced by fairly uniform microwear patterns, while others became more variable (Molleson and Jones, 1991). Some cereal diets left characteristic microwear patterns remarkably similar to those documented in laboratory studies (Gugel, Grupe, and Kunzelmann, 2001). Some changes in food processing, most notably the boiling of foods, led to a marked decrease in the amount of microwear at some sites (Molleson, Jones, and Jones, 1993). However, those changes also left microwear researchers with a bit of a dilemma: how to distinguish the microwear signatures of meat eating versus the consumption of cooked food (Molleson, Jones, and Jones, 1993; Organ et al., 2005)?

Paleontological analyses have focused on a wide variety of animals, ranging from rodents (Rensberger, 1978, 1982), horses (MacFadden, Solounias, and Cerling, 1999), and ungulates (Solounias and Moelleken, 1992, 1994; Solounias and Hayek, 1993; Solounias and Semprebon, 2002; Merceron et al., 2004, 2005a, 2005b; Merceron and Ungar, in press), to carnivores (Van Valkenburgh, Teaford, and Walker, 1990), tyrannosaurids (Schubert and Ungar, 2005), and conodonts (Purnell, 1995). But within the primates, dental microwear analyses have also led to some major insights. Analyses of Miocene hominoid material have helped document an impressive array of dietary adaptations in the early apes (Teaford and Walker, 1984; Ungar, 1996a; Kay and Ungar, 1997; King, Aiello, and Andrews, 1999). By contrast, analyses of Plio-Pleistocene cercopithecoid material have documented a surprisingly limited array of dietary adaptations in East Africa (Lucas and Teaford, 1994; Leakey, Teaford, and Ward, 2003) but a larger array in South Africa (El-Zaatari et al., 2005), while also yielding insights into the degree of terrestriality in some species (Ungar and Teaford, 1996). Molar microwear analyses have also helped to document the effects of phylogenetic constraints in fossil apes by documenting similar functions in taxa that have undergone shifts in molar morphology through time (Ungar, Teaford, and Kay, 2004).

As might be expected, analyses of human ancestors have focused on whichever fossils are available. For the anterior teeth, qualitative studies have suggested similarities between early hominin incisor wear and that observed on modern primates that routinely employ a great deal of incisal preparation (Puech and Albertini, 1984). Quantitative analyses of *Australopithecus afarensis* suggested incisal microwear similarities with those documented for lowland gorillas or perhaps savanna baboons (Ryan and Johanson, 1989). More detailed analyses of *Paranthropus robustus* and *Australopithecus africanus* (Ungar and Grine, 1991) showed great variabilitiy within each species in most standard microwear measurements. However, the greater density of features on the incisors of *A. africanus* helped to show that *A. africanus* had a heavier emphasis on incisal preparation than in *P. robustus*.

In the molar region, qualitative analyses have raised many possibilities that have been often repeated in the literature. For instance, the robust australopithecines have been characterized as indistinguishable from modern chimpanzees or orangutans (Walker, 1981), perhaps with more abrasive molar wear than in the gracile australopithecines (Puech et al., 1985; Puech, Cianfarani, and Albertini, 1986; Puech, 1986b). By contrast, *Homo habilis* has been characterized as using high occlusal pressures but on foods that can chemically etch the enamel (Puech, Serratrice, and Leek, 1983; Puech, 1986b).

Quantitative analyses have begun to refine these interpretations from many different perspectives. Studies of nonocclusal microwear have focused primarily on more recent, European taxa, such as the Neanderthals, together with specimens now attributed to *Homo heidelbergensis*. Initial analyses portrayed the Neanderthals as more carnivorous than their immediate predecessors, or subsequent *Homo sapiens* (Lalueze Fox and Pérez-Pérez, 1993; Lalueza, Pérez-Pérez, Turbón, 1996). However, subsequent work has raised the possibility of sexual differences in diet in *Homo heidelbergensis* (Pérez-Pérez, Bermúdez de Castro, and Arsuaga, 1999) and a more heterogeneous diet for the Neanderthals, with a shift in food processing in the Upper Paleolithic (Pérez-Pérez et al., 2003).

Quantitative analyses of fossil hominin occlusal microwear began with Grine's pioneering work on the South African australopithecines, where *P. robustus* was shown to exhibit more microwear and more pitting on its molars than did *A. africanus* (Grine, 1981, 1986, 1987; Grine and Kay, 1988; Kay and Grine, 1989). This supported Robinson's ideas of dietary differences among the australopithecines, with the so-called robust forms consuming harder foods that required more variable grinding movements in chewing. Recent work has taken analyses a step further by incorporating samples of australopithecines and early *Homo* from East and South Africa (Ungar, Teaford, and Grine, 2001; Teaford, Ungar, and Grine, 2002a). The work is still being written, but initial results give further credence to Ryan and Johanson's (1989) idea of similarities between *A. afarensis* and lowland gorillas, this time for the molars (Teaford, Ungar, and Grine, 2002b; Grine, Ungar, and Teaford, in press). In conjunction with other morphological data for the australopithecine grade of human evolution (Teaford and Ungar, 2000; Teaford, Ungar, and Grine, 2002a), they also helped to make the distinction between dental capabilities and dental use, as the capability to process certain foods may well have been critically important in certain situations. Meanwhile, analyses of early *Homo* have begun to help sort through the variable assemblage that now encompasses early *Homo*, with

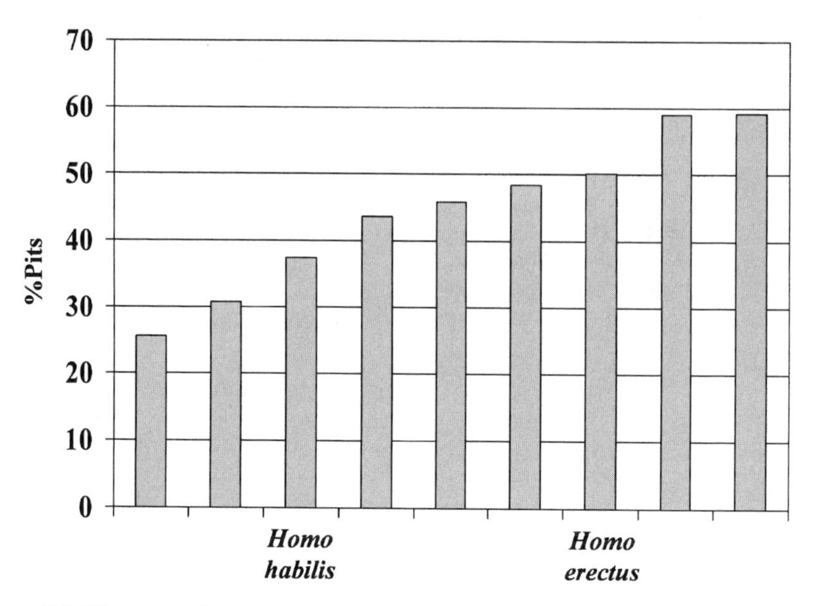

Figure 7.9 Histogram of the incidence of pitting on molars of *Homo habilis* and *Homo erectus* and extant primates. Data from Ungar et al. (2006).

Homo erectus/ergaster showing a higher incidence of pitting on its molars than that found in *H. habilis* (Ungar et al., 2006), suggesting the consumption of tougher or harder food items by the former group, again, as a possible critical fallback food (fig. 7.9).

What Is *Unknown* about Dental Microwear Analyses?

At first glance, with all the work that has been done, it might seem that we know a great deal about dental microwear and diet. In reality, however, all we have are tiny windows into a complex world. Studies of living primates have really only been carried out on two species (*Alouatta palliata* and *Papio hamadryas*) in two habitats, the dry tropical forest of Costa Rica, and the thornbush and savanna grassland of Ethiopia. While those settings certainly have their inherent complexities (e.g., dramatic seasonal changes in rainfall and resource availability), how representative are they of all the other ecological zones in the world? Would dental microwear patterns differ for howlers or baboons in other habitats? Undoubtedly. How might other species share a habitat with *Alouatta* or *Papio* (or other species), and how would that be reflected in differences in dental microwear? What is the magnitude of seasonal, annual, geographic, and interspecific differences in dental microwear for other species elsewhere in the world? How does the incidence of dental microwear relate to specific abrasives and foods in the wild? Clearly, a massive amount of work has yet to be done on live primates in the wild if we are to use that information to help interpret results from fossil samples.

Similarly, laboratory studies have barely begun to sort through the intricacies of dental microwear formation. As noted earlier, the effects of specific food items have yet to be documented in any systematic fashion, and the effects of foods naturally consumed in the wild have yet to be examined in any detail. As primate diets are normally quite variable, and as dental microwear features can change quite quickly (Teaford and Oyen, 1989a, 1989b; Teaford and Glander, 1991), what are the effects of different food items on overall microwear patterns within a specific diet? Will items that are abrasive, hard, or acidic effectively swamp other microwear patterns? Will the so-called Last Supper phenomenon vary between species or populations? In essence, we must not lose sight of the fact that the microwear pattern on a given tooth is essentially a complex summary of past events, often with multiple signals superimposed on each other. Some wear processes, like acid etching, leave fairly distinctive patterns but can have varying effects on teeth, depending on the length of exposure of the tooth to the acid (Teaford, 1994). Other wear processes, like abrasion, may yield evidence of the effects of foods themselves, or the effects of extraneous abrasives adhering to foods (Teaford, 1988a; Pastor, 1993; Merceron et al., 2004a; Nystrom, Phillips-Conroy, and Jolly, 2004). Still other processes, like adhesion, may yield wear patterns similar to those caused by small hard objects (Radlanski and Jäger, 1989; Teaford and Runestad, 1992; Rafferty et al., 2002). When animals have varied diets, incorporating foods of different properties, it is easy to see where the effects of each food may be hard to decipher. Again, much more work needs to be done.

Analyses of museum material have probably been pushed closer to their limits than studies of living animals but only because there are relatively few museums where the associated collection data are of sufficient detail to aid the documentation of geographic or seasonal differences in diet and dental microwear (Teaford and Robinson, 1989; Merceron, Viriot, and Blondel, 2004). Moreover, there are virtually no collections of primate material for which dietary information has been collected before the animals were collected. Thus virtually all studies of museum samples are limited in their resolution by the lack of detailed dietary information for the animals in question. Similarly, analyses of archaeological and paleontological material are limited by the size and extent of collections (and sometimes, in the case of fossils, access to them) and by the associated information for those collections (e.g., geological information, paleoecological interpretations, presence of associated cultural remains, etc.).

Still, when fossils are found, they yield their clues indiscriminately, forcing us to ask questions about subtle intraspecific differences in diet, questions that most analyses of museum material cannot begin to answer. Similarly, "what you see is what you get" in fossils, in terms of dental samples. So preservation may be poor, postmortem wear may be rampant, and, for some collections, we may even have an overabundance of certain tooth types for which we have no analyses of modern material (noting that most analyses to date have focused on incisors or molars). In the face of such problems and possibilities, innumerable questions still need to be answered. For instance, what is the relationship between the biomechanical demands of processing certain foods and the generation of microwear patterns? Can the structural capabilities of bone, or variations in enamel properties, be correlated with variations in microwear pattern within and between jaws? For that matter, what *is* the relationship between microwear patterns between upper and lower jaws? We know their patterns are roughly similar (Teaford and Walker, 1984), but can analyses of upper-lower microwear integration shed new light on jaw movements and food processing in mammals? What about the microwear of other dental materials like dentin? Until now, investigators have shied away from it, mainly because it is hard to clean without introducing artificial microwear patterns. However, because it is softer than enamel, might it be an indicator of even subtler diet distinctions?

Despite all these questions, when all is said and done, the biggest challenge facing dental microwear analyses is a methodological one. Standard SEM analyses are difficult, costly, time consuming, and (most notably) subjective, in that the "measurement" of certain microwear features depends on the recognition of "landmarks" that may be defined differently by different researchers. Even with the use of semiautomated, computerized, digitizing routines (Ungar, Simon, and Cooper, 1990), interobserver error rates are often unacceptably high (Grine, Ungar, and Teaford, 2002). Thus, measurements computed by different researchers should probably not be compared directly. When this is coupled with the fact that most analyses have used relatively small samples (even for species with variable diets), the net effect is that dental microwear analyses have barely begun to live up to their potential.

Recently, workers have begun to address this issue through the use of two new approaches, lower magnification work by light microscopy (where features are categorized into a limited number of sizes; Solounias and Semprebon, 2002; Semprebon et al., 2004) and a higher magnification combination of confocal microscopy and scale-sensitive fractal analysis (Ungar et al., 2003; Scott et al., 2005). The former holds the promise of quick, cheap analyses of larger samples, and published tests of interobserver error hint at better replicability than standard SEM-based analyses (Semprebon et al., 2004), although measures of error rates are not presented in a form that is comparable between studies. However, the technique still requires significant training to master, and interobserver error rates have yet to be checked for categorizations of features into different sizes—the very distinctions that hold the most promise of documenting subtle dietary differences (Godfrey et al., 2004). Moreover, because it works at low magnifications, it may miss information provided by finer microwear features. For instance, the assertion by Godfrey and colleagues (2004) that there were no dietary differences between species of *Megaladapis* may be nothing more than a reflection of the limitations of the technique (if one chooses to ignore the statistical impossibility of proving a *lack* of differences between samples). Finally, the low-magnification technique may only be able to detect the most obvious effects of postmortem wear. Thus, it may ultimately be restricted to making gross dietary distinctions, such as those between browsers and grazers. Only further work with larger samples will tell.

Meanwhile, the confocal work holds promise of providing quick, objective characterizations of entire wear surfaces (in three dimensions no less) with no need to identify or categorize specific wear features. Thus, it would seem to be the closest thing available to putting a specimen in and getting useful numbers out. To emphasize this dramatic shift from standard landmark-based measurements, the developers have referred to it as "dental microwear texture analysis" (Scott et al., 2005). However, it is still a work in progress and, as a result, should be approached cautiously. New analytical software is still in development, a database for future interpretations is still being gathered, and comparisons with data generated by previous techniques are still being completed. Initial results, though, are promising, yielding insights into diet variability in the South African australopithecines, including the possible importance of critical fallback foods (Scott et al., 2005), ultimately holding promise, like in the lower magnification work, of analyses of much larger samples. Only with analyses of larger samples will we truly begin to appreciate the variability of diet and dental microwear in modern and fossil primates. With the added capability of providing data in three dimensions, dental microwear texture analysis also raises the possibility of answering innumerable new questions concerning topics such as the volume of tooth loss in different wear regimes and the depth of enamel removed by certain abrasives or bite forces, thus giving hope for even better correlations between tooth use and wear patterns.

What Is *Unknowable* about Dental Microwear Analyses?

As an optimist, I'm always hesitant to admit that some things are unknowable, largely because I feel that future analyses will always open new, unanticipated possibilities for interpretation. Still, there are some obvious limits to dental microwear analyses. For instance, we can only analyze whatever affects dental materials. Thus, if something has no direct, or indirect, effect on enamel (or dentin), then we will not be able to detect it. As a result, the consumption of certain soft foods may remain invisible, and the relative importance of different foods within overall diets may also be indecipherable because certain foods, or abrasives on foods, effectively overwhelm other evidence. Since dental microwear analysis is ultimately dependent on physical or chemical effects on teeth, it will also yield no direct evidence of the nutritional contributions of food items to the overall diet.

Also, for past populations, we are limited to working on only those specimens preserved in the fossil or archaeological record. Does that mean that we will never obtain samples from certain environments? Probably. Does it mean that we will never obtain samples from certain seasons? Possibly. Does it mean that we will never accurately bracket the range of variation of dietary habits of some species? Probably. In essence, future analyses of fossils will always be limited by the limits of our geological resolution. Thus, while it would be wonderful to be able to document differences within and between populations of prehistoric species, how close to actual biological populations can we come in the fossil record? For the earliest hominins, we probably cannot come very close; for more recent hominins, perhaps we can come much closer. So, for some fossil samples, we may speculate about topics like sexual differences in diet or age-related differences, but if our fossil specimens are thousands, or hundreds of thousands, of years apart, we may be limited to only the most obvious of differences.

Unfortunately, other major sources of information for dental microwear analysis may be vanishing before our eyes, as huge tracts of the earth's environment disappear or are damaged beyond repair. In the process, habitats and organisms of crucial importance for future microwear interpretations may be lost, effectively leaving certain questions unanswerable. For instance, how will we be able to document the effects of the intricate niche partitioning in different ecosystems if some of those ecosystems are no longer with us?

In conclusion, while some opportunities are dwindling and some pieces of information may remain invisible, we always need to be open to new opportunities, as the effects of some foods, or the means of documenting them, may be hard to anticipate. Each method of data collection and each piece of evidence has its strengths and weaknesses. Dental microwear analysis is certainly no exception; it definitely has its limitations. But it also has the potential to give us direct glimpses of the past. As such, it can tell us about how teeth were actually used rather than what they were evolutionarily capable of doing. So we need to better understand its strengths and weaknesses. New methods raise new hopes of doing so. Of course, in the long run, the picture we are trying to decipher is incredibly complicated. So we also need to consider every piece of evidence, be it dental microwear, or otherwise. With a little luck and foresight, we will have the good fortune to contribute to a better understanding of the origin and evolution of human diet, among many other things!

Acknowledgments I would like to thank (1) Peter Ungar for inviting me to participate in the symposium The Evolution of Human Diet: The Known, the Unknown, and the Unknowable and (2) The Sloan Foundation and the University of Arkansas for making that session possible. I would also like to thank all of the symposium participants for their insightful discussions, before, during, and after the sessions. Special thanks go to Peter Ungar and Fred Grine for their constant feedback and continuing collaborative efforts in the study of the evolution of human diet. Peter and an anonymous reviewer added significant improvements to this manuscript. This chapter has benefited from the help of innumerable colleagues and curators from around the world, and it also would not have been possible without the support of the National Science Foundation whose help is gratefully acknowledged.

References

Ainamo, J., 1971. Prenatal occlusal wear in guinea pig molars. *Scand. J. Dent. Res.* 79, 69–71.

Bax, J.S., and Ungar, P.S., 1999. Incisor labial surface wear striations in modern humans and their implications for handedness in Middle and Late Pleistocene hominids. *Int. J. Osteoarchaeol.* 9, 189–198.

Boyde, A., 1964. The structure and development of mammalian enamel. PhD diss., University of London.

Bullington, J. 1991. Dental microwear of prehistoric juveniles from the lower Illinois River Valley. *Am J. Phys. Anthropol.* 84, 59–74.

Butler, P.M., 1952. The milk molars of Perissodactyla, with remarks on molar occlusion. *Proc. Zool. Soc. Lond.* 121, 777–817.

Daegling, D.J., and Grine, F.E., 1999. Terrestrial foraging and dental microwear in *Papio ursinus*. *Primates* 40, 559–572.

Danielson, D.R., and Reinhard, K.J., 1998. Human dental microwear caused by calcium oxalate phytoliths in prehistoric diet of the lower Pecos region, Texas. *Am J. Phys. Anthropol.* 107, 297–304.

Dennis, J.C., Ungar, P.S., Teaford, M.F., and Glander, K.E., 2004. Dental topography and molar wear in *Alouatta palliata* from Costa Rica. *Am J. Phys. Anthropol.* 125, 152–161.

Dirks, W., Reid, D.J., Jolly, C.J., Phillips-Conroy, J.E., and Brett, F.L., 2002. Out of the mouths of baboons: Stress, life history, and dental development in the Awash National Park hybrid zone, Ethiopia. *Am J. Phys. Anthropol.* 118, 239–252.

El-Zaatari, S., Grine, F.E., Teaford, M.F., and Smith, H.F., 2005. Molar microwear and dietary reconstructions of fossil Cercopithecoidea from the Plio-Pleistocene deposits of South Africa. *J. Hum. Evol.* 49, 180–205.

Godfrey, L.R., Semprebon, G.M., Jungers, W.L., Sutherland, M.R., Simons, E.L., and Solounias, N., 2004. Dental use wear in extinct lemurs: evidence of diet and niche differentiation. *J. Hum. Evol.* 47, 145–170.

Gordon, K.D., 1982. A study of microwear on chimpanzee molars: Implications for dental microwear analysis. *Am. J. Phys. Anthropol.* 59, 195–215.

Gordon, K.D., 1984a. The assessment of jaw movement direction from dental microwear. *Am J. Phys. Anthropol.* 63, 77–84.

Gordon, K.D., 1984b. Hominoid dental microwear: Complications in the use of microwear analysis to detect diet. *J. Dent. Res.* 63, 1043–1046.

Gordon, K.D., 1984c. Taphonomy of dental microwear. II. *Am J. Phys. Anthropol.* 63, 164–165.

Gordon, K.D., 1988. A review of methodology and quantification in dental microwear analysis. *Scanning Microsc.* 2, 1139–1147.

Grine, F.E., 1981. Trophic differences between "gracile" and "robust" australopithecines: A scanning electron microscope analysis of occlusal events. *S. Afr. J. Sci.* 77, 203–230.

Grine, F.E., 1986. Dental evidence for dietary differences in *Australopithecus* and *Paranthropus*: A quantitative analysis of permanent molar microwear. *J. Hum. Evol.* 15, 783–822.

Grine, F.E., 1987. Quantitative analysis of occlusal microwear in *Australopithecus* and *Paranthropus*. *Scanning Microsc.* 1(2), 647–656.

Grine, F.E., and Kay, R.F., 1988. Early hominid diets from quantitative image analysis of dental microwear. *Nature* 333, 765–768.

Grine, F.E., Ungar, P.S., and Teaford, M.F., 2002. Error rates in dental microwear quantification using scanning electron microscopy. *Scanning* 24, 144–153.

Grine, F.E., Ungar, P.S., and Teaford, M.F., in press. Molar microwear in *Praeanthropus afarensis*: Evidence for dietary stasis through time and under diverse paleoecological conditions. *J. Hum. Evol.*

Gugel, I.L., Grupe, G., and Kunzelmann, K.H., 2001. Simulation of dental microwear: Characteristic traces by opal phytoliths give clues to ancient human dietary behavior. *Am J. Phys. Anthropol.* 114, 124–138.

Harmon, A.M., and Rose, J.C., 1988. The role of dental microwear analysis in the reconstruction of prehistoric diet. In: Kennedy, B.V., and Lemoine, G.M. (Eds.), *Diet and Subsistence: Current Archaeological Perspectives*. Archaeological Association of the University of Calgary, Calgary, Alberta, pp. 267–272.

Hrdlička, A., 1945. *The Aleutian and Commander Islands and Their Inhabitants*. Wistar Institute of Anatomy & Biology, Philadelphia.

Kay, R.F., 1975. The functional adaptations of primate molar teeth. *Am J. Phys. Anthropol.* 42, 195–215.

Kay, R.F., and Grine, F.E., 1989. Tooth morphology, wear and diet in *Australopithecus* and *Paranthropus* from southern Africa. In: Grine, F.E. (Ed.), *The Evolutionary History of the Robust Australopithecines*. Aldine de Gruyter, New York, pp. 427–444.

Kay, R.F., and Hylander, W.L., 1978. The dental structure of mammalian folivores with special reference to primates and Phalangeroidea (Marsupialia). In: Montgomery, G.G. (Ed.), *The Biology of Arboreal Folivores*. Smithsonian Institution Press, Washington, DC, pp. 173–192.

Kay, R.F., and Ungar, P.S., 1997. Dental evidence for diet in some Miocene catarrhines with comments on the effects of phylogeny on the interpretation of adaptation. In: Begun D.R., Ward C., and Rose M. (Eds.), *Function, Phylogeny and Fossils: Miocene Hominoids and Great Ape and Human Origins*. Plenum Press, New York, pp. 131–151.

Kelley, J., 1986. Paleobiology of Miocene hominoids. PhD diss., Yale University.

Kelley, J., 1990. Incisor microwear and diet in three species of *Colobus*. *Folia Primatol.* 55, 73–84.

King, T., Aiello, and L.C., Andrews, P., 1999a. Dental microwear of *Griphopithecus alpani*. *J. Hum. Evol.* 36, 3–31.

King, T., Andrews, P., and Boz, B., 1999b. Effect of taphonomic processes on dental microwear. *Am J. Phys. Anthropol.* 108, 359–373.

Lalueza Fox, C., 1992. Dental striation pattern in Andamanese and Veddahs from skulls collections of the British Museum. *Man in India* 72, 377–384.

Lalueza Fox, C., and Pérez-Pérez, A., 1993. The diet of the Neanderthal child Gibralter 2 (Devils' Tower) through the study of the vestibular striation pattern. *J. Hum. Evol.* 24, 29–41.

Lalueza Fox, C., Pérez-Pérez, A., and Juan, J., 1994. Dietary information through the examination of plant phytoliths on the enamel surface of human dentition. *J. Archaeol. Sci.* 21, 29–34.

Lalueza C., Pérez-Pérez A., and Turbón D., 1996. Dietary inferences through buccal microwear analysis of middle and upper Pleistocene human fossils. *Am J. Phys. Anthropol.* 100, 367–387.

Leakey, M.G., Teaford, M.F., and Ward, C.W., 2003. Cercopithecidae from Lothagam. In: Leakey, M.G., and Harris, J.M. (Eds.), *Lothagam: Dawn of Humanity in Eastern Africa.* Columbia University Press, New York, pp. 201–248.

Lucas, P.W., 1979. The dental-dietary adaptations of mammals. *Neues Jahrb. Geol. Palaeontol.* 8, 486–512.

Lucas P.W., 1991. Fundamental physical properties of fruits and seeds in the diet of Southeast Asian primates. In: Ehara, A., Kimura, T., Takenaka, O., and Iwamoto, M. (Eds.), *Primatology Today*. Elsevier, Amsterdam, pp. 125–128.

Lucas, P.W., 2004. *Dental Functional Morphology: How Teeth Work.* Cambridge University Press, New York.

Lucas, P.W., and Teaford, M.F., 1994. The functional morphology of colobine teeth. In: Oates, J., and Davies, A.G. (Eds.), *Colobine Monkeys: Their Evolutionary Ecology*. Cambridge University Press, New York, pp. 173–203.

Lucas, P.W., and Teaford, M.F., 1995. Significance of silica in leaves to long-tailed macaques (*Macaca fascicularis*). *Folia Primatol.* 64, 30–36.

MacFadden, B.J., Solounias, N., and Cerling, T.E., 1999. Ancient diets, ecology, and extinction of 5-million-year-old horses from Florida. *Science* 283, 824–827.

Mannerberg, F., 1960. Appearance of tooth surface as observed in shadowed replicas. *Odontol. Rev.* 11, Suppl. no. 6, 114 pp.

Merbs, C.F., 1968. Anterior tooth loss in Arctic populations. *Southwest. J. Anthropol.* 24, 20–32.

Merceron, G., Blondel, C., Brunet, M., Sen., S., Solounias, N., Viriot, L., and Heintz, E., 2004. The Late Miocene paleoenvironment of Afghanistan as inferred from dental microwear in artiodactyls. *Palaeogeogr. Palaeoclimatol. Palaeoecol.* 207, 143–163.

Merceron, G., Bonis, L. de, Viriot, L., and Blondel, C., 2005a. Dental microwear of fossil bovids from northern Greece: Paleoenvironmental conditions in the eastern Mediterranean during the Messinian. *Palaeogeogr. Palaeoclimatol. Palaeoecol.* 217, 173–185.

Merceron, G., Bonis, L. de, Viriot, L., and Blondel, C., 2005b. Dental microwear of the late Miocene bovids of northern Greece: Vallesian/Turolian environmental changes and disappearance of *Ouranopithecus macedoniensis*. *Bull. Soc. Geol. Fr.* 176, 491–500.

Merceron, G., and Ungar P.S., in press. Dental microwear and palaeoecology of bovids from the Early Pliocene of Langebaanweg, Western Cape province, South Africa. *S. Afr. J. Sci.*

Merceron, G., Viriot, L., and Blondel, C., 2004. Tooth microwear pattern in roe deer (*Capreolus capreolus* L.) from Chizé (western France) and relation to food composition. *Small Rumin. Res.* 53, 125–132.

Mills, J.R.E., 1955. Ideal dental occlusion in the primates. *Dent. Pract.* 6, 47–61.

Mills, J.R.E., 1963. Occlusion and malocclusion of the teeth of primates. In: Brothwell, D.R. (Ed.), *Dental Anthropology*. Pergamon Press, New York, pp. 29–52.

Molleson, T., and Jones, K., 1991. Dental evidence for dietary change at Abu Hureyra. *J. Archaeol. Sci.* 18, 525–539.

Molleson, T., Jones, K., and Jones, S., 1993. Dietary change and the effects of food preparation on microwear patterns in the Late Neolithic of Abu Hureyra, northern Syria. *J. Hum. Evol.* 24, 455–468.

Morel, A., Albuisson, E., and Woda, A., 1991. A study of human jaw movements deduced from scratches on occlusal wear facets. *Arch. Oral Biol.* 36, 195–202.

Nagel, U., 1973. A comparison of anubis baboons, hamadryas baboons and their hybrids at a species border in Ethiopia. *Folia Primatol.* 19, 104–165.

Noble, V.E., and Teaford, M.F., 1995. Dental microwear in Caucasian American *Homo sapiens*: Preliminary results. *Am J. Phys. Anthropol.* Suppl. 20, 162.

Nystrom, P., Phillips-Conroy, J.E., and Jolly, C.J., 2004. Dental microwear in anubis and hybrid baboons (*Papio hamadryas*, sensu lato) living in Awash National Park, Ethiopia. *Am J. Phys. Anthropol.* 125, 279–291.

Oliveira, E.V., 2001. Micro-desgaste dentário em alguns Dasypodidae (Mammalia, Xenarthra). *Acta Biol. Leopold.* 23, 83–91.

Organ, J.M., Teaford, M.F., and Larsen, C.S., 2005. Dietary inferences from dental occlusal microwear at Mission San Luis de Apalachee. *Am J. Phys. Anthropol.* 128, 801–811.

Pastor R.F., 1992. Dietary adaptations and dental microwear in Mesolithic and Chalcolithic South Asia. Special issue, *J. Hum. Evol.* 2, 215–228.

Pastor R.F., 1993. Dental microwear among inhabitants of the Indian subcontinent: A quantitative and comparative analysis. PhD diss., University of Oregon.

Pastor, R.F., and Johnston, T.L., 1992. Dental microwear and attrition. In: Kennedy, K.A.R. (Ed.), *Human Skeletal Remains from Mahadaha: A Gangetic Mesolithic Site.* Cornell University Press, Ithaca, NY, pp. 271–304.

Pérez-Pérez, A., Bermúdez de Castro, J.M., and Arsuaga, J.L., 1999. Nonocclusal dental microwear analysis of 300,000-year-old *Homo heidelbergensis* teeth from Sima de los Huesos (Sierra de Atapuerca, Spain). *Am J. Phys. Anthropol.* 108, 433–457.

Pérez-Pérez, A., Espurz, V., Bermúdez de Castro, J.M., de Lumley, M.A., and Turbón, D., 2003. Non-occlusal dental microwear variability in a sample of Middle and Late Pleistocene human populations from Europe and the Near East. *J. Hum. Evol.* 44, 497–513.

Pérez-Pérez, A., Lalueza, C., and Turbón, D., 1994. Intraindividual and intragroup variability of buccal tooth striation pattern. *Am J. Phys. Anthropol.* 94, 175–187.

Peters, C.R., 1982. Electron-optical microscopic study of incipient dental microdamage from experimental seed and bone crushing. *Am J. Phys. Anthropol.* 57, 283–301.

Phillips-Conroy J.E., 1978. Dental variability in Ethiopian baboons: An examination of the anubis-hamadryas hybrid zone in the Awash National Park, Ethiopia. PhD diss., New York University.

Phillips-Conroy, J.E., and Jolly, C.J., 1986. Changes in the structure of the baboon hybrid zone in the Awash National Park, Ethiopia. *Am J. Phys. Anthropol.* 71, 337–350.

Phillips-Conroy, J.E., Jolly, C.J., and Brett, F.L., 1991. Characteristics of hamadryas-like male baboons living in anubis baboon troops in the Awash hybrid zone, Ethiopia. *Am J. Phys. Anthropol.* 86, 353–368.

Phillips-Conroy, J.E., Bergman, T., and Jolly, C.J. 2000. Quantitative assessment of occlusal wear and age estimation in Ethiopian and Tanzanian baboons. In: Whitehead, P.F., and Jolly, C.J. (Eds.), *Old World Monkeys*. Cambridge University Press, Cambridge, pp. 321–340.

Puech, P.-F., 1977. Usure dentaire en anthropolgie étude par la technique des répliques. *Rev. d'Odonto-Stomatol.* 6, 51–56.

Puech, P.-F., 1984a. À la recherche du menu des premiers hommes. *Cahiers Lig. Préhist. Protohist.* 1, 46–53.

Puech, P.-F., 1984b. Acidic food choice in *Homo habilis* at Olduvai. *Curr. Anthropol.* 25, 349–350.

Puech, P.-F., 1986a. Dental microwear features as an indicator for plant food in early hominids: A preliminary study of enamel. *Hum. Evol.* 1, 507–515.

Puech, P.-F., 1986b. Tooth microwear in *Homo habilis* at Olduvai. *Mem. Mus. Hist. Nat., Paris* Ser. C 53, 399–414.

Puech, P.-F., and Albertini, H., 1984. Dental microwear and mechanisms in early hominids from Laetoli and Hadar. *Am J. Phys. Anthropol.* 65, 87–91.

Puech P.-F., Cianfarani, F., and Albertini, H., 1986. Dental microwear features as an indicator for plant food in early hominids: a preliminary study of enamel. *Hum. Evol.* 1, 507–515.

Puech, P.-F., and Prone A., 1979. Reproduction experimentale des processes d'usure dentaire par abrasion: Implications paleoecologique chex l'Homme fossile. *C. R. Acad. Sci. Paris* 289, 895–898.

Puech, P.-F., Prone A., and Albertini H., 1981. Reproduction expérimentale des processus d-altération de la surface dentaire par friction non abrasive et non adhésive: Application à l'étude de alimentation de L'Homme fossile. *C. R. Acad. Sci. Paris* 293, 729–734.

Puech, P.-F., Prone, A., and Kraatz, R., 1980. Microscopie de l'usure dentaire chez l'homme fossile: bol alimnetaire et environnement. *C. R. Acad. Sci. Paris* 290, 1413–1416.

Puech, P.-F., Prone, A., Roth, H., and Cianfarani, F., 1985. Reproduction expérimentale de processus d'usure des surfaces dentaires des Hominides fossiles: Conséquences morphoscopiques et exoscopiques avec application à l'Hominidé I de Garusi. *C. R. Acad. Sci. Paris* 301, 59–64.

Puech, P.-F., Serratrice, C., and Leek, F.F., 1983. Tooth wear as observed in ancient Egyptian skulls. *J. Hum. Evol.* 12, 617–629.

Purnell, M.A., 1995. Microwear in conodont elements and macrophagy in the first vertebrates. *Nature* 374, 798–800.

Radlanski R.J., and Jäger A., 1989. Zur mikromorphologie der approximalen Kontaktflächen und der okklusalen Schliffacetten menschlicher Zähne. *Dtsch. Zahnärztl. Z.* 44, 196–197.

Rafferty, K.L., Teaford, M.F., and Jungers, W.L., 2002. Molar microwear of subfossil lemurs: Improving the resolution of dietary inferences. *J. Hum. Evol.* 43, 645–657.

Raphael, K., Marbach, J., and Teaford, M.F., 2003. Is bruxism severity a predictor of oral splint efficacy in patients with myofascial face pain? *J. Oral Rehabil.* 30, 17–29.

Rensberger, J.M., 1978. Scanning electron microscopy of wear and occlusal events in some small herbivores. In: Butler, P.M., and Joysey, K.A. (Eds.), *Development, Function and Evolution of Teeth*. Academic Press, New York, pp. 415–438.

Rensberger, J.M., 1982. Patterns of change in two locally persistent successions of fossil Aplodontid rodents. In: Kurtén, B. (Ed.), *Teeth: Form, Function, and Evolution*. Columbia University Press, New York, pp. 323–349.

Rensberger, J.M., 1986. Early chewing mechanisms in mammalian herbivores. *Paleobiology* 12, 474–494.

Rose, J.C., 1984. Bioarchaeology of the Cedar Grove Site. In: Trubawitz, N (Ed.), *Cedar Grove*. Arkansas Archeological Research Series No. 23, Fayetteville, pp. 227–256.

Rose, J.C., and Marks, M.K., 1985. Bioarchaeology of the Alexander Site. In: Hemmings, E.T. and House J.H. (Eds.), *The Alexander Site, Conway County, Arkansas*. Arkansas Archeological Research Series no. 24, Fayetteville, pp. 76–98.

Rose, K.D., Walker, A., and Jacobs, L.L., 1981. Function of the mandibular tooth comb in living and extinct mammals. *Nature* 289, 583–585.

Ryan, A.S., 1979. Wear striation direction on primate teeth: A scanning electron microscope examination. *Am J. Phys. Anthropol.* 50, 155–168.

Ryan, A. S., 1981. Anterior dental microwear and its relationship to diet and feeding behavior in three African primates (*Pan troglodytes troglodytes*, *Gorilla gorilla gorilla*, and *Papio hamadryas*). *Primates* 22, 533–550.

Ryan, A.S., and Johanson, D.C., 1989. Anterior dental microwear in *Australopithecus afarensis*: Comparisons with human and nonhuman primates. *J. Hum. Evol.* 18, 235–268.

Schmidt, C.W., 2001. Dental microwear evidence for a dietary shift between two nonmaize-reliant prehistoric human populations from Indiana. *Am J. Phys. Anthropol.* 114, 139–145.

Schubert, B.W., and Ungar, P.S., 2005. Wear facets and enamel spalling in tyrannosaurid dinosaurs. *Acta Palaeontol. Polonica* 50, 93–99.

Scott, R.S., Ungar, P.S., Bergstrom, T.S., Brown, C.A., Grine, F.E., Teaford, M.F., and Walker, A., 2005. Dental microwear texture analysis shows within-species diet variability in fossil hominins. *Nature* 436, 693–695.

Semprebon, G.M., Godfrey, L.R., Solounias, N., Sutherland, M.R., and Jungers, W.L., 2004. Can low-magnification stereomicroscopy reveal diet? *J. Hum. Evol.* 47, 115–144.

Shkurkin, G.V., Almquist, A.J., Pfeihofer, A.A., and Stoddard, E.L., 1975. Scanning electron microscopy of dentition: Methodology and ultrastructural morphology of tooth wear. *J. Dent. Res.* 54, 402–406.

Silcox, M.T., and Teaford, M.F., 2002. The diet of worms: An analysis of mole dental microwear and its relevance to dietary inference in primates and other mammals. *J. Mammal.* 83, 804–814.

Solounias, N., and Hayek, L.C., 1993. New methods of tooth microwear analysis and application to dietary determination of two extinct antelopes. *J. Zool. Lond.* 229, 421–445.

Solounias, N., and Moelleken, S.M.C., 1992. Tooth microwear analyses of *Eotragus sansaniensis* (Mammalia: Ruminantia), one of the oldest known bovids. *J. Vert. Paleontol.* 12, 113–121.

Solounias, N., and Moelleken, S.M.C., 1994. Differences in diet between two archaic ruminant species from Sansan, France. *Hist. Biol.* 7, 203–220.

Solounias, N., and Semprebon, G., 2002. Advances in the reconstruction of ungulate ecomorphology with application to early fossil equids. *Am. Mus. Novit.* 3366, 1–49.

Strait, S.G., 1993. Molar microwear in extant small-bodied faunivorous mammals: An analysis of feature density and pit frequency. *Am J. Phys. Anthropol.* 92, 63–79.

Sugawara, K., 1979. Sociological study of a wild group of hybrid baboons between *Papio anubis* and *Papio hamadryas* in the Awash Valley, Ethiopia. *Primates* 20, 21–56.

Szmulewicz, M.N., Andino, L.M., Reategui, E.P., Woolley-Barker, T. Jolly, C.J., Disotell, T.R., and Herrera, R.J., 1999. An Alu insertion polymorphism in a baboon hybrid zone. *Am J. Phys. Anthropol.* 109, 1–8.

Teaford, M.F., 1988a. A review of dental microwear and diet in modern mammals. *Scanning Microsc.* 2, 1149–1166.

Teaford, M.F., 1988b. Scanning electron microscope diagnosis of wear patterns versus artifacts on fossil teeth. *Scanning Microsc.* 2, 1167–1175.

Teaford, M.F., 1991. Dental microwear what can it tell us about diet and dental function? In: Kelley, M.A., and Larsen, C.S. (Eds.), *Advances in Dental Anthropology*. Alan R. Liss, New York, pp. 341–356.

Teaford, M.F., 1993. Dental microwear and diet in extant and extinct *Theropithecus*: Preliminary analyses. In: Jablonski, N.G. (Ed.), *Theropithecus: The Life and Death of a Primate Genus*. Cambridge University Press, Cambridge England, pp. 331–349.

Teaford, M.F., 1994. Dental microwear and dental function. *Evol. Anthropol.* 3, 17–30.

Teaford, M.F., 2002. Dental enamel microwear analysis. In: Hutchinson, D.L. (Ed.), *Foraging, Farming and Coastal Biocultural Adaptation in Late Prehistoric North Carolina*. University Press of Florida, Gainesville, pp. 169–177.

Teaford, M.F., and Byrd, K.E., 1989. Differences in tooth wear as an indicator of changes in jaw movement in the guinea pig (*Cavia porcellus*). *Arch. Oral Biol.* 34, 929–936.

Teaford, M.F., and Glander, K.E., 1991. Dental microwear in live, wild-trapped *Alouatta palliata* from Costa Rica. *Am J. Phys. Anthropol.* 85, 313–319.

Teaford, M.F., and Glander, K.E., 1996. Dental microwear and diet in a wild population of mantled howlers (*Alouatta palliata*). In: Norconk, M., Rosenberger, A., and Garber, P. (Eds.), *Adaptive Radiations of Neotropical Primates*. Plenum Press, New York, pp. 433–449.

Teaford, M.F., Larsen, C.S., Pastor, R.F., and Noble, V.E., 2001. Pits and scratches: Microscopic evidence of tooth use and masticatory behavior in La Florida. In: Larsen C.S. (Ed.), *Bioarchaeology of Spanish Florida: The Impact of Colonialism*. University Press of Florida, Gainesville, pp. 82–112.

Teaford, M.F., Lucas, P.W., Ungar, P.S., and Glander, K.E., 2006. Mechanical defenses in leaves eaten by Costa Rican *Alouatta palliata*. *Am J. Phys. Anthropol.* 129, 99–104.

Teaford, M.F., and Lytle, J.D., 1996. Diet-induced changes in rates of human tooth microwear: a case study involving stone-ground maize. *Am J. Phys. Anthropol.* 100, 143–147.

Teaford, M.F., and Oyen, O.J., 1989a. Differences in the rate of molar wear between monkeys raised on different diets. *J. Dent. Res.* 68, 1513–1518.

Teaford, M.F., and Oyen, O.J., 1989b. *In vivo* and *in vitro* turnover in dental microwear. *Am J. Phys. Anthropol.* 80, 447–460.

Teaford, M.F., and Robinson, J.G., 1989. Seasonal or ecological differences in diet and molar microwear in *Cebus nigrivittatus*. *Am. J. Phys. Anthropol.* 80, 391–401.

Teaford, M.F., and Runestad, J.A., 1992. Dental microwear and diet in Venezuelan primates. *Am J. Phys. Anthropol.* 88, 347–364.

Teaford, M.F., and Tylenda, C.A., 1991. A new approach to the study of tooth wear. *J. Dent. Res.* 70, 204–207.

Teaford, M.F., and Ungar, P.S., 2000. Diet and the evolution of the earliest human ancestors. *Proc. Natl. Acad. Sci.* 97, 13506–13511.

Teaford, M.F., Ungar, P.S., and Grine, F.E., 2002a. Fossil evidence for the evolution of human diet. In: Ungar, P.S., and Teaford, M.F. (Eds.), *Human Diet: Its Origins and Evolution.* London, Bergen & Garvey, Westport, CT, pp. 143–166.

Teaford, M.F., Ungar, P.S., and Grine, F.E., 2002b. Molar microwear and diet of *Praeanthropus afarensis*: Preliminary results from the Denan Dora Member, Hadar Formation, Ethiopia. *Am J. Phys. Anthropol.* Suppl. 34, 154.

Teaford, M.F., and Walker, A., 1983. Dental microwear in adult and still-born guinea pigs (*Cavia porcellus*). *Arch. Oral Biol.* 28, 1077–1081.

Teaford, M.F., and Walker, A., 1984. Quantitative differences in dental microwear between primate species with different diets and a comment on the presumed diet of *Sivapithecus*. *Am J. Phys. Anthropol.* 64, 191–200.

Ungar, P.S., 1990. Incisor microwear and feeding behavior in *Alouatta seniculus* and *Cebus olivaceus*. *Am. J. Primatol.* 20, 43–50.

Ungar P.S., 1994. Incisor microwear of Sumatran Anthropoid primates. *Am J. Phys. Anthropol.* 94, 339–363.

Ungar, P.S., 1996a. Dental microwear of European Miocene catarrhines: Evidence for diets and tooth use. *J. Hum. Evol.* 31, 335–366.

Ungar, P.S., 1996b. Relationship of incisor size to diet and anterior tooth use in sympatric Sumatran anthropoids. *Am. J. Primatol.* 38, 145–156.

Ungar, P.S., 2004. Dental topography and diets of *Australopithecus afarensis* and early *Homo*. *J. Hum. Evol.* 46, 605–622.

Ungar, P.S., Brown, C.A., Bergstrom, T.S., and Walker, A., 2003. Quantification of dental microwear by tandem scanning confocal microscopy and scale-sensitive fractal analyses. *Scanning* 25, 185–193.

Ungar, P.S., and Grine, F.E., 1991. Incisor size and wear in *Australopithecus africanus* and *Paranthropus robustus*. *J. Hum. Evol.* 20, 313–340.

Ungar, P.S., Grine, F.E., Teaford, M.F., and El Zaatari, S., 2006. Dental microwear and diet in African early *Homo*. *J. Hum. Evol.* 50, 78–95.

Ungar, P.S., Simon, J.-C., and Cooper, J.W., 1990. A semiautomated image analysis procedure for the quantification of dental microwear. *Scanning* 13, 31–36.

Ungar, P.S., and Spencer, M.A., 1999. Incisor microwear, diet, and tooth use in three Amerindian populations. *Am J. Phys. Anthropol.* 109, 387–396.

Ungar, P.S., and Teaford, M.F., 1996. Preliminary examination of non-occlusal dental microwear in anthropoids: Implications for the study of fossil primates. *Am J. Phys. Anthropol.* 100, 101–113.

Ungar, P.S., Teaford, M.F., Glander, K.E., and Pastor, R.F., 1995. Dust accumulation in the canopy: A potential cause of dental microwear in primates. *Am J. Phys. Anthropol.* 97, 93–99.

Ungar, P.S., Teaford, M.F., and Grine, F.E., 2001. A preliminary study of molar microwear of early *Homo* from East and South Africa. *Am J. Phys. Anthropol.* Suppl. 32, 153.

Ungar, P.S., Teaford, M.F., and Kay, R.F., 2004. Molar microwear and shearing crest development in Miocene catarrhines. *Anthropologie* 42, 21–35.

Van Valkenburgh, B., Teaford, M.F., and Walker A., 1990. Molar microwear and diet in large carnivores: Inferences concerning diet in the sabretooth cat, *Smilodon fatalis*. *J. Zool. Lond.* 222, 319–340.

Walker, A., 1981. Diet and teeth, dietary hypotheses and human evolution. *Philos. Trans. R. Soc. Lond.* B 292, 57–64.

Walker, A., 1984. Mechanisms of honing in the male baboon canine. *Am J. Phys. Anthropol.* 65, 47–60.

Walker, A., Hoeck, H.N., and Perez, L., 1978. Microwear of mammalian teeth as an indicator of diet. *Science* 201, 908–910.

Walker, P.L., 1976. Wear striations on the incisors of cercopithecoid monkeys as an index of diet and habitat preference. *Am J. Phys. Anthropol.* 45, 299–308.

Young, W.G., Brennan, C.K.P., and Marshall, R.I., 1990. Occlusal movements of the brushtail possum, *Trichosurus vulpecula*, from microwear on the teeth. *Aust. J. Zool.* 38, 41–51.

Young, W.G., and Robson, S.K., 1987. Jaw movement from microwear on the molar teeth of the koala *Phascolarctos cinereus*. *J. Zool. Lond.* 213, 51–61.

8

Icarus, Isotopes, and Australopith Diets

MATT SPONHEIMER
JULIA LEE-THORP
DARRYL DE RUITER

Early hominin diets can be studied in many ways, such as dental allometry/morphology, dental microwear, and archaeology (e.g., Robinson, 1954; Isaac, 1971; Grine, 1981). Although these and other techniques are commonly used to reconstruct "diet," we rarely question the degree to which they are actually capable of doing so. For instance, in the strictest sense, dental morphology cannot tell us about the totality of an organism's diet, but it can tell us about the foods that proved challenging to that organism's ancestors, whether they were consumed daily, seasonally, or only during short periods of extreme privation. Thus, while there certainly is an important relationship between dental morphology and diet, it is far less direct than researchers often assume. Dental microwear, in turn, does reveal something of what an individual ingested; yet, it probably tells us mostly about the mechanical properties of these foods and the exogenous grit with which they are associated. A main point of this disquisition is not to suggest that these tools and others are not useful because they patently are useful and important, but rather that we often talk about reconstructing "diet," when we are in fact illuminating only facets of the complicated interactions between an organism and its environment that broadly constitute diet.

In this chapter, we will discuss a recently developed technique called *stable carbon isotope analysis* that has allowed us to illumine another facet of our ancestors' diets. In the past decade, stable isotope analysis has revealed much about the diets of a variety of hominins, including anatomically modern humans (Richards et al., 2001), Neanderthals (Fizet et al., 1995; Richards et al., 2000; Bocherens, Billiou, and Mariotti, 2001), and australopiths (Lee-Thorp, van der Merwe, and Brain, 1994; Sponheimer and Lee-Thorp, 1999c; van der Merwe et al., 2003; Sponheimer et al., 2005). However, these advances have also served to highlight just how much remains unknown about early hominin diets in general, as well as the inherent limitations of stable isotope analysis. Like all paleodietary techniques, it does some things very well and other things not so well. We hope to spell these things out in the following pages. Due to space constraints, our particular focus will be on stable carbon isotopes and australopith diets, and just what they allow us to say without flying too high.

The Known

Methodological Background

The stable carbon isotope studies of australopiths are founded on our knowledge of photosynthesis in plants, in particular the fixation of atmospheric CO_2 (see Ehleringer and Monson, 1993, for a review of photosynthetic pathways). In tropical savannas, trees, bushes, shrubs, and forbs use the C_3 photosynthetic pathway, while tropical grasses and some sedges use the C_4 pathway. Another pathway known as crassulacean acid metabolism (CAM) is found primarily in succulent plants such as cacti and euphorbias. However, because CAM plants make up a small part of most ecosystems, we will ignore them at present, although we will return to them a bit later. Because of both anatomical and biochemical differences, C_3 plants discriminate more heavily against ^{13}C during photosynthesis than do C_4 plants. As a result, C_3 plants have highly depleted $^{13}C/^{12}C$ ratios, which are expressed as δ values in parts per thousand (‰) relative to the Pee Dee Belennite (PDB) standard (a Cretaceous cephalopod, *Belemnitella americana*, which is highly enriched in ^{13}C, resulting in most terrestrial plants and animals having negative carbon isotope ratios). C_4 plants, on the other hand, are much less depleted in ^{13}C (Smith and Epstein, 1971; Vogel, Fuls, and Ellis, 1978). This contrast can be seen in figure 8.1, which shows the $\delta^{13}C$ values of C_3 (trees and forbs) and C_4 plants (grasses) in Kruger National

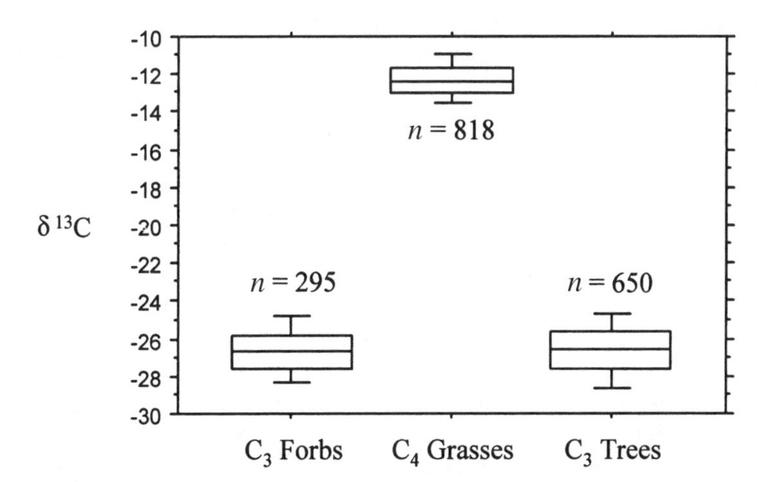

Figure 8.1 $\delta^{13}C$ of trees, forbs, and grasses in Kruger National Park, South Africa. The boxes represents the twenty-fifth to seventy-fifth percentiles (with the medians as horizontal lines) and the whiskers show the tenth to ninetieth percentiles. Note that the C_3 plants are highly distinct from the C_4 plants, with no overlap in their carbon isotope compositions.

Park, South Africa. Although there is some variability within the C_3 and C_4 groups, it tends to be fairly small within any given savanna ecosystem (Codron et al., 2005); however, $\delta^{13}C$ of C_3 vegetation, in particular, can change significantly between sere, open environments (relatively enriched $\delta^{13}C$), and damp, closed canopy forests (relatively depleted $\delta^{13}C$; Vogel, 1978b; Ehleringer and Cooper, 1988; van der Merwe and Medina, 1991). Thus, plant (and animal) stable isotope values can serve as environmental indicators.

We also know that the stable carbon isotopes in plants are passed down into the tissues of animals that eat them. Thus, the tissues of animals that eat C_3 plants, such as giraffes, have very different $\delta^{13}C$ than those that eat C_4 grasses, such as zebra (e.g., Ambrose and DeNiro, 1986; Lee-Thorp, Sealy, and van der Merwe, 1989). However, the relationship between dietary and tissue stable isotope compositions varies between tissues. For instance, muscle and hair tend to be enriched by about 1–3‰ compared with diet, while fat is depleted by a few parts per thousand compared with diet (Vogel, 1978a; Tieszen and Fagre, 1993; Sponheimer, Lee-Thorp, et al., 2003). Yet, from a paleontological perspective, we are only concerned with the hard tissues that constitute the vast majority of the fossil record. The carbon isotopes in bone and enamel mineral readily provide dietary information in modern animals, but only enamel mineral has proved reliable with fossils (e.g., Lee-Thorp and van der Merwe, 1987; Cerling, Harris, and Leakey, 1999; Sponheimer, 1999). Bone is highly organic and poorly crystalline, which leaves it very susceptible to diagenetic processes, while enamel has very little organic content and is highly crystalline, making it essentially "prefossilized" and much less susceptible to postmortem alteration (see LeGeros, 1991). For this reason, enamel is preferentially used for paleontological paleodietary studies and is the only material we will discuss. Studies of large mammals in the field and in semicontrolled settings have shown that enamel mineral is enriched by about 13‰ compared with dietary $\delta^{13}C$ (Lee-Thorp, Sealy, and van der Merwe, 1989; Cerling and Harris, 1999; Balasse, 2002). Thus, a browser eating a typical C_3 diet with a $\delta^{13}C$ value of −27‰, will have tooth enamel $\delta^{13}C$ of about −14‰. We will discuss this relationship between dietary and enamel $\delta^{13}C$ in the "Unknown" section below.

Australopith Diets

Several studies of australopith diet using stable carbon isotopes (Lee-Thorp, van der Merwe, and Brain, 1994; Sponheimer and Lee-Thorp, 1999c; van der Merwe et al., 2003; Sponheimer et al., 2005) have been conducted. These studies were explicitly designed to test hypotheses about australopith diets that had been derived using other techniques; for instance, *Australopithecus* was a consumer of fleshy fruits and leaves (Grine, 1986; Grine and Kay, 1988). Thirty-seven australopith specimens from Swartkrans, Kromdraai, Sterkfontein, and Makapansgat have now been analyzed, the data for which are summarized in figure 8.2. Analysis of variance and post hoc tests of the combined datasets demonstrate that the fossil C_3 consumers (e.g., giraffes) and fossil C_4 consumers (e.g., zebra) have very different $\delta^{13}C$ (Scheffé, $P < 0.0001$), just as is the case today. This shows that diagenesis does not obscure the dietary carbon isotope signal. The data also show that both *Australopithecus africanus* and *Paranthropus robustus* are highly distinct from the C_3 browsers and

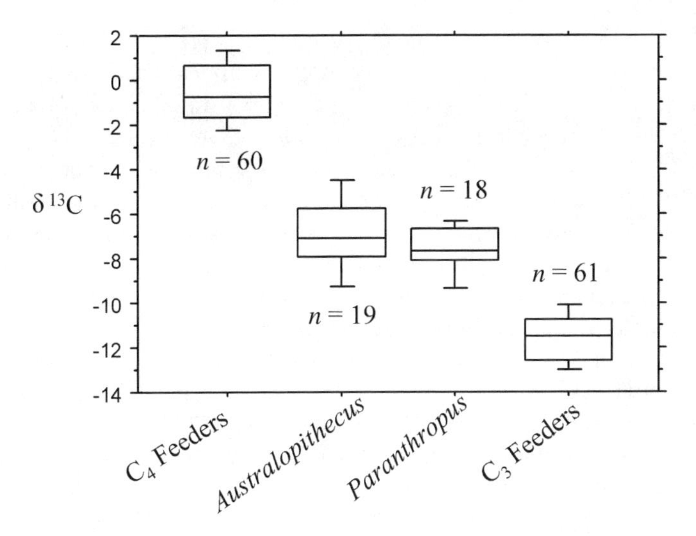

Figure 8.2 $\delta^{13}C$ for *Australopithecus africanus* and *Paranthropus robustus* specimens, as well as C_3 plant consumers (browsing/frugivorous bovids and giraffids) and C_4-plant consumers (grazing bovids and equids). The boxes represents the twenty-fifth to seventy-fifth percentiles (with the medians as horizontal lines) and the whiskers show the tenth to ninetieth percentiles. Given the size of this data set, there can be no doubt that australopith $\delta^{13}C$ is highly distinct from that of associated browser/frugivores.

C_4 grazers with which they are associated (Scheffé, $P < 0.0001$), although they are not different from each other. This indicates that both australopiths consumed foods enriched in ^{13}C, such as C_4 grasses, C_4 sedges, or animals that ate these foods, which is quite surprising because prior studies suggested that *Australopithecus* ate fleshy fruits and leaves and that *Paranthropus* consumed small, harder foods such as nuts (e.g., Grine, 1986; Grine and Kay, 1988). The isotope results are certainly consistent with the consumption of these foods, but they also suggest that australopith diets were rather more complex and probably included significant quantities of foods that few considered to be serious possibilities (but see Jolly, 1970; Dunbar, 1983). Still, it is important to acknowledge that the australopith $\delta^{13}C$ is highly variable, far more so than most modern and fossil taxa in South Africa (e.g., Lee-Thorp, van der Merwe, and Brain, 1994; Sponheimer, Robinson, et al., 2003; van der Merwe et al., 2003). Two or three of the thirty-seven australopiths analyzed had nearly pure C_3 diets, while several were extremely enriched in ^{13}C, with $\delta^{13}C$ values within the range of the grass-eating baboon *Theropithecus oswaldi* (Lee-Thorp, van der Merwe, and Brain, 1989; Sponheimer et al., 2005).

We also know that modern chimpanzees (*Pan troglodytes*) do not typically eat detectable quantities of C_4 foods. This is hardly surprising for chimps in heavily forested environments where few if any C_4 resources are available, but it also holds true for those in semiarid woodland environments where C_4 foods are extremely abundant (Schoeninger, Moore, and Sept, 1999; Carter, 2001). This is not to say that we will never find a chimpanzee with a significant percentage of C_4 foods in its diet, but only that while this is the norm for australopiths, it would be the exception for chimpanzees. Moreover, because many australopiths probably inhabited environments similar to those of today's woodland chimpanzees (McGrew, Baldwin, and Tutin, 1981; Reed, 1997), this strongly suggests that different dietary adaptations, rather than environmental differences, are responsible for the disparity in their consumption of ^{13}C-enriched foods. In other words, it appears likely that, even in identical woodland savanna habitats, chimpanzees would continue to consume their favored forest foods to the degree that was possible, while australopiths would extensively supplement these with C_4 resources. Although this is a reasonable hypothesis, given the available data, it represents a step away from what we know into the rather murkier territory of the unknown.

Nevertheless, australopiths are not the only primates for which we have direct evidence of significant C_4 consumption. A recent study of baboon (*Papio ursinus*) feces showed that populations in marginal environments may eat up to 50% ^{13}C-enriched foods such as grasses and CAM plants, while those in more hospitable woodland environments eat only about 10% C_4 vegetation (Codron, 2003). Furthermore, some of the *Papio* and *Parapapio* specimens from Swartkrans and Sterkfontein show significant C_4 consumption. *Theropithecus*, of course, has been shown to be heavily dependent on C_4 vegetation (Lee-Thorp, van der Merwe, and Brain, 1989), as are some specimens of the colobine monkey *Cercopithecoides williamsi* (Luyt, 2001; Codron, 2003). Thus, despite the great difference between australopiths and contemporaneous ungulates, they do not stand alone from a primatological perspective—only apart from extant apes. This is a crucial point, as it is commonly postulated that a wide suite of behavioral and dietary adaptations were shared by chimpanzees and australopiths (e.g., Wrangham, 1987; Stanford, 1999). While this "chimpanzee paradigm" is likely true to some extent, given the remarkable genetic similarity between *Pan* and extant hominins (e.g., Goodman et al., 1998), the carbon isotope data suggest that other primates, such as the papionins, which often consume significant quantities of C_4 foods and, like australopiths, are highly isotopically heterogeneous (Codron, 2003), might make better dietary analogs. This is not to say that papionin diets are, or were, like those of australopiths but only that both taxa exhibit a dietary flexibility and willingness to use C_4 resources not paralleled in chimpanzees. After all, australopith craniodental and locomotor adaptations as well as physical environments and biotic communities were, in aggregate, highly distinct from those of all living taxa (e.g., Grine, 1981; Stern and Susman, 1983; Reed, 1997). Thus, australopith diets may have been quite dissimilar in character from those of all living primates, although there is reason to believe there was considerable dietary overlap with extant hominoids (e.g. Grine, 1986; Grine and Kay, 1988; Ungar, 2004).

The Unknown

Methodological Background

There is still much that we do not know about stable isotope distributions in many modern ecosystems, and this is especially the case for modern primates, on which there have only been a handful of published stable isotope studies (Schoeninger, Iwaniec, and Glander, 1997; Schoeninger, Iwaniec, and Nash, 1998; Schoeninger, Moore, and Sept, 1999; Cerling, Hart, and Hart, 2004). In addition, we cannot meaningfully distinguish between diets as different as frugivory, folivory, and carnivory in forest ecosystems, because all are ultimately based on on C_3 plants. Some data, however, suggest we might be able to distinguish between these diets, at least in some cases. Fruits, for instance, tend to be slightly enriched in ^{13}C (~1.5‰) compared with leaves, which might allow us to distinguish between frugivory and folivory (Chodron et al., 2005). Indeed, Carter (2001) found that frugivorous C_3 consumers are slightly enriched in ^{13}C compared with C_3 folivores; yet, Cerling, Hart, and Hart (2004) found no difference in the $\delta^{13}C$ of folivores and frugivores. Both of these studies were quite small, however, and more data are clearly needed. Similarly, animals that feed high in the canopy would be expected to be enriched compared with those that feed nearer to the ground, as upper canopy foods are relatively enriched in ^{13}C, but there are far too few data to confirm this (see Schoeninger, Iwaniec, and Glander, 1997; Schoeninger, Iwaniec, and Nash, 1998).

We also need to know more about the relationship between dietary and enamel $\delta^{13}C$. As mentioned earlier, large mammal enamel apatite is enriched in ^{13}C by about 13‰ compared to diet. Controlled-feeding studies of rodents, however, show their bone mineral to be enriched by only 10‰ (Ambrose and Norr, 1993; Tieszen and Fagre, 1993). Strangely, we can only speculate as to why this is the case. It may be because the rodent studies used bone mineral rather than enamel or because of some difference in rodent digestive physiology, but the truth is that we simply do not know. Although this is not a problem for the hominin isotope studies (as hominins are most decidedly not rodents), it does underscore that there is much yet to be learned. Moreover, there is even minor disagreement about the proper diet-enamel fractionation for larger mammals. Lee-Thorp, Sealy, and van der Merwe (1989) conducted a field study that suggested that enamel was enriched by about 12‰ or possibly a bit more. And while a recent controlled-feeding experiment with pigs also argued for an enrichment of 12.5‰ (Young, 2002), two semicontrolled studies suggested an enrichment between 13.4‰ and 14.1‰ (Cerling and Harris, 1999; Balasse, 2002). Although the difference between 12‰ and 14‰ is fairly small, it is potentially important, especially if it turns out there are systematic differences between taxa with disparate digestive physiologies. One possibility, for instance, is that ruminants that produce and expel great amounts of highly ^{13}C-depleted methane would end up with blood bicarbonate and enamel slightly more enriched in ^{13}C than animals that produce less methane, such as primates (Cerling and Harris, 1999; Hedges and van Klinken, 2000). If this is the case, then it might mean that most modern primates and early hominins have a diet-enamel fractionation of about 12‰, but that the ruminants have a fractionation of 14‰. This would mean that we cannot directly compare the $\delta^{13}C$ of ruminants and hominins, but would rather have to adjust the values of the hominins by about 2‰ to make them comparable. This would then mean that we have underestimated the amount of ^{13}C-enriched foods that early hominins consumed, and that 50% would be closer to correct than the 35% to 40% we currently accept (Sponheimer et al., 2005). Indeed, there is some reason to believe that this is the case, as a recent field study suggested that fractionation in primates is slightly more than 1‰ less than that for ungulates (Cerling, Hart, and Hart, 2004). Clearly, more research in this area is required. However, there is little doubt that much of the unknown in this area can be revealed with appropriate field and experimental studies.

Since we are on the topic of the unknown, it might be worth addressing a few tangential, yet methodologically related, issues: what might we derive from oxygen and nitrogen isotopes in australopiths? We routinely produce oxygen isotope ($\delta^{18}O$) data during analysis for carbon isotopes, as the isotope abundances are derived from CO_2 gas. And while students of human evolution are no doubt familiar with the use of oxygen isotopes for paleoclimatic reconstruction (especially in foraminifera, e.g., Prentice and Denton, 1988), few realize that mammalian oxygen isotope compositions may also tell us something about an animal's diet and thermophysiological adaptations. For example, mammal's that get most of their water through food (like many browsing bovids) tend to be enriched in ^{18}O because evapotranspiration leaves leaf water relatively enriched in ^{18}O compared with meteoric water, which is the primary water source for regular drinkers (Kohn, Schoeninger, and Valley, 1996; Sponheimer and Lee-Thorp, 1999b). In contrast, animals that eat roots and meat tend to be depleted in ^{18}O (Sponheimer and Lee-Thorp, 1999b; Sponheimer and Lee-Thorp, 2001). Furthermore, there is some theoretical and empirical basis for believing that folivore $\delta^{18}O$ is higher than that of frugivores, once again due to the evaporative enrichment that occurs in leaves (Carter, 2001; Sponheimer and Lee-Thorp, 2001; Cerling, Hart, and Hart, 2004). Alas, comprehensive intertaxonomic studies of mammalian oxygen isotope compositions within modern ecosystems have not been carried out (but see Kohn et al., 1996; Sponheimer and Lee-Thorp, 2001; Smith, Sharp, and Brown, 2002 for limited attempts); most data have been collected so as to eschew this ecologically induced variability because it makes paleoclimatic reconstruction difficult, if not impossible. Thus, even though we have oxygen isotope data for australopiths and other fossil taxa, we are not yet in a position to interpret them meaningfully. It is worth noting that *Australopithecus* from Sterkfontein is significantly depleted in ^{18}O compared with *Paranthropus* from Swartkrans (Sponheimer et al., 2005; fig. 8.3). This might indicate that the prevailing

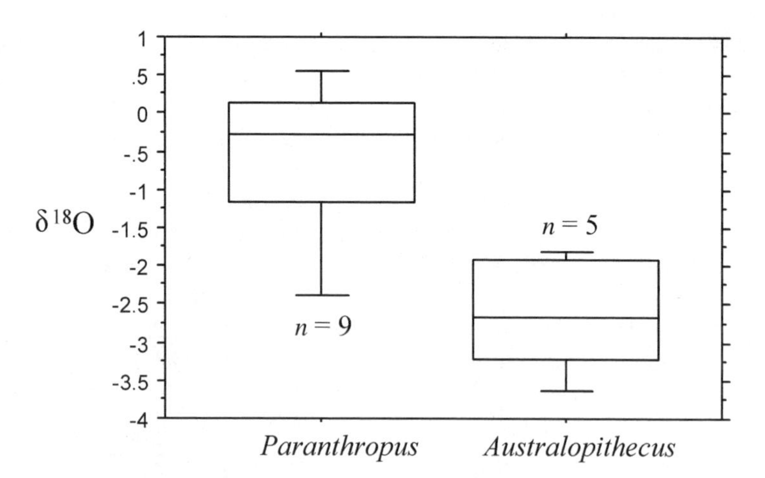

Figure 8.3 $\delta^{18}O$ of 2.5 Ma *Australopithecus africanus* and 1.8 Ma *Paranthropus robustus*. The boxes represent the twenty-fifth to seventy-fifth percentiles (with the medians as horizontal lines) and the whiskers show the tenth to ninetieth percentiles. The relatively depleted $\delta^{18}O$ of *Australopithecus* at 2.5 Ma compared to *Paranthropus* at 1.8 Ma may indicate either increasing aridity in the Sterkfontein Valley over time ecological differences in these taxa.

environmental conditions for *Paranthropus* were more arid than those for *Australopithecus* or it might also reveal a fundamental ecological or thermophysiological difference. At present, we cannot meaningfully distinguish between these possibilities.

Now, let us briefly turn to nitrogen isotopes. Nitrogen isotope compositions ($\delta^{15}N$) are known to increase by about 3‰ up every step in the food chain and can therefore be used as trophic level indicators (e.g., Minigawa and Wada, 1984; Schoeninger and DeNiro, 1984; Ambrose and DeNiro, 1986). Indeed, nitrogen isotope analysis of Neanderthals has been used to good effect, showing that they are enriched in ^{15}N, which is most consistent with a high degree of carnivory (Fizet et al., 1995; Richards et al., 2000; Bocherens, Billiou, and Mariotti, 2001). We have been unable to analyze nitrogen isotopes in australopiths, however, because they reside in organic material (e.g., collagen) that is rarely preserved for tens of thousands, much less millions of years. There is, nevertheless, tantalizing evidence that autochthonous organic material (osteocalcin and enamelins) can survive relatively intact for many millions of years (Glimcher, 1990; Collins et al., 2000). Thus, it is possible that recent advances in mass spectrometry that allow analysis of extremely small samples and individual compounds may make it possible to obtain biological nitrogen isotope data from australopiths in the next few years—but then again, maybe not.

Australopith Diets

We now return to stable isotopes and australopith diets. We possess overwhelming evidence that australopiths consumed foods that are enriched in ^{13}C such as C_4 grasses, C_4 sedges, or animals that ate these plants, but the stable isotope studies have created a new unknown, namely, which of these foods were actually utilized? Answering this question is important, as the use of these different resources might have a variety of physiological, social, and behavioral implications. For instance, if australopiths had a grass-based diet similar to the modern gelada baboon (*Theropithecus gelada*), this would almost certainly indicate that their diets were less nutrient-dense than those of modern apes, possibly placing important limitations on burgeoning hominin brains and sociality (Aiello and Wheeler, 1995; Milton, 1999). The converse, that australopiths ate diets rich in animal foods, would indicate a leap in dietary quality over modern apes, which could have been a crucial step toward hominin encephalization, the development of stone tool industries, and increased social complexity (Milton, 1999). Similarly, it has been suggested that consuming the underground storage organs of plants like C_4 sedges could represent an increase in dietary quality over that of extant great apes because, when they are edible, they are lower in dietary fiber than ape fallback foods (Conklin-Brittain, Wrangham, and Smith, 2002).

Yet, we are still uncertain as to which ^{13}C-enriched foods were consumed by australopiths. Part of the difficulty is that we do not know enough about the potential variety and availability of ^{13}C-enriched foods, although we have been working to rectify this gap (see Sponheimer et al., 2005). Recent work has shown, for instance, that in some savanna woodland environments, sedges are a less likely C_4 resource than was previously supposed. While it is certainly true that the underground parts of *some* sedges (e.g., *Cyperus papyrus*, *Cyperus esculentus*) would make an attractive food for tool-wielding hominins, at least seasonally, it was found that there are few edible C_4 sedges in South African woodland environments (Peters and Vogel, 2005; Sponheimer et al., 2005), although they are abundant in perennial wetlands like the Okavango Delta (Ellery et al., 1995). Indeed, only 28% of the sedges in Kruger National Park use C_4 photosynthesis (Sponheimer et al., 2005). Thus, while sedges could certainly have contributed, it is unlikely that their consumption alone could account for the observed ^{13}C-enrichment of australopiths. However, other early hominin habitats, such as the wetlands of the Eastern Lacustrine Plain at Olduvai Gorge (Hay, 1976; Deocampo et al., 2002), might have been better sources of edible C_4 sedges. The extremely enriched $\delta^{13}C$ of *Paranthropus boisei* might indicate that C_4 sedges were locally abundant and heavily used (van der Merwe, Cushing, and Blumenschine, 1999). Puech and colleagues (1986) have also suggested that the dental microwear of early East African hominins is consistent with the consumption of such foods.

Animals are another potential ^{13}C-enriched food. Of course, animal foods can be many different things, including large and small vertebrates, invertebrates, birds' eggs, and even honey. These foods can also be acquired in a variety of ways, including active hunting of large game, passive scavenging, and gathering of insects and eggs. Although chimpanzees are known to hunt a variety of small vertebrates such as red colubus monkeys (*Piliocolobus badius*) and blue duiker (*Cephalophus monticola*), these are pure C_3 consumers (Teleki, 1981; Goodall, 1986). Therefore, intake of such foods could not contribute to the C_4 component of australopith diets. More likely sources of the reported C_4 signal include small grass-eating taxa, such as hyraxes (*Procavia* spp.), cane rats (*Thryonomys swinderianus*), young antelope (Bovidae), and arthropods. Baboons are known to eat grass-eating grasshoppers (Acrididae) almost exclusively during temporary gluts (Hamilton, 1987). Grass-eating termites represent another intriguing possibility, particularly given recent studies suggesting that bone tools from Swartkrans were used to extract termites from mounds (Backwell and D'Errico, 2001). Stable isotope studies of termites in African savanna environments have shown that they could have contributed to the australopiths' ^{13}C enrichment (Boutton, Arshad, and Tieszen, 1983; Sponheimer et al., 2005). While termites range from nearly pure C_3 to pure C_4 consumers, the vast majority of savanna termites, even in densely wooded riverine microhabitats, consume significant proportions of C_4 foods. In fact, termites throughout Kruger National Park eat 35% C_4 vegetation on average (Sponheimer et al., 2005). Thus, termite consumption by australopiths in woodland savanna and even in riverine forest would be expected to impart some C_4 carbon to consumers. Nevertheless, the fact that so few termites have a pure C_4 signal makes it unlikely that termite consumption alone was the source of the strong C_4 signal of australopiths because it would require a diet of nearly 100% termites. Alternatively, if the hominins selectively preyed on grass-specialist harvester termites (*Trinervitermes*, *Hodotermes*) with virtually pure C_4 diets, a diet of about 35% to 40% termites would be sufficient to produce the observed hominin carbon isotope ratios. This scenario, however, is highly unlikely because these C_4 termites are much less common than those with mixed C_3/C_4 diets in woodlands today; and while harvester termites are more abundant in open grasslands and during acute droughts (Braack

and Kryger, 2003), there is no reason to believe that australopiths frequented such open environments or that drought conditions were so preponderant. Moreover, while *Trinervitermes* builds highly visible aboveground nests (mounds), *Hodotermes* does not, making it much less conspicuous on the landscape (Carruthers, 1997; Stuart and Stuart, 2000). Thus, it is possible and even likely that termites contributed in some way to the unusual $\delta^{13}C$ values of australopiths, but other C_4 resources were almost certainly consumed in considerable quantities.

As far as the consumption of mammals goes, we know that many with C_4 signatures were available for consumption by australopiths (Lee-Thorp, van der Merwe, and Brain, 1994; Sponheimer and Lee-Thorp, 1999c). However, most evidence also suggests that C_3 mammals were very common, especially at Makapansgat and Sterkfontein (Luyt and Lee-Thorp, 2003; Sponheimer and Lee-Thorp, 2003). Thus, as with our discussion of sedges and termites, unless the hominins were eating very large quantities of these animals, or concentrating on C_4 mammals alone, mammals are not a likely source of the ^{13}C enrichment on their own. In short, what we now know about isotopic compositions of ^{13}C-enriched foods suggests that none, save C_4 grasses themselves, appear to be capable of producing a high degree of ^{13}C enrichment on their own. Indeed, from an isotopic perspective, it seems far more likely that several C_4 foods were consumed in tandem, but this conclusion is based on a lack of evidence for the consumption of one ^{13}C-enriched food, rather than direct isotopic evidence for the consumption of a variety of resources.

We also do not know about possible seasonal differences in C_4 consumption. Our traditional sampling protocols provide us with stable isotope abundances for enamel that is formed over many months and, in some cases, years (see Sponheimer et al., 2005). Consequently, the stable isotope data are generally telling us something about an individual's "average" diet. It would be a mistake, therefore, to assume that ^{13}C-enriched resources were important foods for australopiths year round. Primate diets are known to change from season to season, sometimes profoundly so (e.g., Altmann and Altmann, 1970; Dunbar, 1983; Wrangham, Conklin-Brittain, and Hunt, 1998). As such, it might well be that C_4 resources were largely fallback foods that made up the bulk of australopith diets during the dry season when preferred resources were no longer available or when their nutritional quality was no longer acceptable. Then again, australopiths might have consumed ^{13}C-enriched resources all year, perhaps C_4 grass and sedge underground storage organs (USOs) during the dry season and grass seeds during the rains. We simply do not know, which is quite unfortunate, as it is potentially a very important point for understanding the relationship between environmental/climatic change and hominin evolution. We could, in theory, sample for stable isotopes along the growth axes of teeth and, in doing so, trace the diets of individuals over time (e.g., Balasse et al., 2002; Passey and Cerling, 2002); but such sampling is far more destructive than our traditional bulk analyses and as such has not been attempted on rare hominin specimens. Nonetheless, preliminary work using a laser ablation technique to measure stable carbon isotopes along the growth axes of teeth has been promising and may allow us to obtain temporal data while producing minimal damage (Sponheimer, de Ruiter, and Lee-Thorp, 2004). Thus, this particular "unknown" is fast becoming "knowable," and should soon become "known."

On a side note, this high-resolution analysis should allow us to address other previously unassailable questions, such as "When were australopiths weaned?" as carbon and oxygen isotope values are known to change during the weaning process (Wright and Schwarcz, 1998; Witt and Ayliffe, 2001). Thus, incremental sampling along an australopith M_1 is likely to provide evidence of the isotopic shift that accompanies weaning and contribute to our nascent understanding of australopith life history. Moreover, it has been suggested that the metabolic disturbance associated with weaning increased susceptibility to predation among hominins, which in turn could significantly bias fossil assemblages (White, 1978). Consequently, age of weaning data might also have significant taphonomic implications.

To make one last foray into the unknown, we would like to quickly discuss what stable carbon isotopes do not tell us at present about australopith paleoenvironments. As discussed earlier, we know that the $\delta^{13}C$ of C_3 vegetation is sensitive to a variety of environmental parameters, which makes enamel $\delta^{13}C$ a potential source of paleoenvironmental information. However, making the jump from the $\delta^{13}C$ of fossils to that of paleovegetation is much more complicated than many researchers, even within the stable isotope community, appreciate. There are several reasons for this, the first of which is the old bête noire of stable isotope paleodietary studies—diagenesis. It has been shown repeatedly that diagenesis does not obscure the primary dietary signal in enamel apatite (e.g., Lee-Thorp and van der Merwe, 1987; Lee-Thorp, van der Merwe, and Brain, 1994; Cerling, Harris, and Leakey, 1999). This does not mean, however, that enamel is altogether unaffected by diagenesis (e.g., Michel, Ildefonse, and Morin, 1995; Kohn, Schoeninger, and Barker, 1999; Sponheimer and Lee-Thorp, 1999a; Schoeninger et al., 2003). In fact, there is good reason to believe carbon isotope values are altered, albeit only slightly. This is best evidenced by the isotopic separation between C_4 grazers and C_3 browsers. In modern savanna ecosystems, they tend to be different by about 13‰, but this difference shrinks to ~11‰ at South African fossil sites (M. Sponheimer, unpublished data), which probably reflects some small degree of diagenesis. At South African australopith sites, grazer and browser carbon isotopes values are probably pulled slightly together because the matrix in which they are interred has intermediary $\delta^{13}C$ values. This change affects *all* fauna, and although it is slight, it is nonetheless real. While it does not affect our ability to compare taxa to each other in any significant way (as all boats are lifted by the same isotopic tide), it does make reconstructing paleovegetation $\delta^{13}C$ more difficult.

A second problem is the aforementioned uncertainty with regard to the relationship between dietary and enamel $\delta^{13}C$. Take a hypothetical fossil browser like a kudu with a $\delta^{13}C$ value of –12‰. If we assume that enamel is enriched by +12‰ over diet (Lee-Thorp, Sealy, and van der Merwe, 1989), it would translate to a paleovegetation value of about –24‰. If we subtract a further 1.5‰ to adjust for the fossil fuel effect (combustion of ^{13}C-depleted fossil fuels has lowered the $\delta^{13}C$ of atmospheric CO_2 in the past century; see Friedli et al., 1986), we then have a value of –25.5‰, which is typical of dry areas with less than 500 mm of rainfall a year. What if, however, the kudu became enriched by +1.5‰ because of diagenesis, and what if the proper diet to enamel relationship is 14‰? After correcting for the fossil fuel effect, this would translate to a paleovegetation value of –29‰ and would be more indicative of a well-watered closed woodland or forest—quite a different picture indeed! A final difficulty is that stable isotope studies are subject to the same taphonomic biases that bedevil all faunal environmental reconstructions and which can lead to assemblages not accurately reflecting the living community at any one moment in time (e.g., Brain, 1981; Lyman, 1994). Nonetheless, our point here is not to say that any attempt to reconstruct paleoenvironments from enamel $\delta^{13}C$ is bankrupt from the start, but only that we must be cognizant of these limitations.

The Unknowable

The boundaries between the known, the unknown, and the unknowable are often fuzzy or ephemeral. At least part of this problem is semantic: We tend to couple the words "known" and "true"; yet, one can "know" something that in hindsight is untrue. People knew for eons that the sun traveled around the earth, scientists knew for decades the humans had forty-eight chromosomes (as do chimpanzees), but alas, knowing was not quite what it was cracked up to be. Indeed, people can know a great many things, including mutually incompatible notions. This does not mean, of course, that all such notions are equally true (and this is assuming that *The X-Files* is the ultimate epistemological arbiter and therefore "the truth is out there") but rather that while as scientists we are always seeking some form of external truth that is "out there," what we know at any given moment stands in an uncertain relationship with this external truth. Thus, in practice, it might well be argued that we know very little about the diets of early hominins (ironically, perhaps, both more and less than we knew fifty years ago). What we do have is a great many reasonable hypotheses for which there are greater or lesser degrees of empirical and theoretical support. There can be little doubt that some fifty years hence, many will look back at what we know today and shake their heads with wry amusement. We are probably on the right track about many things, but like all of our predecessors, we are undoubtedly incorrect in many ways—even in some areas that we feel fairly secure—perhaps our "chimpcentric" notions of early hominin ecology included?

However, this difficulty with the known and the unknown aside, it might be that the distinction between the unknown and the unknowable is even more fluid and tenuous. Imagine taking a fiery brand out of a campfire on a dark night and then twirling it with superhuman speed before you. Those looking at this spectacle from a few hundred meters away will see nothing but a ring of fire. Sure enough, the ring of fire itself would be illusory, as the entire ring would never exist at any given moment. Yet, from their perspective, the ring is all that they would, and ever could, see. The truth that they are looking on a twirling brand rather than a ring of fire would be forever unknowable to them—so long, that is, that they looked on the phenomenon with human eyes. With high-speed photography, telephoto lenses, and slow-motion playback, however, they could discern the truth. The basic point here is that the unknowable is not some land that is separated from the unknown by a massive epistemological discontinuity, but that the unknowable is often defined by the "eyes" with which you observe the phenomenon at hand. And with changes in technology and in our thinking (paradigm shifts), the "eyes" available to scientists are in constant flux. One hundred years ago, who would have imagined that we would be able to analyze carbon locked in a hominin's tooth that was derived from a fig consumed on a breezy afternoon or that we could look on a tooth's microscopic pits and scratches to reveal the contents of an ancestor's last supper? These things would have been unthinkable a century ago, but science has given us the eyes of mass spectrometry and electron microscopy, and we surely see all the better into the distant past for them. No doubt, technology will continue to advance, and will make it possible to address many questions about the diets of early hominins that would now be relegated to the realm of science fiction.

Thus, we are hesitant to talk about the unknowable, except perhaps in the sense of what is unknowable given current technology. If we are further restrictive, however, and use "unknowable" in the sense of what we cannot know from stable isotopes alone, then the realm of the unknowable is considerable indeed. For one, it would be impossible to ever know which C_4 resources australopiths used because of insurmountable equifinality problems. In other words, there are many diets that can conceivably produce the same carbon isotope signature. Therefore, there is no one-to-one relationship between any stable isotope signature and a given diet (e.g., crabs, bovids, fruits). This point must not be overlooked, although it frequently is in practice, lest our paleodietary interpretations fly (metaphorically) beyond our data. It would be impossible, for instance, for stable carbon isotopes of tooth enamel carbonate to ever distinguish between a purely vegetarian hominin that ate 50% C_3 fruits and 50% C_4 grass and a highly predacious hominin that ate nothing but C_3- and C_4-derived mammal flesh in roughly equal proportions. Stable isotopes will never be up to the task of reconstructing early hominin "menus" (e.g., 60% figs; 20% bovid burgers, 10% grass roots, 5% honey, 5% Coco Puffs), which is rather unfortunate, because "Eat Like an Australopith" has the ring of an international bestseller.

Fortunately, however, stable isotope studies do not operate in a vacuum. We can always use insights gleaned from nutritional analysis, dental microwear, elemental analysis, and other areas to aid in interpreting stable isotope data. Moreover, advances in these other areas may allow us to better interpret carbon isotope data in the future. For instance, perhaps microwear will be able to conclusively rule out the consumption of C_4 grass seeds and roots. If this were to happen, it would likely mean that both C_4 sedges and animal foods were important resources, especially for those hominin specimens that are most enriched in ^{13}C, as it would be very difficult to attain *Theropithecus*-like $\delta^{13}C$ through the consumption of sedges or animal foods alone.

In the end, stable carbon isotope analysis has provided us with significant insights into the diets of early hominins. Ten years ago, very few paleoanthropologists would have countenanced the idea that australopiths consumed significant quantities of grasses, sedges, or animal foods, but stable isotope studies have forced us to confront these possibilities. However, as important as these studies have been in forcing us to broaden the potential australopith dietary, we still need to improve our knowledge of stable isotope distributions in modern ecosystems and better our understanding of the relationship between dietary and enamel stable isotope compositions if we are to get the most out of stable carbon isotope studies. But this caveat aside, stable isotopes still have a great deal to tell us about early hominin diets. To date, the South African australopiths are the only early hominins that have been analyzed for stable carbon isotopes to a significant extent, which leaves entire continents and vast periods of time ripe for investigation with this technique. There can be little doubt that stable carbon isotope analysis, especially when performed as part of a larger, integrated paleodietary investigation, will reveal much about the diets of these hominins.

References

Aiello, L.C., and Wheeler, P., 1995. The expensive tissue hypothesis. *Curr. Anthropol.* 36, 199–221.

Altmann, S.A., and Altman, J., 1970. *Baboon Ecology*. University of Chicago Press, Chicago.

Ambrose, S.H., and DeNiro, M.J., 1986. The isotopic ecology of East African Mammals. *Oecologia* 69, 395–406.

Ambrose, S.H., and Norr, L., 1993. Experimental evidence for the relationship of the carbon isotope ratios of whole diet and dietary protein to those of bone collagen and carbonate. In: Lambert J.B., and Grupe, G. (Eds.), *Prehistoric Human Bone: Archaeology at the Molecular Level*. Springer, Berlin, pp. 1–37.

Backwell, L.R., and d'Errico, F., 2001. Evidence of termite foraging by Swartkrans early hominids. *Proc. Natl. Acad. Sci.* 98, 1358–1363.

Balasse, M., 2002. Reconstruction dietary and environmental history from enamel isotopic analysis: Time resolution of intra-tooth sequential sampling. *Int. J. Osteoarchaeol.* 12, 155–165.

Balasse, M., Ambrose, S.H., Smith, A.B., and Price, T.D., 2002. The seasonal mobility model for prehistoric herders in the south-western Cape of South Africa assessed by isotopic analysis of sheep tooth enamel. *J. Archaeol. Sci.* 29, 917–932.

Bocherens, H., Billiou, D., and Mariotti, A., 2001. New isotopic evidence for dietary habits of Neandertals from Belgium. *J. Hum. Evol.* 40, 497–505.

Boutton, T.W., Arshad, M.A., and Tieszen, L.L., 1983. Stable isotope analysis of termite food habits in East African grasslands. *Oecologia* 59, 1–6.

Braack, L., and Kryger, P., 2003. Insects and savanna heterogeneity. In: du Toit, J.T, Rogers, K.H., and Biggs, H.C. (Eds.), *The Kruger Experience: Ecology and Management of Savanna Heterogeneity*. Island Press, Washington DC, pp. 263–275.

Brain, C.K., 1981. *The Hunters or the Hunted?* University of Chicago Press, Chicago.

Carruthers, V., 1997. *The Wildlife of Southern Africa*. Southern Book Publishers, Halfway House.

Carter, M.L., 2001. Sensitivity of stable isotopes (^{13}C, ^{15}N, and ^{18}O) in bone to dietary specialization and niche separation among sympatric primates in Kibale National Park, Uganda. PhD diss., University of Chicago.

Cerling, T.E., and Harris, J.M., 1999. Carbon isotope fractionation between diet and bioapatite in ungulate mammals and implications for ecological and paleoecological studies. *Oecologia* 120, 347–363.

Cerling, T.E., Hart, J.A., and Hart, T.B., 2004. Stable isotope ecology in the Ituri Forest. Oecologia 138, 5–12.

Cerling, T.E., Harris, J.M., and Leakey, M.G., 1999. Browsing and grazing in elephants: The isotope record of modern and fossil proboscideans. *Oecologia* 120, 364–374.

Codron, D.M., 2003. Dietary Ecology of Chacma Baboons (*Papio ursinus*) and Pleistocene Cercopithecoidea in Savanna Environments of South Africa. MSc thesis, University of Cape Town.

Codron, J., Codron, D., Sponheimer, M., Lee-Thorp, J., Bond, W.J., de Ruiter, D., and Grant, R., 2005. Taxonomic, anatomical, and spatio-temporal variations in stable carbon and nitrogen isotopic composition of plants from an African savanna. *J. Archaeol. Sci.* 32, 1757–1772.

Collins, M.J., Gernaey, A.M., Nielsen-Marsh, C.M., Vermeer, C., and Westbroek, P., 2000. Osteocalcin in fossil bones: Evidence of very slow rates of decomposition from laboratory studies. *Geology*, 28, 1139–1142.

Conklin-Brittain, N.L., Wrangham, R.W., and Smith, C.C., 2002. A two-stage model of increased dietary quality in early hominid evolution: the role of fiber. In: Ungar, P.S., and

Teaford, M.F. (Eds.), *Human Diet: Its Origin and Evolution*. Bergin & Garvey, Westport, pp. 61–76.

Deocampo D.M., Blumenschine R.J., and Ashley G.M., 2002. Wetland diagenesis and traces of early hominids, Olduvai Gorge, Tanzania. *Quatern. Res.* 57, 271–281.

Dunbar, R.I.M., 1983. Theropithecines and hominids: Contrasting solutions to the same ecological problem. *J. Hum. Evol.* 12, 647–658.

Ehleringer, J.R., and Cooper, T.A., 1988. Correlations between carbon isotope ratio and microhabitat in desert plants. *Oecologia* 76, 562–566.

Ehleringer, J.R., and Monson, R.K. (1993). Evolutionary and ecological aspects of photosynthetic pathway variation. *Annual Review of Ecology and Systematics* 24, 411–439.

Ellery W. N., Ellery K., Rogers K. H., and McCarthy T. S., 1995. The role of *Cyperus papyrus* L. in channel blockage and abandonment in the northeastern Okavango Delta, Botswana. *Afr. J. Ecol.* 33, 2549.

Fizet, M., Mariotti, A., Bocherens, H., Lange-Badré, B., Vandermeersch, B., Borel, J.P., and Bellon, G., 1995. Effect of diet, physiology and climate on carbon and nitrogen isotopes of collagen in a late Pleistocene anthropic paleoecosystem (France, Charente, Marillac). *J. Archaeol. Sci.* 22, 67–79.

Friedli, H., Lotscher, H., Oeschger, H., Siegenthaler, U., and Stauffer, B., 1986. Ice core record of the $^{13}C/^{12}C$ ratio of atmospheric CO_2 in the past two centuries. *Nature* 324, 237–238.

Glimcher, M.J., Cohen-Solal, L., Kossiva, D., and de Ricqles, A., 1990. Biochemical analyses of fossil enamel and dentin. *Paleobiology* 16, 219–232.

Goodall, J., 1986. *The Chimpanzees of Gombe*. Cambridge University Press, Cambridge.

Goodman, M., Porter, C.A., Czelusniak, J., Page, S.L., Schneider, H., Shoshani, J., Gunnell, G., and Groves, C.P., 1998. Toward a phylogenetic classification of primates based on DNA evidence complemented by fossil evidence. *Mol. Phylogenet. Evol.* 9, 585–598.

Grine, F.E., 1981. Trophic differences between gracile and robust australopithecines. *S. Afr. J. Sci.* 77, 203–230.

Grine, F.E., 1986. Dental evidence for dietary differences in *Australopithecus* and *Paranthropus*. *J. Hum. Evol.* 15, 783–822.

Grine, F.E., and Kay, R.F., 1988. Early hominid diets from quantitative image analysis of dental microwear. *Nature* 333, 765–768.

Hamilton, W.J., 1987. Omnivorous primate diets and human overconsumption of meat. In: Harris, M., Ross, E.B. (Eds.), *Food and Evolution: Toward a Theory of Human Food Habits*. Temple University Press, Philadelphia, pp. 117–132.

Hay, R. L., 1976. *Geology of the Olduvai Gorge*. University of California Press, Berkeley.

Hedges, R.E.M., and van Klinken, G.J., 2000. "Consider a Spherical Cow"—on Modeling and Diet. In: Ambrose, S., and Katzenberg, M.A. (Eds.), *Biogeochemical Approaches to Paleodietary Analysis*. Kluwer Academic/Plenum, New York, pp. 211–242.

Isaac, G.L., 1971. The diet of early man: Aspects of archaeological evidence from lower and middle Pleistocene sites in Africa. *World Archaeol.* 2, 278–298.

Jolly, C.J., 1970. The seed-eaters: A new model of hominid differentiation based on a baboon analogy. *Man* 5, 5–26.

Kohn, M., Schoeninger, M.J., and Barker, W.W., 1999. Altered states: Effects of diagenesis on fossil tooth chemistry. *Geochim. Cosmochim. Acta* 63, 2737–2747.

Kohn, M.J., Schoeninger, M.J., and Valley, J.W., 1996. Herbivore tooth oxygen isotope compositions: Effects of diet and physiology. *Geochim. et Cosmochim. Acta* 60, 3889–3896.

Lee-Thorp, J.A., Sealy, J.C., and van der Merwe, N.J., 1989. Stable carbon isotope ratio differences between bone collagen and bone apatite, and their relationship to diet. *J. Archaeol. Sci.* 16, 585–599.

Lee-Thorp, J.A., and van der Merwe, N.J., 1987. Carbon isotope analysis of fossil bone apatite. *S. Afr. J. Sci.* 83, 712–715.

Lee-Thorp, J.A., van der Merwe, N.J., and Brain, C.K., 1989. Isotopic evidence for dietary differences between two extinct baboon species from Swartkrans. *J. Hum. Evol.* 18, 183–190.

Lee-Thorp, J.A., van der Merwe, N.J., and Brain, C.K., 1994. Diet of Australopithecus robustus at Swartkrans from stable carbon isotopic analysis. *J. Hum. Evol.* 27, 361–372.

LeGeros, R.Z., 1991. *Calcium Phosphates in Oral Biology and Medicine*. Karger, Paris.

Luyt, C.J., 2001. Revisiting palaeoenvironments from the hominid-bearing Plio-Pleistocene sites: New isotopic evidence from Sterkfontein. MSc thesis, University of Cape Town.

Luyt, J., and Lee-Thorp, J.A., 2003. Carbon isotope ratios of Sterkfontein fossils indicate a marked shift to open environments ca. 1.7 Ma. *S. Afr. J. Sci.* 99, 271–273.

Lyman, R.L., 1994. *Vertebrate Taphonomy*. Cambridge University Press, Cambridge.

McGrew, W.C., Baldwin, P.J., and Tutin, C.E., 1981. Chimpanzees in a hot, dry and open habitat: Mt. Assirik, Senegal, West Africa. *J. Hum. Evol.* 10, 227–244.

Michel, V., Ildefonse, P., and Morin, G., 1995. Chemical and structural changes in Cervus elephas tooth enamels during fossilization (Lazaret Cave): A combined IR and XRD Rietveld analysis. *Appl. Geochem.* 10, 145–159.

Milton, K., 1999. A hypothesis to explain the role of meat-eating in human evolution. *Evol. Anthropol.* 8, 11–21.

Minagawa, M., and Wada, E., 1984. Step-wise enrichment of ^{15}N along food chains: further evidence and the relationship between $\delta^{15}N$ and animal age. *Geochim. Cosmochim. Acta* 48, 1135–1140.

Passey, B.H., and Cerling, T.E., 2002. Tooth enamel mineralization in ungulates: Implications for recovering a primary isotopic time-series. *Geochim. Cosmochim. Acta* 66, 3225–3234.

Peters, C.R., and Vogel, J.C., 2005. Africa's wild C_4 plant foods and possible early hominid diets. *J. Hum. Evol.* 48, 219–236.

Prentice, M.L., and Denton, G.H., 1988. The deep-sea oxygen isotope record, the global ice sheet system and hominid evolution. In Grine, F.E. (Ed.), *Evolutionary History of the "Robust" Australopithecines.* Aldine de Gruyter, New York, pp. 383–403.

Puech P.F., Cianfarani F., and Albertini H., 1986. Dental microwear features as an indicator for plant food in early hominids: A preliminary study of enamel. *Hum. Evol.* 1, 507–515.

Reed, K., 1997. Early hominid evolution and ecological change through the African Plio-Pleistocene. *J. Hum. Evol.* 32, 289–322.

Richards, M.P., Pettett, P.B., Stiner, M.C., and Trinkaus, E., 2001. Stable isotope evidence for increasing dietary breadth in the European mid-Upper Paleolithic. *Proc. Natl. Acad. Sci.* 98, 6528–6532.

Richards, M.P., Pettett, P.B., Trinkaus, E., Smith, F.H., Paunovic, M., and Karavanic, I., 2000. Neanderthal diet at Vindija and Neanderthal predation: The evidence from stable isotopes. *Proc. Natl. Acad. Sci.* 97, 7663–7666.

Robinson, J.T., 1954. Prehominid dentition and hominid evolution. *Evolution* 8, 324–334.

Schoeninger, M.J., and DeNiro, M.J., 1984. Nitrogen and carbon isotopic composition of bone collagen from marine and terrestrial animals. *Geochim. Cosmochim. Acta* 48, 625–639.

Schoeninger, M.J., Hallin, K., Reeser, H., Valley, J.W., and Fournelle, J., 2003. Isotopic alteration of mammalian tooth enamel. *Int. J. Osteoarchaeol.* 13, 11–19.

Schoeninger, M.J., Iwaniec, U.T., and Glander, K.E., 1997. Stable isotope ratios indicate diet and habitat use in New World monkeys. *Am. J. Phys. Anthropol.* 103, 69–83.

Schoeninger, M.J., Iwaniec, U.T., and Nash, L.T., 1998. Ecological attributes recorded in stable isotope ratios of arboreal prosimian hair. *Oecologia* 113, 222–230.

Schoeninger, M.J., Moore, J., and Sept, J.M., 1999. Subsistence strategies of two savanna chimpanzee populations: the stable isotope evidence. *Am. J. Primatol.* 49, 297–314.

Smith, B.N., and Epstein, S., 1971. Two categories of $^{13}C/^{12}C$ ratios for higher plants. *Plant Physiol.* 47, 380–384.

Smith, K.F., Sharp, Z.D., and Brown, J.H., 2002. Isotopic composition of carbon and oxygen in desert fauna: investigations into the effects of diet, physiology, and seasonality. *J. Arid Environ.* 52, 419–430.

Sponheimer, M., 1999. Isotopic ecology of the Makapansgat Limeworks fauna. PhD diss., Rutgers University.

Sponheimer, M., de Ruiter, D., and Lee-Thorp, J., 2004. Seasonality and Australopithecine Diets: New High-Resolution Carbon Isotope Data. *Am. J. Phys. Anthropol.* Suppl. no. 38, 12, 186.

Sponheimer, M., and Lee-Thorp, J., 2003. Using bovid carbon isotope data to provide paleoenvironmental information. *S. Afr. J. Sci.* 99, 273–275.

Sponheimer, M., and Lee-Thorp, J.A., 1999a. The alteration of enamel carbonate environments during fossilisation. *J. Archaeol. Sci.* 26, 143–150.

Sponheimer, M., and Lee-Thorp, J.A., 1999b. The ecological significance of oxygen isotopes in enamel carbonate. *J. Archaeol. Sci.* 26, 723–728.

Sponheimer, M., and Lee-Thorp, J.A., 1999c. Isotopic evidence for the diet of an early hominid, *Australopithecus africanus. Science* 283, 368–370.

Sponheimer, M., and Lee-Thorp, J.A., 2001. The oxygen isotope composition of mammalian enamel carbonate: A case study from Morea Estate, Mpumalanga Province, South Africa. *Oecologia* 126, 153–157.

Sponheimer, M., Lee-Thorp, J., de Ruiter, D., Codron, D., Codron, J., Baugh, A.T., and Thackeray, F., 2005. Hominins, sedges, and termites: New carbon isotope data from the Sterkfontein valley and Kruger National Park. *J. Hum. Evol.* 48, 301–312.

Sponheimer, M., Lee-Thorp, J., de Ruiter, D, Smith, J., van der Merwe, N., Reed, K., Ayliffe, L., Heidelberger, C., and Marcus, W., 2003. Diets of southern African Bovidae: stable isotope evidence. *J. Mammal.* 84, 471–479.

Sponheimer, M., Robinson, T., Ayliffe, L., Passey, B., Roeder, B., Shipley, L., Lopez, E., Cerling, T., Dearing, D., and Ehleringer, J., 2003. An experimental study of carbon-isotope fractionation between diet, hair, and feces of mammalian herbivores. *Can. J. Zool.* 81, 871–876.

Stanford, C., 1999. *The Hunting Apes*. Princeton University Press, Princeton.

Stern, J.T., and Susman, R.L., 1983. The locomotor anatomy of *Australopithecus afarensis. Am. J. Phys. Anthropol.* 60, 279–317.

Stuart, C., and Stuart, T., 2000. *A Field Guide to the Tracks and Signs of Southern and East African Wildlife*. Stuik Publishers, Cape Town.

Teleki, G., 1981. The omnivorous diet and eclectic feeding habits of chimpanzees in Gombe National Park, Tanzania. In: Harding, R.S.O., and Teleki, G. (Eds.), *Omnivorous Primates*. Columbia University Press, New York, pp. 303–343.

Tieszen, L.L., and Fagre, T., 1993. Effect of diet quality and composition on the isotopic composition of respiratory CO2, bone collagen, bioapatite, and soft tissues. In: Lambert, J.B., Grupe, G. (Eds.), *Prehistoric Human Bone: Archaeology at the Molecular Level.* Springer, Berlin, pp. 121–155.

Ungar, P., 2004. Dental topography and diets of *Australopithecus afarensis* and early *Homo. J. Hum. Evol.* 46, 605–622.

van der Merwe, N.J., Cushing, A.E., and Blumenschine, R.J., 1999. Stable isotope ratios of fauna and the environment of paleolake Olduvai. *J. Hum. Evol.* 34, A24–A25.

van der Merwe, N.J., and Medina, E., 1991. The canopy effect, carbon isotope ratios and foodwebs in Amazonia. *J. Archaeol. Sci.* 18, 249–259.

van der Merwe, N.J., Thackeray, J.F., Lee-Thorp, J.A., and Luyt, J., 2003. The carbon isotope ecology and diet of *Australopithecus africanus* at Sterkfontein, South Africa. *J. Hum. Evol.* 44, 581–597.

Vogel, J.C., 1978a. Isotopic assessment of the dietary habits of ungulates. *S. Afr. J. Sci.* 74, 298–301.

Vogel, J.C., 1978b. Recycling of carbon in a forest environment. *Oecol. Plantar.* 13, 89–94.

Vogel, J.C., Fuls, A., and Ellis, R.P., 1978. The geographical distribution of kranz grasses in South Africa. *S. Afr. J. Sci.* 74, 209–215.

White, T.D., 1978. Early hominid enamel hypoplasia. *Am. J. Phys. Anthropol.* 49, 79–84.

Witt, G.B., and Ayliffe, L.K., 2001. Carbon isotope variability in the bone collagen of red kangaroos (*Macropus rufus*) is age dependent. *J. Archaeol. Sci.* 28, 247–252.

Wright, L.E., and Schwarcz, H.P., 1998. Stable carbon and oxygen isotopes in human tooth enamel: Identifying breastfeeding and weaning in prehistory. *Am. J. Phys. Anthropol.* 106, 1–18.

Wrangham, R.W., 1987. The significance of African apes for reconstructing human social evolution. In Kinzey, W.G. (Ed.), *Primate Models of Hominid Evolution*. SUNY Press, Albany, pp. 51–71.

Wrangham, R.W., Conklin-Brittain, N.L., and Hunt, K.D., 1998. Dietary response of chimpanzees and cercopithecines to seasonal variation in fruit abundance. I. Antifeedants. *Int. J. Primatol.* 19, 949–970.

Young, S., 2002. Metabolic mechanisms and isotopic investigation of ancient diets with an application to human remains from Cuello, Brazil. PhD diss., Harvard University.

9

Reconstructing Early Hominin Diets

Evaluating Tooth Chemistry and Macronutrient Composition

MARGARET J. SCHOENINGER

The elevated health risks associated with high-fat, energy-dense, and possibly protein-dense diets underscore the need to determine the natural diet for humans (Gaulin and Konner, 1977; Eaton, Shostak, and Konner, 1988), if one actually exists. The fossil record indicates that our early apelike ancestors ate largely fruit and leaves (Milton, 1984). By 1.8 million years ago, however, *Homo erectus* looked similar to us from the neck down and was hunting/scavenging significant quantities of meat from large animals in addition to gathering plant foods and, when available, seasonally, smaller fauna such as insects (Shipman and Walker, 1989). This change involves a transition from diets based largely on simple sugars, complex carbohydrates, and plant proteins to diets that include animal fat and protein (Schoeninger et al., 2001b). As other chapters in this volume indicate, data from living primates, fossil primates, modern human foragers, and bone chemistry all have been used to elucidate when various dietary changes occurred throughout human evolution. For each of these approaches, however, we must assess the quality of the data and, based on that assessment, clearly identify what we know versus what is, perhaps, unknown but knowable. This chapter will address a limited subset of the approaches available for early diet reconstruction (tooth chemistry and macronutrient composition of foods eaten by an extant group of human foragers). When placed in context with data from other approaches, a general picture of early hominin diet appears.

Bone and Tooth Compositional Analysis: The Known and the Unknown but Knowable

A common approach to habitat reconstruction is to analyze the species composition of fossil bones and teeth recovered from early hominin sites (e.g., Kappelman et al., 1997; Reed, 1997; Semaw et al. 2005). A complementary approach focuses on the mineral compositions of these bones and teeth to reconstruct paleoclimate (Bryant et al., 1996), paleodiet (Schwarcz and Schoeninger, 1991), and paleoecology (Morgan, Kingston, and Marino, 1994). For the former approach, we must establish that all of the fauna compared as a single assemblage truly represent a life assemblage rather than some artifact of the fossilization process. For the latter approach, we must determine that the fossil materials have undergone no chemical or isotopic alteration of the original composition or that all analyzed samples have been subjected to the same alteration. In a series of studies, my colleagues and I analyzed fossil teeth from a single excavation in Locality 261-1 (M. Leakey, personal communication, 1996) at the Pliocene site of Allia Bay, east Lake Turkana, Northern Kenya, which produced several specimens of the early biped, *Australopithecus anamensis* (Coffing et al., 1994; Leakey et al., 1995). We analyzed tooth enamel (following the procedures outlined in Koch, Tuross, and Fogel, 1997) because it is generally assumed to be resistant to diagenetic alteration (Koch, Tuross, and Fogel, 1997) because of its structure. Hydroxylapatite crystallites in enamel are over six times larger than those in bone, producing a relatively smaller surface area, with lower solubility and less substitution, and no remodeling during life (LeGeros, 1981). Only a few studies report effects, and these are minimal (Lee-Thorp, 2000; but see Sponheimer and Lee-Thorp, 2006). To establish baseline conditions in our study, we analyzed modern teeth and a single, Holocene-aged tooth collected nearby the fossil site (Barthelme, 1985). The fossil material that we analyzed was deposited in an inactive channel when flooding occurred along the main channel of the 3.9 Myr. Omo River. The complete faunal collection suggested a mixture of forest, grassland, and bushland habitats (Leakey et al., 1995).

We considered the possibility that the fauna could represent deposits from multiple floods that occurred over tens or hundreds of years (Potts, 1998) and that the faunal diversity might represent the compression of sequential habitats. If true, different fossil teeth could show different reactions to the burial environments. Further, the surface of enamel, the area of the enamel-dentine junction that often separates during surface weathering, and the surface of cracks that form in enamel during weathering could show different elemental compositions than the tooth's interior. Using transmission electron microscope imaging, ion microprobe analysis, electron microprobe spot analysis, and x-ray mapping, we showed higher levels of manganese, iron, silica, rare earth elements, uranium, fluorine, and strontium in fossil enamels from the site when compared with modern teeth from the same area (Kohn, Schoeninger, and Barker, 1999; and see the "Unknowable" section). The pattern of the data suggested contamination by fine-grained secondary oxyhydroxides, low levels of clay incorporation, and chemical alteration of the original apatite. These results indicate that use of strontium and/or barium concentration for diet reconstruction would produce inaccurate results at this site (contra Sillen, 1992, for fossils recovered in South Africa).

Subsequent analysis with cathodoluminescence microscopy demonstrated that the alterations occurred in zones in some teeth, whereas they were distributed throughout the teeth in other teeth from the same deposit (Schoeninger et al., 2003). Under cathodoluminescence, our modern and Holocene samples all appeared blue to blue violet. In contrast, the Pliocene samples clustered loosely into three different color groups. The first group had interiors that were dark and some had very bright outer margins that were white, yellow, or pink. The second group had an uneven intensity of luminescence that glowed green or yellow with some outer margins that glowed brightly in various colors. The third group consisted of a single tooth that was uniformly bright orange. The appearance of the enamel surface when viewed under ordinary light was not associated with any of these luminescent patterns.

Extensive elemental analyses revealed that all of the Pliocene samples had calcium to phosphorus ratios that were similar to modern teeth (Schoeninger et al., 2003). Even though this ratio is often used for establishing the integrity of tooth composition, our intensive study showed that the brightly colored zones identified by cathodoluminescence had elevated levels of fluorine as well as variable levels of manganese, silica, and iron. In addition, the vast majority of the Pliocene samples had higher carbon dioxide concentrations than those measured in the modern and Holocene samples. Clearly, these fossil enamel samples had been altered from their original state.

Next, we investigated whether the carbon and oxygen isotope ratios of the teeth had also been altered from biological levels before using these ratios to compare the Pliocene ecology of the region with that of today. These ratios in the tissues of living primates correlate with several ecological variables. For example, the $\delta^{13}C$ values in hair samples from extant primates living in varying ecological habitats record the extent of canopy cover (see fig. 9.1). Those from closed-canopy forests have significantly lower values than those from more open habitats (Schoeninger, Iwaniec, and Glander, 1997; Schoeninger, Iwaniec, and Nash, 1998; Schoeninger, Moore, and Sept, 1999). Tooth enamel $\delta^{13}C$ values in ruminants from closed-canopy habitats are also lower than those from animals living in more open habitats (Cerling, Harris, and Leakey, 1999). If the carbon isotope values within the fossil teeth from the Allia Bay site retain their biological signatures, it should be possible to identify aspects of the ecology of the site at the time the animals were alive. First, we checked the integrity of the samples for isotopic analyses by performing isotope analysis on the color zones previously identified by cathodoluminescence (Schoeninger et al., 2003). The thick bright yellow zones along the outer edges of otherwise dark sections were removed and analyzed independently of the dark interiors. Additional fragments were separated mechanically into interior and exterior samples that were analyzed separately. All of the browsing fauna had exterior enamel that has a higher $\delta^{13}C$ value than their interior enamel. In contrast, all of the grazing fauna and mixed feeders, with the exception of two bovid samples, had exterior enamel that had lower $\delta^{13}C$ values than their interior enamel. In other words, the enamel exteriors of browsing and grazing fauna are more similar to each other than are the enamel interiors. The pattern of alteration suggested that the surface of the enamels was altered in the direction of sediments that surrounded the enamel during fossilization (see Lee-Thorp, 2000, for another example).

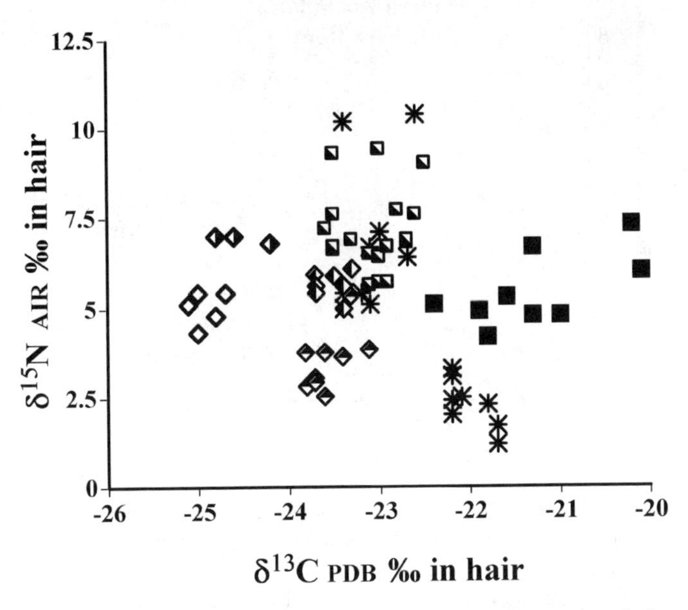

Figure 9.1 The $\delta^{13}C$ values in hair from six primate species including prosimians, monkeys and apes vary in correlation with forest canopy cover. Primates from closed canopy, evergreen forests are 3‰–5‰ more negative (i.e., –25‰) than those from open, dry canopies (i.e., –22‰ to –20‰).

The magnitude of the change, however, is much larger in some cases than previously recognized, and it varies between fragments. In some cases, the alteration is great enough to misinterpret the animal's feeding strategy where the external enamel indicated a mixed feeder, while the interior enamel indicated a browser or a grazer. Even the smaller amounts of alteration, however, would affect interpretation of the ecology in which the animals lived if it had not been identified and removed.

When we removed the altered enamels from the fossils from Allia Bay, most of the fauna have lower $\delta^{13}C$ values in their enamel interiors than is true of similar fauna today and of fauna from the later site of Swartkrans in South Africa (see fig. 9.2; Schoeninger, Reeser, and Hallin, 2003). This suggests that the animals fed in regions that were more closed (i.e., forested) than today. In contrast, if we had used data from the enamel exteriors, we would have concluded that the fossil environments were similar to today.

Once the samples were screened and limited to those showing minimal alteration, the oxygen isotope data in these fossil teeth were used to assess, in a very general way, rainfall levels at the time the animals were alive (Schoeninger, Reeser, and Hallin, 2003). On the basis of oxygen isotope data from benthic foraminifera (fig. 66 in Hoefs, 1997), the oxygen isotope composition of the middle Pliocene ocean should have been approximately 2‰ to 4‰ less positive than today. Because the source of water to the Omo River was the same for the region of the Pliocene Allia

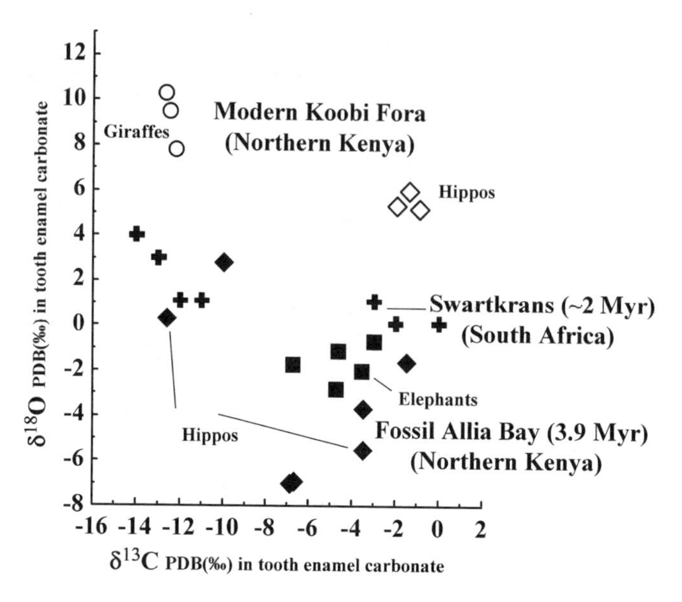

Figure 9.2 The $\delta^{18}O$ and $\delta^{13}C$ values in fossil teeth from east (Allia Bay) and South (Swartkrans) Africa show that both fossil sites were wetter with more canopy cover than the dry scrubland of today's northern Kenya regions. Allia Bay was probably wetter and more closed that Swartkrans. Redrawn from Schoeninger, Reeser, and Hallin, 2003.

Bay site as for today's Allia Bay, the fossil enamel carbonates should be 2‰–4‰ lower than modern enamels if the ecology of the region was the same as today. In contrast to expectations, the fossil enamels have $\delta^{18}O$ values that are 10‰–12‰ lower than the values in modern enamels. Such a difference indicates higher rainfall levels on the Ethiopian plateau during the time that these animals lived in the Turkana basin than is true today. As a result, groundwater levels were probably higher and greater amounts of water would have been available for plant growth and for the animals living on those plants.

These oxygen isotope data, in concert with the more negative carbon isotope data in the fossil fauna when compared with modern fauna in the region, indicate that the environment was better watered during the middle Pliocene than today. In comparison with isotope data on browsing and grazing fauna recovered from Member 2 at Swartkrans (Sponheimer and Lee-Thorp, 1999), Allia Bay appears to have been wetter with more woodland than in the much younger site in South Africa. Also supporting this view are the paleosol carbonates from Kanapoi (Wynn, 2000) and the nearby Tugen Hills sequence (Kingston, 1999), where the data overlap those of modern woodland savanna. The specific composition of the plant community in the Turkana basin cannot be determined from our stable isotope data; but trees recovered nearby (Feibel, Harris, and Brown, 1991), are nut and seed bearers. Although such trees are now restricted to the Kenyan coast, our results are consistent with previous suggestions of a fluvial corridor connecting the Lake Turkana basin with the coast during the middle Pliocene (Feibel, Harris, and Brown, 1991). Further, they suggest that miombo-type woodland (i.e., open tree cover with grass beneath) may have existed in the region at the time that bipedal hominins lived in the region.

More detailed studies of the compositions of fossil tooth enamels from sites like Allia Bay could elucidate climate and ecological changes that occurred within the basin over time. Perhaps greater detail could be determined than was possible in our summarized studies. For example, greater use of cathodoluminescence coupled with elemental and isotopic analyses could identify groups of fossils that had experienced different processes. We know that in synthetic and geological apatites, manganese and silica-containing minerals activate a yellow luminescence, rare earth elements activate various colors (pink, violet, red, etc.) whenever they substitute for calcium, and a light blue color results from defects in the apatite lattice, cesium replacement of calcium and/or low levels of carbonate substitution (Barbarand and Pagel, 2001). Dark areas in geological apatites associate with the presence of iron and of manganese in the oxide form (Filippelli and Delaney, 1993). Further, some manganese oxides form under anaerobic conditions, while others form under aerobic conditions. These patterns of uptake and alteration must indicate specific burial environments and histories. We could use the diagenetic alteration in fossil teeth to reconstruct the ecological history of a fossil site (see Karkanas et al., 2000, for a similar approach). Once we identify particular environments and/or histories, we could focus our analyses on faunal groups where all the fossilized animals experienced the same extent and type of diagenetic process. Such similarity would suggest that they had been buried at similar times and hence had been alive at the same or nearly the same time. The studies at Allia Bay suggest that the faunal collection excavated at 261-1 does not represent a single point in time; but with further analyses, it should be possible to identify representatives from generalized points in time. Then, the faunal compositions would truly be indicative of the environments at particular points rather than averaging across times.

Food Composition Analysis: The Known and Unknown but Knowable

The early work of Glynn Isaac (1971) stimulated many projects by his students and others in which the wild plant and animal foods of Africa serve as referential models in the reconstruction of diets of early australopithecines and hominins, (for examples, see Peters, 1987; Bunn and Ezzo, 1993; Sept, 2001, among many others). In proposing hypotheses that involve dependence on these plants and animals, it is critical to have accurate data on their nutrient compositions rather than relying on data from plant domesticates or from analyses designed to analyze domestic animal fodder rather than human foods (Van Soest, 1963). To address these concerns, two studies (Murray et al., 2001; Schoeninger et al., 2001a) undertook the analyses of several foods that compose a large portion of the diet among Hadza foragers living east of Lake Eyasi in northern Tanzania (Woodburn, 1968; Hawkes, O'Connell, and Blurton Jones, 1991). The region is semiarid with marked wet and dry seasons similar to that where the 3.9 Myr hominins lived at Allia Bay in northern Kenya. In other words, the modern day Hadza live in a "mosaic" habitat of open grassy plains, woodland/savanna, and brushy/wooded riverbeds. In general, their dry-season diet consists of large animals, tubers, baobab fruit pulp and seeds, some berries, and honey, which is unavailable some years but plentiful in others (A.N. Crittenden, personal communication, 2004). Their wet-season diets consist of small animals (hornbills, hyraxes, and dikdik), tubers, many fruits, baobab seeds, and honey.

Previous studies had produced inconsistent compositional data on several of these foods, especially the tubers (Vincent, 1985a) and the baobab (see various publications referenced in Murray et al., 2001). In addition, the available data on honeys was limited to refined domestic honeys from which all larvae had been removed. In contrast, early hominins probably ate liquid honey and bee larvae at all stages of larval development (Skinner, 1991). We analyzed fourteen indigenous foods, including honey (with variable amounts of larvae) from three different types of bees, six different species of fruits (five berries and one stone fruit species), ground baobab seed, baobab fruit pulp, and three different species of tubers. The honey and fruits are eaten as collected with no additional preparation. The baobab seeds are chewed when young but when mature they are commonly removed from the fruit pulp by hand, crushed using a stone, and then winnowed to remove the seed coats, which contain a trypsin inhibitor (Addy and Eteshola, 1984). The Hadza either mix the flour with water or eat it dry. The baobab fruit pulp is scraped from the hard green shell of the baobab fruit and separated from the seeds by hand. The dry, chalky pulp, which contains significant levels of calcium (Prentice et al., 1993) and vitamin C (Carr, 1955), is eaten directly or mixed with a little water. The tubers are sometimes eaten directly as they are removed from the ground after they are stripped of their outer barklike coating. The majority of tubers, however, are roasted for a short time on an open fire and then eaten after the barklike coating is removed. Most of these foods and these processing methods, with the exception of the fire roasting, would have been available to early hominins in savanna woodlands like the ones surrounding 3.9 Myr Allia Bay.

The analyses show clearly that honeys are foods high in energy and that when larvae are present, they are also high in protein. The energy values of the honeys were similar to one another ($N = 6$, average = 4,208 kcal/kg dry weight, standard deviation = 145 kcal/kg) and to 490 domestic honeys (White et al., 1962). In contrast to the refined domestic honeys, however, the Hadza honeys contained higher levels of protein (20–40 g/kg dry weight) and fat (15–80 g/kg dry weight). More analyses of these honeys and their larvae are needed, but the data point to the potential importance of bee larvae and honey as protein and energy sources for early hominins (McGrew, 2001). Some South African honeys (Lee-Thorp et al., 1987) show stable carbon isotope values that are high enough (−19‰ when corrected for the recent 2‰ shift in atmospheric CO_2) to account for some of the values reported for the South African australopithecines (Sponheimer and Lee-Thorp, 1999; van der Merwe et al., 2003). Although termites also have carbon isotope values that could account for the hominin data (Sponheimer et al., 2005), their chitinous exoskeletons probably result in lower energy and protein density in comparison with bee larvae. Clearly, these and other insects. which can be collected seasonally in quantities large enough to meet the nutrient needs of large-bodied primate species (Fleagle, 1999), deserve greater attention (McGrew 2001).

The energy levels of the berries are similar to one another ($N = 5$, average = 3,302 kcal/kg dry berry flesh, standard deviation = 96 kcal/kg). On a dry-weight basis, the berry flesh appear to be a good source of energy; but they all contain much larger seeds than our domestic berries, which lowers their energy density per kilogram of whole, wet berry. Although berries are important sources of energy for the Hadza when available (Marlowe, 2006), large quantities must be eaten to meet energy requirements. The baobab pulp and the flesh of the stone fruit are similar to each other ($N = 2$, average = 2,175 kcal/kg dry weight) and lower than the berries. The baobab, however, is available for a longer period throughout the year than the berries.

The baobab seed flour is high in both energy (4,540 kcal/kg dry weight) and protein (363 g/kg dry weight). At 36% total protein, it is significantly higher than common agricultural grains like sorghum (11% protein) and millet (12% protein) and is roughly equivalent to domestic legumes (38% protein; Glew et al., 1997). The baobab seed flour also has a better amino acid quality than all agricultural plants, except beans (Glew et al., 1997). Baobab seed protein is inadequate in only three amino acids, and minimal amounts of insects or other animal meats would probably meet complete early hominin amino acid requirements.

Throughout Africa today, the tubers of many types of domestic and wild plants provide important energy sources for human foragers (Vincent, 1985b) as well as for small-scale agriculturists (Newman, 1975). Recently, tubers have appeared prominently in several different scenarios about human evolutionary history, including the Grandmother hypothesis (O'Connell, Hawkes, and Blurton Jones, 1999), hypotheses involving the use of cooking (Wrangham et al., 1999), possible non-C_3 foods eaten by South African australopithecines (Peters and Vogel, 2005; Sponheimer et al. 2005), and as fallback foods (Laden and Wrangham, 2005). In some of these, the nutrient composition of domestic tubers is used as a referent for reconstructing the expected energy contribution from tubers used by early hominins. Yet, there are reasons to expect that the composition of the wild tubers used by the Hadza differs from the common agricultural tubers. Most of the wild tubers belong to the Leguminosae (Vincent, 1985a, 1985b) whereas most of the cultivated tubers belong to the Convolvulaceae. Further, the wild tubers grow much more deeply in the soil (up to 3 m), whereas the cultivated tubers grow superficially. In terms of appearance, agricultural tubers vary somewhat across species, although all have fairly thin outer coverings. In contrast, the wild tubers vary dramatically in superficial appearance and have extremely thick, almost barklike external coverings. Each must be peeled before consumption, and the weight of the peel can account for nearly 30% of the total wet weight of the tuber (Schoeninger et al., 2001a).

In addition, some previously published nutrient analyses of the tubers eaten by the Hadza use a method (neutral detergent method published by van Soest in 1963 for use on animal fodders) that incorporates some of the digestible starch into the fraction measured as fiber and fails to include soluble fiber (Marlett, 1990). Because not all nutrient components are accounted for using this type of analysis, it is impossible to estimate energy density accurately. The previous analyses also analyzed all of the tubers rather than just the edible portion. The Hadza typically chew the tubers for thirty seconds to three minutes before expectorating significantly large quids. When included in the nutrient analysis, these quids can lead to overestimation of the overall energy density because the carbohydrate portion is calculated by difference rather than determined through direct analysis. Under such conditions, fiber contributes to the calculation of energy at the rate for digestible carbohydrate (4 kcal/g dry weight). In contrast, fiber, which is fermented rather than digested, actually contributes only 1–2 kcal/g dry weight (Kritchevsky and Bonfield, 1997). In addition, the overall size of the hindgut in modern humans severely constrains the total intake of fiber (Milton, 1999). This particular constraint may not have affected the austalopithecines, and the potential significance of this is discussed next.

Our analyses (Schoeninger et al., 2001a) show that the energy density of wild tubers is less than half that of the domestic tubers. Across four species of domestic tubers that are indigenous to Africa, the average energy density is 3,840 kcal/kg of dry weight (±85kcal/kg dry weight) with no inedible portion. Across five additional domestic species that are not indigenous to Africa but which are commonly cultivated there today, the average energy density is extremely similar (3,862 kcal/kg dry weight ±57 kcal/kg dry weight). In marked contrast, the three species of wild tubers yielded 2,143 kcal/kg dry weight with striking variation across the species (±772 kcal/kg dry weight). There was substantial variation in the amount of edible tuber (that portion not expectorated). The two legume species were less than 50% edible, whereas the nonlegume was about 90% edible. This last tuber, however, was almost 90% water, which probably accounts for why the Hadza say that it does not fill them up (Schoeninger et al., 2001a). Significant variation in energy density also occurs within the most commonly ingested species (//ekwa) as well as across the three species. Five //ekwa samples yielded an average of 1,462 kcal per kg dry weight (±775 kcal/kg dry weight).

The variation in the amount of edible fraction and the associated variation in energy density are rather surprising as the Hadza women do not appear to discard tubers on the basis of size. Perhaps the older tubers have the higher amount of inedible fraction; a new, more extensive study of the nutrient composition of Hadza tubers promises to answer some of these questions (A.N. Crittenden, personal communication, 2005). In any case, the wild tubers provide approximately half the energy of cultivated tubers while, at the same time, requiring large energy investments on the part of Hadza women. These observations support proposals that these tubers are "starvation," or fallback, foods rather than highly ranked preferred foods (Laden and Wrangham, 2005). If the australopithecines had hindguts similar in relative size to those of modern gorillas, they could have obtained significant energy via fermentation of fiber in these tubers. In a diet of 2,250 kcal, modern gorillas ingest around 600–800 grams of fiber (estimated from Conklin-Brittain et al., 1998), which could provide 1,200–1,600 kcal of energy. A diet of Hadza tubers in which the whole tuber is swallowed (see fig. 9.3, "Hadza no quids") could provide 900–1,800 kcal of energy through fermentation in the hindgut. Such an investment in fermentation, however, would probably be associated with restricted mobility, such as in modern gorillas. This scenario contrasts with others that emphasize the advantages of bipedality in hominin mobility (Isbell and Young, 1996). Such a focus on fermentation would also restrict the ingestion of other foods because the tubers would fill the gut. If tubers were the main food taken during harsh times, the resulting overall calorie intake per day would be low.

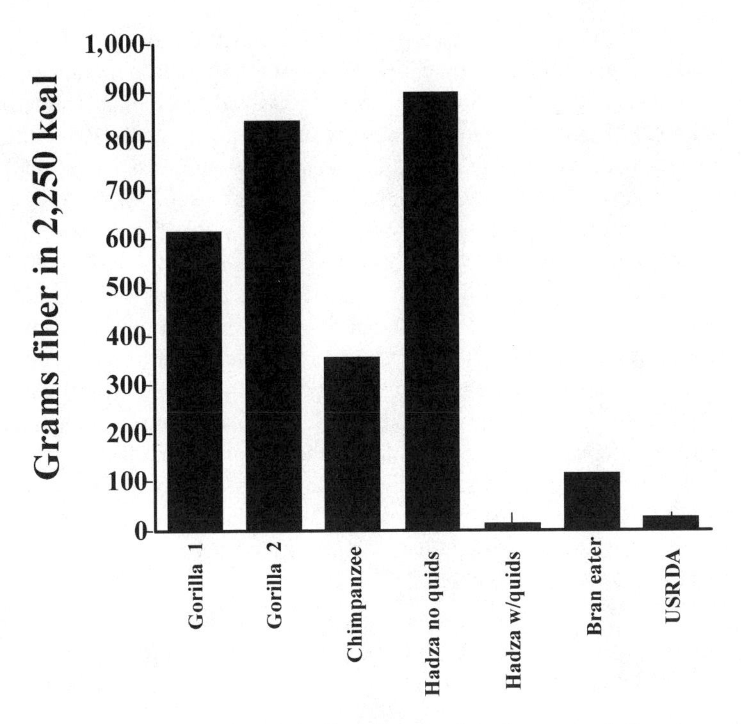

Figure 9.3 Daily intake of indigestible fiber (in grams) in the reported diets (normalized to 2,250 kcal/day) in gorilla and chimpanzee (estimated from Conklin-Brittain, Wrangham, and Hunt, 1998), modern human foragers (from Schoeninger et al., 2001a), and modern humans eating bran cereal. United States Recommended Daily Allowances (USRDA) is 25 g of fiber/day. "Hadza no quids" refers to a diet where whole tubers are chewed and swallowed; those "w/quids" indicated that a fibrous mass is expectorated.

Conclusions

In combination, the enamel and nutrient composition studies provide data for choosing among the several scenarios suggested for early hominin adaptations as well as indicating avenues for future research to address that all-so-elusive unknown but knowable. The element compositions of teeth from the single excavation at Allia Bay, northern Kenya, clearly show variable diagenetic alteration. This argues strongly against the common assumptions that fossils taken from single deposits within sites represent a single assemblage. Rather, it points to the necessity for dealing with each fossil sample individually, demonstrating through screening methods like cathodoluminescence that they are really comparable before doing isotopic or trace element analyses.

When the enamel samples were screened to analyze only those portions of each sample in which the tooth enamel had remained unaltered, the carbon and oxygen isotope values showed some general patterns. The data were all more negative than expected from the assumed composition of the Pliocene ocean and atmospheric carbon dioxide. This indicates that the region surrounding the site at Pliocene Allia

Bay was wetter with greater tree canopy cover than exists today. Use of the elemental data could address the "unknown" by identifying individual faunal assemblages within the complete faunal assemblage. These individual assemblages should represent those animals that lived at single points in time or at least, closely related points in time. The study also argues that the diagenetic history of a fossil assemblage, elucidated through extensive chemical analyses, would be useful in teasing apart the various, and as yet unknown, ecologies represented by single fossil sites.

The nutrient composition analyses show that in environments similar to those occupied by early Pliocene and, quite possibly, late Miocene (Semaw et al., 2005) australopithecines, most energy and protein needs can be met through the consumption of hard seeds through much of the year. If we assume that some species of early australopithecine is ancestral to the hominin lineage, this argues in favor of earlier hard foods proposals (Jolly, 1970; Kay, 1981; Peters, 1987) as a significant adaptation in the competitions between the early australopithecines and the clades including the ancestors of extant chimpanzees, bonobos, and gorillas (Schoeninger et al., 2001). If, however, the early australopithecines had limited mobility and gastrointestinal tracts similar to gorillas, they could have obtained minimal energy amounts through the fermentation of fiber from tubers.

The Unknowable

For any scientist, this is the hardest section to address. So many areas once considered unknowable have become known, with hardly any remembrance of ignorance in times past. Advances in theoretical orientation or technology continue to elucidate our understanding of the natural world in unanticipated and productive ways. As stated by one of my anonymous reviewers "the unknowable is so temporally contingent that it is really a philosophical question" rather than a scientific one. Just as an example, before the nineteenth century, many chemists believed that molecular structures could never be known. Yet, Kukule's and, independently, Couper's proposals regarding valence in addition to experimental work on reactivity showed this assumption to be false. How different our world is today with this knowledge!

Yet, there are times when the data we recover can be interpreted in multiple ways. Will we ever know the "true" interpretation? Is the "true" interpretation unknowable? A good example from my own work is the concentration of various elements in fossil teeth. The levels of several elements in the fossil teeth from Allia Bay fall within biological levels reported from modern teeth collected elsewhere. Does this mean that the levels in the fossil teeth from Allia Bay are the result of biology and can, therefore, be interpreted as evidence of diet? Or, alternatively, are the Allia Bay concentrations the result of diagenetic alteration? The Allia Bay fossil teeth differ from modern teeth recovered in the region today so perhaps it is only fortuitous that the final concentrations fall within those measured in modern fauna from other regions. At this point, I choose to err on the side of caution whenever diagenesis is as likely as a biological process. At the same time, I recognize that the alternative explanation cannot be rejected completely.

Will some independent signature of biology be discovered someday? Will the next generations of scientists be stymied by the same uncertainty that produces many of today's disagreements? Possibly there is something beyond isotopes or elemental concentrations, something we cannot envision today, that will conclusively identify results from biology versus those from chemistry. Perhaps some new technological method will separate the two sets; or, more likely, perhaps there is some signature of cellular involvement that is not yet recognized. I can only hope this will be the case, while at the same time, I regret that such a discovery probably belongs to the next generations rather than my own.

References

Addy, E.O., and Eteshola, E., 1984. Nutritive value of a mixture of tigernut tubers (*Cyperus esculentus* L.) and baobab seeds (*Adansonia digitata* L.). *J. Sci. Food Agric.* 35, 437–440.

Barbarand, J., and Pagel, M., 2001. Cathodoluminescence study of apatite crystals. *Am. Mineral.* 86, 473–484.

Barthelme, J.W., 1985. *Fisher-Hunters and Neolithic Pastoralists in East Turkana, Kenya.* B.A.R., Oxford.

Bryant, J.D., Froelich, P N., Showers, W.J., and Genna, B.J., 1996. Biologic and climatic signals in the oxygen isotope composition of Eocene-Oligocene equid enamel phosphate. *Palaeogeogr. Palaeoclimatol. Palaeoecol.* 126, 75–89.

Bunn, H.T., and Ezzo, J.A., 1993. Hunting and scavenging by Plio-Pleistocene hominids: Nutritional constraints, archaeological patterns, and behavioural implications. *J. Archaeol. Sci.* 20, 365–398.

Carr, W.R., 1955. Ascorbic acid content of baobab fruit. *Nature* 176, 1273.

Cerling, T.E., Harris, J.M., and Leakey, M.G., 1999. Browsing and grazing in elephants: The isotope record of modern and fossil proboscideans. *Oecologia* 120, 364–374.

Coffing, K., Feibel, C., Leakey, M., and Walker, A., 1994. Four-million-year-old hominids from East Lake Turkana, Kenya. *Am. J. Phys. Anthropol.* 93, 55–65.

Conklin-Brittain, N.L., Wrangham, R.W., and Hunt, K.D., 1998. Dietary response of chimpanzees and cercopithecines to seasonal variation in fruit abundance. II. Macronutrients. *Int. J. Primatol.* 19, 971–998.

Eaton, S.B., Shostak, M., and Konner, M., 1988. *The Paleolithic Prescription: A Program of Diet and Exercise and a Design for Living.* Harper & Row, New York.

Feibel, C.S., Harris, J.M., and Brown, F. H., 1991. Neogene paleoenvironments of the Turkana basin. In: Harris, J.M. (Ed.), *Koobi Fora Research Project.* Vol. 3: *Stratigraphy, Artiodactyls and Paleoenvironments.* Clarendon Press, Oxford, 321–346.

Filippelli, G.M., and Delaney, M.L., 1993. The effects of manganese(II) and iron(II) on the cathodoluminescence signal in synthetic apatite. *J. Sed. Petrol.* 63, 167–173.

Fleagle, J.G., 1999. *Primate Adaptation and Evolution.* Academic Press, San Diego, CA.

Gaulin, S.J.C., and Konner, M., 1977. On the natural diet of primates, including humans. In: Wurtman, R.J., and Wurtman, J.J. (Eds.), *Nutrition and the Brain.* Vol. 1. Raven Press, New York, 2–86.

Glew, R.H., VanderJagt, D.J., Lockett, C., Grivetti, L.E., Smith, G.C., Pastuszyn, A., and Millson, M., 1997. Amino acid, fatty acid, and mineral composition of 24 indigenous plants of Burkina Faso. *J. Food Compos. Anal.* 10, 205–217.

Hawkes, K., O'Connell, J.F., and Blurton Jones, N.G., 1991. Hunting income patterns among the Hadza: Big game, common goods, foraging goals and the evolution of the human diet. *Philos. Trans. R. Soc. Lond.* B 243–251.

Hoefs, J., 1997. *Stable Isotope Geochemistry.* Springer, New York.

Isaac, G., 1971. The diet of early man: Aspects of archaeological evidence from lower and Middle Pleistocene sites in Africa. *World Archaeol.* 2, 278–299.

Isbell, L.A., and Young, T.P., 1996. The evolution of bipedalism in hominids and reduced group size in chimpanzees: Alternative responses to decreasing resource availability. *J. Hum. Evol.* 30, 389–397.

Jolly, C.J., 1970. The seed-eaters: A new model of hominid differentiation based on a baboon analogy. *Man* 5, 5–26.

Kappelman, J., Plummer, T., Bishop, L., Duncan, A., and Appleton, S., 1997. Bovids as indicators of Plio-Pleistocene paleoenvironments in East Africa. *J. Hum. Evol.* 32, 229–256.

Karkanas, P., Bar-Yosef, O., Goldberg, P., and Weiner, S., 2000. Diagenesis in prehistoric caves: The use of minerals that form "in situ" to assess the completeness of the archaeological record. *J. Archaeol. Sci.* 27, 915–929.

Kay, R.F., 1981. The nut-crackers: A new theory of the adaptations of the Ramapithecinae. *Am. J. Phys. Anthropol.* 55, 141–151.

Kingston, J.D., 1999. Environmental determinants in early hominid evolution: Issues and evidence from the Tugen Hills, Kenya. In: Andrews, P., and Banham, P. (Eds.), *Late Cenozoic Environments and Hominid Evolution*. Geological Society, London, pp. 69–84.

Koch, P.L., Tuross, N., and Fogel, M.L., 1997. The effects of sample treatment and diagenesis on the isotopic integrity of carbonate in biogenic hydroxylapatite. *J. Archaeol. Sci.* 24, 417–429.

Kohn, M.J., Schoeninger, M.J., and Barker, W.W., 1999. Altered states: Effects of diagenesis on fossil tooth chemistry. *Geochim. Cosmochim. Acta* 63, 2737–2747.

Kritchevsky, D., and Bonfield, C. (Eds.), 1997. *Dietary Fiber In Health And Disease*. Plenum Press, New York.

Laden, G., and Wrangham, R.W., 2005. The rise of hominids as an adaptive shift in fallback foods: Plant underground storage organs (USOs) and australopith origins. *J. Hum. Evol.* 49, 482–498.

Leakey, M.G., Feibel, C.S., McDougall, I., and Walker, A., 1995. New four-million-year-old hominid species from Kanapoi and Allia Bay, Kenya. *Nature* 376, 565–571.

Lee-Thorp, J.A., 2000. Preservation of biogenic carbon isotopic signals in Plio-Pleistocene bone and tooth mineral. In: Ambrose, S. H., and Katzenberg, M. A. (Eds.), *Biogeochemical Approaches to Paleodietary Analysis*. Kluwer Academic/Plenum Publishers, New York, pp. 89–115.

Lee-Thorp, J.A., Lanham, J.L., Wenner, D., and van der Merwe, N.J., 1987. A carbon isotope survey of South African honey. *S. Afr. J. Sci.* 83, 186.

LeGeros, R.Z., 1981. Apatites in biological systems. *Prog. Cryst. Growth Charact.* 4, 1–45.

Marlett, J.A., 1990. Analysis of dietary fiber in human foods. In: Kritchevsky, D., Bonfield, C., and Anderson, J. W. (Eds.), *Dietary Fiber*. Plenum, New York, pp. 3–48.

Marlowe, F.W., 2006. Central place provisioning: The Hadza as an example. In: Hohmann,G., Robbins, M., and Boesch, C. (Eds.), *Feeding Ecology in Apes and Other Primates*. Cambridge University Press, Cambridge, pp. 357–375.

McGrew, W.C., 2001. The other faunivory: Primate insectivory and early human diet. In: Stanford, C.B., and Bunn, H.T. (Eds.), *Meat-Eating and Human Evolution*. Oxford University Press, New York, pp. 160–178.

Milton, K., 1984. The role of food-processing factors in primate food choice. In: Rodman, P.S., and Cant, J.G.H. (Eds.), *Adaptations for Foraging in Nonhuman Primates: Contributions to an Organismal Biology of Prosimians, Monkeys, and Apes*. Columbia University Press, New York, pp. 249–279.

Milton, K., 1999. A hypothesis to explain the role of meat-eating in human evolution. *Evol. Anthropol.* 8, 11–21.

Morgan, M.E., Kingston, J.D., and Marino, B.D., 1994. Carbon isotopic evidence for the emergence of C4 plants in the Neogene from Pakistan and Kenya. *Nature* 367, 162–165.

Murray, S.S., Schoeninger, M.J., Bunn, H.T., Pickering, T.R., and Marlett, J.A., 2001. Nutritional composition of some wild plant foods and honey used by Hadza foragers of northern Tanzania. *J. Food Compos. and Anal.* 14, 3–13.

Newman, J.L., 1975. Dimensions of Sandawe diet. *Ecol. Food Nutr.* 4, 33–39.

O'Connell, J.F., Hawkes, K., and Blurton Jones, N.G., 1999. Grandmothering and the evolution of *Homo erectus*. *J. Hum. Evol.* 36, 461–485.

Peters, C.R., 1987. Nut-like oil seeds: food for monkeys, chimpanzees, humans, and probably ape-men. *Am. J. Phys. Anthropol.* 73, 333–363.
Peters, C.R., and Vogel, J.C., 2005. Africa's wild C4 plant foods and possible early hominid diets. *J. Hum Evol.* 48, 219–236.
Potts, R., 1998. Variability selection in hominid evolution. *Evol. Anthropol.* 7, 81–96.
Prentice, A., Laskey, M.A., Shaw, J., Hudson, G.J., Day, K.C., Jarjou, L.M.A., Dibba, B., and Paul, A.A., 1993. The calcium and phosphorus intakes of rural Gambian women during pregnancy and lactation. *Brit. J. Nutr.* 69, 885–896.
Reed, K.E., 1997. Early hominid evolution and ecological change through the African Plio-Pleistocene. *J. Hum. Evol.* 32, 289–322.
Schoeninger, M.J., Bunn, H.T., Murray, S.S., Pickering, T.R., and Marlett, J.A., 2001a. Nutritional composition of tubers used by Hadza foragers of northern Tanzania. *J. Food Compos. and Anal.* 14, 15–25.
Schoeninger, M.J., Bunn, H.T., Murray, S.S., Pickering, T., and Moore, J., 2001b. Meat-eating by the fourth African ape. In: Stanford, C.B., and Bunn, H.T. (Eds.), *Meat-Eating and Human Evolution*. Oxford University Press, New York, 179–195.
Schoeninger, M.J., Hallin, K., Reeser, H., Valley, J.W., and Fournelle, J., 2003. Isotopic alteration of mammalian tooth enamel. *Int. J. Osteoarchaeol.* 13, 11–19.
Schoeninger, M.J., Iwaniec, U.T., and Glander, K.E., 1997. Stable isotope ratios monitor diet and habitat use in New World Monkeys. *Am. J. Phys. Anthropol.* 103, 69–83.
Schoeninger, M.J., Iwaniec, U.T., and Nash, L.T., 1998. Ecological attributes recorded in stable isotope ratios of arboreal prosimian hair. *Oecologia* 113, 222–230.
Schoeninger, M.J., Moore, J., and Sept, J.M., 1999. Subsistence strategies of two "savanna" chimpanzee populations: The stable isotope evidence. *Am. J. Primatol.* 47, 297–314.
Schoeninger, M.J., Reeser, H., and Hallin, K., 2003. Paleoenvironment of *Australopithecus anamensis* at Allia Bay, East Turkana, Kenya: Evidence from mammalian herbivore enamel stable isotopes. *J. Anthropol. Archaeol.* 22, 200–207.
Schwarcz, H., and Schoeninger, M.J., 1991. Stable isotope analyses in human nutritional ecology. *Yearb. Phys. Anthropol.* 34, 283–321.
Semaw, S., Simpson, S.W., Quade, J., Renne, P.R., Butler, R.F., McIntosh, W.C., Levin, N., Dominguez-Rodrigo, M., and Rogers, M.J., 2005. Early Pliocene hominids from Gona, Ethiopia. *Nature* 433, 301–305.
Sept, J., 2001. Modeling the edible landscape. In: Stanford, C.B., and Bunn, H.T. (Eds.), *Meat-Eating and Human Evolution*. Oxford University Press, New York, 73–98.
Shipman, P., and Walker, A., 1989. The costs of becoming a predator. *J. Hum. Evol.* 18, 373–392.
Sillen, A., 1992. Strontium-calcium ratios (Sr/Ca) of *Australopithecus robustus* and associated fauna from Swartkrans. *J. Hum. Evol.* 23, 495–516.
Skinner, 1991. Honey and bee larvae. *J. Hum. Evol.* 20, 493–503.
Sponheimer, M., and Lee-Thorp, J.A., 1999. Isotopic evidence for the diet of an early hominid, *Australopithecus africanus*. *Science* 283, 368–370.
Sponheimer, M., and Lee-Thorp, J.A. 2006. Enamel diagenesis at South African Australopith sites: Implications for paleoecological reconstruction with trace elements. *Geochim. et Cosmochim. Acta,* in press.
Sponheimer, M., Lee-Thorp, J., de Ruiter, D., Codron, D., Codron, J., Baugh, A.T., and Thackeray, F., 2005. Hominins, sedges, and termites: New carbon isotope data from Sterkfontein valley and Kruger National Park. *J. Hum. Evol.* 48, 301–312.
van der Merwe, N.J., Thackeray, J.F., Lee-Thorp, J.A., and Luyt, J., 2003. The carbon isotope ecology and diet of *Australopithecus africanus* at Sterkfontein, South Africa. *J. Hum. Evol.* 44, 581–597.
Van Soest, P.J., 1963. Use of detergent in the analysis of fibrous feeds. II. A rapid method for the determination of fiber and lignin. *J. AOAC Int.* 46, 829–835.
Vincent, A., 1985a. Plant foods in savanna environments: A preliminary report of tubers eaten by the Hadza of northern Tanzania. *World Arch.* 17, 132–148.
Vincent, A., 1985b. Wild tubers as a harvestable resource in the East African savannas: Ecological and ethnographic studies. PhD thesis, University of California, Berkeley.

White, J.W.J., Riethof, M.L., Subers, M.H., and Kushnir, I., 1962. *Composition of American Honeys*. U.S. Government Printing Office, Washington, DC.

Woodburn, J.C., 1968. An introduction to Hadza ecology. In: Lee, R., and DeVore, I. (Eds.), *Man the Hunter*. Aldine de Gruyter, Chicago, 49–55.

Wrangham, R.W., Jones, J.H., Laden, G., Pilbeam, D., and Conklin-Brittain, N., 1999. The raw and the stolen: Cooking and the ecology of human origins. *Curr. Anthropol.* 40, 567–594.

Wynn, J.G., 2000. Paleosols, stable carbon isotopes, and paleoenvironmental interpretation of Kanapoi, Northern Kenya. *J. Hum. Evol.* 39, 411–432.

PART III

THE ARCHAEOLOGICAL RECORD

10

Zooarchaeology and the Ecology of Oldowan Hominin Carnivory

ROBERT J. BLUMENSCHINE
BRIANA L. POBINER

Zooarchaeology is the study of animal-body and trace fossils found in association with hominin body or trace fossils. Zooarchaeology in the nineteenth century served to assign antiquity to stone artifacts associated with extinct animals and to evaluate the paleoenvironmental setting of archaeological occurrences (Daniels, 1975). The current role of zooarchaeology has expanded to include evaluations of the physical and biotic processes responsible for the formation of paleoanthropological assemblages, including most centrally the suite of hominin behaviors and ecological interactions recorded by the animal remains. For the dietary ecology of Quaternary hominins, zooarchaeology is the principle source of evidence for documenting the origin and evolution of larger-mammal carnivory.

Larger-mammal carnivory signals a major niche expansion within the hominin lineage. The likely frugivorous ancestral diet (e.g., Andrews and Martin, 1991) that probably included small animals (<5 kg body weight) was supplemented by foods from animals of similar and far greater body size than hominins. The novel stone-tool slicing technology of the Oldowan Industry enabled this niche shift (e.g., Isaac, 1983). It involved often distant transport of stone materials to carcass-processing locales. Carcass foods were also likely transported at times from acquisition sites to places affording safer processing and consumption (Isaac, 1983; Peters and Blumenschine, 1995, 1996). Stone-tool-assisted butchery of larger mammals is highly distinctive from the small-animal faunivory of nonhuman primates, limiting the usefulness of predation by chimpanzees and baboons for modeling its ecology and evolutionary consequences (cf. Stanford, 1996).

Oldowan hominins were apparently the first primate to enter the larger-carnivore guild, composed most prominently of a diverse series of felids, hyaenids, and canids (Turner, 1990; Werdelin and Lewis, 2005), as well as crocodiles and vultures.

Oldowan hominins likely remained prey for most guild members, but they were now also potential food competitors for some. This ecological transformation can be linked to the evolution of the hominin brain and social systems (cf. Washburn and Lancaster, 1968; Isaac, 1978; Foley and Lee, 1991; Aiello and Wheeler, 1995; Foley, 2001).

The zooarchaeological record makes three unique contributions to early hominin dietary reconstruction. First, butchery marks on bone allow the realized carnivorous niche of hominins to be itemized. These dietary trace fossils record the edible tissues hominins extracted from animals of specifiable size and sometimes species. Such specificity about paleodiet composition cannot be ascertained currently through the morphology or chemistry of hominin body fossils.

Second, co-occurring butchery-marked and tooth-marked bone is informative about modes of carcass acquisition and the carnivore species with which hominins interacted over food directly or indirectly. Temporal changes in these ecological traits define the evolving role of hominins in the larger carnivore guild.

Third, the relative commonness of fossils bearing feeding traces of the carnivore guild allows hominin dietary ecology to be examined in paleolandscape context (Blumenschine and Peters, 1998). The suite of carnivores associated with landscapes supporting patchy tree cover theoretically afforded hominins varying degrees of predator encounter risk and competition for carcass foods. These ecostructural factors, which have been modeled for the Plio-Pleistocene Olduvai lake basin (Peters and Blumenschine 1995, 1996), likely affected the amounts, sources, and processing of carcass foods by hominins in different landscape settings (Blumenschine and Peters, 1998). The zooarchaeological record in paleolandscape context can be used to test these theoretical constraints on early hominin carnivory (Blumenschine et al., 2006.).

In this chapter, we evaluate the state of zooarchaeological knowledge about larger-animal carnivory in Oldowan hominins, an informal ichnospecies designation that includes possibly several species who made and/or used stone artifacts assigned to the Oldowan Industry, dating from approximately 2.6–1.6 Ma (Plummer, 2004). The zooarchaeology of the Developed Oldowan and later industries is not considered here. We will address four aspects of Oldowan hominin carnivory that together define the emerging role of hominins in the larger carnivore guild:

1. Diet composition, focusing on larger mammals.
2. Carcass processing equipment and techniques.
3. The relative importance of larger mammal foods to Oldowan hominin diets.
4. The ecology of larger mammal carnivory by Oldowan hominins, emphasizing modes of carcass acquisition and potential interactions with large carnivores.

Methodological Considerations: The Known, the Unknown, and the Unknowable

Assigning aspects of hominin carnivory to "the known, the unknown, and the unknowable" depends on how well these can be linked to the production of dietary traces (cf. Binford, 1981; Gifford-Gonzalez, 1991). Zooarchaeological traces of diet include bone modifications and the taxonomic, skeletal part, and age composition of fossil assemblages. The significance of these traces of ancient diets can be ascertained by reference to direct observations of modern processes that leave similar traces. Such observations are referred to as *actualistic*, or *neotaphonomic*.

Zooarchaeological assemblages are often taphonomically complex, products of numerous biological, chemical, and physical processes in addition to hominin dietary behavior. Traces of hominin diet are imparted on bones during the very brief period that they bear or contain food resources. Stone-tool processing of carcasses inflicts feeding traces on bone during dismemberment, skinning, defleshing, and extraction of within-bone nutrients. The great quantity of nutrients concentrated in larger mammal carcasses prolongs their resource life, thus attracting a potentially wide range of vertebrate and invertebrate carnivores. Each consumer has the potential to modify, destroy, or disperse bones bearing hominin butchery traces or to add bones to an assemblage. Once the soft-tissue nutrients of a carcass are exhausted, its degreased bones are subject to a further array of microbial, physical, and chemical processes during burial, fossilization (Hedges, 2002), and reexposure. For these reasons, taphonomic complexity can limit dietary inference and bias it if undetected.

The taphonomic complexity of an assemblage also adds ecological information to zooarchaeological interpretations (Behrensmeyer and Kidwell, 1985; Gifford-Gonzalez, 1991). Feeding traces of larger carnivores are especially informative about prehistoric hominin feeding ecology because they record intraguild interactions that took place during the brief resource life of a carcass (figs.10.1 and 10.2).

Neotaphonomic observations provide a method for benefiting from the taphonomic complexity of zooarchaeological assemblages. Neotaphonomic observations that establish process-trace links for carnivory and distinguish these from other potential trace-mimicking processes can be used to build models of hominin diet and intraguild feeding interactions. Neotaphonomic models that achieve realism (demonstrating contextual similarity with the past) and precision (replicating trace discrimination) are best suited for discerning the behavioral and ecological traces in taphonomically complex assemblages. Extinctions, unfamiliarity with relevant modern processes, Holocene degradation of modern landscape analogs, and a poor understanding of the prehistoric system limit a model's realism. Available instrumentation, small control samples, poor observational controls, and insufficient training in trace discrimination limit its precision.

Some classes of zooarchaeological evidence impart more precision to diet model building than others. Hominin dietary interpretations based on assemblage-level traces, such as species, skeletal element, and age profiles, are subject to interpretive ambiguity arising from equifinalities in process-trace relationships (e.g., Marean et al., 1992; Marean, 1995, 1998; Bartram and Marean, 1999; Marean and Assefa, 1999; Pobiner and Blumenschine, 2003). Because the dietary trace is the assemblage itself, or some subset of the assemblage, one must assume that all elements of the sample are attributable to hominin accumulation and modification. Such taphonomic singularity would pertain only if hominins maintained exclusive access to carcass-processing locales during the time that their food debris remained nutritionally attractive to other carnivores and if no other agents subsequently added or subtracted bones from the locale. These conditions appear unlikely to apply to Oldowan localities. Deemphasis of the effect of nonhominin agencies, a common occurrence

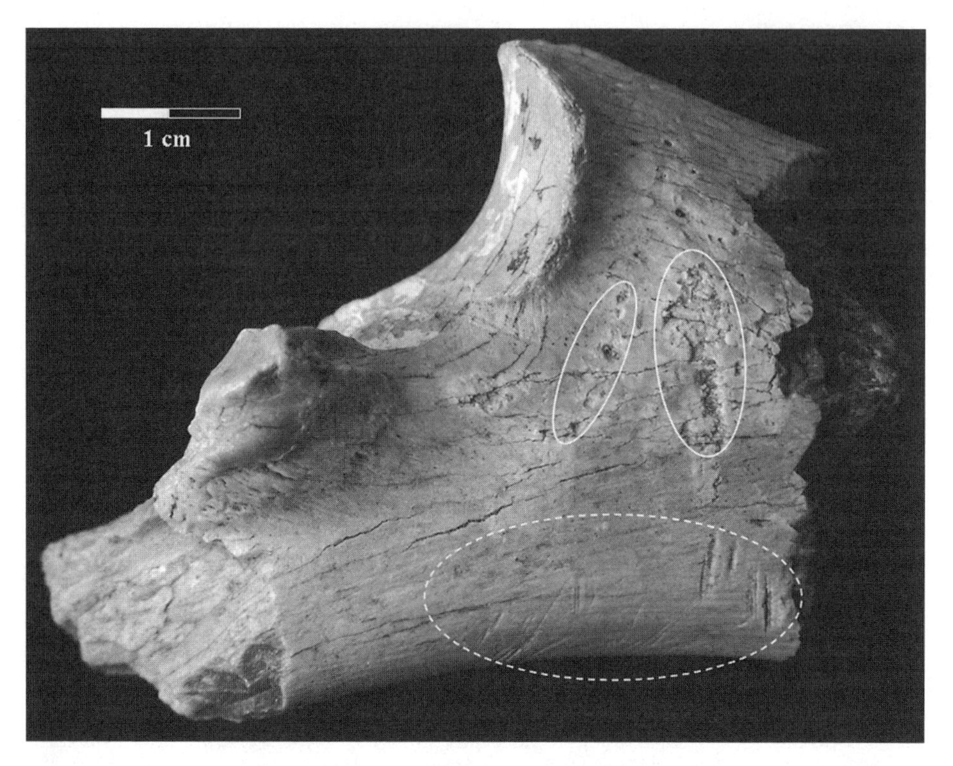

Figure 10.1 Fossil proximal ulna of a medium-sized bovid from the surface of Bed I, Olduvai Gorge, Tanzania. The specimen has been cut-marked repeatedly with one or more stone tools (narrow linear marks encircled by dashed line) and tooth marked conspicuously by a large carnivore (pits and broad scores encircled by solid lines). Such evidence records potential intra-guild interactions, the nature of which is critical to an understanding of hominin evolutionary biology.

in zooarchaeology, suppresses the ecological information encoded in taphonomically complex assemblages.

Dietary interpretations based on the modification of individual bones can take interpretive advantage of taphonomic complexity. Patterns of gross gnawing on bones by carnivores can indicate the body size and jaw strength of the carnivore (Haynes, 1983; Pobiner and Blumenschine, 2003; Pobiner, in prep.). Notches on bone fracture edges produced by carnivore teeth can in many cases be distinguished from those inflicted by hammerstone impact (Capaldo and Blumenschine, 1994). Fracture angles produced by hominins and carnivores on green-broken limb bones are also often distinctive (Pickering et al., 2005). Marks on bone surfaces are the most agent-specific form of bone modification. Stone tool butchery traces (cut marks, chop marks, percussion marks, scraping marks) are morphologically distinct from tooth marks inflicted by carnivores, and both classes of marks are distinct from surface marks produced by rodent gnawing, root etching, insects, and trampling (Bunn, 1981; Potts and Shipman, 1981; Shipman and Rose, 1984; Behrensmeyer, Gordon, and Yanagi, 1986; Watson and Abbey, 1986; Blumenschine and Selvaggio, 1988, 1991). Given

Figure 10.2 Spotted hyena feeding on a Thomson's gazelle, Serengeti, Tanzania. Observations of carcass consumption by modern carnivores coupled with controlled butchery using stone tools provide a basis for understanding the trace fossil signatures of the evolving role of hominins in the larger carnivore guild.

analyst experience with collections of modern bones modified under controlled conditions (Blumenschine, Marean, and Capaldo, 1996), marks on bone surfaces provide a reliable inventory of many of the biotic processes that influenced a zooarchaeological assemblage, including those potentially interacting with hominins over carcass foods (fig. 10.1).

Given these methodological considerations, we consider a "known" zooarchaeological inference about diet to be one based on a fossil trace that has been shown by independent observations to be reproduced by an observed modern process and demonstrated to be distinct from similar traces produced by other modern processes. "Knowns" will always be subject to some uncertainty if only because extinct processes may remain undiscovered. Despite being interpretively most reliable, tooth marking, cut marking, and percussion marking together have been reported systematically for only a few Oldowan bone assemblages from Olduvai Gorge and East Turkana. As well, the total known sample of butchery-marked bone in Oldowan contexts is very small, the vast majority deriving from FLK 22 (*Zinjanthropus* level), Bed I, Olduvai Gorge (table 10.1). A new analysis by Capaldo and colleagues (in prep.; Blumenschine et al., 2006) suggests that butchery marking is more common than indicated previously (Bunn, 1982) at FLKN 1-2, also in Bed I Olduvai. Otherwise, butchery marking is uncommon in the other Oldowan assemblages that

Table 10.1 Occurrence of Cut-Marked and Percussion-Marked Specimens in Those Oldowan Assemblages for Which Butchery Marking Has Been Investigated

Assemblage	No. Specimens Analyzed for Surface Marks	No. Butchery Marked	Source
Olduvai Gorge (Beds I/Lower II)			
FLK 22	2,787 (excluding teeth)	252 cut marked	Bunn and Kroll, 1986
	731 (long bones)	200 percussion marked	Blumenschine, 1995
FLKN 1-2	Not specified	Unspecified number of cut-marked specimens[a]	Bunn, 1982
	1,067 (long bones)	304 percussion marked	Capaldo in Blumenschine et al., 2006
FLKN 6	622	Several probable cut-marked *Elephas* specimens,[b] plus 5 cut-marked specimens from other animals	Bunn, 1982
FLKNN 2	393	1 cut marked	Bunn, 1982
HWKE 1 & 2	884	5 cut marked and 3 percussion marked	Monahan, 1996
OLAPP Trench 57	72 (long bones)	3 cut marked, 6 percussion marked	Blumenschine et al., 2003
East Turkana (KBS Member)			
FxJj 1	245[c]	0 (but poor bone surface preservation)	Bunn, 1997
FxJj 3	363[c]	0 (but poor bone surface preservation)	Bunn, 1997
FxJj 10	5-6	0 (but poor bone surface preservation)	Isaac and Harris, 1997
Lokalalei 1A (W. Turkana)	415	2 possibly cut marked, 1 possibly percussion marked	Kibunjia, 1994
Gona			
DAN2	Not specified[d]	5 cut-marked surface specimens	Domínguez-Rodrigo et al., 2005
EG13	Not specified	1 cut-marked surface specimen	Domínguez-Rodrigo et al., 2005
OGS-6	Not specified	1 cut-marked surface specimen	Semaw et al., 2003
WG9	Not specified[e]	2 cut-marked surface specimens	Domínguez-Rodrigo et al., 2005
Bouri	Not specified	3 cut-marked surface specimens, 1 with percussion damage	de Heinzelin et al., 1999
Sterkfontein Member 5[f]	29,038	2 cut-marked	Pickering, 1999; Pickering et al., 2000

Note: Butchery marks have not been reported for—and likely have not been sought in—the following Oldowan assemblages: those from Shungura Member E (Howell et al., 1987) and Member F (Merrick and Merrick, 1987) of the Lower Omo; Senga 5 (Harris et al., 1987); Lokalakei 2C (Roche et al., 1999); A.L. 666 from Hadar (Kimbel et al., 1996); and Ain Hanech (Sahnouni and de Heinzelin, 1998). J. Ferraro (personal communication) reports from an ongoing analysis the presence of cut- and percussion-marked specimens from Kanjera South (Plummer et al., 1999).

[a] "Cut marks are considerably less abundant than at FLK *Zinjanthropus* [FLK 22], although well preserved examples are present, especially limb bones of *Parmularius altidens*" (Bunn, 1982, p. 135).

[b] Glue used as preservative on *Elephas* bones inhibited confident mark identification by Bunn.

[c] Number of specimens identified to at least Mammalia, presumably all of which were examined by Bunn for surface modifications.

[d] "Hundreds of [surface] stone tools and fossil bone specimens" (Domínguez-Rodrigo et al., 2005, p. 114).

[e] "Several well-preserved [surface] fossil bone fragments" (Domínguez-Rodrigo et al., 2005, p. 114).

[f] Includes two assemblages, the STW53 Breccia and the Oldowan Infill.

have been examined taphonomically (table 10.1). The paucity of evidence for Oldowan hominin butchery may be real, leaving FLK 22 as an unexplained behavioral anomaly. Alternatively, it may reflect the excellent preservation of bone surfaces in the assemblage (e.g., Bunn and Kroll, 1986; Blumenschine, 1995) and the limited to nonexistent investigation of bone modification in other assemblages (table 10.1).

We consider an "unknown" but knowable zooarchaeological inference about diet to be one based on a distinctive fossil trace that can be investigated neotaphonomically but has yet to meet the qualifications of the "knowns." A knowable zooarchaeological inference would also include any aspect of prehistoric diet that could potentially produce a yet-to-be recognized trace.

Finally, we consider as unknowable those aspects of dietary ecology that cannot be conceived to leave a fossil trace. These aspects of diet must be evaluated theoretically.

Zooarchaeological Inferences about Early Hominin Carnivory

Diet Composition

The range of food items consumed is a basic parameter of an organism's dietary niche. Prey characteristics such as body size, habitat preference, population structure, biomass, and antipredator defense mechanisms reflect the consumer's acquisition and processing capabilities. Interspecific competition can restrict diet breadth. The species and tissue types consumed by Oldowan hominins partly define the niche space they occupied within the larger carnivore guild.

Taxonomic lists are provided routinely for Oldowan assemblages, but an inventory of the taxa that bear butchery marking—the only assemblage subset that securely indicates hominin food items—has not been compiled. Tables 10.2 and 10.3 list the ungulates present in East African Oldowan assemblages that have yielded butchery-marked bone, indicating those taxa that are reported to be butchery marked.

Fourteen of the ungulate genera and species from Oldowan localities are reported to be butchery marked (table 10.2). All but three of these species are represented by butchery-marked specimens at FLK 22. Assuming that the butchery marking signals extraction of edible tissues (as opposed to only skinning, disarticulation, or bone breakage for nondietary reasons), hominins consumed foods from animals ranging in adult body size from approximately 10 kg to more than 2,500 kg and representing grazing and browsing species associated with a range of vegetation settings (table 10.2). Given that most butchery-marked specimens are fragmentary and can be identified no more specifically than to Mammalia (table 10.3), the number of taxa in the larger mammal diet of Oldowan hominins was probably greater than that indicated in table 10.2. This could be taken to indicate that the carnivorous niche of Oldowan hominins was very broad, although FLK 22, from which the preponderance of evidence derives, may be unrepresentative.

Table 10.2 Ungulate Genera and Species at East African Oldowan Localities

Size/Taxon	FLK 22 CM	FLK 22 PM	BM Elsewhere	No BM at Any Site	Notes on Diet, Vegetation Cover Preferences, and Water Dependency
Size 1 (ca. 10–15 lbs.)					
Antidorcas cf. *recki*[a] (bovid)	X	X			Fossil femora = open to light cover.[b] Fossil teeth = C_4 grazer.[c] Fossil jaws = grass feeder.[d]
Size 2 (50–250 lbs.)					
Gazella janenschi (bovid)				X	Modern *Gazella* range from browsers to grazers.[e]
Aepyceros sp.				X	Modern *Aepyceros* = C_3/C_4 mixed feeder.[e]
Size 3 (250–750 lbs.)					
Kolpochoerus limnetes (suid)	X				Fossil teeth = C_4 grazer.[f] More water dependent than *Metridiochoerus*.[f]
Metridiochoerus andrewsi (suid)	X				Fossil teeth = C_4 grazer.[f] Less water dependent than *Kolpochoerus*.[f]
Metridiochoerus modestus				X	See *Metridiochoerus andrewsi.*
Notochoerus sp. (suid)				X	Fossil teeth of *Notochoerus euilus & N. scotti* = C_4 grazers.[f]
Tragelaphus strepsiceros (bovid)	X	X			Fossil tragelaphine femora = light to heavy cover.[b] Modern *Tragelaphus strepsiceros* = C_3 browser.[e]
Tragelaphus nakuae (bovid)				X	Six other modern tragelaphines are also C_3 browsers.[e]
Tragelaphus pricei (bovid)				X	
Parmularius altidens (bovid)	X	X	FLKN 6, FLKN 1-2		Fossil teeth = C_4 grazer.[c] Fossil jaw = grass feeder.[d]
Parmularius rugosus (bovid)				X	
Kobus sigmoidalis (bovid)	X	X			Fossil teeth = C_4 grazer.[c] Fossil jaw = grass feeder.[d]
Kobus kob (bovid)				X	Fossil femora = forest vegetation.[b] Modern *K. kob* = C_4 grazer/ hypergrazer.[e]
Connochaetes cf. *gentryi*[g] (bovid)	X	X			Fossil jaws = short grass feeder.[d] Modern *Connochaetes* = C_4 grazer/hypergrazer.[e]
Hipparion sp. (equid)			Bouri		
Size 4 (750–2000 lbs.)					
Megalotragus kattwinkeli (bovid)				X	Fossil teeth = C_4 grazer.[c]
Syncerus cf. *acoelotus*[h] (bovid)	X				Modern *Syncerus* = C_4 grazer/ hypergrazer.[e]
Hippotragus gigas (bovid)				X	Hippotragine fossil femora = open vegetation.[b] Modern *Hippotragus equinus & H. niger* = C_4 grazer/ hypergrazer.[e]
Pelorovis sp. (bovid)				X	Fossil femora = heavy cover.[b]
Oryx sp. (bovid)	X				Modern *Oryx* = C_4 grazer.[e]
Giraffa sp. A (giraffid)	X				

Table 10.2 (*Continued*)

Size/Taxon	FLK 22 CM	FLK 22 PM	BM Elsewhere	No BM at Any Site	Notes on Diet, Vegetation Cover Preferences, and Water Dependency
Size 5 (2000–6000 lbs.)					
Giraffa sp. B (giraffid)	X				
Sivatherium sp. (giraffid)				X	
Hippopotamus cf. *gorgops*			FLKN 6		
Hexaprotodon sp. (hippopotamid)				X	
Size 6 (> 6000 lbs.)					
Elephas recki (elephantid)			FLKN 6 (possibly)		Fossil teeth = C_4 grazer.[i]
Deinotherium cf. *bozasi*				X	Fossil teeth = C_3 browser.[i]

Note: Localities for which data on butchery marking are available indicating those that have and have not been reported to be butchery marked (BM). FLK 22, from which the preponderance of evidence for cut-marked (CM; Bunn 1982, Table 4.30) and percussion-marked (PM; Blumenschine, 1995 and unpublished data) specimens derive, is treated individually. See Table 10.1 for references to other assemblages. Four species reported from Bouri (*Beatragus whitei,* cf. *Numidocarpa crassicornis*, cf. *Rabaticeras arambourgi, Damaliscus ademassui;* de Heinzelin et al., 1999) are of uncertain body size; none of these are butchery marked.

[a] Identified as *Antidorcas* sp. at Bouri.
[b] Kappelman et al. (1997).
[c] Blumenschine et al. (2003).
[d] Spencer (1997).
[e] Cerling et al. (2003).
[f] Harris and Cerling (2002).
[g] Identified as *Connochaetes* sp. at FLK 22.
[h] Identified as *Syncerus* sp. at Bouri.
[i] Cerling et al. (1999).

Two species other than ungulates have been reported to be butchery marked in Oldowan assemblages. One is the hedgehog, *Erinaceous broomi*, represented by three cut-marked mandibles from FLKN 5 (Bed I, Olduvai) and interpreted to indicate skinning in preparation for consumption (Fernandez-Jalvo, Andrews, and Denys, 1999). The second is the hominin specimen STW 53c from Sterkfontein Member 5 (Pickering, White, and Toth, 2000). Butchery marking has not been reported on birds, amphibians, reptiles, or fish (Stewart, 1994) in any Oldowan assemblage.

Some unknown, but knowable, aspects of larger vertebrate diet composition include the contribution of birds and other nonmammals. The brain and pulps of the nasal cavity and mandible may have been sought regularly, but taphonomic analyses of cranial and mandibular specimens have yet to be published (e.g., Capaldo et al., in prep.). Oldowan hominins may have eaten grease but, lacking boiling technology, this may have been accomplished by gnawing softer cancellous bone portions of small animals or by chewing pounded, pulverized cancellous bone from larger animals. As yet, hominin tooth marks cannot be distinguished from those inflicted by carnivores.

Unknowable aspects of the larger mammal composition of Oldowan hominin diets include the species consumed only in nondepositional settings and consumption of viscera and blood, which would not be expected to leave distinctive traces on bone.

Table 10.3 Ungulate Taxa Identified Above the Genus Level from FLK 22 and Other East African Oldowan Localities That Have Been Reported to be Butchery Marked

Size/Taxon	FLK 22		Butchery-Marked Elsewhere
	CM	PM	
Size 1 (ca. 10–50 lbs.)			
Bovidae gen. and sp. indet. (bovid)			WG9[a]
Mammalia gen. and sp. indet.			DAN2
Size 1/2			
Mammalia gen. and sp. indet.			LA1 (possibly)
Size 2 (50–250 lbs.)			
Antilopini gen. and sp. indet. (bovid)	X	X	
Bovidae gen. and sp. indet. (bovid)			LA1 (possibly)
Mammalia gen. and sp. indet.			LA1 (possibly)
Size 3 (250–750 lbs.)			
Alcelaphini gen. and sp. indet. (bovid)			Bouri,[b] FLKN 6
Bovidae gen. and sp. indet.			FLKN 6, EG13[b]
Suidae gen. and sp. indet.	X		
Equidae gen. and sp. indet.	X	X	OGS-6, FLKNN 2
Mammal fam., gen., sp. indet.			DAN2 WG9[b]
Size 4 (750–2000 lbs.)			
Bovidae gen. and sp. indet.			Bouri[c]
Mammal size and taxon indet.			DAN2 OLAPP Trench 57 HWKE 1&2

Note: See table 10.1 for references to assemblages. CM = cut-marked; PM = percussion-marked.
[a] Reported in source publication as a "small" bovid.
[b] Reported in source publication as "medium-sized."
[c] Reported in source publication as a "large" bovid.

Carcass Processing Equipment and Techniques

The equipment used by Oldowan hominins to process carcasses is understood broadly from modern replication of butchery marks using stone tools (e.g., Shipman and Rose, 1983). The knifelike edge of unmodified stone flakes was used to skin carcass parts, as documented by cut marks on metapodials (Wilson, 1982), and to slice flesh from bone, as documented by cut marks away from joints (e.g., Binford, 1981; Potts and Shipman, 1981; Bunn and Kroll, 1986; fig. 10.1). Bone disarticulation may be indicated by cut marks at joints (Shipman, 1986). While some authors support the argument linking cut marks on the midshafts of limb bones to defleshing whole muscle masses (e.g., Domínguez-Rodrigo, 1997; Bunn, 2001; see Nilssen [2000] for a thorough review), we caution that additional neotaphonomic observations are needed before cut mark location can be used to infer specific butchery actions and yields (Gifford-Gonzalez, 1989; Egeland, Hill, and Byerly, 2002; B. L. Pobiner, personal observation).

Rounded, coarse-grained hammerstones and anvil stones were used to crack into long bone marrow cavities. This is indicated by the common presence of fine patches of microstriae associated with percussion marks made by grainy material (Blumenschine and Selvaggio, 1988) and by the similar incidence and size distribution of percussion-marked midshaft fragments in modern assemblages generated by hammerstone-on-anvil breakage and the FLK 22 assemblage (Blumenschine, 1995). Had hominins at FLK 22 not used an anvil stone in conjunction with a hammerstone, one would expect a lower, though yet-to-be determined, incidence of percussion-damaged midshafts.

Chopping marks, made by unifacially or bifacially flaked cores, are observed very rarely on long bones from Oldowan contexts (R. J. Blumenschine and B. L. Pobiner, personal observations). Choppers are less effective for marrow extraction than rounded hammerstones because they shatter bone at the impact sites, rather than creating longitudinal fractures to expose the whole marrow cavity. Chopping, however, may have been an effective way to expose the brain.

Unknown but potentially decipherable aspects of carcass processing include the specific stone materials used for various butchery tools, and whether sharp bone flakes or other bone tools were used (e.g., Shipman, Fisher, and Rose, 1984; Shipman, 1989). Use of carcass soft tissues for nonfood purposes is probably not knowable. Potts and Shipman (1981) suggest that cut marks on bovid metapodials indicate tendon use at Olduvai sites, although skinning and removal of periosteum to allow marrow extraction is a more likely cause (e.g., Wilson, 1982). Skin and viscera may have been used for various nondietary purposes, but processing of these tissues is unlikely to leave distinct traces. Likewise, meat drying may have been practiced by Oldowan hominins, but it is unassociated with a distinctive trace (Bartram, 1993).

The Dietary Importance of Larger Mammals

The contribution of larger mammal foods to Oldowan hominin diets is a long-standing issue (e.g., Isaac and Crader, 1981) with a number of components. These include the frequency and quantity of nutrients consumed from different carcass tissue types and species relative to other foods, as well as variability in these parameters by season, landscape setting, and consumer sex and age. While many of these aspects of diet are unknown, with some likely remaining unknowable, the zooarchaeological record provides some reliable and relevant information on the topic.

Oldowan hominins at FLK 22 and FLKN 1-2 broke long bones of size 1–4 mammals in direct proportion to their estimated gross caloric yield from marrow fat (Blumenschine, 1991; Blumenschine and Madrigal, 1993; see table 10.2 for definition of larger mammal size classes). The strongest, most significant correlations are obtained using direct determinations of caloric yield from modern bovids that had suffered minimal-to-no marrow fat depletion, indicating hominin access to carcasses of nutritionally unstressed animals. Long bone abundance of size 3–4 mammals at FLK 22 is also correlated positively and significantly to the net yield of marrow bones (based on Lupo's [1998] estimates of caloric yield per unit processing time; Madrigal and Blumenschine, 2000). At least for FLK 22, where percussion marking has been analyzed, the incidence of percussion-marked long bone fragments indicates that hominins broke most of the marrow bones recovered (Blumenschine, 1995). As well, Bunn estimates long bone abundances using midshaft fragments, such that the values should be affected minimally, if at all, by carnivore deletion of hammerstone-generated long bone end fragments, following neotaphonomic findings (e.g., Blumenschine, 1988; Marean and Spencer, 1991). Here, long bones may have been introduced to the locality by carnivores, hominins, or natural deaths in abundances differing from those recovered, but mainly those that were hammerstone broken were not deleted completely by scavenging carnivores.

The findings that Oldowan hominins made marrow processing decisions partly on the basis of gross and net energy yield suggests that carcass processing decisions in general may have taken account of the energy yields of other classes of foods. The net yield of defleshed marrow bones from small and medium-sized bovids ranges from approximately 1,000–10,000 kcal/h processing time (R. J. Blumenschine, unpublished data). Assuming opportunistic encounter of carcasses, these net yields are comparable to or higher than those for nonlarger mammal food items harvested by tropical hunter-gatherers (e.g., the Ache [Hawkes, Hill, and O'Connell, 1982] and the Alyawara [O'Connell and Hawkes, 1981]). The net yields of extracting the brain and flesh are even higher, in some cases, by almost two orders of magnitude (R. J. Blumenschine, unpublished data). Following optimal diet breadth models (e.g., MacArthur and Pianka, 1966), food items in the optimal diet set are expected to be consumed whenever encountered. The question of the regularity of animal food intake by Oldowan hominins therefore is about the prehistoric availability of larger mammal carcasses, which is dependent on a variety of ecological variables (e.g., mode of acquisition, degree of competition). Regardless, the absolute high yield of marrow alone suggests that hominins would exploit it whenever encountered, assuming low risk: the twelve major long bones from a non-fat-depleted adult wildebeest yield about 3,000 kcal (table 10.4), an amount that would have easily satisfied

Table 10.4 Food Yields from Selected Adult Bovids and Carcass Portions

Food Item	Marrow g	Marrow Kcal[a]	Flesh kg
Thomson's gazelle (size 1)			
1 hindlimb	19.6	164	1.6
1 forelimb[b]	10.2	85	0.7
Whole carcass[c]	59.6	498	9.1
Grant's gazelle (size 2)			
1 hindlimb	66.7	558	4.7
1 forelimb[b]	34.5	229	2.4
Whole carcass[c]	202.4	1,694	30.2
Wildebeest (size 3)			
1 hindlimb	104.7	876	9.4
1 forelimb[b]	74.8	626	7.4
Whole carcass[c]	359.0	3,004	71.0

Note: Flesh weights are from Blumenschine and Caro (1986, appendix 1, mean values for all adults). Marrow wet weights are adult means from Blumenschine and Madrigal (1993, table 10.2). The estimated daily caloric expenditure for *Homo habilis* ranges from ca. 1,200–2,700 kcal (Leonard and Robertson, 1997).

[a] kcal = 93% of wet marrow weight in grams (assumes maximal fat levels, as in unstressed animals) times 9.

[b] Includes scapula for flesh weight.

[c] For marrow, includes the 12 major limb bones only.

the daily caloric requirements of an adult Oldowan hominin (Leonard and Robertson, 1997) with a low investment of processing time (less than thirty minutes; Blumenschine 1986b, 1991).

The amounts of flesh hominins extracted and presumably consumed from larger mammal carcasses can likely be resolved. A current debate centers on whether the frequency and anatomical patterning of cut marks on larger mammal bones indicates defleshing of large muscle masses or only small scraps of flesh. Citing a similar overall proportion of cut-marked bone at FLK 22 to that found in Holocene assemblages, and the presence of cut marks on bones with high flesh yields, Bunn (2001; Bunn and Kroll, 1986) argues that Oldowan hominins defleshed intact carcass parts, including whole limbs. Bunn, however, has not supported this assertion with butchery trials that vary the amounts of flesh present. The presumption is that large quantities of flesh were consumed. For example, a single hindlimb of an adult Grant's gazelle (size 2) yields almost 5 kg of flesh, while that of an adult wildebeest (size 3) yields about 10 kg of flesh; an entire wildebeest carcass yields approximately 70 kg of flesh (table 10.4).

Binford (1986, 1988) argues, however, that the cut marks at FLK 22 were inflicted while extracting relatively small quantities of near-bone flesh scraps abandoned by nonhominin carnivores. Cheetah and lions typically abandon tens to hundreds of grams of flesh scraps on individual skeletal parts distributed widely over a carcass (Blumenschine, 1986b; but see Domínguez-Rodrigo, 1999).

Butchery trials conducted by Domínguez-Rodrigo (1997) suggest that bones with more flesh present before butchery are more likely to be cut marked. Bunn (2001) notes that no simple relationship exists between cut-mark frequency and the amount of flesh removed by Hadza hunter-gatherers; he suggests that cut-mark frequency is directly proportional to the strength of muscle attachments. Other studies show no relationship between the number of cut marks on a bone and either the amount of flesh removed (Pobiner and Braun, 2005) or the number of slicing strokes employed (Egeland, 2003). Further butchery studies are needed to identify the relationship between the incidence, orientation, and anatomical placement of cut marks and the amount of flesh removed.

Potential seasonality in carcass consumption should also be detectable. Speth and Davis (1976) use the presence of tortoises in some Olduvai assemblages to suggest that they accumulated during the wet season. Other indicators of season include cementum bands on ungulate teeth (Lieberman, 1994) and the presence of migratory bird species (e.g., Matthiesen, n.d.). These, however, have not been applied to the question of seasonality in Oldowan hominin carnivory.

The Ecology of Oldowan Hominin Carnivory

The relative contribution of hunting and scavenging to Oldowan hominin carnivory is an ecological question about the emerging role of hominins within the larger carnivore guild of the sub-Saharan African Plio-Pleistocene. During the Oldowan in East Africa, this guild included the lion, leopard, cheetah, and forms ancestral to the modern spotted, brown, and striped hyenas; it also included the extinct saber-toothed cats (*Homotherium, Megantereon, Dinofelis*), a wolf-sized canid (*Canis africanus*), and a hypercarnivorous hyaenid (*Chasmoporthetes*; Werdelin and Lewis,

2005). Crocodiles (*Crocodylus lloidi*; Tchernov, 1986) and vultures were among the nonmammalian members of the guild. Four issues are relevant to the role of Oldowan hominins in the guild: (1) the timing of hominin access to carcasses relative to that of other carnivores; (2) the identity of the carnivores with which hominins interacted directly or indirectly over carcass foods; (3) the nature and intensity of these intraguild interactions; and (4) the hominin technological, behavioral and social responses to these interactions.

Early emergence of effective predatory capabilities is implied by the hypothesis that Oldowan hominins hunted larger mammals. If hunting accounts for the full size and ecological range of butchery-marked animals (table 10.2), capabilities comparable to today's top African predators (lions, crocodiles) would be indicated (cf. Schaller, 1972; Pooley, 1989). Such capabilities imply that hominins possessed effective weaponry and used cooperative tactics to gain regular access to and a major dietary contribution from larger mammal foods. Such capabilities were assumed in some earlier reconstructions (e.g., Dart, 1953), but decisive zooarchaeological test implications are still lacking.

Early emergence of dominant parasitizing capabilities within the guild is implied by the hypothesis that Oldowan hominins obtained larger mammal carcasses by driving predators from kills. Such confrontational, or "power" (Bunn and Ezzo, 1993; Bunn, 2001), scavenging would likely require that hominins parasitized both solitary and social carnivores. Claims that hominins acquired fully fleshed carcass parts in this way (Bunn and Ezzo, 1993; Domínguez-Rodrigo, 1997, 2002; Bunn, 2001), including those typically consumed very early in feeding episodes (e.g., upper hindlimbs; Blumenschine, 1986a), imply confrontation with predators very soon after a kill was made. As with predation, confrontational parasitism implies regular access to very large quantities of carcass foods (table 10.4), as well as the possession of effective weaponry and coordinated group tactics.

A more peripheral, low-ranking role within the larger carnivore guild is implied by the hypothesis that hominins scavenged carcasses permanently abandoned by initial consumers. Two forms of such passive scavenging have been hypothesized. One involves parasitizing solitary predators that had abandoned a carcass temporarily, as in the hypothesized case of hominins scavenging kills cached in trees by leopards between feeding bouts (Cavallo and Blumenschine, 1989). Here, substantial quantities of flesh as well as marrow and brain tissue can be obtained without confronting the predator.

Another hypothesized form of passive scavenging involves kills permanently abandoned by the predator due to satiation or anatomical limitations to extracting remaining food. Felid kills remaining with near-bone flesh scraps, all marrow, and the brain are the main example hypothesized by Blumenschine (1986b, 1987), based on the minimum amount of food remaining on cheetah kills of size 1 adult ungulates and lion kills of size 3 adult and juvenile ungulates. Passive scavenging of abandoned kills involves competition with other potential scavenging species, including small carnivores, vultures, and other birds for flesh scraps, and hyaenids and possibly large canids for marrow and brain tissue. Hominin competition with these scavengers could have involved direct interspecies contests over food. Hominins may also have been effective indirect competitors with hyaenids for felid kills abandoned during the day in riparian woodland settings, given the delayed discovery of these carcasses by spotted hyenas in the modern Serengeti (Blumenschine, 1986b, 1987). Passive scavenging is commonly assumed to involve opportunistic and infrequent access to nutritionally marginal amounts of pathogenic, putrefied animal foods (e.g., Ragir, Rosenberg, and Tierno, 2000). However, observations of scavenging opportunities from over 250 fresh carcasses in the modern Serengeti region coupled with the high gross and net yields of common leftovers such as marrow and the brain indicate otherwise (table 10.4; Blumenschine, 1986b, 1987; Cavallo and Blumenschine, 1989; Blumenschine and Cavallo, 1992).

Oldowan hominins also remained a prey species for larger carnivores. Adult hominins were likely vulnerable to all large felid, canid, hyaenid, and crocodilian predators, and young hominins were additionally vulnerable to smaller predators. The hypothesized capabilities of predatory and confrontational scavenging hominins suggest lower vulnerability to predation than does passive scavenging. In especially the latter case, avoidance of potential predators may have been an important determinant of Oldowan hominin land use (Peters and Blumenschine, 1995, 1996).

Neotaphonomic simulations of hominin butchery and bone breakage coupled with observations of carnivore consumption of whole carcasses and butchered food refuse provide a means for testing the alternative hypotheses about carcass acquisition by hominins. These neotaphonomic studies provide expectations for the assemblage-wide frequencies of larger mammal bone bearing tooth marks and/or butchery marks that vary with the sequence of carnivore and hominin access to carcass parts. Blumenschine (1988, 1995) proposed three models of carcass access based on bovid long bones. These include one for defleshing and hammerstone breakage of marrow bones by hominins only; another for defleshing and marrow extraction by carnivores only; and a third for hominin defleshing and marrow extraction followed by carnivore (mainly spotted hyena) consumption of long bone ends retaining bone grease. The hominin-only and hominin-followed-by-carnivore models provide test implications for the hunting and early confrontational scavenging hypotheses that have been replicated independently by Capaldo (1997, 1998) and Marean (in Blumenschine and Marean, 1993). A fourth model was added by Selvaggio (1994, 1998) that involves defleshing by felids followed by removal of flesh scraps and marrow extraction with stone tools, and finally, in some cases, consumption of bones retaining grease by spotted hyenas. This model provides test implications for the passive scavenging hypothesis. The incidence of percussion- and tooth-marked bone can distinguish early access to carcasses by hominins (hunting and confrontational scavenging) from late access to largely defleshed carcasses permanently abandoned by initial consumers.

These models have been applied to only four Oldowan assemblages, all from Bed I Olduvai (Blumenschine, 1995; Selvaggio, 1998; Blumenschine et al., 2006.; Capaldo et al., in prep.). Only the data for larger mammal long bones from FLK 22 has been published fully.

Two critical facts emerge for the FLK 22 analysis. One noted earlier, that hominins broke most or all marrow bones at FLK 22, is based on an incidence of percussion-marked specimens statistically indistinguishable from the hominin-only and hominin-followed-by-carnivore models. The second is that the incidence of tooth-marked long bone midshaft fragments (58%) is statistically distinct from and intermediate to the rates of tooth-marking in the hominin-followed-by-carnivore model (11%) and the carnivore-only model (83%). These results were hypothesized to indicate that hominins at FLK 22 broke marrow bones that had been largely defleshed by felids who left tooth marks on the still intact long bone shafts (Blumenschine, 1995). Selvaggio's (1998) finding of similar (65%) tooth mark frequencies on long bone shafts in her carnivore-to-hominin simulations fail to reject this hypothesis, suggesting that hominins scavenged long bones from permanently abandoned felid kills, removed remaining scraps of flesh, and extracted marrow. This inference is consistent with Shipman's (1986) finding of eight long bone specimens from Bed I Olduvai that bear a carnivore tooth mark overlain by a stone-tool cut mark, indicating hominin access after carnivore feeding. Selvaggio (1998) also finds that the model involving initial bulk defleshing by felids predicts the frequency of cut-marked bone at FLK 22. Domínguez-Rodrigo (1997) claims that the frequency of cut marking on different long bones indicates hominin access at FLK 22 to fully fleshed bones, but this inference is based on small neotaphonomic samples, the results of which have not been replicated independently (Blumenschine et al., 2006). Domínguez-Rodrigo (1999) also asserts that lions do not produce large numbers of tooth marks on long bone midshafts during flesh consumption, but this conclusion is based on bones that had not been collected and cleaned, and on which he states that only the most conspicuous tooth marks could be detected (Domínguez-Rodrigo 1999, p. 380).

While passive scavenging by hominins is indicated by the FLK 22 assemblage, S.D. Capaldo's results from FLKN 1-2 show that frequencies of tooth-marked long bone midshaft fragments are predicted by the hominin-followed-by-carnivore model (Blumenschine et al., 2006). This result suggests that hominins at FLKN 1-2 may have had access to larger quantities of flesh than available on thoroughly defleshed predator kills and underscores the likelihood that carcass foods were acquired from multiple sources in a variety of ways.

The identity of the carnivores with which hominins interacted directly or indirectly over carcass foods is largely unknown but knowable to perhaps the level of family and/or body-size class. The apparent deletion of hammerstone-generated long bone ends of size 3 and larger animals is strongly indicative of scavenging of hominin food refuse by large bone-cracking carnivores, including *Crocuta* or possibly *Canis africanus*. Progress has been made recently toward identifying the carnivores that may have provided scavengeable food to hominins and the major predators of hominins. Following on studies of gross bone damage by specific carnivores (Brain, 1981; Haynes, 1983), Pobiner and Blumenschine (2003) demonstrate a scaling in bone damage and destruction with increasing carcass size by modern carnivores of distinct bone-eating capabilities (lion, cheetah, spotted hyena). This result suggests that general taxon-specific patterns of skeletal element and portion survival can be diagnosed in the fossil record, given knowledge of the bone-modification capabilities of fossil carnivores (e.g., Emerson and Radinsky, 1980; Marean, 1989; Marean and Ehrhardt, 1995; Van Valkenburgh, Teaford, and Walker, 1990; Anyonge, 1996; Biknevicius, Van Valkenburgh, and Walker, 1996).

The taxon-specific morphology of carnivore tooth marks is also being explored. On the basis of comparisons with the size of tooth marks made by known modern carnivores, Selvaggio and Wilder (2001) conclude that more than one carnivore taxon modified bones at FLK 22. Domínguez-Rodrigo and Piqueras (2003) show that the diameter of tooth pits reflects carnivore body size generally. Njau and Blumenschine (2006) demonstrate the unique morphology of some tooth marks inflicted by Nile crocodiles and have identified crocodile-modified bone in Oldowan contexts from Olduvai Gorge. Pobiner (in prep.) is conducting a neotaphonomic study that aims to establish taxon specificity of larger mammalian carnivore gross bone damage and destruction, as well as the frequency, anatomical location, and morphology of tooth marks.

Aspects of hominin responses to intraguild interactions in different landscape settings are also potentially knowable. For example, the ratio of long bone shaft to long bone end specimens is a measure of hyena ravaging of bone assemblages that varies with the degree of competition for carcass foods (Blumenschine, 1989; Marean et al., 1992; Blumenschine and Marean, 1993). High levels of ravaging occur in contexts where hyenas are relatively abundant, such as less-wooded landscapes where expected predation risk for hominins is also high (Peters and Blumenschine, 1995, 1996). Theoretically, hominins might have transported carcass parts from dangerous settings to those affording relatively low predation risk and carried large modified or unmodified stones for use as defensive weapons against potential predators (Blumenschine and Peters, 1998). The long bone end : shaft ratio has been shown to correlate to the density and composition of Oldowan stone artifact assemblages across landscapes of the eastern lowermost Bed II Olduvai Basin (Blumenschine et al., in press; Blumenschine et al., 2006) in theoretically expected ways.

Carcass acquisition in dangerous locales might expectedly be limited to older subadult and prime adult male hominins, although direct zooarchaeological traces of such sex and age limitations are unlikely to be discerned (Blumenschine and Peters, 1998). Likewise, the size of carcass foraging groups, and the potential sharing of carcass foods with individuals not involved in carcass acquisition (cf. Isaac, 1978) seems unknowable for the Oldowan through direct zooarchaeological traces, remaining in the theoretical realm (cf. Marshall, 1994).

Conclusions

Zooarchaeology provides conclusive evidence that Oldowan hominins extracted edible tissues from a wide size and ecological range of larger mammals that sustained the diverse carnivore guilds of the sub-Saharan African Plio-Pleistocene. Observations of fresh carcass consumption by modern African carnivores show that marrow from long bones, the brain, and scraps of flesh would have been the most commonly available carcass foods (Blumenschine, 1986a, 1986b) unless hominins hunted routinely. Marrow, providing a rich source of fat, was extracted using a hammerstone-on-anvil technique preferentially from long bones with higher gross and net caloric yields. The brain, a concentrated source of protein and fat, was probably consumed, although published efforts to detect percussion marks on cranial bones are lacking. Knifelike stone flakes were used to remove flesh, but the amount of flesh Oldowan hominins consumed is uncertain. Even minimal flesh yields, such as the scraps that are commonly abandoned by felids, would provide substantial protein while incurring little extraction time and effort and relatively low risk of injury or predation.

Access to more fully fleshed carcasses would provide hominins with extremely large amounts of protein, even from size 1 larger mammals (table 10.4).

The amount of flesh to which hominins had access defines a zooarchaeological objective central to evaluating the nature of the emerging role of Oldowan hominins in the larger carnivore guild. This role, whether peripheral and subordinate or central and dominant, is a knowable ecological datum of great relevance to early hominin evolutionary biology. A more peripheral and subordinate role for hominins as passive scavengers of abandoned felid kills is the most conservative hypothesis that has yet to be rejected. It is consistent with the apparent lack of effective weaponry for subduing prey or driving predators from kills, and the small body size of Oldowan hominins. It is also consistent with the apparent paucity of direct, butchery-mark evidence for carnivory. In combination with a developing understanding of the carcass acquisition capabilities of Plio-Pleistocene carnivores, zooarchaeology provides paleoanthropology with an opportunity to test the nature of hominin emergence into—and eventual domination of—the larger carnivore guild.

Acknowledgments We are grateful to Travis Pickering for comments on a draft of this chapter, and to Peter Ungar for inviting us to write it.

References

Aiello, L.C., and Wheeler, P., 1995. The expensive-tissue hypothesis: The brain and digestive system in human and primate evolution. *Curr. Anthropol.* 36, 199–221.

Andrews, P., and Martin, L., 1991. Hominoid dietary evolution. *Philos. Trans. R. Soc. Lond.* B 334, 199–209.

Anyonge, W., 1996. Microwear on canines and killing behavior in large carnivores: Saber function in *Smilodon fatalis*. *J. Mammal.* 77(4), 1059–1067.

Bartram, L., 1993. Perspectives on skeletal part profiles and utility curves from Eastern Kalahari ethnoarchaeology. In: Hudson, J., (Ed.), *From Bones to Behavior*. University of Southern Illinois Press, Carbondale, pp. 115–137.

Bartram, L.E., and Marean, C.W., 1999. Explaining the "Klasies pattern": Kua ethnoarchaeology, the Die Kelders Middle Stone Age archaeofauna, long bone fragmentation and carnivore ravaging. *J. Archaeol. Sci.* 26, 9–29.

Behrensmeyer, A.K., Gordon, K.D., and Yanagi, G.T., 1986. Trampling as a cause of bone surface damage and pseudo-cutmarks. *Nature* 319, 768–771.

Behrensmeyer, A.K., and Kidwell, S. M., 1985. Taphonomy's contribution to paleobiology. *Paleobiology* 11, 105–119.

Biknevicius, A.R., Van Valkenburgh, B., and Walker, J., 1996. Incisor size and shape: Implications for feeding behaviors in saber-toothed "cats." *J. Vert. Paleontol.* 16(3), 510–521.

Binford, L.R., 1981. *Bones: Ancient Men and Modern Myths*. Academic Press, New York.

Binford, L.R., 1986. Comment on Bunn and Kroll's "Systematic butchery by Plio-Pleistocene hominids at Olduvai Gorge." *Curr. Anthropol.* 27, 444–446.

Binford, L.R., 1988. Fact and fiction about the Zinjanthropus floor: Data, arguments and interpretations. *Curr. Anthropol.* 29, 123–135.

Blumenschine, R.J., 1986a. Carcass consumption sequences and the archaeological distinction of scavenging and hunting. *J. Hum. Evol.* 15, 639–659.

Blumenschine, R.J., 1986b. *Early Hominid Scavenging Opportunities: Implications of Carcass Availability in the Serengeti and Ngorongoro Ecosystems*. BAR International Series 283, Oxford.

Blumenschine, R.J., 1987. Characteristics of an early hominid scavenging niche. *Curr. Anthropol.* 28, 383–407.

Blumenschine, R.J., 1988. An experimental model of the timing of hominid and carnivore influence on archaeological bone assemblages. *J. Archaeol. Sci.* 15, 483–502.

Blumenschine, R.J., 1989. A landscape taphonomic model of the scale of prehistoric scavenging opportunities. *J. Hum. Evol.* 18, 345–371.

Blumenschine, R.J., 1991. Hominid carnivory, foraging strategies, and the socioeconomic function of early archaeological sites. *Philos. Trans. R. Soc. Lond.* B 334, 211–221.

Blumenschine, R.J., 1995. Percussion marks, tooth marks, and experimental determinations of the timing of hominid and carnivore access to long bones at FLK *Zinjanthropus*, Olduvai Gorge, Tanzania. *J. Hum. Evol.* 29, 21–51.

Blumenschine. R.J., and Caro, T.M., 1986. Unit flesh weights of some East African bovids. *Afr. J. Ecol.* 24, 273–286.

Blumenschine, R.J., and Cavallo, J.A., 1992. Scavenging and human evolution. *Sci. Am.* 247, 90–96.

Blumenschine, R.J., and Madrigal, T.C., 1993. Variability in long bone marrow yields of East African ungulates and its zooarchaeological implications. *J. Archaeol. Sci.* 20, 555–587.

Blumenschine, R.J., and Marean, C.W., 1993. A carnivore's view of archaeological bone Assemblages. In: Hudson, J., (Ed.), *From Bones to Behavior*. University of Southern Illinois Press, Carbondale, pp. 273–300.

Blumenschine, R.J., Marean, C.W., and Capaldo, S.D., 1996. Blind tests of interanalyst correspondence and accuracy in the identification of cut marks, percussion marks, and carnivore tooth marks on bone surfaces. *J. Archaeol. Sci.* 23, 493–507.

Blumenschine, R.J., Masao, F.T., and Peters, C.P., in press. Broad-scale landscape traces of Oldowan hominid land use at Olduvai Gorge, and the Olduvai Landscape Paleoanthropology Project. In: Mapunda, B.B.B., and Msemwa, P. (Eds.), *Salvaging the Cultural Heritage of Tanzania*. University of Dar Es Salaam Press.

Blumenschine, R.J., and Peters, C.R., 1998. Archaeological predictions for hominid land use in the paleo-Olduvai Basin, Tanzania, during lowermost Bed II times. *J. Hum. Evol.* 34, 565–607.

Blumenschine, R.J., Peters, C.R., Capaldo, S.D., Andrews, P., Njau, J.K., and Pobiner, B.L., 2006. Vertebrate taphonomic perspectives on Oldowan hominin land use in the Plio-Pleistocene Olduvai basin, Tanzania. In: Pickering, T., Schick, K., and Toth, N. (Eds.), *African Taphonomy: A Tribute to the Career of C. K. "Bob" Brain*. Bloomington: CRAFT Press (Indiana University). In press.

Blumenschine, R.J., Peters, C.R., Masao, F.T., Clark, R.J., Deino A.L., Hay, R.L., Swisher, C.C., Stanistreet, I.G., Ashley, G.M., McHenry, L.M., Sikes, N.E., van der Merwe, N.J., Tactikos, J.C., Cushing, A.E., Deocampo, D.M., Njau, J.K., and Ebert, J.I., 2003. Late Pliocene *Homo* and hominid land use from western Olduvai Gorge, Tanzania. *Science* 299, 1217–1221.

Blumenschine, R.J., and Selvaggio, M.M., 1988. Percussion marks on bone surfaces as a new diagnostic of hominid behaviour. *Nature* 333, 763–765.

Blumenschine, R. J., and Selvaggio, M.M., 1991. On the marks of marrow bone processing by hammerstones and hyenas: Their anatomical patterning and archaeological implications. In: Clark, J.D. (Ed.), *Cultural Beginnings*. Dr. R. Habelt GMBH, Bonn, pp. 17–32.

Brain, C.K., 1981. *The Hunters or the Hunted? An Introduction to African Cave Taphonomy.* University of Chicago Press, Chicago.

Bunn, H.T., 1981. Archaeological evidence for meat-eating by Plio-Pleistocene hominids from Koobi Fora and Olduvai Gorge. *Nature* 291, 547–577.

Bunn, H.T., 1982. Meat-eating and human evolution: Studies of the diet and subsistence patterns of Plio-Pleistocene hominids. PhD diss., University of California, Berkeley.

Bunn, H.T., 1997. The bone assemblages from the excavated sites. In: Isaac, G. (Ed.), *Koobi Fora Research Project*. Vol. 5: *Plio-Pleistocene Archaeology*. Clarendon Press, Oxford, pp. 402–458.

Bunn, H.T., 2001. Hunting, power scavenging, and butchering by Hadza foragers and by Plio-Pleistocene *Homo*. In: Stanford, C.B., and Bunn, H.T. (Eds.), *Meat Eating and Human Evolution*. Oxford University Press, Oxford, pp. 199–281.

Bunn, H.T., and Ezzo, J.A., 1993. Hunting and scavenging by Plio-Pleistocene hominids: Nutritional constraints, archaeological patterns, and behavioral implications. *J. Archaeol. Sci.* 20, 365–398.

Bunn, H.T., and Kroll, E.M., 1986. Systematic butchery by Plio/Pleistocene hominids at Olduvai Gorge, Tanzania. *Curr. Anthropol.* 27, 431–452.

Capaldo, S.D., 1997. Experimental determinations of carcass processing by Plio-Pleistocene hominids and carnivores at FLK 22 (*Zinjanthropus*), Olduvai Gorge, Tanzania. *J. Hum. Evol.* 33, 555–597.

Capaldo, S.D., 1998. Simulating the formation of dual-patterned archaeofaunal assemblages with experimental control samples. *J. Archaeol. Sci.* 25, 311–330.

Capaldo, S.D., and Blumenschine, R.J., 1994. A quantitative diagnosis of notches made by hammerstone percussion and carnivore gnawing on bovid long bones. *Am. Antiq.* 59, 724–748.

Capaldo, S.D., Johnson, P., and Pante, M.C., in prep. Experimental determinations of carcass processing by hominids and carnivores, and intersite variability at FLK 22 (*Zinjanthropus*), FLKN 1-2, FLKN 3, and FLKNN 3, Olduvai Gorge, Tanzania.

Cavallo, J.A., and Blumenschine, R.J., 1989. Tree-stored leopard kills: Expanding the hominid scavenging niche. *J. Hum. Evol.* 18, 393–399.

Cerling, T.E., Harris, J.M., and Leakey, M.G., 1999. Browsing and grazing in elephants: The isotope record of modern and fossil proboscideans. *Oecologia* 120, 364–374.

Cerling, T.E, Harris, J.M., and Passey, B.H., 2003. Diets of East African bovids based on stable isotope analysis. *J. Mammal.* 84, 456–470.

Daniels, G., 1975. *A Hundred and Fifty Years of Archaeology*. Harvard University Press, Cambridge.

Dart, R.A., 1953. The predatory transition from ape to man. Int. Anthropol. Ling. Rev. 1, 201–219.

de Heinzelin, J., Clark, D., White, T., Hart, W., Renne, P., WoldeGabriel, G., Beyene, Y., and Vrba, E., 1999. Environment and behavior of 2.5-million-year-old Bouri hominids. *Science* 284, 625–629.

Domínguez-Rodrigo, M., 1997. Meat-eating by early hominids at the FLK 22 *Zinjanthropus* site, Olduvai Gorge (Tanzania): An experimental approach using cut-mark data. *J. Hum. Evol.* 33, 669–690.

Domínguez-Rodrigo, M., 1999. Flesh availability and bone modifications in carcasses consumed by lions: Palaeoecological relevance in hominid foraging patterns. *Palaeogeogr. Palaeoclimatol. Palaeoecol.* 149, 373–388.

Domínguez-Rodrigo, M., 2002. Hunting and scavenging by early humans: the state of the debate. *J. World Prehist.* 16, 1–54.

Domínguez-Rodrigo, M., Pickering, T.R., Semaw, S., and Rogers, M.J., 2005. Cutmarked bones from Pliocene archaeological sites at Gona, Ethiopia: Implications for the function of the world's oldest stone tools. *J. Hum. Evol.* 48, 109–121.

Domínguez-Rodrigo, M., and Piqueras, A, 2003. The use of tooth pits to identify carnivore taxa in tooth-marked archaeofaunas and their relevance to reconstruct hominid carcass processing behaviours. *J. Archaeol. Sci.* 20, 1385–1391.

Egeland, C.P., 2003. Processing intensity and cutmark creation: An experimental approach. *Plains Anthropol.* 48, 39–51.

Egeland, C.P., Hill, M.G., and Byerly, R.M., 2002. Archaeological vs. experimental manifestations of large mammal butchery: Increasing the reliability of inferences of prehistoric subsistence behavior. Society for American Archaeology, Denver, CO.

Emerson, S.B., and Radinsky, L., 1980. Functional analysis of sabertooth cranial morphology. *Paleobiology* 6, 295–312.

Fernandez-Jalvo, Y., Andrews, P., and Denys, C., 1999. Cut marks on small mammals at Olduvai Gorge Bed-I. *J. Hum. Evol.* 36, 587–589.

Foley, R., 2001. The evolutionary consequences of increased carnivory in hominids. In: Stanford, C.B., and Bunn, H.T. (Eds.), *Meat Eating and Human Evolution*. Oxford University Press, Oxford, pp. 305–331.

Foley, R.A., and Lee, P.C., 1991. Ecology and energetics of encephalization in hominid evolution. *Philos. Trans. R. Soc. Lond.* B 334, 223–232.

Gifford-Gonzalez, D., 1989. Shipman's shaky foundations. *Am. Anthropol.* 91, 180–186.

Gifford-Gonzalez, D., 1991. Bones are not enough: Analogues, knowledge, and interpretive strategies in zooarchaeology. *J. Anthrop. Archaeol.* 10, 215–254.

Harris, J. M., and Cerling, T.E., 2002. Dietary adaptations of extant and Neogene African suids. *J. Zool. Lond.* 256, 45–54.

Harris, J.W.K., Williamson, P., Verniers, J., Tappen, M., Stewart, K., Helgren, D., de Heinzelin, J., Boaz, N., and Bellomo, R., 1987. Late Pliocene hominid occupation in Central Africa: the setting, context, and character of the Senga 5A site, Zaire. *J. Hum. Evol.* 16, 701–728.

Hawkes, K., Hill, K., and O'Connell, J.F., 1982. Why hunters gather: Optimal foraging and the Ache of eastern Paraguay. *Am. Ethnol.* 9, 379–398.

Haynes, G., 1983. A guide for differentiating mammalian carnivore taxa responsible for gnaw damage to herbivore limb bones. *Paleobiology* 9, 164–172.

Hedges, R.E.M., 2002. Bone diagenesis: An overview of processes. *Archaeometry* 44, 319–328.

Howell, F.C., Haesaerts, P., and de Heinzelin, J., 1987. Depositional environments, archeological occurrences and hominids from Members E and F of the Shungura Formation (Omo Basin, Ethiopia). *J. Hum. Evol.* 16, 665–700.

Isaac, G.L., 1978. The food-sharing behavior of proto human hominids. *Sci. Am.* 238, 90–108.

Isaac, G.L., 1983. Early stages in the evolution of human behavior: The adaptive significance of stone tools. Sixtieth Kroon Lecture. Amsterdam, Stichting Nederlands Museum voor Anthropologie en Prehistorie.

Isaac, G.L., and Harris, J.W.K., 1997. Sites stratified within the KBS tuff: Reports. In: Isaac, G.L. (Ed.), *Koobi Fora Research Project*. Vol 5: *Plio-Pleistocene Archaeology*. Clarendon Press, Oxford, pp. 71–114.

Isaac, G.L., and Crader, D.C., 1981. To what extent were early hominids carnivorous? An archaeological perspective. In: Harding, R.S.O., and Teleki, G. (Eds.), *Omnivorous Primates: Gathering and Hunting in Human Evolution*. Columbia University Press, New York, pp. 37–103.

Kappelman, J., Plummer, T., Bishop, L., Duncan, A., and Appleton, S., 1997. Bovids as indicators of paleoenvironments in East Africa. *J. Hum. Evol.* 32, 229–256.

Kibunjia, M., 1994. Pliocene archaeological occurrences in the Lake Turkana basin. *J. Hum. Evol.* 27, 159–171.

Kimbel, W.H., Walter, R.C., Johanson, D.D., Reed, K.E., Aronson, J.L., Assefa, Z., Marean, C.W., Eck, G.G., Bobe, R., Hovers, E., Rak, Y., Vondra, C., Yemane, T., York, D., Chen, Y., Evensen, N.M., and Smith, P.E., 1996. Late Pliocene Homo and Oldowan tools from the Hadar Formation (Kada Hadar Member), Ethiopia. *J. Hum. Evol.* 31, 549–561.

Leonard, W.R., and Robertson, M.L., 1997. Comparative primate energetics and hominid evolution. *Am. J. Phys. Anthropol.* 102, 265–281.

Lieberman, D., 1994. The biological basis for seasonal increments in dental cementum and their application to archaeological research. *J. Archaeol. Sci.* 21, 525–539.

Lupo, K.D., 1998. Experimentally derived extraction rates for marrow: implications for body part exploitation strategies of Plio-Pleistocene hominid scavengers. *J. Hum. Evol.* 33, 657–675.

MacArthur, R.H., and Pianka, E. R., 1966. On optimal use of a patchy environment. *Am. Nat.* 100, 603–609.

Madrigal, T.C., and Blumenschine, R.J., 2000. Preferential processing of high return rate marrow bones by Oldowan hominids: A comment on Lupo. *J. Archaeol. Sci.* 27, 739–741.

Marean, C.W., 1989. Sabertoothed cats and their relevance for early hominid diet and evolution. *J. Hum. Evol.* 18, 559–582.

Marean, C.W., 1995. Of taphonomy and zooarchaeology. *Evol. Anthropol.* 4, 64–72.

Marean, C.W., 1998. A critique of the evidence for scavenging by Neandertals and early modern humans: New data from Kobeh Cave (Zagros Mountains, Iran) and Die Kelders Cave 1 Layer 10 (South Africa). *J. Hum. Evol.* 35, 111–136.

Marean, C.W., and Assefa, Z., 1999. Zooarchaeological evidence for the faunal exploitation behavior of Neandertals and early modern humans. *Evol. Anthropol.* 8, 22–37.

Marean, C.W., and Ehrhardt, C.L., 1995. Paleoanthropological and paleoecological implications of the taphonomy of a sabertooth's den. *J. Hum. Evol.* 29, 515–547.

Marean, C.W., and Spencer, L.M., 1991. Impact of carnivore ravaging on zooarchaeological measures of element abundance. *Am. Antiq.* 56, 645–658.

Marean, C.W., Spencer, L.M., Blumenschine, R.J., and Capaldo, S.D., 1992. Captive hyena bone choice and destruction, the schlepp effect, and Olduvai archaeofaunas. *J. Archaeol. Sci.* 19, 101–121.

Marshall, F., 1994. Food sharing and body part representation in Okiek faunal assemblages. *J. Archaeol. Sci.* 21, 65–77.

Matthiesen, D.G., n.d.. Prodromus of the paleoecology of the Plio-Pleistocene fossil birds from the early man sites of Omo and Hadar, Ethiopia, and Olduvai Gorge, Tanzania. Paper presented at the Sixth International Council for Archaeozoology, Washington DC, 1990.

Merrick, H.V., and Merrick, J.P.S., 1987. Archaeological occurrences of earlier Pleistocene age from the Shungura Formation. In: Coppens, Y., Howell, F. C., Isaac, G. Ll., and Leakey, R. E. F. (Eds.), *Earliest Man and Environments in the Lake Rudolf Basin*. University of Chicago Press, Chicago, pp. 574–584.

Monahan, C.M., 1996. New zooarchaeological data from Bed II, Olduvai Gorge, Tanzania: Implications for hominid behavior in the Early Pleistocene. *J. Hum. Evol.* 31, 93–128.

Nilssen, P.J., 2000. An actualistic butchery study in South Africa and its implications for reconstructing hominid strategies of carcass acquisition and butchery in the Upper Pleistocene and Plio-Pleistocene. PhD diss., University of Cape Town.

Njau, J. K., and Blumenschine, R. J., 2006. A diagnosis of crocodile feeding traces on larger mammal bone, with examples from the Plio-Pleistocene Olduvai Basin, Tanzania. *J. Hum. Evol.* 50, 142–162.

O'Connell, J.F., and Hawkes, K., 1981. Alywara plant use and optimal foraging theory. In: Winterhalder, B., and Smith, E. A. (Eds.), *Hunter-Gatherer Foraging Strategies: Ethnographic and Archeological Analyses*. University of Chicago Press, Chicago, pp. 99–125.

Peters, C.P., and Blumenschine, R.J., 1995. Landscape perspectives on possible land use patterns for early hominids in the Olduvai basin. *J. Hum. Evol.* 29, 321–362.

Peters, C.P., Blumenschine, R.J., 1996. Landscape perspectives on possible land use patterns for early hominids in the Olduvai basin, Tanzania. Part II: Expanding the landscape models. In: Magori, C., Saanane, C.B., and Schrenk, F. (Eds.), *Four Million Years of Hominid Evolution in Africa: Papers in Honour of Dr. Mary Douglas Leakey's Outstanding Contribution in Paleoanthropology, Kaupia 6*. Darmstadter Beitrage zur Naturgeschichte, Darmstadt, pp. 175–221.

Pickering, T.R., 1999. Taphonomic interpretations of the Sterkfontein early hominid site (Gauteng, South Africa) reconsidered in light of recent evidence. PhD diss., University of Wisconsin–Madison.

Pickering, T.R., Domínguez-Rodrigo, M., Egeland, C.P., and Brain, C.K., 2005. The contribution of limb bone fracture patterns to reconstructing early hominid behavior at Swartkrans Cave (South Africa): Archaeological application of a new analytical method. *Int. J. Osteoarchaeol.* 15, 247–260.

Pickering, T.R., White, T.D., and Toth, N., 2000. Cutmarks on a Plio-Pleistocene hominid from Sterkfontein, South Africa. *Am. J. Phys. Anthropol.* 111, 579–584.

Plummer, T., 2004. Flaked stones and old bones: Biological and cultural evolution at the dawn of technology. *Yearb. Phys. Anthropol.* 47, 118–164.

Plummer, T., Bishop, L.C., Ditchfield, P., and Hicks, J., 1999. Research on late Pliocene Oldowoan sites at Kanjera South, Kenya. *J. Hum. Evol.* 36, 151–170.

Pobiner, B.L., in prep. A taphonomic perspective on Oldowan hominid carnivory: Evidence from Olduvai Gorge and Koobi Fora. PhD diss., Rutgers University.

Pobiner, B.L., and Blumenschine, R.J., 2003. A taphonomic perspective on the Oldowan hominid encroachment on the carnivoran paleoguild. *J. Taph.* 1, 115–141.

Pobiner, B.L., and Braun, D.R., 2005. Strengthening the inferential link between cutmark frequency data and Oldowan hominid behavior: Results from modern butchery experiments. *J. Taph.* 3, 107–119.

Pooley, A.C., 1989. Food and feeding habits. In: Ross, C. A., and Garnett, S. (Eds.), *Crocodiles and Alligators*. Facts on File, New York, pp. 76–91.

Potts, R.B., and Shipman, P., 1981. Cutmarks made by stone tools on bones from Olduvai Gorge, Tanzania. *Nature* 291, 577–580.

Ragir, S., Rosenberg, M., and Tierno, P., 2000. Gut morphology and the avoidance of carrion among chimpanzees, baboons, and early hominids. *J. Anthropol. Res.* 56, 477–512.

Roche, H., Delagnes, A., Brugal, J.P., Feibel, C., Kibunjia, M., Mourre, V., and Texier, P.-J., 1999. Early hominid stone tool production and technical skill 2.34 Myr ago in West Turkana, Kenya. *Nature* 399, 57–60.

Sahnouni, M., and de Heinzelin, J., 1998. The site of Ain Hanech revisited: new investigations at this Lower Pleistocene site in northern Algeria. *J. Archaeol. Sci.* 25, 1083–1101.

Schaller, G.B., 1972. *The Serengeti Lion: A Study of Predator-Prey Relations*. University of Chicago Press, Chicago.

Selvaggio, M.M., 1994. Carnivore tooth marks and stone tool butchery marks on scavenged bones: Archaeological implications. *J. Hum. Evol.* 27, 215–228.

Selvaggio, M.M., 1998. Evidence for a three-stage sequence of hominid and carnivore involvement with long bones at FLK *Zinjanthropus*, Olduvai Gorge, Tanzania. *J. Archaeol. Sci.* 25, 191–202.

Selvaggio, M.M., and Wilder, J., 2001. Identifying the involvement of multiple carnivore taxa with archaeological bone assemblages. *J. Archaeol. Sci.* 28, 465–470.

Semaw, S., Rogers, M.J., Quade, J., Renne, P., Butler, R., Domínguez-Rodrigo, M., Stout, D., Hart, W., Pickering, T., and Simpson, S., 2003. 2.6-Million-year-old stone tools and associated bones from OGS-6 and OGS-7, Gona, Afar, Ethiopia. *J. Hum. Evol.* 45, 169–177.

Shipman, P., 1986. Scavenging or hunting in early hominids: theoretical frameworks and tests. *Am. Anthropol.* 88, 27–43.

Shipman, P., 1989. Altered bones from Olduvai Gorge, Tanzania: Techniques, problems, and implications of their recognition. In: Bonnichsen, R., and Sorg, M. (Eds.), Bone Modification. Center for the Study of the First Americans, Orono, Maine, pp. 247–258.

Shipman, P., Fisher, D.C., and Rose, J.J., 1984. Mastodon butchery: microscopic evidence of carcass processing and bone tool use. *Paleobiology* 10, 358–365.

Shipman, P., and Rose, J., 1983. Early hominid hunting, butchering and carcass-processing behaviors: Approaches to the fossil record. *J. Anthropol. Archaeol.* 2, 57–98.

Shipman, P., and Rose, J., 1984. Cutmark mimics on modern and fossil bovid bones. *Curr. Anthropol.* 25, 116–117.

Spencer, L.M., 1997. Dietary adaptations of Plio-Pleistocene Bovidae: Implications for hominid habitat use. *J. Hum. Evol.* 32, 201–228.

Speth, J.D., and Davis, D.D., 1976. Seasonal variability in early hominid predation. *Science* 192, 441–445.

Stanford, C.B., 1996. The hunting ecology of wild chimpanzees: Implications for the evolutionary ecology of Pliocene hominids. *Am. Anthropol.* 98, 96–113.

Stewart, K.M., 1994. Early hominid utilization of fish resources and implications for seasonality and behavior. *J. Hum. Evol.* 27, 229–245.

Tchernov, E., 1986. Evolution of the Crocodiles in East and North Africa. Cahiers de Paleontologie, Travaux de Paleontologies Est-Africaine, CNRS, Paris, pp. 1–65.

Turner, A., 1990. The evolution of the guild of larger terrestrial carnivores in the Plio- Pleistocene of Africa. *Geobios* 23, 349–368.

Van Valkenburgh, B., Teaford, M.F., and Walker, A., 1990. Molar microwear and diet in large carnivores: inferences concerning diet in the sabretooth cat, *Smilodon fatalis*. *J. Zool. Lond.* 222, 319–340.

Washburn, S.L., and Lancaster, C.S., 1968. *Man the Hunter*. Aldine, Chicago.
Watson, J.A.L., and Abbey, H.M., 1986. The effects of termites (Isoptera) on bone: Some archaeological implications. *Sociobiology* 11, 245–254.
Werdelin, L., and Lewis, M.E., 2005. Plio-Pleistocene Carnivora of eastern Africa: Species richness and turnover patterns. Zool. J. Linn. Soc. 144, 121–144.
Wilson, M.C., 1982. Cut marks and early hominids: evidence for skinning. *Nature* 298, 303.

11

Meat Made Us Human

HENRY T. BUNN

The role of meat in the evolution of hominin diet and behavior is controversial. During the later reign of the "Man the Hunter" paradigm, fossil and archaeological evidence from Plio-Pleistocene localities in southern and eastern Africa seemed at face value to corroborate significant involvement with meat and hunting early in human evolution (Dart, 1957; Isaac, 1971; Leakey, 1971). Leakey and Isaac interpreted the archaeological evidence as generally reminiscent of the behavioral adaptations of human foragers. Isaac (1971, 1978) attributed to Plio-Pleistocene *Homo* a human-like adaptation for the procurement and consumption of meat, including gender-based division of labor, food transport, and food sharing, which he called the *home-base*, or *central-place* model. Given the lingering image of Man the Hunter, it is perhaps unsurprising that these findings were challenged. If males usually provide the meat in foraging societies, and if meat was pivotal in the evolution of human diet and behavior, then what was the contribution of females? Alternative models based on analogies to nonhuman primates defined a significant role for females, particularly for australopithecines predating the known archaeological record (e.g., Zihlman and Tanner, 1978).

When Brain (1967, 1969) introduced a taphonomic approach into paleoanthropology, the research landscape changed profoundly. By developing an interpretive baseline through observations of modern bone-related processes, particularly carnivore feeding and resulting bone damage, Brain was able to show that patterns in the fossil bone assemblages from complex breccias in South African caves, particularly Swartkrans, were mainly from feeding by large carnivores, rather than from pronounced carnivorous proclivities of *Australopithecus africanus*, as Dart had envisioned. In East Africa, paleontologically oriented taphonomic studies by Behrensmeyer (1975) and Hill (1975) identified the potential of and the need for taphonomic research at the Plio-Pleistocene archaeological sites. By 1977, to test and possibly falsify his own home-base model, Isaac already had a team of researchers in the field at Koobi Fora, Kenya, conducting taphonomic studies of the bone assemblages and more broadly, site-formation studies of the complex, overlapping processes affecting not only bones but also stone tools and other evidence found at the archaeological sites. Corroboration of aspects of the home-base model came from discoveries and analyses of stone-tool cut marks and percussion notches on bones of Plio-Pleistocene age from the archaeological site of FxJj 50 at Koobi Fora, along with conjoining stone tools and bones and the tight spatial distribution of the conjoining pieces (Bunn et al., 1980), bones with cut marks from the FLK Zinj archaeological site at Olduvai Gorge (Bunn, 1981; Potts and Shipman, 1981), and microwear on stone tools from the cutting of meat and of soft plants and sawing or scraping of wood at FxJj 50 (Keeley and Toth, 1981). Hominins clearly had a significant, though not exclusive, role in the accumulation and use of the bones and stone tools for meat and marrow consumption.

Subsequent taphonomic research in Plio-Pleistocene contexts, involving many researchers, differing analytical approaches, different data sets, and different interpretations, has continued productively (Isaac, 1984; Bunn, 1991; Domínguez-Rodrigo, 2002; Plummer, 2005). In this chapter, I shall focus on taphonomic evidence from the largest and best preserved Plio-Pleistocene archaeological site, FLK Zinj, to illustrate how bone assemblages have been studied, what basic undisputed information has resulted, and how alternative analytical approaches (my own and other researchers') have sought to interpret evidence of hominin diet and foraging behavior. These results will then be placed in an evolutionary framework to consider the strengths and weaknesses of the proposition that meat made us human. I contend that the procurement and consumption of meat by early *Homo erectus* is knowable from the Plio-Pleistocene archaeological record and created selective pressures for the evolution of human behavior.

Taphonomic Perspectives on the FLK Zinj Archaeological Site

Taphonomy and the Knowable in Archaeological Research

Taphonomy can be used constructively as a tool for determining how bone assemblages form, or it can be used as a weapon against unwanted evidence and interpretations. Taphonomic (sometimes labeled as actualistic, middle-range, or site-formation) observations of cause-and-effect relationships can be made in modern settings to define the dynamics of bone assemblage formation. Before ever looking at an ancient fossil assemblage, taphonomic knowledge can be stated as testable hypotheses. Comparable patterning (the effect) in an ancient bone assemblage can, with appropriate uniformitarianist reasoning, enable researchers to infer the taphonomic agent (the cause) responsible. Nothing is novel about this approach in science or more particularly in archaeology (e.g., Binford, 1968, 1972). Detectives, judges, and juries all practice somewhat similar logic, and when the absolute proof by direct visual observation is lacking, decisions or scientific interpretations compensate with disclaimers of "beyond reasonable doubt." Taphonomic interpretations of the formation of any bone assemblage are really probability statements, identifying dominant rather than exclusive taphonomic agents. Taphonomic research is challenging because animal carcasses and the bones contained within them are an attractive food source to multiple agents and processes, and some of the patterning in an ancient assemblage is not unambiguously attributable to just one taphonomic cause. The dynamics forming a bone assemblage are complex, and the interplay of multiple taphonomic agents and processes is the norm. Since the discovery of Plio-Pleistocene cut marks a quarter-century ago, an interpretive baseline has developed along two main trajectories focused on the procurement and processing of animal carcasses for food: (1) observations of human forager behavior and (2) observations of carnivore behavior. Both have contributed significantly to ongoing reconstructions of subsistence behavior in Plio-Pleistocene hominins and human evolution.

Taphonomic Knowledge of the FLK Zinj Bone Assemblage

The FLK Zinj archaeological site at Olduvai Gorge, is one of the largest excavations (~300 m^2) and apparently one of the most completely excavated archaeological sites known from the Plio-Pleistocene (Leakey, 1971). Bones and stone tools have been found in a discrete, four-inch-thick layer in sediments, which accumulated within a kilometer of the margin of a fluctuating, saline lake. An air-fall ash abruptly buried the stone tools and bones, some of which actually extended into the ash layer (Leakey, 1971; Hay, 1976; R.L. Hay, personal communication, 1991). Leakey (1971) reported a total of 2,470 Oldowan stone artifacts, (more than 90% were sharp-edged flakes and flake fragments and the remainder a mix of core tools) and a total of 3,510 bone specimens of which 1,090 were considered identifiable to taxon or skeletal part; some had been gnawed by carnivores. Following what was then standard practice, paleontologists studied the more complete and diagnostic specimens and provided taxonomic identifications and some quantitative data. Leakey (1971) compiled the taxonomic lists and skeletal percentages and described them as "merely a preliminary report, to be amplified." She regarded the FLK Zinj site as a prime example of a hominin living floor, or occupation site, which had been produced by the transport, use, and discard of stone tool and bone remains by hominins, followed by hyena gnawing and scavenging of some leftover bone food remains.

With the encouragement of Mary Leakey, my analysis of the archaeological bones from FLK Zinj began in 1977. My aim was to conduct the first (and still only) taphonomic analysis of *all* of the bone specimens from the site, modeled after Brain's taphonomic research. Efforts to reconstruct the unique taphonomic history of the FLK Zinj bone assemblage would be enhanced by analysis of all bone specimens, rather than a subset of bones selected because they are taxonomically diagnostic to genus or species. My first inspection of the bones from FLK Zinj revealed how special an opportunity this site provides for paleoanthropology. The preservation of surface detail on the original bone surfaces is unparalleled for the Plio-Pleistocene and so good that one can readily recognize and confidently and accurately distinguish, with unaided eyes, the telltale signatures of artifact-induced cut marks, carnivore gnaw marks, rodent gnaw marks, and root marks. Examining the bone surfaces with a hand lens (10×) or microscope (~20–40×) reveals another dimension of taphonomic evidence on the bones. Taphonomic analysis of the entire bone assemblage has revealed strong patterns of skeletal representation, of bone damage from hominin butchery and from gnawing by other animals, and other results (Bunn, 1981, 1982; Bunn and Kroll, 1986, 1988).

The FLK Zinj bone assemblage is larger than the preliminary report had indicated, including approximately 3,500 skeletally identifiable mammalian specimens, more than 40,000 skeletally nonidentifiable mammalian specimens, more than 16,000 microfaunal specimens, several hundred bird specimens, and some uncommon specimens of other nonmammalian taxa. The MNI (minimum number of individuals) estimate for larger mammals is at least forty-eight: twenty-nine bovids, five equids, five suids, two giraffes, one hyenid, two small carnivores, two hominins, one hippo (tooth fragment), and one elephant (tooth fragment).

Comprehensive analysis of skeletal proportions, involving specimen-by-specimen comparisons of all skeletally identifiable pieces indicates that FLK Zinj, is dominated by hemimandibles and by meaty and marrow-rich limb elements (humerus, radioulna, femur, tibia). Nonmeaty, but still marrow-rich, metapodials are less abundant, as are other skeletal elements. Crania (including isolated maxillary teeth), ribs, and vertebrae are poorly represented. Methodology means everything here. Incorporating all specimens into the analysis, and specifically using anatomical landmarks on limb shaft fragments to determine limb proportions, reveals (incontrovertibly) the actual pattern of hemimandible/meaty limb abundance in the assemblage (Bunn, 1982, 1986, 1991; Bunn and Kroll, 1986, 1988). Paleontological work on selected subsets of the assemblage and its derivatives (Binford, 1981; Potts, 1988) obscure, rather than reveal, the actual skeletal pattern.

Patterns of bone damage include: (1) fracture and fragmentation and (2) marks inflicted into surfaces. The bones are highly fragmented, and fracture morphology indicates that the fragmentation occurred predominantly when the bones were fresh, rather than during fossilization. A high degree of fragmentation is a common feature of humanly broken bones at more recent archaeological sites, in contrast to the relatively more complete bone elements characteristic of many bone accumulations generated by large carnivores. Although suggestive of breakage by hominins, bone fragmentation alone is not conclusive, because other taphonomic agents can fragment bones. In the FLK Zinj bone assemblage, abundant percussion notches and large bone flakes, which were caused by hominins using hammerstones to break open marrow-rich limbs and hemimandibles, establish hominins as the dominant taphonomic agent causing bone fragmentation. Refitting of these specimens yielded more than one hundred sets of conjoining bones (predominantly limb shaft specimens), and figure 11.1 shows the distribution pattern of refitted sets across the site, indicating minimal disturbance of the spatial pattern following hominin discard of the butchered bones (Bunn, 1982; Kroll and Isaac, 1984). Quantification of bone fragmentation (Bunn, 1989), and particularly quantification of fracture morphology (Capaldo and Blumenschine, 1994), strengthens the conclusion that hominins caused most of the breakage of marrow-rich bones.

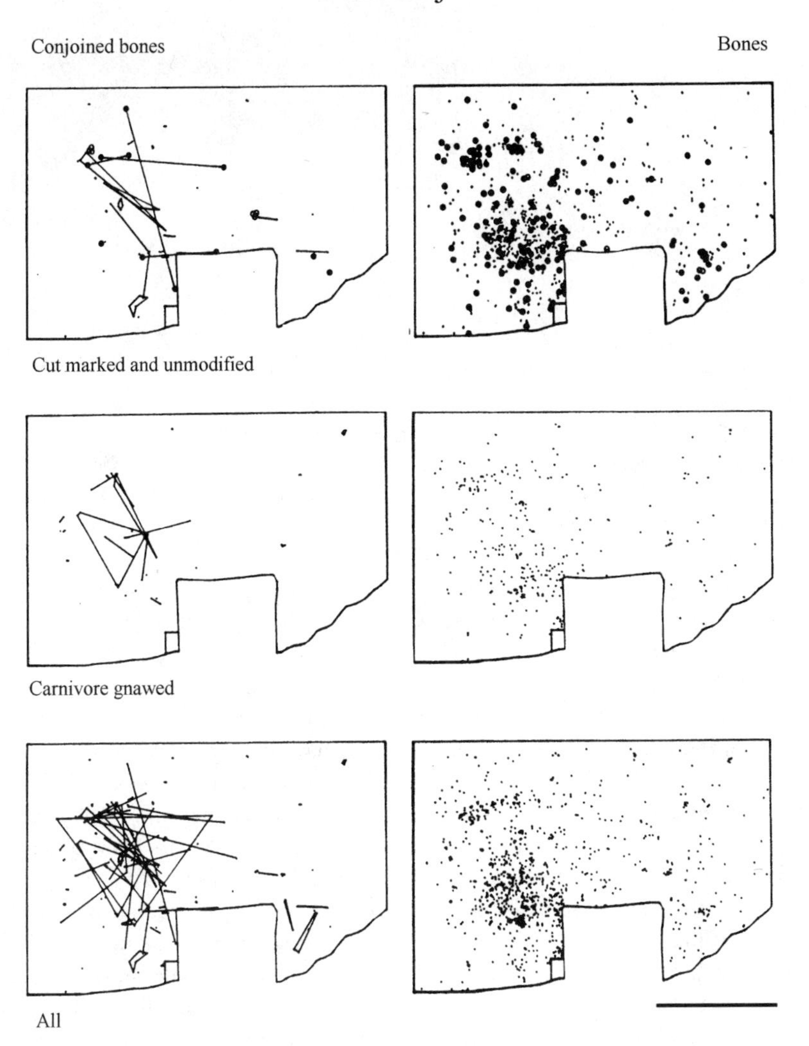

Figure 11.1 Excavation by Mary Leakey at FLK Zinj, Olduvai Gorge. Image elements courtesy of Ellen Kroll. Scale bar = 5 m.

Marks inflicted in bone surfaces are abundant and varied in the FLK Zinj assemblage. The mere presence of cut marks (fine slicing grooves) produced by hominins using sharp stone flakes in butchery puts to rest the concern that for taphonomic reasons, the bone assemblage might have no functional association with hominin behavior. But investigating what that hominin behavioral role might have been requires more evidence. Documentation of more than 200 bone specimens with cut marks in the assemblage provides a more secure foundation. The total number of cut-marked bone specimens is actually closer to 300, but discussion here refers to the 200 or so specimens that have been identified by both macroscopic and microscopic criteria to

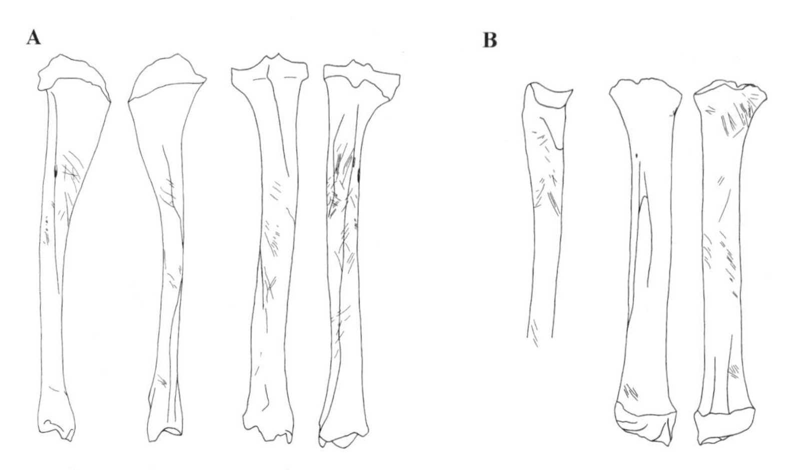

Figure 11.2 (*A*) Composite cut marks on tibia specimens from FLK Zinj. Cut marks on fragmentary tibia specimens are shown on anterior, posterior, lateral, and medial views of a whole tibia. (*B*) Composite cut marks on radius specimens from FLK Zinj. Cut marks on fragmentary radius specimens are shown on anterior, posterior, and distal views of a whole radius.

meet the requirements of all researchers involved. The cut marks are distributed on various skeletal elements, and, given knowledge of animal anatomy and butchery, specific behavioral meaning can be derived from the locations and orientations of the cut marks. A buffalo hyoid bone, for example, has cuts that were inflicted during removal of the tongue. Bovid metapodial shafts have cuts from skinning those nonmeaty elements before breakage to access marrow. A few dismembering (disarticulating) cut marks occur on some but not many of the surviving limb epiphyses.

Most of the cut marks, however, are clustered on once-meaty limb bone shafts near major muscle attachments. Figure 11.2 shows three or four views (anterior, medial, posterior, lateral) each of whole tibia and radius elements, onto which *all* cut marks from the individual, fragmented tibia and radius specimens in the FLK Zinj assemblage have been transferred to scale and in correct anatomical orientation. Thus, cut marks on bone specimens from all sizes of animals and from left and right sides of the skeleton, have all been transferred onto one set of drawings. On the tibia, an obvious cluster of cut marks occurs on the posterior, proximal shaft where the major calf muscles attach. Cuts on the nonmeaty distal, posterior shaft result from cutting of the "Achilles tendon" (i.e., tendon of the gastrocnemius muscle). Similar reasoning applies to the anterior surface of the radius. Bunn (2001) has composite illustrations of cut marks on humerus and femur specimens from FLK Zinj. The composite pattern of defleshing cut marks on the FLK Zinj limb elements indicates that meat was a primary target of hominin butchers, and the spatial distribution of the cut-marked bones spans the excavation area (fig. 11.1).

Carnivore gnaw marks are abundant and require systematic interpretations of which carnivore inflicted them and when in the sequence they were inflicted. Some gnaw marks are large, hyena-tooth-sized grooves and punctures, but most of the gnawed specimens exhibit marks that are very small and probably from a different and much smaller carnivore. Jackals and/or viverrids (family of small carnivores including civets, genets, mongooses) are potentially responsible for the smaller carnivore gnawing damage. The anatomical distribution of carnivore gnawing is noteworthy. Gnawing damage occurs on the outer surfaces of limb bones, but also on the inner, medullary surfaces of hammerstone-broken bones. The spatial distribution of carnivore-gnawed bones spans the excavation area (fig. 11.1). Rodent gnaw marks are small and were made by small, mouse-sized rodents, not by much-larger porcupines. The probable role of small rodents was in the gnawing of bones discarded at the site, rather than any significant transport of bones to the site.

Reconstruction of the Taphonomic History of the FLK Zinj Bone Assemblage

Judging from the multiple taphonomic agents involved with the bone assemblage, FLK Zinj was a popular location on the Olduvai landscape during formation of the site. It is feasible to reconstruct the most probable sequence of site formation and the taphonomic history of the bone assemblage with considerable resolution because of the exceptional bone preservation and the careful excavation work by Mary Leakey.

Transport of stones and bones by hominins to the site is strongly implicated. The amount of shattered quartz debitage alone indicates that hominins flaked quartz artifacts at FLK Zinj after transporting tabular raw material available from nearby outcrops. The fossiliferous layer is only four inches thick and only marginally thicker than the bones buried in it. Yet, the vast majority of the bone surfaces are fresh to lightly weathered (i.e., weathering stage 0 to early stage 2 of Behrensmeyer, 1978), indicating a maximum of several years of subaerial exposure and weathering, rather than decades, centuries, or millennia (Bunn and Kroll, 1987). The concentration of at least fourty-eight individual larger mammals in this temporally restricted bone assemblage is anomalous relative to the distribution of bones on the modern and (samples of) the ancient landscape and indicates that carcass remains were transported to the location (Behrensmeyer, 1983; Blumenschine and Masao, 1991; Bunn, Kroll, and Bartram, 1991; Tappen, 1995). Taphonomic agents, other than hominins, however, are capable of concentrating bones of multiple animals, but these typically leave telltale signatures of their involvement (e.g., porcupines, hyenas, fluvial processes), which are absent at FLK Zinj. Because of the extensive evidence of hominin butchery of carcasses at FLK Zinj, it is highly probable that hominins repeatedly transported carcass portions to the site for further butchery and consumption of meat and fat.

The acquisition of edible carcass portions from elsewhere on the landscape also has implications for hominin behavior involved in forming FLK Zinj. Alternative possibilities for acquisition include hunting, power scavenging, and passive scavenging, or a combination of those, which would likely have operated together in a foraging adaptation prioritizing animal tissues. Successful carcass acquisition usually requires some combination of lethal weaponry, physical and intellectual prowess, and luck. Because of inherent complexities in distinguishing among those, it is more reasonable to seek reconstruction of the most probable dominant pattern(s) rather than some highly detailed account of each prey animal's fate (which may be unknowable). Two aspects of this reconstruction can be considered: (1) How did hominins acquire carcasses? (2) How much edible meat and fat did hominins acquire? Answers to those questions are knowable within reasonable probability limits because the regular sequence by which large carnivores consume carcasses (thereby potentially removing them from hominin access) is known (e.g., Kruuk, 1972; Schaller, 1972; Blumenschine, 1987; Domínguez-Rodrigo, 1999) and because the skeletal profile at FLK Zinj is known (Bunn, 1982; Bunn and Kroll, 1986, 1988).

A nonhunting reconstruction of early hominin foraging would place them in the rare position as an exclusive scavenger. The very skeletal elements that carnivores consume first are abundant at FLK Zinj, and they also exhibit abundant defleshing cut marks; this combined pattern of skeletal elements and defleshing cut marks indicates early access by hominins to intact carcasses (Bunn and Kroll, 1986, 1988; Bunn and Ezzo, 1993; Bunn, 2001). No weaponry from the Plio-Pleistocene is preserved (e.g., wooden spears or clubs), and consequently, it is difficult to prove that hominins killed any of the animals at FLK Zinj beyond reasonable doubt. Yet, hunting of the smaller, gazelle-sized animals is more probable than scavenging from the kills of large carnivores, simply because small carcasses can be consumed within minutes by lions or hyenas, which, at best, offers a fleeting opportunity for others to scavenge. Insofar as hunting would have increased the primary access to small carcasses, which hominins undoubtedly had, it is probable beyond reasonable doubt that hunting by hominins provided the gazelle-sized carcasses at FLK Zinj, rather than scavenging from lions or hyenas. Power scavenging from more retiring predators, such as cheetahs or leopards, could also provide intact gazelle-sized carcasses, as could repeated good luck in finding carcasses from natural causes other than predation before other carnivore scavengers (which were better adapted than hominins at finding carcasses during the day and night) did so.

In the absence of weaponry, whether hominins hunted or power scavenged the predominantly larger sizes (i.e., waterbuck, wildebeest) of animals butchered at FLK Zinj is uncertain. As with the smaller animals, the combined abundant evidence of (1) the very limb elements eaten quickly by primary predators and (2) the defleshing cut marks on those limb elements indicate beyond reasonable doubt that hominins regularly had early access to those larger carcasses. As a dominant though not exclusive strategy, power scavenging (i.e., interference competition involving aggressively driving predators from their kills) would have provided hominins with this early access to intact or nearly intact larger carcasses, which the skeletal and cut-mark patterns demonstrate that they had. Though less likely, more good luck could have provided the larger carcasses through passive scavenging ahead of a suite of carnivores possessing greater speed, better vision and smell, and an adaptation for scavenging throughout the day and night. Either way, strong patterns in the bone assemblage at FLK Zinj indicate that hominins obtained large quantities of meat and fat from those larger carcasses.

To carry edible meat and fat from acquisition points to FLK Zinj, carcasses (particularly large, size group 3 animals) would have required field butchery into transportable units: whole limbs and segments of the axial skeleton. Unless hominin group size (unknowable except by modern analogy) was sufficiently large (~10 or more individuals) to readily transport field-butchered carcasses, decisions prioritizing portions for transport and for abandonment would have been made as well. Judging from the bone assemblage recovered at FLK Zinj, hominins regularly prioritized for transport the meaty limb units and fat-rich hemimandibles (also meaty, if tongue and throat flesh are grouped with them), over the remaining axial portions. The availability of fully fleshed limbs indicates, in relation to carnivore feeding from carcasses before hominin acquisition, that other axial meat and fat and even some internal viscera would have been available to hominins. The paucity of vertebrae and crania, for example, makes it difficult to know what use hominins made of those portions. Possibilities include (1) defleshing significant loin meat from the vertebral column and transporting meat but not vertebrae to FLK Zinj; (2) field butchery and transport of vertebral portions to FLK Zinj, followed by removal from the site by scavengers; (3) breakage of crania and consumption of brains at death sites without bone transport. A similar possibility exists for any viscera obtained at death sites. Thus, if taphonomic factors have reduced the evidence of the parts transported, then even more meat and more fat were probably consumed than is indicated by direct bone evidence.

The abundance and strong patterning of defleshing cut marks on limb elements from FLK Zinj documents systematic, thorough butchery for meat by knowledgeable butchers. Similarly, abundant percussion notches and other evidence of hammerstone breakage of marrow bones documents thorough butchery for marrow fat, even extending to the breakage of some phalanges for marrow. Given this keen interest, it is unlikely that hominins would have been satisfied by the minor amount of total carcass fat (~2%) that marrow bones provide. More than 95% of total carcass fat in a typical tropical ungulate comes from other fat-rich sources, including brains, thoracic and abdominal viscera, and intramuscular fat, rather than from bone marrow. That is why pressure persisted on hominins to obtain intact carcasses, rather than the remnants available through passive scavenging (Bunn and Ezzo, 1993; Bunn, 2001).

During a daytime visit at FLK Zinj to process and consume meat and fat from a carcass, a group of cooperating hominins probably shared their food significantly before discarding bones and abandoning the site for a more secure sleeping location elsewhere. The amount of meat and fat available from a gazelle-sized or larger carcass would take hours to eat even by multiple hominins sharing that day's solution to the group's foraging needs. Scavenging hyenas and other carnivores probably had unhindered nocturnal access to the day's freshly discarded bones, which they exploited, thereby producing the documented depletion of limb epiphyses in the bone assemblage at FLK Zinj.

The logic favoring nightly abandonment of FLK Zinj and similar sites comes from observations of longer-term Hadza and shorter-term Kua San base camps. By their sustained presence at some base camps for weeks at a time, the Hadza unintentionally protect many limb epiphyses as they lose grease and nutritional appeal to potential scavengers, whereas at shorter-term camps of several days, the Kua San leave unprotected much fresher bone with adhering edible tissues (Bunn and Kroll, 1988; Bunn, Bartram, and Kroll, 1988). Loss of cancellous bones, and particularly of limb epiphyses, from shorter-term sites is much more pronounced. Thus, the loss of most fat-rich epiphyses from FLK Zinj suggests recurrent abandonment of the location by hominins. This does not, however, rule out the possibility of hominin presence at or near FLK at night, particularly if future research provides more convincing evidence of hominin control of fire. Cancellous bones do, of course, retain diminishing grease for days and weeks. The evidence, however, favors more intermittent, rather than more continuous use of FLK and similar Plio-Pleistocene sites as daytime home bases. The bone assemblage as a whole, accumulated through repeated daytime visits by hominins during the several-year time interval preceding the abrupt burial of the site.

After each abandonment of the FLK Zinj site by hominins, freshly discarded bones with contained grease and adhering remnants of flesh and fat evidently attracted hyenas and other, smaller scavengers. While searching through the butchered bones for the most appealing food items, hyenas gnawed some bones at the location and removed others, particularly greasy limb epiphyses, from the location. Most of the gnawing was done by small carnivores, probably jackals or even smaller viverrids. Analysis of the microfauna at a similar Bed I site (FLKN) indicates viverrid coprolites as the probable source of the microfaunal bones (Andrews, 1983) and hence, the presence of live viverrids at the site. Following diet-breadth logic (1) the small live rodents (microfauna) gnawing bones at FLK Zinj provided a potential food source for smaller carnivores (viverrids) and/or (2) butchered bones with adhering remnants of flesh and fat, food items that evidently were unappealing to larger hyenas, still provided an appealing food source for smaller carnivores (viverrids or canids). As reported a quarter-century ago, the mean width of 146 linear gnaw marks on bones from FLK Zinj is 0.70 mm (Bunn, 1981). Although few quantitative data on taxon-specific gnaw mark dimensions were then available, knowledge of various carnivore tooth sizes and shapes alone indicated that carnivores much smaller than hyenas probably produced most of the gnaw marks. Pickering et al. (2004) have recently reported a mean breadth of jackal gnaw marks of <1 mm, contrasted to hyena gnaw marks two to three times larger (Domínguez-Rodrigo and Piqueras, 2003). The most parsimonious taphonomic sequence forming the FLK Zinj bone assemblage is thus:

1. Repeated acquisition by hominins of intact carcasses through hunting of smaller carcasses and power scavenging (and/or less likely hunting) of larger carcasses (i.e., carnivore-exploitation strategy).
2. Transport of field-butchered portions to the FLK Zinj location.
3. Further butchery, sharing, and consumption of substantial quantities of meat and fat at FLK Zinj during the day.
4. Hyena removal of fresh limb epiphyses and some hyena gnawing of butchered bones; more intensive gnawing by small carnivores (jackals, viverrids) and by small rodents of butchered bones.

The fresh to lightly weathered (i.e., early stage 2 of Behrensmeyer, 1978) surfaces of the overwhelming majority of bones at FLK Zinj indicates that accumulation occurred during a geologically brief several-year period before burial of the site by an air-fall volcanic ash.

Other views of the taphonomic sequence at FLK Zinj differ in the reconstruction of hominin and carnivore interactions by depicting hominins as passive scavengers for marrow bones and scraps of meat from kills by lions that had already been defleshed and abandoned by the primary predators (e.g., Binford, 1981; Blumenschine, 1987; Marean et al., 1992; Selvaggio, 1994; Capaldo, 1997). They favor a hominin strategy of carnivore avoidance instead of carnivore exploitation. Several assumptions underlie these interesting actualistic studies: (1) the frequency of gnawing and bone destruction by captive and wild spotted hyenas can be used to accurately reveal the taphonomic sequence at FLK Zinj; (2) the frequencies of gnaw marks on eipiphyses, near-epiphyses, and midshafts in the actualistic observations and in the FLK Zinj assemblage show that limb bones were defleshed by large carnivores prior to hominin access; (3) if the paucity of vertebrae in the FLK Zinj assemblage is attributed to removal by hyenas, then hominins did not selectively transport meaty limb portions by hominins to the site. The underlying assumptions are problematic.

First, smaller carnivores, not hyenas, produced most of the linear gnaw marks on the bones, so there is no reason to assume that the interpretive framework provided by actualistic hyena gnawing is relevant to the frequency or anatomical location of gnawing by small carnivores on the FLK Zinj bones. Second, reporting the frequency of gnaw marks by fragmentary limb categories (epiphysis, near-eipiphysis, midshafts) overlooks the biasing effect caused by differential fragmentation of the specimens (Domínguez-Rodrigo, 1997; Bunn, 2001). By this methodology, a lengthy limb specimen that was classified as an epiphysis portion might easily include portions of epiphysis, near-epiphysis, and midshaft; a gnaw mark on the midshaft portion of that specimen would be counted as a gnawed epiphysis specimen. If increased fragmentation separated the midshaft portion as a distinct specimen, however, the same gnaw mark would be counted as a gnawed midshaft specimen, even though the original position of the gnawing was identical. An alternate approach that avoids the biasing effect of fragmentation involves graphic presentation of surface damages on composite drawings of whole limb elements. For cut marks, this has been achieved both manually (see Bunn, 2001) and with computer software (Abe et al., 2002), and the same approach needs to be done with gnaw marks at FLK Zinj. A recent reexamination of FLK limb specimens concludes that most of the linear grooves identified as carnivore gnaw marks by Blumenschine and colleagues actually result from biochemical decay, not carnivore gnawing, which casts further doubt on any significant role of carnivores in defleshing carcasses or in gnawing bones at FLK Zinj (Domínguez-Rodrigo and Barba, 2006).

Third, studies sampling bone distributions on modern landscapes in ecological settings comparable to Olduvai indicate that associations of more than approximately three animals in an area the size of FLK Zinj result from transport or from detectable catastrophic events (Behrensmeyer, 1983; Bunn, Kroll, and Bartram, 1991; Tappen, 1995). The MNI estimate at FLK Zinj is more than fifteen times greater than the sample likely to result from in situ death, unless considerably more time for site formation was involved. Both taphonomic and geological evidence, however, preclude that, which yields the reconstruction that carcass portions were transported to FLK Zinj. The dynamics of carcass transport are complex and have been discussed elsewhere (e.g., Bunn, Bartram, and Kroll, 1988; O'Connell, Hawkes, and Blurton Jones, , 1988, 1990; Bunn, 1993; Lyman, 1994). The salient point is that carcass size, fat content, and adhering meat would certainly influence bone selection by hominins and by hyenas for consumption and for transport. Regarding hominin decision making, transport distance varies inversely with the number of bones and amount of carcass selected for transport. For a specific carcass size, if transport distance to FLK Zinj was relatively short, more would have been transported, and vice versa. Field butchery into transportable portions would have been required for the size 3 carcasses predominant at FLK Zinj. But why would hominins have carried backbones from acquisition locations to FLK Zinj? For hominins lacking the technology to boil cancellous vertebrae for fat, what was the adaptive value in transporting those portions (Bunn, Bartram, and Kroll, 1988; Bunn 2001)? Most of the loin meat (>90% or more) can easily be defleshed from the backbone and carried away without requiring transport of the bones. If, however, hominins transported vertebral portions of large carcasses to FLK Zinj (in addition to the meaty limbs), followed by taphonomic loss of vertebrae from the site, then access to even more meat from intact carcasses, and more dedicated use of meat, is indicated.

The Unknown and the Unknowable

Even though Plio-Pleistocene archaeological evidence establishes hominin use of large mammal carcasses for meat and marrow with considerable resolution, there are some rather basic aspects of this reconstruction that are unknown or unknowable. Given the uncertain taxonomic affinities of key hominin fossils and the known presence in the East African Rift Valley of several hominin taxa around the Plio-Pleistocene boundary, it is difficult to know whether one hominin species produced all of the Oldowan archaeological record, as exemplified here by FLK Zinj, and if so, which species that was. It is difficult to know, except by modern analogy, what hominin group size, composition, and organization characterized the Plio-Pleistocene. And it is difficult to know the full range of hominin foraging adaptations of the time from examination of just one particularly large and well-preserved site. How important was meat in hominin diet, and how did that vary from place to place?

Alternative Interpretations of the Evolutionary Trajectory of Hominin Diet

Tubers and/or Meat?

Despite all that is known and all that is knowable from FLK Zinj and other Plio-Pleistocene archaeological sites, an alternative view suggests that meat was not a significant factor in Plio-Pleistocene diet or in providing selective evolutionary pressure for more humanlike foraging behavior. Wrangham and colleagues (1999) argue that because of butchery evidence at 2.5 Mya (de Heinzelin et al., 1999), a dietary change to increased meat consumption cannot have been the principal cause in the appearance of *H. erectus* more than one-half million years later. In other words, they contend that too much time passed following the earliest evidence of butchery to define a cause-and-effect relationship between increased meat in the diet and the evolution of *H. erectus*. Instead, Wrangham and colleagues claim that a dietary change to increased tuber consumption enabled the evolution of *H. erectus*. Claims of supporting evidence have been questioned elsewhere (Bunn, 1999).

The postulated shift to tuber consumption needs to be considered in relation to documented physical and metabolic changes in the hominins themselves, such as increased brain and body size and decreased cheek tooth and gut size. Foraging for tubers would not have selected for most of these, at least not if realistic estimates of the nutritional yields of wild tubers are considered (Schoeninger et al., 2001a, 2001b). Only by doubling or tripling the actual energetic value of wild tubers can O'Connell, Hawkes, and Blurton Jones (2002) describe them as an attractive energy source relative to meat. So, does it follow that increased meat and marrow in hominin diet made us human independently of other components of hominin lifeways, that other aspects of foraging strategy, social organization, biology, and diet immediately melted away in favor of a foraging strategy obsessed over meat? Not at all. I contend that in the context of existing foraging adaptations for a plant-based diet, meat is what changed first and what altered the dynamics toward a more humanlike foraging adaptation and diet in *H. erectus.*

Time spent foraging specifically for meat is time not spent accessing and consuming plant foods. (In practice, of course, an individual can do both simultaneously but probably less efficiently; this is meant as a heuristic discussion.) Individuals devoting significant time and energy foraging for animal carcasses, particularly when unsuccessful, which is most of the time judging from ethnography, come up short, with a deficit in the energy budget, much of the time, unless foraging dynamics change. What are some of the possibilities? One possibility is increasing foraging time, which is constrained by day length for a diurnal hominin. Another is increasing emphasis on higher-ranked food items, such as nuts and seeds rich in lipids and protein (Schoeninger et al., 2001b). Yet another is discovering and adding a new plant food, such as tubers, or increasing use of other carbohydrate sources, for example, honey, to the existing diet. Given that female chimps use sexual access as leverage to induce males to share meat (e.g., Mitani and Watts, 2001; Stanford, 2001), how big an intellectual leap would it have been to collect and then use surplus plant foods as leverage to induce meat sharing? Encephalized *Homo* was adapted for more complex decision making than australopithecines or chimps. This is not a particularly complex decision for early *H. erectus* to make. It is called division of labor.

The escalation of meat eating, which appears at the beginning of the Pleistocene, may have been supported by the addition of other energy sources to the diet (Aiello and Wells, 2002), but tubers did not initiate or support the beginning of meat eating in the hominin adaptation. On the basis of a simplistic view of Hadza foraging, a two- to threefold exaggeration of the energetic yields of Hadza tubers, and an incautious, distorted reading of the Plio-Pleistocene archaeological evidence, O'Connell, Hawkes, and Blurton Jones (1999, 2002) imagine that foraging for tubers by female *H. erectus* enabled male hunting and scavenging. For the Hadza, recognizing some relationship between foraging for tubers and for meat seems reasonable, given that those foods are reciprocally shared. But characterizing that relationship evolutionarily as O'Connell, Hawkes, and Blurton Jones do confuses the dynamic by reversing cause and effect and thereby losing a sense of time depth in evolution. How did a tuber-meat relationship develop for the Hadza, or, hypothetically, for ancestral Plio-Pleistocene hominins? No plausible evidence or claim of tuber use and cooking technology, the combined adaptation envisioned by Wrangham and others as an efficient substitute for meat, extends back into the Pliocene. A recent hypothesis by Laden and Wrangham (2005), however, proposes tubers as a pivotal fallback food in the early evolution of australopithecines but without considering how the earliest hominins of the Pliocene and even Miocene would have acquired deeply buried tubers in the first place. Evidence of butchery for meat and marrow, however, extends all the way back to 2.6 Mya (de Heinzelin et al., 1999; Semaw et al., 2003; Domínguez-Rodrigo et al., 2005). Thus, consumption of meat and marrow of large animals greatly preceded any likely tuber foraging. Meat provided the selective pressure. Meat was the evolutionary dynamic driving the adaptation and defining the cause-and-effect relationship. Tubers (the effect), or honey, more likely, may have compensated for foraging time and food value lost in unsuccessful searches for meat (the cause).

However, even if hominins controlled fire as early as some authors claim, tubers need not have played a singularly dominant role in hominin foraging strategy and diet at the Plio-Pleistocene boundary. If hominins regularly controlled fire by 1.64 Mya (Bellomo, 1994) or earlier (although there is no evidence), particularly if the fires were cooking fires, then foraging efficiency may have been broadly increased for various plants and for meat. In that context, a positive energy balance may have enabled more foraging for meat. If tubers were a significant new dietary component at the beginning of the Pleistocene, then an even more pronounced response in foraging effort for meat and marrow would be predicted. Again, the documented inclusion of meat and marrow in the diet of Pliocene hominins started *Homo* on this trajectory. If fire were routinely controlled by *Homo* at the beginning of the Pleistocene, then so-called home bases—intermittent daytime sites—routinely and quickly would have become the residential base camp sites characteristic of human foragers today.

Meat Made Us Human

I contend that Wrangham and colleagues (1999) and O'Connell, Hawkes, and Blurton Jones (1999, 2002) have unrealistically reversed cause and effect in claiming that (1) the exploitation of tubers enabled hominins to invest foraging energy in hunting and scavenging for carcasses, and (2) cooked tubers, not meat, supported the biological changes seen in fossils of early *H. erectus*, including expanded brains, smaller teeth, and more human body proportions (limbs and gut) and body size. To reach that perspective, it is necessary to collapse time depth and view the dietary changes, essential technology, and biological changes as having all appeared simultaneously and abruptly at 1.9 Mya, which is the earliest date for claimed, fragmentary *H. erectus* fossils in East Africa. A date of 1.9 Mya is significantly older than the oldest debated claim for control of fire by hominins, and it is older than the small bone "scratching sticks" from Swartkrans that do not compare well to the heavy-duty digging sticks used by human foragers to access deeply buried tubers. Yet, Wrangham and colleagues needing a source for these behavioral innovations, suggest more tenuously that an even older population of early *Homo* (i.e., *Homo habilis*) would have already possessed the controlled fire and digging sticks essential for the cooked-tuber adaptation. O'Connell, Hawkes, and Blurton Jones imagine that these technological innovations all appeared de novo with *H. erectus*. At any rate, the implication is that a cooked-tuber adaptation was in place simultaneous with the actual evidence of Plio-Pleistocene butchered bones and firmly inferred meat eating.

Aiello and Wheeler (1995) provide, in their expensive-tissue hypothesis, an informative biological framework in which to view the evolution of hominin diet and foraging adaptations. They use physiological and fossil evidence in tracing an evolutionary relationship between encephalization and gut reduction to emphasize the increased metabolic costs of those changes and an essential shift to higher quality food, which, they argue, increased meat consumption provided in the early evolution of the genus *Homo*. Aiello and Wells (2002) assert that the hypothesized dietary shift did not simply involve adding more meat to an apelike (frugivorous) diet but would have required the addition of other foods to compensate for the energetic costs and limitations of increased meat foraging and consumption. In concert with Wrangham and colleagues, Aiello and Wells identify tubers as a likely food that could have balanced the energetic needs of *H. erectus*.

How might evidence from Plio-Pleistocene archaeology build onto these intriguing perspectives of the diet and foraging adaptations of early *Homo*? How human was the diet and foraging behavior of early *Homo*? First, the archaeological evidence in this chapter, from 1.8 to 1.6 Mya, when there is no doubt that early *H. erectus* was present in the East African Rift Valley, documents significant hominin involvement in the repeated transport of carcass portions to favored, central locations (i.e., daytime home bases) for further butchery, sharing, and the consumption of meat and fat. Among Oldowan sites of this age, evidence of foraging behavior is best illustrated by the FxJj 50 archaeological site (Bunn et al., 1980, 1997) at Koobi Fora, where *H. erectus* is present in the fossil sample, and by FLK Zinj at Olduvai, where *H. erectus* has not (yet) been recovered from Bed I deposits. At both field areas, additional, smaller Oldowan sites, each with its own taphonomic history, provide smaller samples of similar behavioral evidence. Taken together, this substantial body of evidence makes it unlikely that early *Homo* was as marginally involved in meat eating as Wrangham and colleagues (1999) and O'Connell, Hawkes, and Blurton Jones (1999, 2002) imply. If meat and tubers were causally related, as Aiello and Wells (2002) argue, with tubers compensating for the energetic inefficiency of meat, then how did that relationship originate? One perspective has tubers driving the adaptation and leading abruptly to *H. erectus* at 1.9 Mya, with tuber gathering then facilitating meat foraging. But that perspective is contradicted as soon as time depth is added and known archaeological evidence of meat eating dating back to 2.6 Mya is considered. Unless one is willing to imagine the existence of technology for accessing and cooking tubers at 2.6 Mya, then tubers, as the cause, did not enable meat, as the effect. The evolutionary trajectory from hominin to humanity, from small-brained australopithecine to encephalized *Homo erectus*, began 2.6 Mya with an interest in meat.

Wrangham and colleagues (1999) reason that because of the evidence of butchery at 2.6 Mya, meat cannot have been influential in the evolution leading to *H. erectus* because *H. erectus* did not evolve abruptly at that early date. O'Connell, Hawkes, and Blurton Jones (2002) adopt that argument and allege that Pliocene evidence extending back to 2.6 Mya is comparable or similar to FLK Zinj. But why does that logic necessarily apply? Why would it not have required an extended interval for a small-bodied hominin with the gut of a frugivore to be transformed through natural selection into a more humanlike omnivore, to convert a potentially risky dietary interest in meat into an ability to obtain it with some reasonable efficiency and regularity? Given the major changes documented in the hominin fossil record between 2.6 and 1.6 Mya, it is logical to predict behavioral change as well. Rather than obscuring this possibility by lumping the archaeology as similar throughout this million-year time interval, it is possible to divide the evidence into a three-phase sequence of behavioral evolution: (1) 2.6–2.5 Mya, (2) 2.3–1.9 Mya, (3) 1.8–1.6 Mya.

1. In situ (?) butchery of several bones with the oldest flaked stone tools. The sample of butchered bones at 2.6–2.5 Mya could not be much smaller and still exist. Yet, it is decisive and highly significant in providing hard evidence that the long-standing argument that flaked stone tools were invented for butchery is correct. Still, it would be incautious to infer too much from the handful of butchered bones. At face value, they do not indicate the transport and concentration of food bones at central locations comparable to evidence from the 1.8–1.6 million year old sites (but see Domínguez-Rodrigo and colleagues (2005) who refer to the Hadar occurrences as Type C sites but do not provide MNI estimates or other conclusive evidence of repeated food transport in the behavioral sense intended by Isaac's (1978) definition of a home base/Type C site). That does not imply a marginal interest in meat and marrow, just a marginal ability to obtain it.

An understanding of the biology of the actual hominin toolmaker(s) between 2.6 Mya and the origin of *H. erectus* would help clarify its foraging adaptations. But achieving that is a problem because of the uncertain taxonomic status of key fossils. This intriguing period, from 2.6 to 1.9 or 1.8 Mya witnessed the early evolution of the genus *Homo* or its direct ancestors. What were those hominins like in body form and foraging adaptations? Judging crudely from partial skeletons of two relevant taxa in East Africa, *Australopithecus garhi* at 2.5 Mya (BOU-VP-12/1; Asfaw et al., 1999) and *Homo habilis* at 1.8 Mya (OH 62; Johanson et al., 1987), they differed from early *H. erectus* in having smaller brains, larger teeth, smaller bodies, and relatively apelike forearms. Notably, Asfaw and colleagues (1999) describe the sequential evolution of hindlimb and forelimb proportions, with *A. garhi* already exhibiting humanlike femoral proportions (relative to humerus) but retaining an apelike brachial index. An elongate forearm also characterizes OH-62 (Johanson et al., 1987). Those limb proportions probably indicate selection for greater terrestrial mobility in the progressively drier and more open habitats of the Late Pliocene, combined with retention of more apelike forelimbs for proficient arboreality (e.g., deMenocal, 1995; Wood and Collard, 1999; Wood and Strait, 2004).

Is that the ancestral *Homo* population predicted by Wrangham and colleagues (1999) to have headed into more open habitats to dig up tubers? If so, what was the adaptive value of the apelike forearms? A more parsimonious view predicts, instead, that if proficient climbing ability characterized that frugivore, then following Aiello and Wells (2002), an energy investment in meat while living on the ground had to be compensated through the addition of more energy. Even a relatively small investment in meat is time and energy not spent being a frugivore. Instead of habitually defying the adaptations of the forelimb and also requiring technological innovations (fire, heavy-duty digging sticks) for which evidence is lacking in the Pliocene and tenuous in the Early Pleistocene, why not use that proficient climbing ability in the food quest to balance the energy budget? In addition to fruit and nuts already used, what other high-energy food is available in the trees? Honey may have sweetened the diet and balanced the energy budget of early *Homo*. It may be as hard to test with direct evidence as tubers, but honey is widely available and accessible using the technology of chimps (McGrew, 1992). The existence of multiple species of non-stinging bees mean that fire and smoke are not required to obtain honey without the discomfort of being stung. Equipping hominins with simple Oldowan stone tools would increase the efficiency of accessing hives beyond the chimp level.

2. Bone-bearing archaeological sites from this time interval (2.3–1.9 Mya) at West Turkana (Kibunjia, 1994) and Koobi Fora (Isaac, 1997) yield MNI estimates in excess of one dozen animals in temporally restricted assemblages and, thus, evidence of food transport. Abundant evidence of hammerstone breakage of limb bones but minimal cut-mark evidence could imply more access to marrow than to meat. Insofar as the first evidence of significant repeated transport of carcass portions for delayed processing and consumption indicates an increased investment in foraging and consumption of animal tissues (relative to the 2.5–2.6 Mya in situ butchery evidence), then it follows that selection would have favored increased body size for mobility in open habitats and for accessing and carrying carcass portions. The hominin fossil record during the time immediately before *Homo erectus* does not permit thorough evaluation of this predicted relationship. The taxonomic status of the complete femur, KNM-ER 1481, now seems unclear. Although originally assigned to the same species as KNM-ER 1470 (Leakey and Leakey, 1978), that is, *Homo rudolfensis*, Wood and Collard (1999) do not do so. What is described as a modern appearance could also indicate early *H. erectus*, but either way, the specimen does indicate the presence of a large, mobile hominin in the East African Rift by 1.9 Mya.

3. Multiple archaeological sites from the 1.8–1.6 Mya interval, including FLK Zinj and FxJj 50, yield much larger concentrations of thousands of stone tools and MNI estimates, ranging up to four dozen large animals in temporally restricted assemblages. Combined with extensive evidence of butchery for meat and marrow, this demonstrates regular access to mostly intact carcasses and repeated transport of portions to favored, central locations (i.e., intermittent, daytime home bases). What was the likely role of tubers? Given a realistic measure of the energetic yields of tubers (Schoeninger et al., 2001a), particularly in a precooking context, it is unlikely that tubers, instead of meat, fueled the adaptation and selection for the craniodental and postcranial pattern that defines the genus *Homo*. Viewing tubers as a fallback food in the diet of early *Homo* may be prudent and parsimonious, given their availability in an apparently open niche.

The social dynamics would have been quite different from the male-theft scenario developed by Wrangham and colleagues (1999). Instead, meat was demonstrably the sought-after, but risky, food item. Why share it? For chimps, the answer is sex (e.g., Mitani and Watts, 2001; Stanford, 2001). Let's not equate sex with tubers. But if the sought-after food item is not available, you still need to eat. If more commitment to terrestriality and meat foraging occurred at the beginning of the Pleistocene and if the higher-energy requirements of early *H. erectus* could not be balanced with honey, then tubers would be a likely addition, providing the same mixed diet as Wrangham and colleagues (1999) describe, just a very different means to get to a more human division of labor in the food quest. So, what was causing the selective pressure and driving this adaptation toward a more humanlike foraging adaptation? MEAT!

Acknowledgments I thank Peter Ungar for inviting me to the conference and for good advice during the preparation of this chapter and all of the conference participants for stimulating discussions about hominin diet.

References

Abe, Y., Marean, C.W., Nilssen, P.J., Assefa, Z., and Stone, E.C., 2002. The analysis of cutmarks on archaeofauna: A review and critique of quantification procedures, and a new image-analysis GIS approach. *Am. Antiq.* 67, 643–663.

Aiello, L.C., and Wells, C.K., 2002. Energetics and the evolution of the genus *Homo*. *Annu. Rev. Anthropol.* 31, 323–338.

Aiello, L.C., and Wheeler, P., 1995. The expensive-tissue hypothesis: The brain and the digestive system in human and primate evolution. *Curr. Anthropol.* 36, 199–221.

Andrews, P., 1983. Small mammal faunal diversity at Olduvai Gorge, Tanzania. In: Clutton-Brock, J., and Grigson, C. (Eds.), *Animals and Archaeology*. 1. *Hunters and Their Prey*. British Archaeological Reports International Series 163, Oxford, pp. 77–85.

Asfaw, B., White, T., Lovejoy, O., Latimer, B., Simpson, S., and Suwa, G., 1999. *Australopithecus garhi*: A new species of early hominid from Ethiopia. *Science* 284, 629–635.

Behrensmeyer, A.K., 1975. The taphonomy and palaeoecology of Plio-Pleistocene vertebrate assemblages east of Lake Rudolf, Kenya. *Bull. Mus. Comp. Zool.* 146, 473–578.

Behrensmeyer, A.K., 1978. Taphonomic and ecologic information from bone weathering. *Paleobiology* 4, 150–162.

Behrensmeyer, A.K., 1983. Patterns of natural bone distribution on recent and Pleistocene land surfaces: Implications for archaeological site formation. In: Clutton-Brock, J., and Grigson, C. (Eds.), *Animals and Archaeology*. 1. *Hunters and Their Prey*. British Archaeological Reports International Series 163, Oxford, pp. 93–106.

Bellomo, R.V., 1994. Methods of determining early hominid behavioral activities associated with the controlled use of fire at FxJj 20 Main, Koobi Fora, Kenya. *J. Hum. Evol.* 27, 173–195.

Binford, L.R., 1968. Archaeological perspectives. In: Binford, S., and Binford, L. (Eds.), *New Perspectives in Archeology*. Aldine Press, Chicago, pp. 5–32.

Binford, L.R., 1972. *An Archaeological Perspective*. Seminar Press, New York.

Binford, L.R., 1981. *Bones: Ancient Men and Modern Myths*. Academic Press, New York.

Blumenschine, R.J., 1987. Characteristics of an early hominid scavenging niche. *Curr. Anthropol.* 28, 383–407.

Blumenschine, R.J., Masao, F.T., 1991. Living sites at Olduvai Gorge, Tanzania? Preliminary landscape archaeology results in the basal Bed II lake margin zone. *J. Hum. Evol.* 21, 451–462.

Brain, C.K., 1967. Hottentot food remains and their bearing on the interpretation of fossil bone assemblages. Scientific Papers of the Namib Desert Research Station, No. 32, 1–11.

Brain, C.K., 1969. The contribution of Namib Desert Hottentots to an understanding of australopithecine bone accumulations. Scientific Papers of the Namib Desert Research Station, No. 39, 13–22.

Bunn, H.T., 1981. Archaeological evidence for meat-eating by Plio-Pleistocene hominids from Koobi Fora and Olduvai Gorge. *Nature* 291, 574–577.

Bunn, H.T., 1982. Meat-eating and human evolution: studies on the diet and subsistence patterns of Plio-Pleistocene hominids in East Africa. PhD diss.,, University of California, Berkeley.

Bunn, H.T., 1986. Patterns of skeletal representation and hominid subsistence activities at Olduvai Gorge, Tanzania, and Koobi Fora, Kenya. *J. Hum. Evol.* 15, 673–690.

Bunn, H.T., 1989. Diagnosing Plio-Pleistocene hominid activity with bone fracture evidence. In: Bonnichsen, R., and Sorg, M., (Eds.), *Bone Modification*. Center for the Study of the First Americans, Orono, Maine, pp. 299–316.

Bunn, H.T., 1991. A taphonomic perspective on the archaeology of human origins. *Annu. Rev. Anthropol.* 20, 433–467.

Bunn, H.T., 1993. Bone assemblages at base camps: A further consideration of carcass transport and bone destruction by the Hadza. In: Hudson, J., (Ed.), *From Bones to Behavior*. Center for Archaeological Investigations, Southern Illinois University, Carbondale, pp. 156–168.

Bunn, H.T., 1999. Comment on The raw and the stolen: Cooking and the ecology of human origins, by Wrangham, R.W., et al. *Curr. Anthropol.* 40, 579–580.

Bunn, H.T., 2001. Hunting, power scavenging, and butchering by Hadza foragers and by Plio-Pleistocene *Homo*. In: Stanford, C.B., and Bunn, H.T., (Eds.), *Meat-Eating and Human Evolution*, Oxford University Press, Oxford, pp. 199–218.

Bunn, H.T., Bartram, L.E., and Kroll, E.M., 1988. Variability in bone assemblage formation from Hadza hunting, scavenging, and carcass processing. *J. Anthropol. Archaeol.* 7, 412–457.

Bunn, H.T., and Ezzo, J.A., 1993. Hunting and scavenging by Plio-Pleistocene hominids: Nutritional constraints, archaeological patterns, and behavioral implications. *J. Archaeol. Sci.* 20, 365–398.

Bunn, H., Harris, J.W.K., Isaac, G., Kaufulu, Z., Kroll, E., Schick, K., Toth, N., and Behrensmeyer, A.K., 1980. FxJj 50: An early Pleistocene site in northern Kenya. *World Archaeol.* 12, 109–136.

Bunn, H.T., and Kroll, E.M., 1986. Systematic butchery by Plio/Pleistocene hominids at Olduvai Gorge, Tanzania. *Curr. Anthropol.* 27, 431–452.

Bunn, H.T., and Kroll, E.M., 1987. On butchery by Olduvai hominids, reply to Potts, R., *Curr. Anthropol.* 29, 96–98.

Bunn, H.T., and Kroll, E.M., 1988. Fact and fiction about the Zinjanthropus floor: Data, arguments, and interpretations, reply to Binford, L.R., *Curr. Anthropol.* 29, 135–155.

Bunn, H.T., Kroll, E.M., and Bartram, L.E., 1991. Bone distribution on a modern East African landscape and its archaeological implications. In: Clark, J.D. (Ed.), *Cultural Beginnings: Approaches to Understanding Early Hominid Lifeways in the African Savanna*. Romisch-Germanisches Zentralmuseum, Forschungsinstitut fur Vor- und Fruhgeschichte, Dr. Rudolf Habelt GMBH, Bonn, pp. 33–54.

Bunn, H.T., Kroll, E.M., Kaufulu, Z., and Isaac, G.Ll., 1997. FxJj 50. In: Isaac, G.Ll. (Ed.), *Koobi Fora Research Project*. Vol. 5: *Plio-Pleistocene Archaeology*. Clarendon Press, Oxford, pp. 192–211.

Capaldo, S.D., 1997. Experimental determinations of carcass processing by Plio-Pleistocene hominids and carnivores at FLK 22 (*Zinjanthropus*), Olduvai Gorge, Tanzania. *J. Hum. Evol.* 33, 555–597.

Capaldo, S.D., and Blumenschine, R.J., 1994. A quantitative diagnosis of notches made by hammerstone percussion and carnivore gnawing on bovid long bones. *Am. Antiq.* 59, 724–748.

Dart, R.A., 1957. The Osteodontokeratic culture of *Australopithecus prometheus*. *Trans. Mus. Mem.* 10, 1–105.

de Heinzelin, J., Clark, J.D., White, T.W., Hart, W., Renne, P., WoldeGabriel, G., Beyene, Y., and Vrba, E., 1999. Environment and behavior of 2.5-million-year-old Bouri hominids. *Science* 284, 625–629.

deMenocal, P.B., 1995. Plio-Pleistocene African climate. *Science* 270, 53–59.

Domínguez-Rodrigo, M., 1997. Meat-eating by early hominids at the FLK 22 *Zinjanthropus* site, Olduvai Gorge (Tanzania): An experimental approach using cut-mark data. *J. Hum. Evol.* 33, 669–690.

Domínguez-Rodrigo, M., 1999. Flesh availability and bone modifications in carcasses consumed by lions: Palaeoecological relevance in hominid foraging patterns. *Palaeogeogr. Palaeoclimatol. Palaeoecol.* 149, 373–388.

Domínguez-Rodrigo, M., 2002. Hunting and scavenging by early humans: the state of the debate. *J. World Prehist.* 16, 1–54.

Domínguez-Rodrigo, M., and Barba, R., 2006. New estimates of tooth mark and percussion mark frequencies at the FLK Zinj site: The carnivore-hominid-carnivore hypothesis falsified. *J Hum. Evol.* 50, 170–194.

Domínguez-Rodrigo, M., and Piqueras, A., 2003. The use of tooth pits to identify carnivore taxa in tooth-marked archaeofaunas and their relevance to reconstruct hominid carcass processing behaviours. *J. Archaeol. Sci.* 30, 1385–1391.

Domínguez-Rodrigo, M., Pickering, T.R., Semaw, S., and Rogers, M.J., 2005. Cutmarked bones from Pliocene archaeological sites at Gona, Afar, Ethiopia: Implications for the function of the world's oldest stone tools. *J. Hum. Evol.* 48, 109–121.

Hay, R.L., 1976. Geology of the Olduvai Gorge: *A Study of Sedimentation in a Semi-Arid Basin.* University of California Press, Berkeley.

Hill, A.P., 1975. Taphonomy of contemporary and late Cenozoic East African vertebrates. PhD diss.,, University of London.

Isaac, G.Ll., 1971. The diet of early man: Aspects of archaeological evidence from Lower and Middle Pleistocene sites in Africa. *World Archaeol.* 2, 278–299.

Isaac, G.Ll., 1978. The food-sharing behavior of protohuman hominids. *Sci. Am.* 238, 90–108.

Isaac, G.Ll., 1984. The archaeology of human origins: Studies of the Lower Pleistocene in East Africa, 1971–1981. *Adv. World Archaeol.* 3, 1–87.

Isaac, G.Ll. (Ed.), 1997. *Koobi Fora Research Project.* Vol. 5: *Plio-Pleistocene Archaeology.* Clarendon Press, Oxford.

Johanson, D.C., Masao, F.T., Eck, G.G., White, T.D., Walter, R.C., Kimbel, W.H., Asfaw, B., Manega, P., Ndessokia, P., and Suwa, G., 1987. New partial skeleton of *Homo habilis* from Olduvai Gorge, Tanzania. *Nature* 327, 205–209.

Keeley, L.H., Toth, N., 1981. Microwear polishes on early stone tools from Koobi Fora, Kenya. *Nature* 293, 464–465.

Kibunjia, M., 1994. Pliocene archaeological occurrences in the Lake Turkana basin. *J. Hum. Evol.* 27, 159–171.

Kroll, E.M., and Isaac, G. Ll., 1984. Configurations of artifacts and bones at early Pleistocene sites in East Africa, in Hietela, H.J. (Ed.), *Intrasite Spatial Analysis in Archaeology.* Cambridge University Press, Cambridge, pp. 4–31.

Kruuk, H., 1972. The *Spotted Hyena: A Study of Predation and Social Behavior.* University of Chicago Press, Chicago.

Laden, G., and Wrangham, R., 2005. The rise of the hominids as an adaptive shift in fallback foods: Plant underground storage organs (USOs) and australopith origins. *J. Hum. Evol.* 49, 482–498.

Leakey, M.D., 1971. *Olduvai Gorge: Excavations in Beds I and II, 1960–1963.* Cambridge University Press, Cambridge.

Leakey, M.G., and Leakey, R.E. (Eds.), 1978. *Koobi Fora Research Project.* Vol. 1: *The Fossil Hominids and an Introduction to Their Context.* Clarendon Press, Oxford.

Lyman, R.L., 1994. *Vertebrate Taphonomy.* Cambridge University Press, Cambridge.

Marean, C.W., Spencer, L.M., Blumenschine, R.J., and Capaldo, S.D., 1992. Captive hyaena bone choice and destruction, the schlep effect and Olduvai archaeofaunas. *J. Archaeol. Sci.* 19, 101–121.

McGrew, W.C., 1992. *Chimpanzee Material Culture: Implications for Human Evolution.* Cambridge University Press, Cambridge.

Mitani, J.C., and Watts, D.P., 2001. Why do chimpanzees hunt and share meat? *Anim. Behav.* 61, 915–924.

O'Connell, J.F., Hawkes, K., and Blurton Jones, N., 1988. Hadza hunting, butchering, and bone transport and their archaeological implications. *J. Anthropol. Res.* 44, 113–161.

O'Connell, J.F., Hawkes, K., and Blurton Jones, N., 1990. Reanalysis of large mammal body part transport among the Hadza. *J. Archaeol. Sci.* 17, 301–316.

O'Connell, J.F., Hawkes, K., and Blurton Jones, N., 1999. Grandmothering and the evolution of *Homo erectus. J. Hum. Evol.* 36, 461–485.

O'Connell, J., Hawkes, K., and Blurton Jones, N., 2002. Meat-eating, grandmothering, and the evolution of early human diets. In: Ungar, P.S., and Teaford, M.F. (Eds.), *Human Diet: Its Origin and Evolution.* Bergin & Garvey, Westport, CT; pp. 49–60.

Pickering, T.R., Domínguez-Ridrigo, M., Egeland, C.P., and Brain, C.K., 2004. Beyond leopards: Tooth marks and the contribution of multiple carnivore taxa to the accumulation of the Swartkrans Member 3 fossil assemblage. *J. Hum. Evol.* 46, 595–604.

Plummer, T., 2005. Discord after discard: Reconstructing aspects of Oldowan hominin behavior. In: Stahl, A.B. (Ed.), *African Archaeology: A Critical Introduction.* Blackwell,

Malden, MA.Potts, R.B., 1988. Early Hominid Activities at Olduvai. Aldine de Gruyter, New York.

Potts, R.B., and Shipman, P.B., 1981. Cutmarks made by stone tools on bones from Olduvai Gorge, Tanzania. *Nature* 291, 577–580.

Schaller, G.B., 1972. *The Serengeti Lion: A Study of Predator-Prey Relations*. University of Chicago Press, Chicago.

Schoeninger, M.J., Bunn, H.T., Murray, S.S., and Marlett, J.A., 2001a. Composition of tubers used by Hadza foragers of Tanzania. *J. Food Composition Analysis* 14, 15–25.

Schoeninger, M.J., Bunn, H.T., Murray, S., Pickering, T., and Moore, J., 2001b. Meat-eating by the fourth African ape. In: Stanford, C.B., and Bunn, H.T., (Eds.), *Meat-eating and Human Evolution*. Oxford University Press, Oxford, pp. 179–195.

Selvaggio, M.M., 1994. Carnivore tooth marks and stone tool butchery marks on scavenged bones: Archaeological implications. *J. Hum. Evol.* 27, 215–228.

Semaw, S., Rogers, M.J., Quade, J., Renne, P.R., Butler, R.F., Domínguez-Rodrigo, M., Stout, D., Hart, W.S., Pickering, T., and Simpson, S.W., 2003. 2.6-Million-year-old stone tools and associated bones from OGS-6 and OGS-7, Gona, Afar, Ethiopia. *J. Hum. Evol.* 45, 169–177.

Stanford, C.B., 2001. A comparison of social meat-foraging by chimpanzees and human foragers. In: Stanford, C.B., and Bunn, H.T., (Eds.), *Meat-Eating and Human Evolution*. Oxford University Press, Oxford, pp. 122–140.

Tappen, M., 1995. Savanna ecology and natural bone deposition. *Curr. Anthropol.* 36, 223–260.

Wood, B., Collard, M., 1999. The human genus. *Science* 284, 65–71.

Wood, B., Strait, D., 2004. Patterns of resource use in early *Homo* and *Paranthropus*. *J. Hum. Evol.* 46, 119–162.

Wrangham, R.W., Jones, J.H., Laden, G., Pilbeam, D., and Conklin-Brittain, N., 1999. The raw and the stolen: Cooking and the ecology of human origins. *Curr. Anthropol.* 40, 567–594.

Zihlman, A.L., Tanner, N., 1978. Gathering and hominid adaptation. In: Tiger, L., and Fowler, H. (Eds.), *Female Hierarchies*. Beresford Book Service, Chicago, pp. 163–194.

12

Lithic Archaeology, or, What Stone Tools Can (and Can't) Tell Us about Early Hominin Diets

JOHN J. SHEA

Stone tools are the most durable residues of hominin behavior. Our ancestors left these "Stone Age visiting cards," as Isaac (1981) called them, on every major land mass humans have inhabited, except Antarctica. Before the advent of cheap and efficient methods for producing metal implements, stone tools were employed in subsistence tasks by all known human societies. Stone tools are also still used as subsistence aids by chimpanzees, our nearest primate relatives. To the extent that we believe major changes in hominin evolution have been accompanied by dietary shifts, it is reasonable to seek clues to these changes in stone-tool design and variability. In seeking these clues, however, we need to be alert to the complicating effects of behavioral variability. Recent stone-tool-using humans exhibit considerable variability, not only in the kinds of tasks for which stone tools are used but also in the choice of technological strategies they deploy in their land-use strategies and subsistence adaptations. To cite one particularly well-phrased example, observed by the archaeologist/soldier T. E. Lawrence in Arabia, during World War I:

> We put a few miles between us and the railway before we sat down to our feast of mutton. We were short of knives, and, after killing the sheep in relay, had recourse to stray flints to cut them up. As men unaccustomed to such expedients, we used them in the eolithic spirit [*i.e., without flaking them -J.S.*]; and it came to me that if iron had been constantly rare we should have chipped our daily tools skillfully as palaeoliths: whilst had we had no metal whatever, our art would have been lavished on perfect and polished stones. (Lawrence, 1935, p. 294)

Since the earliest days of Paleolithic research, prehistorians have inferred a causal connection between stone tools and hominin carnivory (Isaac, 1971). Juxtaposed concentrations animal bones featuring cut marks and percussion fractures together with stone tools have been discovered at numerous sites in Africa and western Asia, dating to more than 1 Myr, including (to name but a few) Olduvai Gorge, Tanzania (Leakey, 1971; Bunn, 1981; Potts and Shipman, 1981; Potts, 1988; Blumenschine, 1995); East Turkana, Kenya (Bunn et al., 1980; Isaac and Isaac, 1997); Olorgesailie, Kenya (Isaac, 1977; Potts, 1989); Kada Gona, Ethiopia (Semaw et al., 2003); and Ubeidiya, Israel, (Bar-Yosef and Goren-Inbar, 1993; Shea, 1999; Gaudzinski, 2004). The question of hunting versus scavenging as the principal mode of meat/fat procurement at these and other sites is still debated (Bunn, 2001; Lupo and O'Connell, 2002; Domínguez-Rodrigo and Pickering, 2003). Nevertheless, the inferred link between early hominins' making stone tools and meat eating/animal fat procurement retains near-universal support among contemporary archaeologists (Schick and Toth, 1993; Stanford and Bunn, 2001).

But were Early Paleolithic stone tools designed first and foremost as subsistence aids? Microwear- and residue-based functional analyses of Early Paleolithic tools from East Turkana and other sites indicate stone tool use for working wood and processing soft plant matter (Keeley and Toth, 1981; Domínguez-Rodrigo et al., 2001). The ethnographic record is also rich in accounts of simple stone tools like those found in Early Paleolithic assemblages being used as a "secondary technology," as tools for making other tools of wood, bone, and other perishable media and as aids to the collection and processing of plant foods (Gould, 1980). Replicative experiments affirm Early Paleolithic tools would have been effective butchery tools (Jones, 1980), but experiments suggest they would have been effective in many other tasks as well (Schick and Toth, 1993).

This chapter examines what stone tools from Late Pliocene and Early Pleistocene contexts (2.7–0.7 Myr) in Africa and Eurasia can tell us about early hominin diets. The overwhelming majority of Early Paleolithic stone tools are simple pebble cores and flake tools. The technological simplicity of Early Paleolithic stone-tool technology has been viewed as a product of relatively simple hominin cognitive capacities. I argue here that this simplicity is deceptive and that it is better explained as a strategic response to the intrinsic variability of early hominin subsistence activities and other tasks requiring stone tools.

It is tempting to see the inception of stone-tool use by hominins as a revolutionary development, but the cognitive capacities underlying stone-tool production are probably evolutionarily primitive among hominins and at least some African hominoids (chimpanzees and bonobos) (Toth et al., 1993; Wynn, 2002). Most Early Paleolithic stone tools were created by controlled conchoidal fracture in brittle isotropic rocks. These rocks are usually fine-grained silicates (chert/flint, quartz) and volcanics (e.g., basalt). Most of the lithic raw materials selected for tool production (or "flintknapping") were procured from sources close to the places where stone tools were discarded (typically less than 5–10 kilometers). Chimpanzee nut-cracking is a plausible evolutionary precursor to stone tool production. Cracking nuts with stone and wood hammers is a widespread chimpanzee subsistence activity (McGrew, 1992), and one that requires a degree of control over fracture properties similar to that involved in flintknapping. Chimpanzees cracking nuts with stone hammers occasionally produce fractured and battered stones that are superficially similar to some components of Early Paleolithic assemblages (Mercader, Panger, and Boesch, 2002). Where Early Paleolithic stone tool production differs from chimpanzee stone breakage is in the former's emphasis on the systematic production of sharp-edged flakes. Controlled experiments suggest that chimpanzees can be taught to fracture stone and to use the sharp-edged flakes that result (Toth et al., 1993), but they do not appear to make and use stone-cutting tools in the wild. That they do not do this suggests early hominin stone-tool production probably involved activities that are not commonly undertaken by living nonhuman primates. Butchery of large mammal carcasses and bulk processing of plant foods might number among these activities but so too might the production of wooden tools with stone implements.

Archaeological Terminology

The term "Early Paleolithic" is increasingly used to subsume both the Eurasian Lower Paleolithic and the African Early Stone Age (Toth and Schick, 2000). Most archaeologists describe Early Paleolithic stone tools in terms of three major "industrial complexes," the Oldowan, the Developed Oldowan, and the Acheulean (Leakey, 1971; Isaac, 1984; Schick and Toth, 1993; Klein, 1999; Plummer, 2004).

The Oldowan Industrial Complex (or "Oldowan Industry") was defined on the basis of the Leakeys' excavations in Beds I–II of Olduvai Gorge Tanzania (Leakey, 1971). Most lithic assemblages formally recognized as Oldowan date to more than 1.5 Myr, but younger assemblages are sometimes likened to the Oldowan as well (Plummer, 2004). Oldowan assemblages are composed primarily of flakes struck from pebble cores and a variety of morphologically distinct core types (i.e., choppers, discoids, core scrapers, and polyhedrons). Retouched flakes are relatively rare and typically take the forms of simple scrapers, notches, denticulates, and awls. Some of the best-documented African Oldowan assemblages occur in Bed I of Olduvai Gorge (Tanzania), the KBS Member of the Koobi Fora Formation at East Turkana (Kenya), Member F of the Omo Shungura Formation and Kada Hadar/Kada Gona (both in Ethiopia), and Member 5 of Sterkfontein Cave (South Africa).

The attribution of later (i.e., post–1.6 Myr) and non–African Early Paleolithic assemblages to the Oldowan is problematic. The Oldowan does not possess uniquely derived technological and typological features other than those relating to the production of flakes from pebble cores by hard-hammer percussion. Any set of circumstances that encourages this mode of stone-tool production (i.e., expedience, local raw material supplies, limited toolmaker skills) will create an Oldowan-like core assemblage, irrespective of age. Consequently, it is probably a mistake to treat the Oldowan as the same kind of industrial entity as the Developed Oldowan, the Acheulean, and other Paleolithic industries that have uniquely derived technological and typological characteristics that are not shared with penecontemporaneous and chronologically sequential stone-tool industries.

Developed Oldowan assemblages appear in Middle Bed II at Olduvai Gorge, circa 1.7 Myr. The main differences between Developed Oldowan assemblages and their predecessors are decreased numbers of choppers, increased numbers of scrapers on flakes, and increased numbers of "pounded pieces" (spheroids and subspheroids). Small numbers of elongated discoidal cores or "proto-bifaces" occur in some Developed Oldowan assemblages. The chief point of technological similarity between the Oldowan and the Developed Oldowan (and a point of contrast with the penecontemporaneous Early Acheulean) is the emphasis on the production of small (<10 cm long) flakes. Developed Oldowan or Developed Oldowan–like assemblages older than 1.0 Myr are known from the Mid-Upper Bed II at Olduvai Gorge, the Okote Member of the Koobi Fora Formation at East Turkana, and Members 1–3 or Swartkrans Cave (South Africa) and Ain Hanech (Algeria) (Leakey, 1971; Isaac, 1984).

Acheulean assemblages appear around 1.6–1.7 Myr and are roughly contemporaneous with the Developed Oldowan. Early Acheulean assemblages differ from Developed Oldowan ones in featuring large numbers of bifaces/heavy cutting tools and an emphasis on the production of large flakes. Acheulean bifaces are large, elongated cores that are symmetrical in at least two dimensions. Their defining technological feature is that they have had a series of flakes detached from both their upper and lower faces. These bifaces are often subdivided into *handaxes* (bifaces with a sharp distal tip), *cleavers* (bifaces with a broad, usually unretouched, edge aligned perpendicularly to the tool's long axis), and *picks* (bifaces with a thick, tapered distal tip) (Bordes, 1961; Debénath and Dibble, 1994). Examples of Early Acheulean assemblages (i.e., ones reliably dated to >0.7 Myr) include those from upper Bed II at Olduvai (Tanzania), Olorgesailie and Peninj (Kenya), Melka Kunturé and Konso (Ethiopia), and Ubeidiya (Israel) (Isaac, 1984; Bar-Yosef, 1994).

Who made these Early Paleolithic industries? It is generally supposed that most Early Paleolithic stone-tool assemblages were produced by species of the genus *Homo*. Stone-tool assemblages become more consistent features of the African paleolandscape after the emergence of *Homo habilis* circa 2.4–1.6 Myr. Yet, earlier occurrences of stone tools are associated with australopithecines and paranthropines. It could be argued that early *Homo* populations whose remains are not preserved at these localities produced these tools. However, studies of paranthropine phalanges suggest a capacity for the precision grip necessary to make and use stone tools (Susman, 1994).

The appearance of the Developed Oldowan and Acheulean after 1.6 Myr coincides rather closely with the emergence of *Homo erectus* (*sensu lato*) and the dispersal of this taxon beyond Africa. Mary Leakey (1971) originally considered the Developed Oldowan and Early Acheulean to be separate industrial traditions, but subsequent research has blurred the distinctions between these assemblage groups. At Ubeidiya, for example, assemblage composition varies widely between "Early Acheulean" and "Developed Oldowan" all the while retaining the same basic repertoire of flintknapping techniques (Bar-Yosef and Goren-Inbar, 1993). While we cannot reject the hypothesis of distinctive "Early Acheulean" and "Developed Oldowan" traditions, the long persistence of these industries and their completely overlapping geographic distribution in Africa and the Near East suggest the alternative possibility, that they are merely the byproducts of situationally variable patterns of lithic raw material use, tool production, and discard behavior, by a single hominin species or a group of closely related hominin species.

Though the Acheulean is often seen as the material culture of *Homo erectus*, it would be a mistake to reflexively equate this industry solely with this hominin, or vice versa. In Europe at least, early representatives of *Homo heidelbergensis* are associated with the Acheulean at Boxgrove (UK) (Gamble, 1994). In China, *H. erectus* fossils are associated with non-Acheulean assemblages at Zhoukoudian and elsewhere (Wu and

Poirer, 1995). Heavy cutting tools like those found in Acheulean assemblages continued to be made in Europe until circa 45 Ka, among assemblages of the Mousterian of Acheulean Tradition associated with Neandertals (Bordes, 1968).

One could be forgiven for seeing the poor fit between hominin fossil species and major groupings of Early Paleolithic stone-tool assemblages as suggesting little can be learned about major changes in human diet from lithic industrial variability. This is not necessarily the case. The poor fit reflects a misapplication of archaeological methods developed for recent prehistoric periods to the immensely greater timescale of the Plio-Pleistocene (Clark and Riel-Salvatore, 2005; Shea, 2005). The named industrial entities that populate the Paleolithic are modeled after the archaeological "cultures" of recent prehistory. Yet, these Paleolithic industries, particularly those of the Early and Middle Paleolithic, exceed the geographic scope and chronological duration of any Holocene archaeological entity or ethnographic/historical culture by several orders of magnitude. Holocene archaeological cultures are defined on the basis of multiple lines of evidence, tool designs, mortuary behavior, and architecture. Early Paleolithic industries, in contrast, are defined primarily on the basis of stone-tool designs among which there is considerable potential for adaptive convergence. Whatever the Oldowan, Developed Oldowan, and Acheulean are, they are not the Stone Age equivalents of recent human "cultures."

Stone Tools as Subsistence Aids: What Is Known and Why We Think We Know It

If, as we suspect, there were dietary differences among Plio-Pleistocene hominins, then the expectation that these differences are reflected in the lithic record depends on the confidence with which we can link stone-tool design variation to particular subsistence tasks. If this link is robust, then seeming noncorrespondence between hominin taxa and archaeological industries is problematical. If the link is weak or equivocal, then we should expect a weak correlation between lithic industries and hominin taxa. I would argue that the latter is precisely what we see from the earliest archaeological sites, through the Early and Middle Paleolithic and into the most recent phases of Stone Age prehistory.

Much of what we can infer about early hominin diet from lithic technology hinges on epistemology, how we know what we think we know about Early Paleolithic technological organization. Archaeological hypotheses about relationships between stone tool design and early hominin subsistence are informed by contextual clues (e.g., spatial associations between stone tools and animal or plant remains), and by three other main sources of information, ethnographic analogy, microwear and residue analysis, and experimentation.

Ethnoarchaeology, observations of recent human stone-tool use by archaeologists, is a productive source of hypotheses about stone-tool function (Hayden, 1979; Gould, 1980; Binford, 1986; Stout, 2002). One of the most valuable insights these studies have produced is that unretouched flakes were often used effectively as tools by ethnographic populations. Lacking these insights, earlier generations of prehistorians treated unretouched flakes as waste (débitage) and often discarded them at excavation sites. Unfortunately, very few recent human societies still use stone tools as primary subsistence instruments. Stone-tool use observed by nonarchaeologists (i.e., cultural anthropologists and explorers) can enlarge this database somewhat, but these latter accounts often suffer from imprecise descriptions that limit their use for all but anecdotal purposes. A further limitation of using recent human stone-tool use to infer the functions of Early Paleolithic tools is that of potentially spurious analogies. Particular stone tools from recent contexts may resemble Early Paleolithic tools morphologically, but they may have been integrated into their respective users' technological adaptations in fundamentally different ways.

When stone tools are used on other materials, their edges and surfaces accumulate wear traces (microfractures, striations, and polishes) (Semenov, 1964). Residues of worked materials can also become attached to stone-tool surfaces. Both microwear and residues have been identified on African Early Paleolithic stone tools (Keeley and Toth, 1981; Sussman, 1987; Domínguez-Rodrigo et al., 2001). Together with cut marks on bone, these wear traces offer the most direct evidence for the use of stone tools in subsistence tasks. One limitation for microwear and residue analysis is that both are only preserved under exceptional conditions. Consequently, generalizations about Early Paleolithic stone tool use based on microwear/residue analysis are often extrapolated from small and nonprobabilistically selected samples. Such samples unquestionably shed light on the uses of particular tools and the formation processes of particular sites, but they have limited value for broader inferences about stone-tool use. If, for example, one found that three-fourths of the unretouched flakes in such a nonprobabilistic sample bore either organic residues or microwear traces referable to butchery, one could not legitimately infer that the preponderance of unretouched flakes in the larger population of such tools from that site were used for butchery. One would be on even shakier ground, methodologically, to infer the functions of unretouched flakes from other Paleolithic sites.

Experimentation is a third source of hypotheses about Early Paleolithic stone tool function. Experimentation with stone tool use is a particularly prominent aspect of African Paleolithic archaeology (Leakey, 1960; Jones, 1980, 1994; Schick and Toth, 1993; Toth, 1997). The experimental database has been enlarged further still by experiments with bonobos trained to make and use stone tools (Toth et al., 1993) and experiments focused on the neurological foundations of stone-tool production (Stout, Toth, and Schick, 2000). One of the limits of experimentation is that it usually only demonstrates the possibility (and sometimes the improbability) of particular stone-tool uses. For example, demonstrating that one can butcher a horse with a replica of an Acheulean handaxe does not prove that a particular handaxe, much less handaxes as a group, were used to butcher horses. Additional evidence is needed to establish the probability that Early Paleolithic stone tools were used in a particular way.

Cognizant of these limitations, what do these sources of information suggest about the nature of Early Paleolithic stone tool uses?

Early Paleolithic Stone Tools Were Used in Subsistence-Related Tasks

Flaked stone tools are unquestionably useful in subsistence-related tasks. The sharp edges of stone tools offer considerable advantages in butchering animal carcasses, and cut marks on animal bones from numerous Early Paleolithic sites clearly affirm this mode of tool use. Every researcher who has used simple flakes in butchery reports they are adequate to cut through the skin, meat, and fat of even the largest African mammals, including elephants. There appears to be a consensus among experimenters (Frison, 1978; Jones, 1980; Schick and Toth, 1993) that large cutting tools offer advantages over smaller flakes in the prolonged tasks involved in butchering relatively large animals. However, relatively little systematic experimentation has been performed that precisely quantifies differences in efficiency (but see Hounsell, 2004; Tactikos, 2005).

Early Paleolithic Stone Tools Required Relatively Little Advance Planning

The overwhelming majority of raw materials represented in the Early Paleolithic come from sources less than 5–10 km distant (Plummer, 2004), a few hour's walk over level terrain. This suggests that the minimum amount of anticipatory planning involved in the creation of Early Paleolithic assemblages was relatively small compared to later periods, where maximum raw material transfers routinely span tens of kilometers (Feblot-Augustins, 1997). Nor do African Early Paleolithic assemblages exhibit the close relationship between raw material quality and increased transport distance characteristic of later prehistoric periods in Africa and elsewhere (McBrearty and Brooks 2000). This finding is significant because it suggests stone tool transport may not have been subject to strong selective pressure to optimize the quality of raw materials being transported. This is not to say that many Early Paleolithic assemblages were made on the spot, with no advanced planning. Refitting studies clearly point to on-site and off-site transport of lithic artifacts (Toth, 1997; Hallos, 2005). Nor is it to assert that there was no preferential selection of high-quality raw materials. Some selectivity is apparent in even the earliest assemblages (Stout et al., 2005). Rather, it is to suggest that particular qualities of Early Paleolithic stone-tool assemblages, predominantly local raw materials and relatively little discrimination of raw material quality, reflect technological strategies emphasizing the provisioning of places and "events" with useful lithic materials. These "events" might be particular short-lived resources (an animal carcass) at which tools were abandoned or strategic caches of raw material in areas where resources requiring stone tools were available on a predictable schedule (Potts, 1988). Such a strategy contrasts with the increased strategic emphasis on provisioning people with mobile toolkits that one sees in the ethnographic record (Binford, 1979) and in later Paleolithic assemblages (Kuhn, 1992).

Making Early Paleolithic Stone Tools Required Relatively Little Effort

Modern flintknappers can replicate nearly any Early Paleolithic stone tool (pebble cores, retouched flakes, Early Acheulean heavy cutting tools) in a matter of minutes, if not seconds. Several researchers who have taught flintknapping have also noted many points of similarity between the artifacts created by novice flintknappers after some initial experience and Early Paleolithic artifacts (N. Toth, personal communication, January 13, 2005). As Toth (1985) has shown, the final typological forms of pebble cores appear to be strongly influenced by the shapes of the pebbles and cobbles from which they were created. These observations suggest that the particular forms of Early Paleolithic tools are not arbitrary, culturally reinforced design choices but rather largely the almost inevitable by-products of least-effort flake production using hard-hammer percussion. There is thus far little evidence to suggest considerable investment of postproduction effort in stone tools, such as attaching them to handles. Most Early Paleolithic stone tools are relatively thick in cross section, which is an obstacle for hafting. The kind of stone-tool-shape standardization usually associated with hafting does not become a regular feature of the African record until after 0.2–0.3 Myr (Ambrose, 2001).

Most Early Paleolithic Stone Tool Types Were Probably Multipurpose Tools

Although some of the names given to various classes of Paleolithic artifacts, such as "point," "scraper," "awl," or "burin," imply specialized functions, there is little evidence to support the hypothesis that archaeologically defined stone-tool types were functionally specialized. Microwear and residue studies of retouched Early and Middle Paleolithic stone-tool types consistently point to multiple modes of use (Keeley and Toth, 1981; Anderson-Gerfaud, 1990; Shea, 1991). Ethnographic observations of flake tool use (e.g., Gould, Koster, and Sontz, 1971) and experiments using replicas of Early Paleolithic stone tools (e.g., Toth, 1997) both point to versatility as an important feature of Early Paleolithic stone-tool design.

If one were searching for an organizing metaphor to describe Early Paleolithic stone tools, "instant technology" would fit the bill. The overwhelming majority of the stone tools in Early Paleolithic assemblages, pebble cores and flakes, are by-products of technological strategies involving short-distance transport of lithic materials to resources, minimum-effort flintknapping (predominantly hard-hammer percussion), short-tool-use lives, and minimal curation, either by resharpening, hafting, or prolonged transport. There is little convincing evidence for functional specialization among these tools, and therefore, we cannot reasonably expect their design variation to shed much light on particular dimensions of early hominin subsistence. For this, we must turn to the not insignificant residual of more complex Early Paleolithic artifact types.

What Is Unknown: Were There Specialized Lithic Subsistence Aids in the Early Paleolithic?

The instant technology model does not explain the totality of Early Paleolithic stone-tool design. Two groups of artifacts, Acheulean handaxes and Developed Oldowan "spheroids," have been viewed by some researchers as specialized lithic subsistence aids, specifically, hunting weapons. Both handaxes and spheroids differ from other Early Paleolithic artifact types in terms of their symmetry and in terms of the considerable amounts of time and labor necessary to create them. If they are such specialized subsistence aids, then the increasing numbers of handaxes and spheroids found in the archaeological record from circa 1.7 Myr onward could point to an increased role for predation in hominin subsistence.

Acheulean Handaxes

Acheulean handaxes were among the first stone artifacts recognized as genuine evidence of Pleistocene human antiquity (Grayson, 1983). Some early reconstructions of prehistoric human lifeways depicted handaxes mounted onto wooden handles, but over the course of the twentieth century, handaxes were increasingly viewed as handheld tools (e.g., Howell, 1968; McPherron, 2000). Twenty years ago, O'Brien (1981) proposed that Acheulean handaxes were used as projectile weapons, thrown like a discus. She claimed that field trials using a weighted plastic replica of a handaxe repeatedly resulted in these tools landing edge-on in the ground, a quality that would be desirable in a projectile weapon. (Discuses were originally used in naval combat.) Though what has come to be called the "killer frisbee" hypothesis has some supporters (Calvin, 2002), the majority of Paleolithic archaeologists are skeptical about Acheulean handaxes having been specifically designed as projectile weapons. Recently, Whittaker and McCall (2001) attempted to replicate O'Brien's throwing experiment using actual handaxes. They found that handaxes did not routinely land on-edge, and that their flight trajectories were highly unpredictable (much like those of discuses used in track and field events). These experiments do not refute the claim that some handaxes or other Acheulean heavy cutting tools may have been used as projectile weapons, but they do challenge the inferred causal connection between Acheulean handaxe design and the aerodynamics of projectile weapons.

If Acheulean handaxes were not projectile weapons, what factors underlie their distinctive designs? Several researchers have noted the superior qualities of Acheulean handaxes in butchery tasks, particularly in skin removal and disarticulation of larger mammals (figs. 12.1 and 12.2). Given that in tropical savannas and woodlands such large mammal carcasses would have been "carnivore magnets," handaxe design may be related to the need for butchery tools that performed quickly and efficiently in a situation where the hominins using these tools would have been exposed to considerable predation risk. In terms of current stone-tool design theory (Bleed, 1986), handaxe design may have reflected efforts to improve the reliability of more generalized lithic butchery aids.

A second possible factor in Acheulean handaxe design may be these tools' role as cores, as sources of flakes. Being relatively broad and thin, handaxes allow a flintknapper to detach flakes that are broader and thinner than those resulting from the reduction of a more spherical pebble core, all the while preserving the functionality of the handaxe/core as an effective tool in its own right. It is possible that some significant aspect of handaxe morphology may reflect the design of such a combination tool/core. If this hypothesis is correct, then we must be alert to the possibility that the forms in which we find Acheulean handaxes may not so much reflect the design targets ("finished artifacts") of prehistoric flintknappers as they

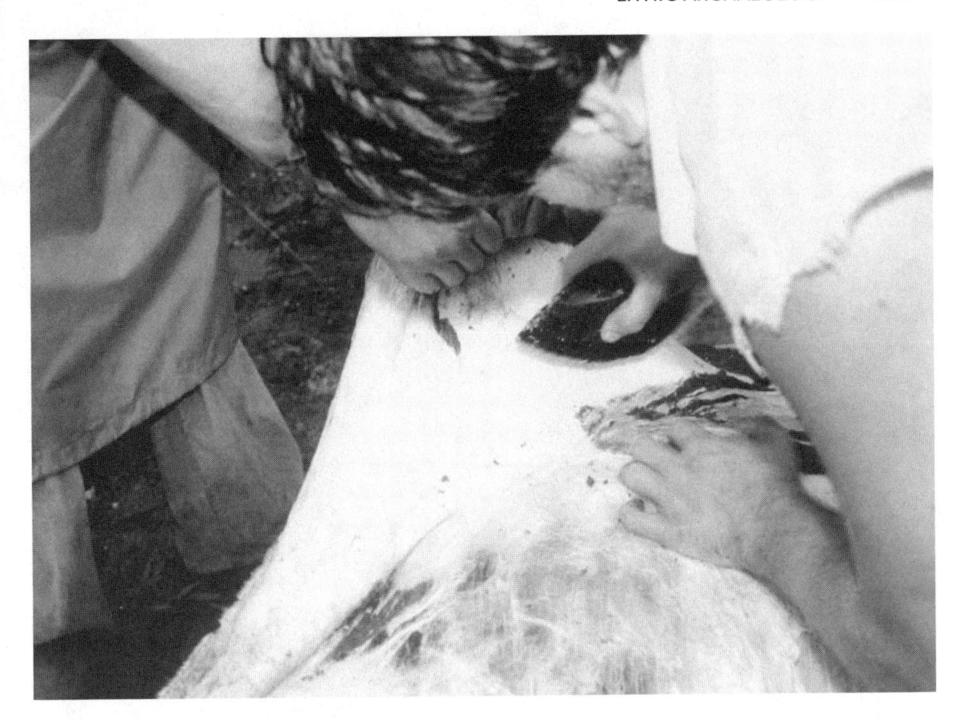

Figure 12.1 The author using a replica of an Acheulean handaxe to remove the skin from a horse.

do patterning in "discard thresholds" (the point at which a tool is judged no longer useful and abandoned) (Davidson and Noble, 1993). The "handaxes as cores" hypothesis may seem implausible if one's impression of these artifacts is shaped largely from illustrations in the archaeological literature. These illustrations are biased in favor of large, thin, symmetrical and carefully flaked artifacts. However, such "trophy specimens" are almost always a minority of the bifacial core tools in any substantial Acheulean assemblage. Most such tools are short, thick, asymmetrical and with exactly the kinds of deep and invasive flake scars that result from flake production.

A third possible explanation for Acheulean handaxe design proposed by Kohn and Mithen (1999) views these artifacts emerging as by-products of hominin sexual selection. A certain measure of handaxes' utility as butchery tools reflects their size and weight, rather than their overall shape. Large flakes work just as well as handaxes of similar size in most tasks, including butchery. Kohn and Mithen argue that the extensive retouch and imposition of symmetry involved in handaxe design reflect male hominins signaling their fitness as potential mates by demonstrating complex cognitive and technological skills in tasks that do not directly influence foraging success. Such "costly signaling" of individual fitness is not uncommon in the annals of ethology and human behavioral ecology (Bird, Smith, and Bird, 2001), but most archaeologists recognize that this intriguing hypothesis of handaxe function is a difficult one to test directly.

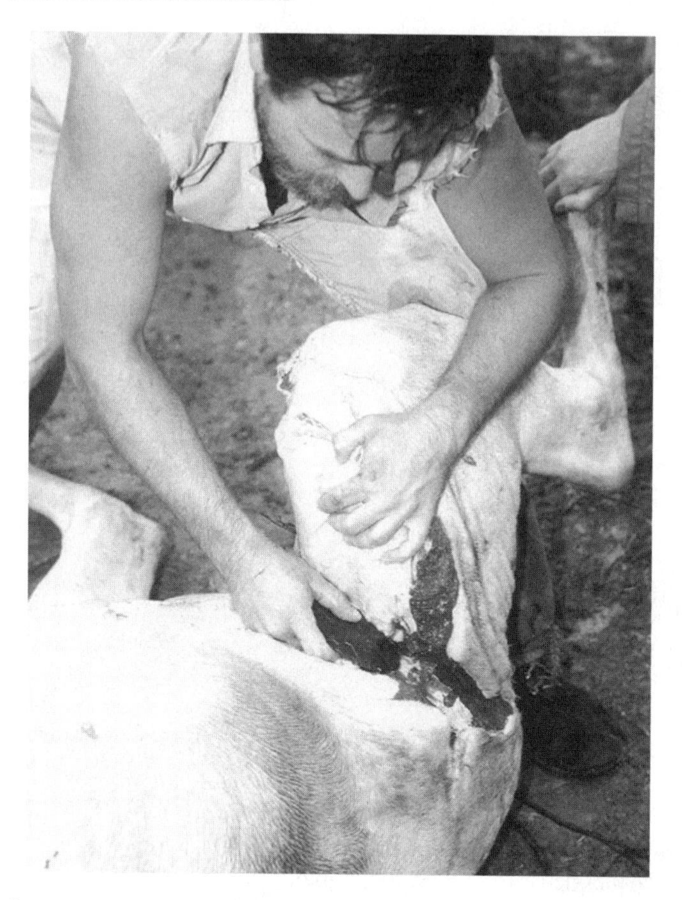

Figure 12.2 The author using a replica of an Acheulean handaxe to detach the hindlimb of a horse.

Spheroids

Louis Leakey (1931) among others (e.g., Clark, 1955, 1959) argued that Early Paleolithic spheroids may have been used as bola stones or "missile stones" for bringing down running prey. Ethnographic bola stones are round pebbles wrapped in leather pouches joined together by a length of string. When thrown, the stones spread out, disabling their target on impact, though not by wrapping themselves around its legs, as is popularly assumed. Such a function for spheroids is not impossible, though it is unlikely that larger examples of this artifact type, some weighing more than a kilogram, were used for this purpose (Willoughby, 1985). The difficulty with accepting Early Paleolithic spheroids as bola stones is that there is as yet no clear evidence to link these tools specifically to this function and to exclude their possible use in other tasks.

As Willoughby (1985) notes, many spheroid-rich assemblages are found near water sources. This has led some researchers to consider an alternative function for spheroids, that of specialized tools for pulverizing soft plant foods. Here, too, however, we lack clear evidence of this use for spheroids beyond their perceived utility. Given that a simple unmodified cobble can function perfectly well as a bola stone or a plant-pulverizing tool, any hypothesis attempting to link spheroids to specific subsistence-related tasks would have to explain the not considerable effort early hominins spent shaping that tool into a sphere. There are few published accounts of the amount of effort involved to create a spheroid deliberately. Out of curiosity, this author once created a spheroid from an angular fragment of oolitic limestone (the kind of rock of which spheroids are made at Ubeidiya) by repetitively pounding it against other large rocks. This task required more than sixty minutes of focused effort and resulted in a spheroid about 15 cm in diameter. If Early Paleolithic toolmakers were even twice as efficient as the author, this would still have been a substantial outlay of time and energy for no clear and obvious improvement in functionality. Although, as with handaxes, one could invoke "costly signaling" as a motive for prehistoric spheroid production.

Schick and Toth (1994) have argued that rather than being tools designed for a specific purpose, spheroids reflect a by-product of rocks with poor conchoidal fracture properties, such as quartz, quartzite, or limestone, being used as hammerstones in flintknapping. As the flintknapper rotates the hammerstone, seeking fresh surfaces, prominent parts of a hammerstone are flaked and abraded away, resulting in a somewhat spherical artifact. Because a spherical hammerstone positions the tips of one's fingers near the point where the stone strikes the core, hammerstones might have been abandoned as they began to approximate a spherical shape. Archaeological support for Schick's and Toth's model of spheroid formation can be seen in the predominance of limestone, quartzite, and quartz among spheroids from Early Paleolithic contexts and the near absence of spheroids from contexts, like those of East Turkana, that are dominated by basalt. Thus viewed, spheroids may not reflect "finished artifacts" designed for a particular purpose but rather patterned discard thresholds for hammerstones.

Variability as a Factor in Early Paleolithic Stone Tool Design

Viewed from our perspective, in which we are surrounded by specialized tools for nearly every activity, the absence of specialized subsistence aids among Early Paleolithic assemblages seems surprising. It should not. Numerous studies of recent human and nonhuman technology have shown technological strategies reflect a complex interplay of factors (Oswalt, 1973; McGrew, 1992). Not the least among these is the *variability* of circumstances requiring tool use.

If hunting and butchery of large animal carcasses were regular and predictable aspects of early hominin subsistence, then one ought to expect early hominins to have devised specialized lithic subsistence aids designed to reduce energy expenditure in these tasks. These might not take the same form as the lithic subsistence aids created by recent humans, but it is reasonable to expect to find Early Paleolithic stone tools whose designs, wear patterns, associated residues, and depositional contexts clearly point to use in particular subsistence-related tasks. That we are unable to identify such tools may reflect the limits of contemporary, archaeological methods, but it is equally plausible that the absence of specialized subsistence aids among Early Paleolithic stone-tool assemblages reflects the intrinsic variability of early hominin subsistence.

The most durable nonlithic residues of early hominin subsistence, cut-marked bones of large terrestrial vertebrates, point toward carnivory. But it is vastly more likely that the early hominin subsistence niche was a broad one, encompassing a wide range of plant and animal food sources. If early hominin subsistence was widely variable, not just at the level of generalizations spanning millions of years, but at the level of day-to-day life, then it is reasonable to expect such variability to be reflected in the choices of early hominin technological strategies.

Pebble-core and flake-tool technology is simultaneously the most simple and versatile technological strategy ever devised by members of the genus *Homo*. Based as it is on direct hard-hammer percussion, it is adaptable to the widest possible range of conchoidally fracturing raw materials. Raw materials that are not amenable to shaping by soft-hammer percussion or pressure flaking because of their toughness, granularity, or other qualities, can be shaped by hard-hammer percussion. A certain amount of technical knowledge is necessary to create pebble cores and flake tools, but it is probably not categorically different from the technical knowledge chimpanzees call on to effectively split nuts with stone or wood hammers. In functional terms, the by-products of pebble-core and flake-tool technology appear adequate to many tasks involving brief uses of stone tools held directly in the hand. If early hominins woke up each morning uncertain of how they would procure their food that day, but confident that they would need stone tools for one reason or another in the near future, their most effective technological strategy would have been to equip themselves with the means to create pebble cores and flake tools, like those that dominate Early Paleolithic assemblages.

Viewing the pebble-core technology as versatile adaptation to subsistence variability may help to explain the late "survival" of Oldowan-like (i.e., pebble-core-based) assemblages on the eastern, northern, and western fringes of *H. erectus*'s range in Eurasia. Although Acheulean assemblages appear to have been well established in the Near East and Indian subcontinent, the earliest archaeological assemblages in Europe, Central Asia, and East Asia do not feature distinctive Acheulean heavy cutting tools (Langbroek, 2004). If early hominins' ecological "foothold" in these frontier regions in Eurasia was less secure than in their African "homeland," it seems reasonable for them to have adjusted their technological strategy to optimize versatility.

Conclusion: From the Known to the Unknown and the Unknowable

We know that Early Paleolithic stone tools were used from about 2.7 Myr onward to aid early hominins in butchering animal carcasses. We know this, however, more from stone-tool cut marks on bone than we do from the analysis of stone tools themselves. We also know, from wear pattern analysis and residue studies, that Early Paleolithic stone tools were used for other purposes not directly related to subsistence, such as woodworking. Consequently, we know that we should not automatically assume that accumulations of stone tools necessarily reflect subsistence activities, much less carnivory, specifically.

We do not know which of the various hominin taxa found in Africa and southern Asia between 2.7–1.7 Myr were the authors of particular tools or Oldowan lithic assemblages. It is possible that rudimentary stone-tool use of the sort resulting in Oldowan assemblages was a strategy practiced by more than one hominin species. It seems reasonable to expect stone-tool assemblages created by different hominin species to have differed from each other (Foley and Lahr, 1997), but whether the scale of these differences is greater than the lithic technological variation associated with any one hominin taxon remains to be seen. This problem remains a potentially fertile field for collaborative work between archaeologists and physical anthropologists.

We do not yet fully understand how the various dimensions of Early Paleolithic stone tool technology were integrated into early hominin land-use strategies. This is a "knowable unknown," an aspect of Early Paleolithic behavioral variability that is remediable by combining detailed reconstruction of paleolandscapes, such as those recently pursued at Olduvai Gorge (Blumenschine et al., 2003) with equally detailed studies of raw material provenience and technological variability (e.g., Tactikos, 2005). Differences in land-use strategies are unlikely to be reflected in traditional archaeological measures of variability, such as relative frequencies of particular artifact types in archaeological assemblages, than they are in structural relationships between raw material provisioning and technological strategies.

A third prominent "known unknown" for the Early Paleolithic is the extent to which stone tools were used to create a secondary technology, wooden tools, for example, that had a more direct role in subsistence. Thus far, the oldest shaped wooden implements are known from contexts in the 0.5–0.8 Myr range, at Gesher Benot Ya'acov, Israel (Goren-Inbar, Werker, and Feibel, 2002), and Schöningen, Germany (Theime, 1997). Somewhat younger preserved wooden implements are known from African Middle Stone Age sites, such as Florisbad, South Africa (Bamford and Henderson, 2003), and Kalambo Falls, Zambia (Clark, 2001). It is not beyond the realm of possibility that future Plio-Pleistocene archaeological research may discover preserved wooden artifacts that shed additional light on Early Paleolithic subsistence technology.

Last, the inception of stone-tool use ranks high among the probable "unknowable unknowns" of Early Paleolithic technology. As T.E. Lawrence's anecdote of stone-tool use in Arabia clearly shows, one does not need to flake stone if sharp-edged stones are available from natural sources. Indeed, there are safety considerations that positively discourage one from doing so. No matter how careful you are, if you knap stone, you will get cut. Absent modern medical techniques for cleaning and stanching wounds, flintknapping would have been a potential source of crippling injuries and lethal infections. Unlike modern recreational flintknappers, who are more like industrial craftsmen than hunter-gatherers, prehistoric humans probably knapped stone as infrequently as possible. If, as studies of chimpanzee stone-tool use suggest (Boesch and Boesch, 1984), the capacity for procuring useful stones and transporting them to places where they are needed is evolutionarily primitive among hominins, there may be a long prologue to the Early Paleolithic. This is not to say that we need to revive the "eolith debate" of the early twentieth century (Grayson, 1986).

Rather, we need to be aware that the recognizably flaked stone tools that populate the oldest Early Paleolithic assemblages may not so much reflect the inception of stone-tool use but rather the increasing need for sharp stone tools on parts of the landscape where they did not naturally occur.

It is frustrating that Early Paleolithic stone tools tell us so little about early hominin diets. But to keep their muteness in perspective, it is worth considering just how much an archaeologist of the remote future could infer about the diets of twentieth century North Americans from our knives, forks, spoons, and chopsticks. First, they would have to be able to filter out the "noise" caused by these implements being used and damaged in tasks unrelated to subsistence. Context, residues, microwear, and some experimentation might establish their role in our subsistence; but, this future archaeologist would certainly not be able to infer the ratio of salads to T-bone steaks in our diet, nor the relative significance of fish versus potatoes. With a little experimentation, however, they would probably observe the versatility of these tools and correctly recognize that they functioned in contexts of wide dietary variability. Recognizing the role that variability may have played in Early Paleolithic stone-tool design is a significant step toward more realistic models of early hominin subsistence.

Acknowledgments I thank Peter Ungar and the staff and students of the Anthropology Department of the University of Arkansas at Fayetteville for organizing the conference for which this chapter was prepared. The quality of this chapter reflects numerous discussions with the conference participants and helpful editorial advice from Geoff Clark. Nick Toth and Joanne Tactikos also supplied valuable insights gleaned from their experiments with Early Paleolithic technology. The opinions expressed in this chapter and any errors are my own.

References

Ambrose, S.H., 2001. Paleolithic technology and human evolution. *Science* 291, 1748–1753.

Anderson-Gerfaud, P.C., 1990. Aspects of behaviour in the Middle Palaeolithic: Functional analysis of stone tools from southwest France. In: Mellars. P.A. (Ed.), *The Emergence of Modern Humans*. Edinburgh University Press, New York, pp. 389–418.

Bamford, M.K., and Henderson, Z.L., 2003. A reassessment of the wooden fragment from Florisbad, South Africa. *J. Archaeol. Sci.* 30, 637–651.

Bar-Yosef, O., 1994. The Lower Paleolithic of the Near East. *J. World Prehist.* 8, 211–265.

Bar-Yosef, O., and Goren-Inbar, N., 1993. *The Lithic Assemblages of 'Ubeidiya*. Hebrew University Institute of Archaeology, Jerusalem.

Binford, L.R., 1979. Organization and formation processes: Looking at curated technologies. *J. Anthropol. Res.* 35, 255–273.

Binford, L.R., 1986. An Alyawara day: Making men's knives and beyond. *Am. Antiq.* 51, 547–562.

Bird, R., Smith, E.A., and Bird, D.W., 2001. The hunting handicap: Costly signaling in human foraging strategies. *Behav. Ecol. Sociobiol.* 50, 9–19.

Bleed, P., 1986. The optimal design of hunting weapons: Maintainability or reliability? *Am. Antiq.* 51, 737–747.

Blumenschine, R.J., 1995. Percussion marks, tooth marks, and experimental determinations of the timing of hominid and carnivore access to long bones at FLK *Zinjanthropus*, Olduvai Gorge, Tanzania. *J. Hum. Evol.* 29, 21–51.

Blumenschine, R.J., Peters, C.R., Masao, F.T., Clarke, R.J., Deino, A.L., Hay, R.L., Swisher, C.C., Stanistreet, I.G., Ashley, G.M., McHenry, L.J., Sikes, N.E., van der Merwe, N.J.,

Tactikos, J.C., Cushing, A.E., Deocampo, D.M., Njau, J.K., and Ebert, J.I., 2003. Late Pliocene Homo and hominid land use from western Olduvai Gorge, Tanzania. *Science* 299, 1217–1221.

Boesch, C., and Boesch, H., 1984. Mental map in wild chimpanzees: An analysis of hammer transports for nut cracking. *Primates* 25, 160–170.

Bordes, F., 1961. Typologie du Paléolithique ancien et moyen. Bordeaux, Delmas.

Bordes, F., 1968. The Old Stone Age. McGraw-Hill, New York.

Bunn, H.T., 1981. Archaeological evidence for meat-eating by Plio-Pleistocene hominids from Koobi Fora and Olduvai Gorge. *Nature* 291, 574–577.

Bunn, H.T., 2001. Hunting, power scavenging, and butchery by Hadza foragers and by Plio-Pleistocene *Homo*. In: Stanford, C.B., and Bunn, H.T. (Eds.), *Meat Eating and Human Evolution*. Oxford University Press, New York, pp. 199–218.

Bunn, H.T., Harris, J.W.K., Kaufulu, Z., Kroll, E., Schick, K., Toth, N., and Behrensmeyer, A.K., 1980. FxJj 50: An early Pleistocene site in northern Kenya. *World Archaeol.* 12, 109–136.

Calvin, W.H., 2002. *A Brain for All Seasons: Human Evolution and Abrupt Climate Change*. University of Chicago Press, Chicago.

Clark, G.A., and Riel-Salvatore, J., 2005. Observations on systematics in Paleolithic Archaeology. In: Hovers, E., and Kuhn, S.L. (Eds.), *Transitions before the Transition*. Plenum/Kluwer, New York, pp. 29–57.

Clark, J.D., 1955. The stone ball: Its associations and use by prehistoric man in Africa. In: Balout, L. (Ed.), *Congrès Panafricain de Préhistoire: Actes de la IIe Session (Alger 1952)*. Arts et Métiers Graphiques, Paris, pp. 403–417.

Clark, J.D., 1959. *The Prehistory of Southern Africa*. London, Penguin.

Clark, J.D., (Ed.), 2001. *Kalambo Falls Prehistoric Site*. Vol. 3: *The Earlier Cultures: Middle and Earlier Stone Age*. Cambridge University Press, New York.

Davidson, I., and Noble, W., 1993. Tools and language in human evolution. In: Gibson, K.R., and Ingold, T. (Eds.), *Tools, Language and Cognition in Human Evolution*. Cambridge University Press, New York, pp. 363–388.

Debénath, A., and Dibble, H.L., 1994. *Handbook of Paleolithic Typology*. Vol. 1: *Lower and Middle Paleolithic of Europe*. University of Pennsylvania Press, Philadelphia.

Domínguez-Rodrigo, M., and Pickering, T.R., 2003. Early hominid hunting and scavenging: A zooarchaeological review. *Evol. Anthropol.* 12, 275–282.

Domínguez-Rodrigo, M., Serralonga, J., Juan-Tresserras, J., Alcala, L., and Luque, L., 2001. Woodworking activities by early humans: A plant residue analysis on Acheulian stone tools from Peninj (Tanzania). *J. Hum. Evol.* 40, 289–299.

Feblot-Augustins, J., 1997. *La circulation des matières premiers au Paléolithique: Synthése de données, perspectives comportementales*. No. 75. Etudes et Recherches Archéologiques de l'Université de Liège, Liège.

Foley, R., and Lahr, M.M., 1997. Mode 3 technologies and the evolution of modern humans. *Cambridge Archaeol.* J. 7, 3–36.

Frison, G.C., 1978. *Prehistoric Hunters of the High Plains*. Academic Press, New York.

Gamble, C., 1994. Time for Boxgrove man. *Nature* 369, 275–276.

Gaudzinski, S., 2004. Subsistence patterns of Early Pleistocene hominids in the Levant–taphonomic evidence from the 'Ubeidiya Formation (Israel). *J. Archaeol. Sci.* 31, 65–75.

Goren-Inbar, N., Werker, E., and Feibel, C.S. (Eds.), 2002. *The Acheulian Site of Gesher Benot Ya'acov, Israel: The Wood Assemblage*. Oxbow Books, Oxford.

Gould, R.A., 1980. *Living Archaeology*. Cambridge University Press, New York.

Gould, R.A., Koster, D.A., and Sontz, A., 1971. The lithic assemblage of the Western Desert Aborigines of Australia. *Am. Antiq.* 36, 149–168.

Grayson, D.K., 1983. *The Establishment of Human Antiquity*. Academic Press, New York.

Grayson, D.K., 1986. Eoliths, archaeological ambiguity, and the generation of "middle-range" research. In: Meltzer, D.J., Fowler, D.D., and Sabloff, J.A. (Eds.), *American Archaeology Past and Future*. Smithsonian Institution Press, Washington DC, pp. 77–133.

Hallos, J., 2005. "15 minutes of fame": Exploring the temporal dimension of Middle Pleistocene lithic technology. *J. Hum. Evol.* 49, 155–179.

Hayden, B., 1979. *Palaeolithic Reflections: Lithic Technology and Ethnographic Excavations among Australian Aborigines*. Australian Institute of Aboriginal Studies, Canberra.
Hounsell, S., 2004. Quantifying stone tool performance: With especial reference to the Oldowan and Acheulean lithic traditions. PhD diss., University of Liverpool.
Howell, F.C., 1968. *Early Man*. Time-Life Books, New York.
Isaac, G.L., 1971. The diet of early man: Aspects of archaeological evidence from Lower and Middle Pleistocene sites in Africa. *World Archaeol.* 2, 278–299.
Isaac, G.L., 1977. *Olorgesailie: Archaeological Studies of a Middle Pleistocene Lake Basin in Kenya*. University of Chicago Press, Chicago.
Isaac, G.L., 1981. Stone Age visiting cards: Approaches to the study of early land-use patterns. In: Hodder, I., Isaac, G., and Hammond, N. (Eds.), *Past in Perspective*. Cambridge University Press, New York, pp. 131–155.
Isaac, G.L., 1984. The earliest archaeological traces. In: Clark, J.D., (Ed.), *The Cambridge History of Africa*. Vol. 1. Cambridge University Press, New York, pp. 157–247.
Isaac, G.L., and Isaac, B. (Eds.), 1997. *Koobi Fora Research Project Series*. Vol. 5: *Plio-Pleistocene Archaeology*. Clarendon Press, Oxford.
Jones, P.R., 1980. Experimental butchery with modern stone tools and its relevance for Palaeolithic archaeology. *World Archaeol.* 12, 153–165.
Jones, P.R., 1994. Results of experimental work in relation to the stone industries of Olduvai Gorge. In: Leakey, M.D., and Roe, D.A. (Eds.), *Olduvai Gorge*. Vol. 5: *Excavations in Beds III, IV and the Masek Beds, 1968–1971*. Cambridge University Press, New York, pp. 254–298.
Keeley, L.H., and Toth, N.P., 1981. Microwear polishes on early stone tools from Koobi Fora, Kenya. *Nature* 293, 464–465.
Klein, R.G., 1999. *The Human Career*. 2nd ed. University of Chicago Press, Chicago.
Kohn, M., and Mithen, S., 1999. Handaxes: Products of sexual selection? *Antiquity* 73, 518–526.
Kuhn, S.L., 1992. On planning and curated technologies in the Middle Paleolithic. *J. Anthropol. Res.* 48, 185–214.
Langbroek, M., 2004. *Out of Africa: An Investigation into the Earliest Occupation of the Old World*. British Archaeological Reports International Series 1244. Archaeopress, Oxford.
Lawrence, T.E., 1935. *Seven Pillars of Wisdom: A Triumph*. Doubleday, Doran, New York.
Leakey, L.S.B., 1931. *The Stone Age Cultures of Kenya Colony*. Cambridge University Press, New York.
Leakey, L.S.B., 1960. *Adam's Ancestors*. 4th ed. Harper & Row, New York.
Leakey, M.D., 1971. *Olduvai Gorge: Excavations in Beds I and II, 1960–1963*. Cambridge University Press, New York.
Lupo, K.D., and O'Connell, J.F., 2002. Cut and tooth mark distributions on large animal bones: Ethnoarchaeological data from the Hadza and their implications for current ideas about early human carnivory. *J. Archaeol. Sci.* 29, 85–109.
McBrearty, S., and Brooks, A.S., 2000. The revolution that wasn't: A new interpretation of the origin of modern human behavior. *J. Hum. Evol.* 39, 453–563.
McGrew, W.C., 1992. *Chimpanzee Material Culture: Implications for Human Evolution*. Cambridge University Press, New York.
McPherron, S.P., 2000. Handaxes as a measure of the mental capacities of early hominids. *J. Archaeol. Sci.* 27, 655–663.
Mercader, J., Panger, M., and Boesch, C., 2002. Excavation of a chimpanzee stone tool site in the African rain forest. *Science* 296, 1452–1455.
O'Brien, E.M., 1981. The projectile capabilities of an Acheulian handaxe from Olorgesailie. *Curr. Anthropol.* 22, 76–79.
Oswalt, W.H., 1973. *Habitat and Technology: The Evolution of Hunting*. Holt, Rinehart & Winston, New York.
Plummer, T., 2004. Flaked stone and old bones: Biological and cultural evolution at the dawn of technology. *Yearb. Phys. Anthropol.* 47, 118–164.
Potts, R., 1988. *Early Hominid Activities at Olduvai*. Aldine de Gruyter, New York.

Potts, R., 1989. Olorgesailie: New excavations and findings in Early and Middle Pleistocene contexts, southern Kenya Rift Valley. *J. Hum. Evol.* 18, 477–484.

Potts, R., Shipman, P., 1981. Cut-marks made by stone tools on bones from Olduvai Gorge, Tanzania. *Nature* 291, 577–580.

Schick, K.D., and Toth, N., 1994. Early Stone Age technology in Africa: A review and case study into the nature and function of spheroids and subspheroids. In: Robert, S.C., and Ciochon, R.L. (Eds.), *Integrative Paths to the Past: Paleoanthropological Advances in Honor of F. Clark Howell.* Prentice-Hall, Englewood Cliffs, NJ, pp. 429–449.

Schick, K.D., and Toth, N.P., 1993. *Making Silent Stones Speak: Human Evolution and the Dawn of Technology.* Simon & Schuster, New York.

Semaw, S., Rogers, M.J., Quade, J., Renne, P.R., Butler, R.F., Domínguez-Rodrigo, M., Stout, D., Hart, W.S., Pickering, T., and Simpson, S.W., 2003. 2.6-Million-year-old stone tools and associated bones from OGS-6 and OGS-7, Gona, Afar, Ethiopia. *J. Hum. Evol.* 45, 169–177.

Semenov, S.A., 1964. *Prehistoric Technology.* Corey Adams Mackay, London.

Shea, J.J., 1991. The behavioral significance of Levantine Mousterian industrial variability. PhD diss., Harvard University.

Shea, J.J., 1999. Artifact abrasion, fluvial processes, and "living floors" at the Early Paleolithic site of 'Ubeidiya (Jordan Valley, Israel). Geoarchaeology 14, 191–207.

Shea, J.J., 2005. The Middle Paleolithic of the Levant: Recursion and convergence. In: Hovers, E., and Kuhn, S.L. (Eds.), *Transitions before the Transition: Evolution and Stability in the Middle Paleolithic and Middle Stone Age.* Plenum/Kluwer, New York, pp. 189–212.

Stanford, C.B., and Bunn, H.T. (Eds.), 2001. *Meat Eating and Human Evolution.* Oxford University Press, New York.

Stout, D., 2002. Skill and cognition in stone tool production: An ethnographic case study from Irian Jaya. *Curr. Anthropol.* 43, 693–722.

Stout, D., Quade, J., Semaw, S., Rogers, M.J., and Levin, N.E., 2005. Raw material selectivity of the earliest stone toolmakers at Gona, Afar, Ethiopia. *J. Hum. Evol.* 48, 365–380.

Stout, D., Toth, N., Schick, K., 2000. Stone tool-making and brain activation: Position emission tomography (PET) studies. *J. Archaeol. Sci.* 27, 1215–1223.

Susman, R., 1994. Fossil evidence for early hominid tool use. *Science* 265, 1570–1573.

Sussman, C., 1987. The results of a microwear study on select artefacts from Olduvai Gorge, Tanzania. *L'Anthropologie* 91, 375–380.

Tactikos, J.C., 2005. A landscape perspective on the Oldowan from Olduvai Gorge, Tanzania. PhD diss., Rutgers University.

Theime, H., 1997. Lower Paleolithic Hunting Spears from Germany. *Nature* 385, 807–810.

Toth, N.P., 1985. The Oldowan reassessed: A close look at early stone artifacts. *J. Archaeol. Sci.* 12, 101–120.

Toth, N.P., 1997. The artifact assemblages in the light of experimental studies. In: Isaac, G.L., and Isaac, B. (Eds.), *Plio-Pleistocene Archaeology.* Vol. 5. Koobi Fora Research Project Series. Clarendon, Oxford, pp. 363–402.

Toth, N., and Schick, K., 2000. Early Paleolithic. In: Delson, E., Tattersall, I., Van Couvering, J.A., and Brooks, A. (Eds.), *Encyclopedia of Human Evolution and Prehistory.* Garland, New York, pp. 225–229.

Toth, N., Schick, K.D., Savage-Rumbaugh, E.S., Sevcik, R., and Rumbaugh, D., 1993. Pan the tool-maker: Investigations into the stone tool-making and tool-using abilities of a bonobo (*Pan paniscus*). *J. Archaeol. Sci.* 20, 81–91.

Whittaker, J.C., and McCall, G., 2001. Handaxe-hurling hominids: An unlikely story. *Curr. Anthropol.* 42, 566–572.

Willoughby, P., 1985. Spheroids and battered stones in the African Early Stone Age. *World Archaeol.* 17, 44–60.

Wu, X., and Poirer, F.E., 1995. *Human Evolution in China: A Metric Description of the Fossils and a Review of the Sites.* Oxford University Press, New York.

Wynn, T., 2002. Archaeology and cognitive evolution. *Behav. Brain Sci.* 25, 389–402.

PART IV

PALEOECOLOGY AND MODELING

13

Theoretical and Actualistic Ecobotanical Perspectives on Early Hominin Diets and Paleoecology

CHARLES R. PETERS

Diet is one direct link between an animal and the multiple environments that it is surrounded by and in part creates (Andrewartha and Birch, 1984; Kuchka, 2001). It is reasonable to assume that the early hominin priority web for resources included potable water, plant foods, arboreal refuge, animal foods, and tool materials in a contingency hierarchy of critical functions (Peters and Blumenschine 1995, 1996). Ecobotanical perspectives are relevant to evaluating all of these resources.

My focus is on Plio-Pleistocene hominins without controlled use of fire as a technology for food procurement and processing. Claims have been made for controlled use of fire by Plio-Pleistocene hominins (e.g., Clark and Harris, 1985; Brain and Sillen, 1988), but I will assume that technology was unavailable.

Other chapters in this volume cover what is known about the diets of australopithecines, early *Homo* and *Paranthropus* in the Pliocene and early Pleistocene. We know very little about these creatures in comparison to historical *Homo sapiens*; it is remarkable that we know as much as we do. Limitations in our paleoanthropological knowledge sometimes invite overly rich interpretations and speculations uninformed in critical ways. Additional frameworks are needed to guide hypothesis construction and temper interpretation. These frameworks explicitly take us into aspects of the unknown and the apparently unknowable. They are theoretical endeavors that rely on patterns in extant systems to explore possibilities in prehistory. Regularities in nature, combined with observations about the past and scientific imagination, lead to an exploratory dialogue between the possible and the critical. A characteristic feature of theoretical science is that it attempts to understand and explain the known world (here the paleoanthropological record) by a postulated invisible world of logical constructions (Liebenberg, 1990). In our case, those constructions are synthesized out of basic environmental data combined with conceptual devices from ecology (Kuchka, 2001). Progressively (following Levins, 1966), we emphasize precision in aspects of the database, generality in its theoretical application, while we attempt some realism in our formulations of hominin-environment relations.

Some of the jobs of theory are (1) clarifying thinking about method; (2) motivating alternative working hypotheses; (3) introducing conceptualizations and model building relevant to the state of development of a discipline; (4) creating new interpretations while evaluating old ones; and (5) helping to push back the boundaries of the unknown and the apparently unknowable.

Method for theory is relatively well developed in ecology. Here "method" refers to principles of inquiry, in contrast to techniques. Philosophy of scientific theory and conceptual terminology used here follow the discussion of Pickett, Kolasa, and Jones (1994; also see Kuchka, 2001). The level of theory that can be attained at this stage in the development of paleoanthropology is relatively unrefined and imprecise. Basic conceptual devices include the concepts of fundamental niche and ecosystem, and the potential interplay "on the ground" between hominin biology, use of technology, and landscape ecostructure/dynamics. The temporal domain is the Plio-Pleistocene and the geographic domain is primarily southern through eastern Africa, including the Zambezian phytochorion, or floristic region. Empirical content includes a number of "facts" about early hominin diet (e.g., at paleontological site *x* there is apparent evidence that butchery of animal *y* occurred with Oldowan stone tool-type *z*): but, as is common with theory still at the intuitive stage of development, confirmed generalizations are often lacking. Derived conceptual devices are primarily limited to hypotheses and theorems. These can, however, be integrated into preliminary stages of model building.

Robustness of hypotheses and theorems in our conceptual models is judged for the most part by Whewell's principle of consilience (Levins, 1966; Kuchka, 2001): disparate lines of information or evidence converge to support the same conclusion. For example, a number of lines of circumstantial and analogical evidence converge/reinforce theoretical conclusions that have been reached about the fundamental trophic niche(s) of the early hominins (see the section that follows below).

We suffer from both limitations to theory and limitations to the prehistoric record. Current limits to theory are rather obvious. Empirical limits are often not as clear. Paleobotanical analyses, for example, are not complete enough to provide reliable vegetation reconstructions for early hominin sites. In the face of this, we can still set out some boundary conditions for our models with modern analogs. In our case, however, there is only a limited role that actualistics can play in model building: present day conditions do not always provide for an adequate understanding of the past. For example, the probably once lush natural vegetation on the footslopes of the mountains in East Africa has been largely destroyed by cultivation and ranching. I doubt a complete picture of the potential natural vegetation can be assembled. Associated hydrological processes have also been transformed. This is another reminder of our inability to simultaneously maximize precision, realism, and generality. Moreover, prehistory reserves for itself elements of surprise. We remain, at this stage in the development of our science, largely ignorant of what life was like in the past. With so many unknowns, the beginning and ends of our story lines are usually unpredictable (a bit of chaos there); we have a better chance understanding the trajectories in-between.

Fundamental Trophic Niche, Potential Competition, Alternative Forms of Omnivory, and Ecophenotypic Specialization: Moving from the Known into the Unknown

Theoretically, the fundamental trophic niche of an animal includes all that it is biomechanically and chemically capable of consuming, while its realized niche is what it actually manages to consume in real-life situations in particular geographic areas, when the full circumstances of environmental variables (including competition) come into play (following Hutchinson, 1957). What plant and animal foods early hominins actually ate is virtually unknown (perhaps, in almost all cases, empirically unknowable). What they were capable of eating and probably ate are more fundamental, theoretically approachable questions. One possible theoretical approach is to find commonalities among closely related extant animals with similar nutritional requirements living in a range of environments that might have been used by early hominins. An analysis and comparison of the undomesticated components of human, chimpanzee, and baboon diets suggests itself as a relevant method. Preliminary findings for plant-food-item types from 461 wild-plant genera exploited by these three primates in eastern and southern Africa (Peters and O'Brien, 1981) indicate the following: fruits, leaves/shoots, and seeds/pods were probably basic components of the fundamental plant food niche(s) of early hominins. Historically, humans and baboons also exploited a taxa-rich variety of wild plants for their rootstocks (underground storage organs), indicating this was probably an additional food-item type exploited by early hominins. A similar analysis could be made of the small wild animals eaten by these three extant primates, and some of those small animals could be added to the fundamental trophic niche(s) of the early hominins. One overall result of this method is a general characterization of their trophic niche(s) as omnivorous. Another general inference is that the abstract and fundamental overlap in diet of these three extant primates implies high potential competition in closely related extinct species of hominins, hominids and large-bodied monkeys (also see Peters and O'Brien, 1982).[1] The easy to exploit fruits are theoretically the plant-food type on which interspecies competition would be focused the most (Peters and O'Brien, 1981).

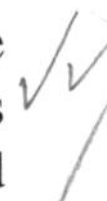

Prior to our work, Leopold and Ardrey (1972) theorized that because of widespread occurrence of toxic chemicals in plant tissues the hominin plant-food diet was greatly restricted before the development of cooking. This is not a robust theorem, in part because of possible losses of physiological capabilities (relaxed selection) and narrowing dietaries after the widespread adoption of cooking. Moreover, Leopold and Ardrey were unfamiliar with Africa's plants as sources of toxins, medicinals (e.g., Watt and Breyer-Brandwijk, 1962), and foods, even its (remarkably valuable) edible wild fruits (see below), about which they drew unwarranted limiting conclusions. They also overlooked the biomechanical effects of cooking food, especially the textural transformations that result in the soft baby-food-like products normally consumed by adults.

Stahl (1984) extended the Leopold-Ardrey approach by ordering some general plant parts into an overall "edibility" ranking on the basis of uncooked toxins and digestion inhibitors, plus protein and soluble carbohydrate (simple sugars, not starch) content. If we take her conclusions (table 1 in Stahl, 1984) at face value, early hominin plant-food diets would theoretically be reduced apparently to flowers, fruit pulp, and a sweet-onion cultigen. Stahl was not familiar with Africa's wild-plant foods, so she was not aware of the nutlike oil seeds, their significant role in traditional plant-food diets, and their theoretical importance for early hominin diets (Peters and Maguire, 1981; Peters and O'Brien, 1984; Peters, 1987a, 1987b; Peters, 1988).[2] However, her concept of quality ranking is useful, if it is enriched with additional categories and we remember its failures and shortcomings. It fails to represent, for example, the concept of an optimal dietary mix.

An alternative hypothetical quality ranking of the wild-plant foods of eastern and southern Africa for somewhat chimpanzee-like early hominins (including earliest *Homo* and *Paranthropus*) is seen in table 13.1 (following Peters and O'Brien, 1984; and O'Brien and Peters, 1991). It is based for the most part on wild-plant foods eaten raw by humans. The ranking hierarchy (strong but only partial) is not strictly linear because of contingent and differential possibilities for feedback among affordances, synergistic effects, and shifting ecological circumstances (e.g., multiyear droughts). Table 13.1 is tentative because, among other things, the nutritional values and deleterious side effects require more work. We do not know how, for example, these factors might vary across locales and regions, depending on soil type or differing climates. Following the suggestion in Peters and O'Brien (1984), some notes on medicinals (based on human usage) have been added. Under "plant type" the information on C_3 and C_4 photosynthetic types is from Peters and Vogel (2005). Most of the other generalizations under "ecology/availability" are based on personal observations and a broad familiarity with the relevant literature. Arguably, the best use for table 13.1 is to highlight some of the theoretical issues about fundamental hominin adaptations.

Notice the theoretical rankings in table 13.1 are dependant on hominin masticatory and/or tool-use specializations that allow them access to foods that are either not easy to process or procure. The most challenging potentially high-ranking foods are the nutlike oil seeds (they have very strong endocarps) and the rootstocks (they are often not easily dug/chopped out of the soil and water). (Notes in table 13.1 on the resistance-strength of nuts and seeds to crushing are based on Peters and Maguire [1981], and Peters [1987a; 1993, which also includes corrections in terminology and notes on keystone species].) All but the most resistant nutlike oil seeds theoretically were probably within the masticatory capabilities of adult male paranthropines (Peters and Maguire, 1981; Peters, 1987a; Demes and Creel, 1988; Lucas, Peters, and Arrandale, 1994; Lucas and Peters, 2000) but not weaning-age infants or young subadults. However, judging by their jaws and cranial architecture, weaning-age paranthropines probably had a significant advantage over "gracile" australopithecines and early *Homo* infants, masticating a variety of the relatively less-challenging foods (Peters, 1979, 1993). Some appreciation for the dramatic differential advantage in potential capability of the young *Paranthropus* can be gained by visual examination of examples of their mandibular dentition (in particular the size of their deciduous molars) in comparison with other hominins (see fig. 13.1). Moreover, theoretically, *Paranthropus* was likely an omnivore (Peters, 1981, 1987a; Peters and Vogel, 2005). This seems an obvious hypothesis in the case of Oldowan-tool using *Homo*, but in the past, some reviewers have expressed unease at the apparently counterintuitive proposition that morphologically specialized paranthropines could have been ecological generalists, that is, that their specialized masticatory apparatus could have increased their fundamental trophic-niche breadth in comparison to gracile australopithecines. Such discomfort may be caused by the confusion stemming from what Robinson and Wilson (1998) have called Liem's paradox: species that are highly specialized in phenotype (e.g., morphology, physiology, and behavior) may act as ecological generalists. This apparent paradox is conceptually resolved when it is realized that (1) some resources are intrinsically easy to use and widely preferred, while others require specialized phenotypic traits on the part of the consumer, and (2) that this resource asymmetry allows optimally foraging (versatile) consumers to evolve phenotypic specializations for more challenging resources without greatly compromising their ability to use otherwise preferred resources (Robinson and Wilson, 1998). There is no trade off: the specialist acts as an ecological generalist whenever easy-to-use high-value resources are available. In the ranking model presented in table 13.1, hominin ecophenotypic specializations make available higher-value, otherwise impractical to use, potential-food resources. Robinson and Wilson (1998) show that theoretically the most extreme phenotypic specializations should occur in the virtual absence of trade-off and point out that a community of such "specialists" can coexist with or even replace phenotypically generalized consumers that are slightly better at using easy-to-use resources but have no specialized abilities to exploit less-easy-to-use potential resources. In a sense, the specialists create new niches. They also note that key evolutionary innovations may allow exceptional degrees of both feeding versatility and phenotypic specialization. They apply this conceptualization to explaining, in part, the radiation of the African lake cichlids. In our case, the ecophenotypic tool-use specializations of early *Homo* (and possibly *Paranthropus*) recognized in table 13.1 can readily be added to this picture of potential trophic versatility, especially considering their morphological correlates in the hands and bipedality (Marzke, 1996).

Table 13.1 also notes the theoretical importance of special forms of bipedality in wetland contexts. This praxis is known to those of us who make botanical collections (on foot) from the interiors of African wetlands. The following notes are based on personal experience.

The structure of clay-dominated hydromorphic soils depends largely on water conditions. When they are dry, they can be very hard and deeply cracked with an angular blocky structure. When permanently waterlogged, and bioturbated, they are almost structureless. Under wet conditions, their field properties include the following. The flat, partially saturated compact soils at the edges of ponded water have well-defined slip-surfaces on which it is very easy to skid and lose balance. These soils show a gradient from adhesive (very sticky and clingy) to easily molded. In contrast, deeply bioturbated, fully saturated highly fluid soils (e.g., hippo muds) can be entrapping because of hydrostatic pressure around feet and lower limbs that have

Table 13.1 Tentative Hypothetical Quality Ranking of the Wild-Plant Foods of Eastern and Southern Africa for Nonfire Using Plio-Pleistocene Hominins with Masticatory and/or Tool-Use Specializations

	Nutritional Values		
Food Type/Value Rank	Fats/oils	Protein	Carbohydrates[a]
First rank			
Nutlike oilseeds	High	High	Low
Mature fleshy fruit (pulp; epicarp?)	—	—	Intermediate to high (sugars)
Second Rank			
Dry fruits (excl. seeds)	—	—	High
Beans (Leguminosae)	Low to intermediate	Intermediate to high	High
Other dicot seeds	Low to intermediate	Low to intermediate	Low to intermediate
Grass seed (wetland species)	—	Intermediate	High
Shallow rootstocks (tubers/bulbs /corms/ rhizomes/fleshy roots)	—	Low	High (starch, some sugars)
Immature leaves/shoots	—	Intermediate	Low
Deep rootstocks (tubers/enlarged roots)	—	Low	Intermediate (starch)
Mushrooms	—	Low	Low to intermediate
Third Rank			
Gum	—	—	High (sugars)
Flowers (incl. pollen)	—	Low to high	Intermediate
Nectar	—	—	High (sugars)
Fourth Rank			
Cambium	—	Low?	High (sugars?)
Young stems/stalks and pith	—	Low to intermediate	Intermediate
Mature leaves malacophyllous (petiole > lamina)	—	Low	Intermediate (starch)
Fifth Rank			
Mature leaves nonmalacophyllous (esp. lamina)	—	Low	—
Bark	—	—	—

Sources: After Stahl 1984 with additional categories; revised from Peters and O'Brien (1984) and O'Brien and Peters (1991) with additional categories and supplemental information from Peters (1990, 1993, 1994, 1996, 1999), Peters et al. (1992), Peters and O'Brien (1994), Peters and Vogel (2005), Watt and Breyer-Brandwijk (1962), Iwu (1993), Van Wyk et al. (1997), and unpublished observations.

[a]Nonstructural carbohydrates.

[b]Does not include travel time and transport costs, nor risk of predation.

Nutritional Values			Deleterious Side Effects	
H_2O	Vitamins	Minerals	Toxicity	Digestion Inhibitors
—	Vit. B (thiamine, niacin)	K, Ca, Mg, P	—	—
High to intermediate	Vit. C (some spp. very high in Vit. C)	K, Ca, Mg	—	—
—	Some Vit. C	Ca, Mg, P, K, Fe, Cu, Zn	—	—
—	Some Vit. B	Ca, Mg, K, Zn, P, Cu, Fe	Low to high	Low to high
—	Vit. B	K, P, Mg, Ca, Fe, Zn, Cu	Low to high	Low to intermediate
—	Vit. B	Ca, Mg, K, P, Fe, Cu, Zn	—	?
High to intermediate	Vit. C, Vit. B	K, Ca, Mg, P, Fe, Cu, Zn, Mn	Variable, common? (incl. medicinals)	± constipation
High	Vit. C, niacin, carotene	Ca, Mg, Fe, K, P, Cu, Zn	Low to high (incl. medicinals)	Low to intermediate
High	—	Ca, Mg	Variable, uncommon? (incl. medicinals)	?
High	Vit. B	K, P, Mg, Cu, Fe, Zn	Low to high	?
—	—	Ca, Mg, K, Fe, Cu, Zn	— (few medicinals)	?
Intermediate	Vit. B, some Vit. C	K, Ca, Mg, P, Cu, Fe, Zn	Allergens?	—
—	—	—	— (few medicinals)	—
Intermediate	?	?	? (incl. medicinals?)	?
High to intermediate	Vit. C?	K, Ca, Mg?, Fe?	Low to intermediate	Low to intermediate
Intermediate to high	?	Ca, K, Mg, P, Fe, Cu, Zn	Intermediate to high (incl. medicinals)	Low to high
Low	?	Ca, K, Mg, P, Fe, Cu, Zn	High (incl. medicinals)	Low to high
—	—	?	Intermediate to high (incl. medicinals)	Intermediate to high

(*continued*)

Table 13.1 (*Continued*)

Food Type/Value Rank	Ecology/Availability		
	Plant Type	Landscape Associations	Wet Season Availability
First rank			
Nutlike oilseeds	Trees: C_3; deciduous > evergreen	Mostly well drained soils: high terraces, valley sides, hills, dunes, plains	High (some available in herbivore dung and regurgitated clusters)
Mature fleshy fruit (pulp; epicarp?)	Trees and shrubs > forbs: C_3; deciduous > evergreen	Mostly well drained soils (incl. river banks)	High
Second Rank			
Dry fruits (excl. seeds)	Trees and shrubs: C_3; deciduous > evergreen	Well-drained soils	Intermediate
Beans (Leguminosae)	Trees > forbs: C_3; deciduous >> evergreen	Mostly well drained soils	Low to intermediate
Other dicot seeds	Trees and shrubs > forbs: C_3 >> C4; deciduous > evergreen	Well-drained soils	High
Grass seed (wetland species)	C_3 and C_4	Floodplains, dambos (vleis), marshlands	High
Shallow rootstocks (tubers/bulbs/corms/rhizomes/fleshy roots)	Mostly monocots: C_3 and C_4	Wetlands > drylands	High
Immature leaves/shoots	Trees, shrubs and forbs: $C_3 > C_4$	Well-drained soils and seasonal wetlands	High
Deep rootstocks (tubers/enlarged roots)	Deciduous trees and forbs: C_3	Well-drained soils	High
Mushrooms	C_3	Woodlands >> bushlands	High
Third Rank			
Gum	Trees > shrubs: C_3; deciduous >> evergreen	Semiarid > mesic	High
Flowers (incl. pollen)	Trees, shrubs, and forbs: C_3	Various	High
Nectar	Trees, shrubs, and forbs: C_3	Well-drained soils	Some
Fourth Rank			
Cambium	Trees and shrubs: C_3	Well-drained soils	—
Young stems/stalks and pith	Trees, shrubs, forbs, and graminoids: $C_3 > C_4$	Various	High
Mature leaves malacophyllous (petiole > lamina)	Trees, shrubs and forbs: $C_3 > C_4$	Various	High
Fifth Rank			
Mature leaves nonmalacophyllous (esp. lamina)	Trees and shrubs: C_3	Well-drained soils	High
Bark	Trees and shrubs: C_3	Well-drained soils	—

Ecology/Availability			
Dry Season Availability	Potential Staples	Procurement[b] Processing Costs	Potential Competition
Low to intermediate	Some (weaning foods) (drought tolerant)	Intermediate to high (resistant shells requiring high crushing/impact forces, e.g., hammer stone)	Low to intermediate (the fruit are keystone spp.)
Low to intermediate	Many (weaning foods)	Low	High (many keystone spp.)
Low	Some (drought tolerant)	Intermediate	High
Intermediate to high	Some	Intermediate (resistant shells requiring moderate to low forces)	Intermediate
Low	Few	Intermediate (resistant shells requiring moderate to low forces)	Intermediate
—	Few	? (shells of low resistance)	Low?
High	Many (incl. weaning foods) (drought tolerant)	Intermediate to high (wetland form of bipedality) (digging implement, ± chopping implement)	Low to high
—	Intermediate	Low	High
Intermediate to low	Some? (drought tolerant)	High (digging implement, ± chopping implement)	Low
—	Some?	Low	High
Some	—	Low	High
Low	Few	Low	High
Few	Few?	Low	High
—	Famine food?	Intermediate (± chopping implement)	Low
—	Some	Low to intermediate	High
—	Famine food	Low to intermediate	Low
Intermediate to high	Famine food	High	Low
—	Famine food	High (± chopping implement)	Low

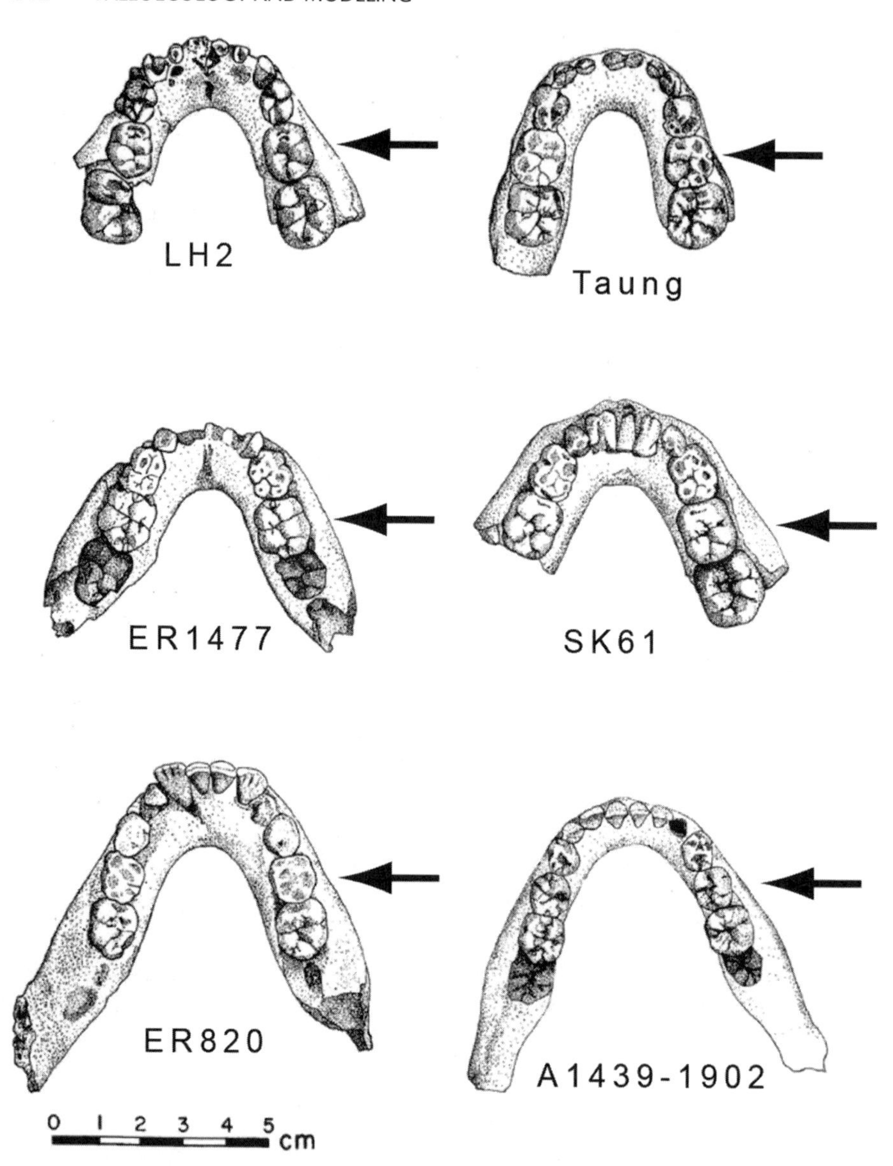

Figure 13.1 Infant mandibles with dm_1 and dm_2 in use. Arrows point to dm_2. LH 2: *Australopithecus afarensis* (M_1 erupting). Taung: *Australopithecus africanus* (M_1 erupted). ER 1477: *Paranthropus boisei* (M_1 unerupted). SK 61: *Paranthropus robustus* (M_1 erupting). ER 820: *Homo ergaster* (M_1 erupting). A1439-1902 (Dart Collection): *Homo sapiens sapiens* (M_1 erupted, M_2 unerupted).

penetrated deep into the matrix. The effect is a resisting force, felt as suction, when attempting to pull the limbs free. A third, equally startling effect that occurs in dambos is microcatastrophic collapse. Surfically bioturbated, moist top soils with a crumbly loose structure can overlie saturated subsoils with chemical undermining from throughflow eluviation and piping, resulting in local surface collapse under foot-fall standing pressures. Limbs may penetrate several centimeters into the matrix, temporarily immobilizing the unwary.

With regard to water depth, vegetation structure and locomotor substrate, in very shallowly flooded marshland ground-level sedge beds (low growing *Cyperus* spp.) are easy to walk across, but because of the give, running and rapid maneuverability are compromised. In somewhat deeper water, closed meter-tall sedge beds (e.g., *Cyperus laevigatus*) can be mattress-like for the broad-footed biped, but balance is difficult to maintain and locomotion is limited in speed to a very slow, awkward walk. Reed (*Phragmites*) and cattail (*Typha*) marsh can present a more complex mosaic, partially submerged in water most of the time. Their reticulated mat of rootstocks has spaces between the subaqueous stems that can impede movement. In deeper water, *Cyperus papyrus* appears simpler, being a larger plant with more uniform stands, but attempting to walk among the stems on top of the rhizomes is more difficult than swimming on the open water side, where the plant is growing out over the water's edge.

In contrast, most of the herbaceous cover in seasonally flooded dambos is not marsh but grassland. In addition, bunchgrasses often dominate. They present small mounds (e.g., 10–15 cm tall) that can not be seen during the wet season when the grass is waist high. Moreover away from the wetter parts of the dambo, especially on the less-flooded margins, small termite mounds can be numerous. Combining the lack of visible ground with the low mounds of bunchgrass, and the low termitaria mounds, not to mention suid diggings, the substrate is surprisingly uneven. The grass canopy blocks view of the ground but is too thin to provide a significant pad or mattress-like effect for locomotion.

Theoretically, environmental ecostructure dominates locomotor requirements (e.g., Youlatos, 2004). In shallow water, firm muds are very slippery. A third (or fourth) leg, in the form of a staff or pole, is sometimes needed by humans. For a large mammal, four big feet are assets (e.g., elephant, hippo, some ungulates, and carnivores). Wet deeper muds are entrapping. Again a third leg prop may be necessary for humans, but in the deep liquid mud of a hippo pool or river delta, humans and dry-land ungulates easily become stuck, and while they usually free themselves after some struggling, they are effectively immobilized for awhile. In general, bipedal movement in mud is a slow, careful affair, with the emphasis on balance in slow motion. Walking in a marsh on a deep mattress-like sedge mat or papyrus root system is even more awkward. Once again, the bipedal emphasis is on balance in slow motion. Trying to move rapidly is quickly exhausting.

Walking through the shallowly flooded margins of a dambo during the wet season one experiences unseen, very uneven ground. Because the grass covers low mounds and depressions, you tend to stumble as you walk. Again, the emphasis in a biped is on maintaining balance, in slow motion. This kind of bipedality is not the kind we are accustomed to. Even scavenging in wetlands, as opposed to plant foraging, would seem not to require our long-legged striding kind of bipedality. Moreover, the plausibility of a wetlands form (or forms) of bipedality forces us to conceptually recognize certain amphibious possibilities, for example, dominantly bipedal in wetlands and partly to dominantly arboreal in adjacent drylands. Those interested in alternative forms of bipedality or bipedality/arboreality will have to examine ecostructure at the facet-to-element levels in relevant landscapes. Theoretically, it appears there is room for amphibious locomotor repertoires in the earliest hominins, within the settings of herbaceous-wetland and woodland mosaics.

Ecosystem Analyses: Forays Deeper into the Unknown and into the Apparently Unknowable

Table 13.1 gives some general notes on seasonality. These are largely based on Peters, O'Brien, and Box (1984) and detailed records such as Peters and Maguire (1981) for Africa's predominantly summer rain climate (fig. 1 in Peters, 1990), as well as personal observations in eastern, south-central, and southern Africa. For the summer rain climate region south of the equator, there is a well-known, pronounced seasonal change in the quality as well as the quantity of types of food items from the full spectrum of food types provided by several different types of plants during the rainy season, to a limited fare in the dry season, consisting mainly of fruits and nut-like oil seeds (relatively scarce) provided by trees, dry beans from the Leguminosae provided by trees (especially, miombo species, e.g., Suzuki, 1969) and rootstocks provided primarily by herbaceous plants. Wild-plant foods from rootstocks are generally available year-round, but those of the geophytes often become bitter in the dry season from water loss and increased concentrations of secondary compounds. Wetland forms do not experience the same degree of seasonal drought stress, and those of perennial wetlands remain fully succulent year-round.

The theoretical importance of dry-season scarcity is in the selection for hominin ecophenotypic specializations that help overcome predicable adversity and open up new possibilities for species-range expansion. Theoretical examples include consumption of nutlike oil seeds at the end of the rainy season to beginning of the dry season, putting on fat reserves that buffer height of the dry-season stress (e.g., Peters, 1987a). Another is the use of perennial wetlands for edible rootstocks, for example, those from *Phragmites* (Peters, 1990, 1994), *Typha* (Peters, 1994), and *Cyperus papyrus* (Peters, 1999).

Severe drought is also a challenge, partially accommodated by the hominin ecophenotypic specializations recognized here. Some of the nut-species trees and dry-fruit species shrubs are dry-climate forms and are able to produce good crops in drought years. Wetlands are usually the refugia of last resort in lowlands (e.g., the Amboseli groundwater marsh in Kenya fed by Mt. Kilimanjaro; see Western, 1997).

That a species may die back at the edge of its range under severe climate conditions is not surprising. What is not so intuitively obvious is that a species may not be demographically viable over most of its geographic range.

The demographic dynamics of a species depends on the mosaic of landscapes and habitats that it occupies across the full range of its distribution. The interactions of local populations across this mosaic determines demographic dynamics at the level of the species. This is true because within this mosaic some landscapes/habitats are demographic sources, while others are demographic sinks (Pulliam, 1988). Geographic areas that are sources are rich enough to support reproductive rates greater than that needed to sustain their local populations. For animals, these areas provide the migrating (surplus) individuals that recolonize adjacent areas where reproductive rates are below replacement (i.e., not self-sustaining). A large (perhaps the largest) fraction of the individuals that make up a species may regularly (or episodically) occur in sink landscapes/habitats, where local reproduction is insufficient to balance local mortality. Here breeding sites may be abundant, but they are of poor quality. Pulliam (1988) has shown that active dispersal from source areas can theoretically maintain large sink populations. If the reproductive surplus of the source areas is large and the reproductive deficit of the sink areas is small, the great majority of the species population may occur in the sink. This might also correspond on a broad timescale to periods of environmental amelioration when species reach their maximum geographic extent. We should also note that, although demographic sources/sinks are usually thought of as different (adjacent) geographic areas, they can be temporally successive characterizations of the same area as it experiences marked flux in the quality of the breeding sites.

In terms of landscape and habitat, for hominins source areas would be characterized by mosaics of extensive woodlands, with local groundwater forest and tall bushland; riparian forest and medium- to small-sized freshwater wetlands; numerous small streams with high water quality and relatively few crocodiles; high productivity and species richness of edible wild plants; and safe sleeping sites.

Areas functioning as demographic sinks for hominins would be characterized by extensive montane and lowland rainforest; extensive semiarid bushland; hot, dry broad valleys; saline and alkaline wetlands; coastal forest with very limited potable water; and increased competition for and/or lower productivity and lower species richness of edible wild plants.

For a large-bodied frugivorous primate, climate is an important parameter, indirectly determining demographic sources and sinks. O'Brien's work (1989 and subsequently, e.g., Field, O'Brien, and Whittaker, 2005) has revealed important predicable relations between climate and the species richness of Africa's woody plants. As an example of this approach applied to woody plants that provide edible plant parts for humans, we compared species richness for areas of approximately 20,000 km^2 centered on Harare, Zimbabwe, and Gaborone, Botswana (Peters and O'Brien, 1994). These two areas are similar in topography and elevation, but the latter is semiarid compared with the subhumid Harare area: Gaborone receives about 60 percent of the average annual precipitation that Harare receives (ca. 500 mm vs. 800 mm), and its potential evapotranspiration is more seasonally variable and extreme than that of Harare. The vegetation of the Gaborone area is mainly leguminous bushland (*Acacia* dominated) in an overall dryland setting, while the woody vegetation of the Harare area is mainly leguminous woodlands (dry miombo, i.e., *Brachystegia* dominated) in a landscape mosaic with herbaceous seasonal wetlands (dambos). The main difference in the edible plant parts from woody species (trees and shrubs) in these two areas is that of fleshy fruits, many of which are keystone food species for a variety of mammals and birds. For Gaborone, thirty-three woody species provide fleshy fruits, compared with eighty-one species for Harare. This difference in species richness is also reflected in dry-season fruit species: six and two versus sixteen and thirteen species for the early and late dry season, respectively. The Gaborone area has one nutlike oil seed species, compared with two species for the Harare area. Theoretically, without wetlands, semiarid climate areas were probably marginal habitats for early hominins, sometimes occupied but functioning as demographic sinks. Further climate and woody plant distribution studies point to potential hominin source areas.

Even reliably fruiting tree species can vary dramatically in individual and stand productivity year to year. Species richness in edible fruit taxa is important insurance for relatively high annual habitat productivity, even in well-watered summer rain environments. We examined the geographic distribution of species richness for wild-fruit trees and shrubs across the region of southern Africa. There is a strong west-to-east gradient of increasing edible fruit providing species similar to that of woody plants as a whole and in accord with changes in climate and vegetation (O'Brien and Peters, 1998). Again, the analysis relies on a grid-cell matrix that divides southern Africa into equal-area units, each about 20,000 km^2. The species richness of edible fruit taxa varies from grid cells with only one species (heart of the Kalahari) to 194 species (eastern Zimbabwe). For the region as a whole, there are 301 known woody species that provide edible fruit. Figure 13.2 shows the geo-

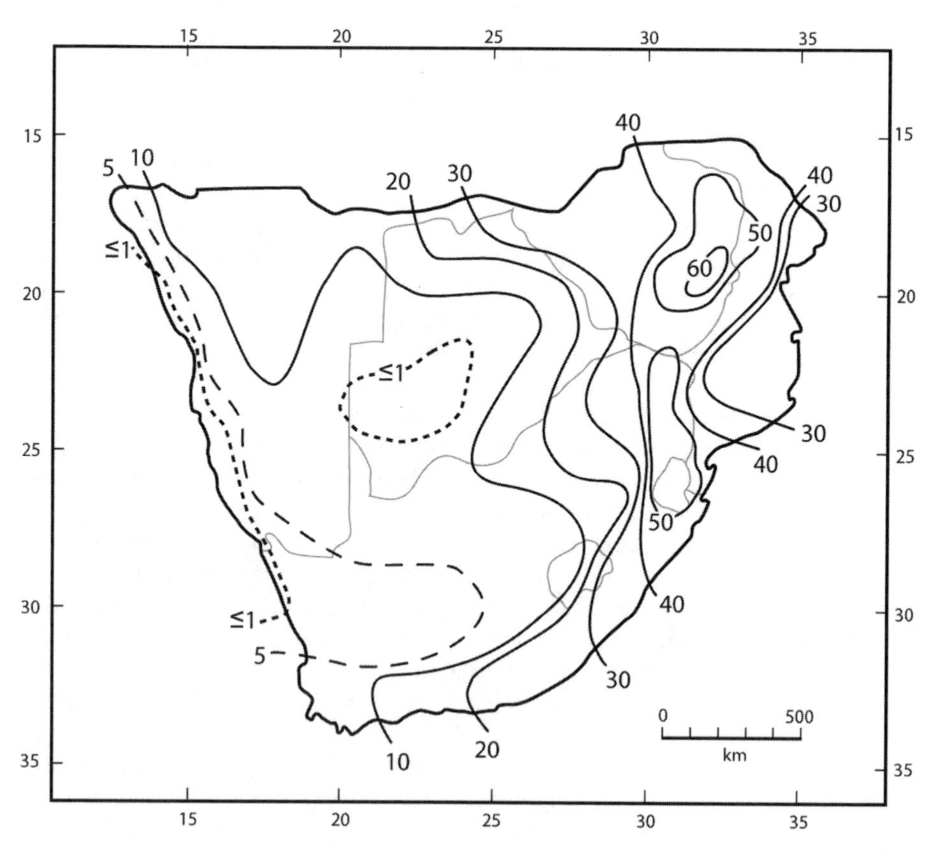

Figure 13.2 Geographic pattern of increasing edible-fruit species richness for woody plants (trees and shrubs) based on their percentage per grid cell (see text) out of a total of 301 edible-fruit-producing woody species in the southern Africa region. Redrawn from O'Brien and Peters (1998).

graphic pattern of percentage in edible-fruit species richness across the region. The gradient of increasing species richness is associated with increased length of the growing season and topographic relief. It reaches its maximum in two areas, the eastern highlands of Zimbabwe and the highlands of the eastern Transvaal, South Africa, and Swaziland. The disjunction between these two areas corresponds to the dry climate trough of the Limpopo River Valley. During the dry season, the Limpopo River dwindles to a stream or a discontinuous series of pools, and its broad valley is virtually a waterless wasteland. At the height of the rainy season in wet years, the crocodile "infested" river would also inhibit large mammal movement between these two highland areas. This partial barrier is not as strong as that of the Zambezi River (the northeast border of the map in fig. 13.2), whose valley creates a disjunction between the eastern highlands of Zimbabwe and those of Malawi to the north. The highlands of Malawi continue up along the southern extension of the Great Western Rift Valley of east Africa to the southern highlands of Tanzania, where the highland system bifurcates (with addition disjunctions) in association primarily with the Great Eastern and Western Rift Valleys of the East African Plateau (fig. 13.3). The theoretical significance of all of this is that the highland system running along the eastern portion of Africa is more like a string of pearls than a highway. It is not a corridor so much as a discontinuous set of probable demographic source areas whose theoretical significance is currently unknown. Their possible significance is explained below, noting that it is their footslopes and associated adjacent wetland land-systems that are of greatest interest to us and that they probably constituted on the East African Plateau oases-like demographic source areas in what may have been largely a demographic sink subregion for hominin species in the Pliocene.

O'Brien and Peters (1999) hypothesized that the East African Plateau was a demographic sink for hominin species in the Pliocene and that south-central Africa was the hominin demographic source area for both eastern and southern Africa. This broad geographic heartland of south-central Africa had the requisite characteristics of a phytochorion (the Zambezian) with probably the highest diversity of wild-plant foods in Africa, diverse landscapes, and complex vegetation mosaics (including dambo grasslands) across a broad geographic area. At a finer scale, some of these characteristics are present on the East African Plateau, closer to the fossil sites that provide the paleoanthropological record, and they increase the likelihood that a hierarchy of geographic and ecosystem relationships shaped the early course of hominin evolution. Below, I reconsider the potential importance of the East African Plateau for our theoretical understanding of the trophic ecology and regional demographics of the early hominins.

The footslopes to toeslopes of well-watered highlands in the subhumid tropic and subtropics theoretically offer all of the characteristics in the profile of the ideal early hominin demographic source area, previously summarized, except perhaps for the wetlands. The water-quality factor afforded by footslope streams is a very important consideration in the face of waterborne zoonoses (especially debilitating to infants and young children) associated with contaminated drinking water (see appendix in Peters and Blumenschine, 1996).

As it turns out, the toeslopes and adjacent piedmont plains of the volcanic mountains of east Africa do present landscape mosaics with important wetlands. This is

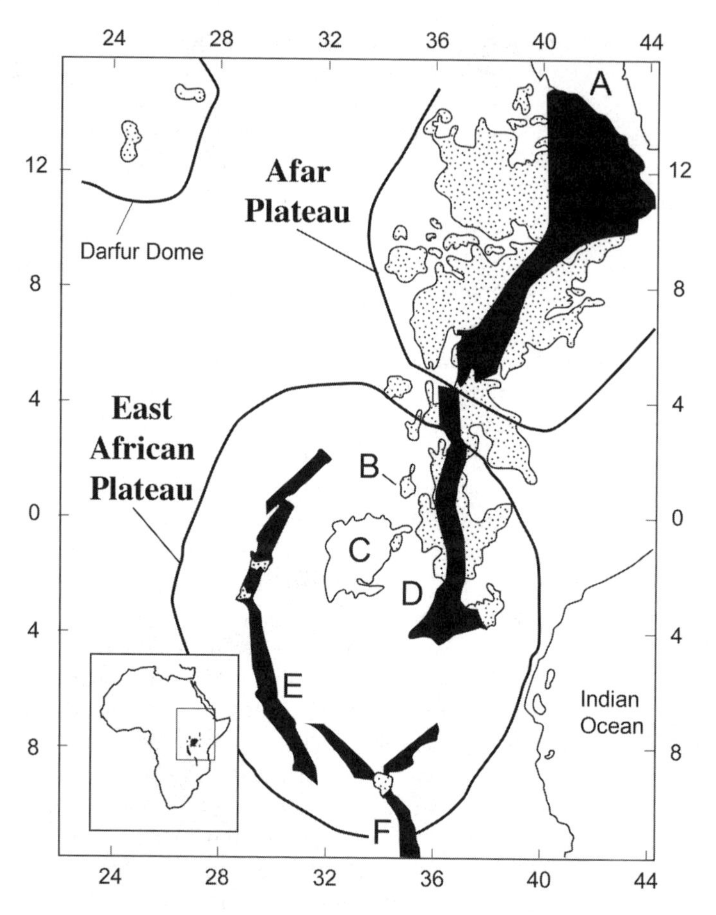

Figure 13.3 East African Rift Valley region showing location of the uplifted domes, or major plateaus. (A) Red Sea. (B) Mt. Elgon volcanic area. (C) Lake Victoria. (D) location of the Serengeti plains. (E) Lake Tanganyika located to the left. (F) Lake Malawi located to the right. Bold lines enclose elevations greater than 800 m. Cenozoic rifting superimposed on the Afar and East African plateaus indicated by black shading. Cenozoic volcanics indicated by stippling. Insert of Africa shows the location of the region, highlighting in black Lake Victoria and surrounding Rift Valley lakes. Redrawn and modified from Ebinger et al. (1989, their fig. 1).

most easily seen along the southwest edge of Mt. Kilimanjaro, although the wetlands there are being developed (destroyed) at a rapid rate. Here, we see, in what remains of the natural landscape, an alternation of groundwater-seepage marshes (with *Typha* and *Cyperus papyrus*), small dambos, and deeply incised channels (from upland streams). Dambos are of particular theoretical interest here, and they may not be a type of landform known to those who are familiar with Africa only to a limited degree.

Dambos are seasonally waterlogged (for the most part), ill-defined floodplains, channelless (unincised) drainageways forming the headwaters of incipient valleys in areas of low relief. The term, now standard in geomorphology, originated in Zambia/

Figure 13.4 Broadscale landforms of Africa. Hatched areas are dominated by the (re-)exposed and broadly leveled ancient cystalline basement, lithologies Pre-Cambrian in origin. Stippled areas are lava uplands, Cenozoic in origin, except for South Africa, where Mesozoic lavas have been re-uplifted. The unmarked remainder (with minor exceptions at this scale) represents erosional plains on Paleozoic and Mesozoic sedimentaries or Cenozoic depositional plains. Simplified from Butzer (1976, fig. 15-11): notes follow Pritchard 1979. Overlain bold lines enclose the distribution of dambos in Africa, based on Acres et al. (1985, their fig. 2). See text for additional information.

Malawi: they are called *mbugas* in Tanzania and *vleis* in South Africa. They occur widely in Africa (see fig. 13.4), primarily within the climate zones having present-day average annual rainfalls of 600–1,500 mm, with a four- to six-month dry season, and they are mostly associated with the gently undulating, broadly leveled ancient crystalline basement lithologies (between 1,000 m and 1,500 m elevation), which are Pre-Cambrian in origin (Acres et al., 1985). They reach their greatest densities across the Zambezian region, with a tongue of relatively high-density occurrence extending north through the Mt. Elgon area on the East African Plateau (fig. 13.5). They occur with less density (sporadically) across the Afar Plateau and Sudanian

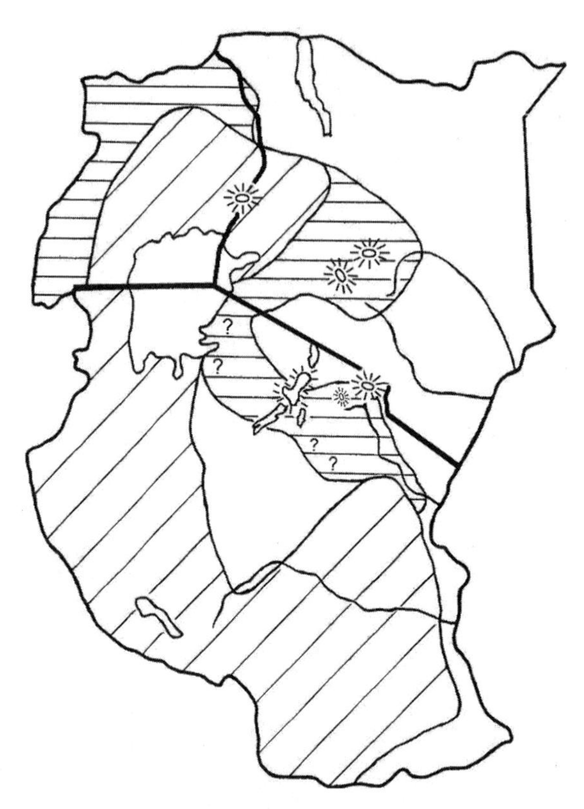

Figure 13.5 Dambo occurrence in East Africa (Uganda, Kenya, and Tanzania). Excerpted and redrawn, with modifications, from Acres et al. (1985, their fig. 2). Horizontal hatching denotes areas of sporadic dambo occurrence; oblique hatching denotes main areas of dambo occurrence. Mt. Elgon is located north of Lake Victoria on the Uganda-Kenya border.

region (Acres et al., 1985). In the formation processes of landscape evolution they represent a continuum from wetland depressions (including pans), through channelless drainageways, to final replacement by fully integrated fluviatile systems (McFarlane, 1995a, 1995b).

Under the current subhumid to semiarid climatic regimes grasses make the greatest contribution to the biomass of most dambos; sedges are second. Sedges become co-dominant to dominant in wetter conditions. Some of these species provide edible plant parts: seeds, stem/leaf bases, and especially shallow rootstocks. A variety of rootstock-food plants belonging to the petaliferous monocots are also present, including terrestrial orchids in wetter environments. Herbaceous vegetation usually dominates because few trees can tolerate poorly drained (commonly clay) soils and/or extended periods of inundation by standing or very slowly moving water. Examples of partially wooded dambos are *Syzygium* swamps in higher rainfall areas, and *Acacia drepanolobium* "bushland" on low-lying, seasonally waterlogged clay soils in semiarid east Africa. During the late Cenozoic, intensification of seasonality in rainfall and decreasing overall precipitation probably resulted in grasses

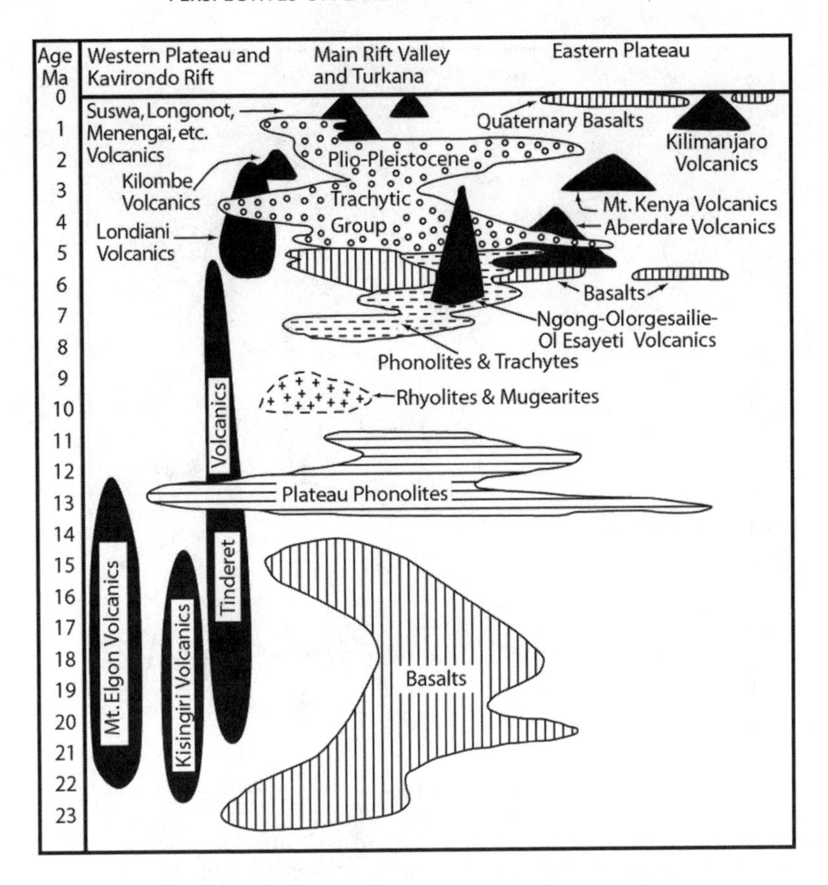

Figure 13.6 Sequence age ranges of the major central volcano groups (black shading) and plateau formations (various patterns of shading) for Kenya and its borders. Redrawn and modified slightly from Baker et al. (1971, their fig. 3).

increasingly replacing marsh in dambos (also along lake margins, in riverine floodplains and in pans).

Unfortunately, because of their economic importance, dambos have been and are being overused and destroyed across most of Africa. Before their destruction, they constituted the major natural grasslands of tropical/subtopical Africa.[3] If early hominins exploited grasslands in any major way, dambos were probably the primary grasslands to which they were adapted. Ecosystem analysis allows us to carry this theoretical line of inquiry a few steps further.

The tongue of main occurrence in dambo distribution extends north through the East African Plateau and surrounds Mt. Elgon (fig. 13.5). Mt. Elgon is an ancient massif (fig. 13.6). Because of its size, fertile soils, and surrounding wetland and woodland landscape mosaic, this ecosystem complex is hypothesized here to have been the most important demographic source area for hominins in east Africa during the (Mio-) Pliocene and early Pleistocene. A full exploration of the implications of this hypothesis can not be undertaken here, but we can take a close look at this landscape mosaic

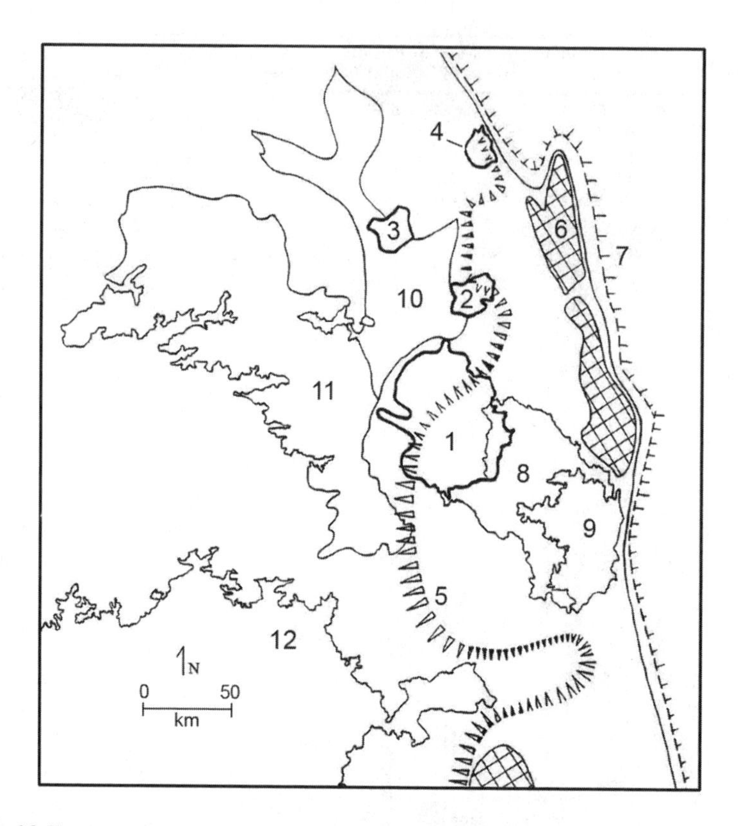

Figure 13.7 Mount Elgon area. The numbers identify the following: (1) Mt. Elgon (4,321 m); (2) Mt. Kadam (3,068 m); (3) Mt. Napak (2,537 m); (4) Mt. Moroto (3,083 m); (5) extent of the west-facing eastern limiting scarp of the Kyoga surface at the start of Miocene volcanic activity; (6) areas higher and older than the Kitale plateau; (7) the Turkana-Elgeyo fault scarp of the Eastern Rift Valley; (8) the Kitale landscape system; (9) the Eldoret landscape system; (10) the Sebei (Karamoja Plain) landscape system; (11) the Amuria landscape system; (12) Lake Victoria. Based on Bishop and Trendall (1966–67, plate 14) plus Ollier et al. (1967) and Scott et al. (1972).

and suggest theoretical lines of inquiry that can probably help us better understand early hominin ecophenotypic trophic adaptations. The Mt. Elgon area landscapes, surrounded by a subregion that was functioning largely as a demographic sink, ultimately connected via the dambo corridor to the Zambezian heartland in a hierarchy of space-time relationships, also might help us understand aspects of both diversification and stability in hominin evolution across the Pliocene and early Pleistocene.

First, we need to be able to picture the land systems and vegetation ecostructure to postulate patterns of selection for common and divergent hominin forms of locomotion and food acquisition. The Mt. Elgon area and the two dambo/floodplain-rich landscape systems immediately adjacent on the monsoon side are depicted in figures 13.7–13.10 and tables 13.2 and 13.3. Mt. Elgon together with the adjacent Kitale

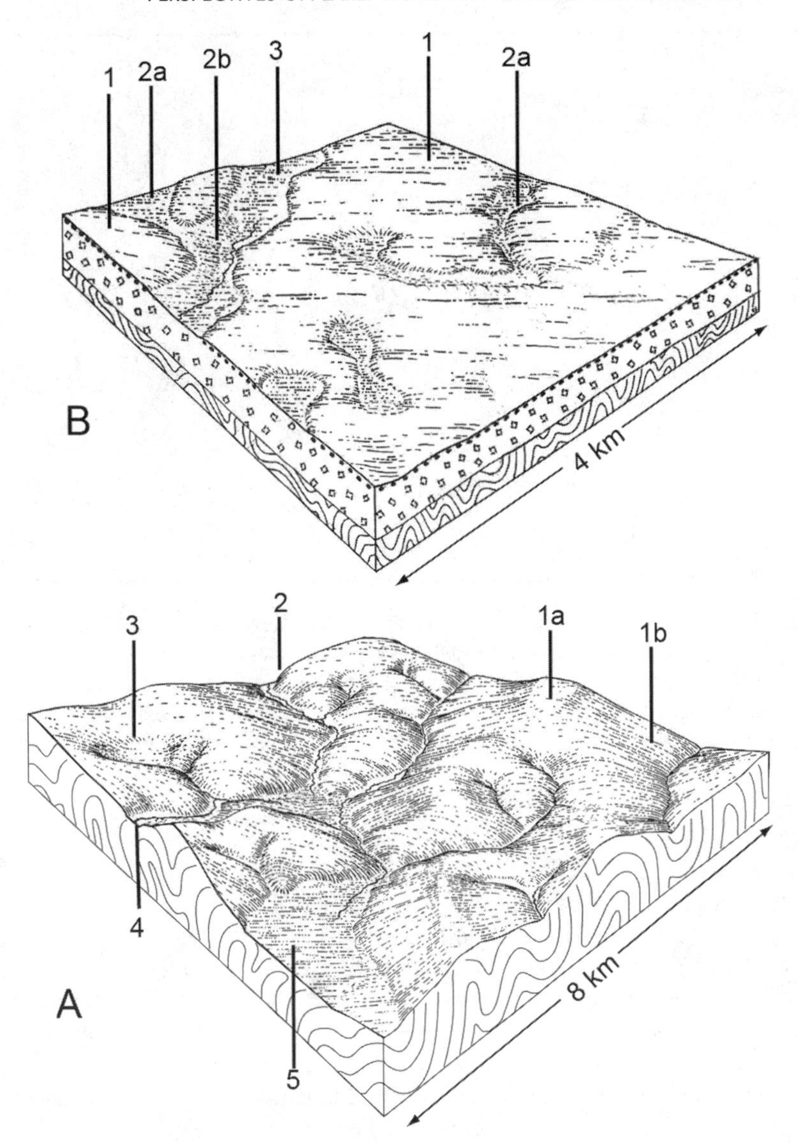

Figure 13.8 Landscape systems to the southeast of Mt. Elgon (see fig. 13.7). (*A*) Idealized block diagram of the Kitale landscape system (the numbered landscape facets are described in table 13.2). (*B*) Idealized block diagram of the Eldoret landscape system (the numbered landscape facets are described in table 13.3). Modified from Scott et al. (1970–71).

and Eldoret land systems illustrate how unique neighboring landscapes can result in a keystone landscape complex.

The landscape facets of the Kitale and Eldoret land systems are detailed in tables 13.2 and 13.3. Judging by the 1970 *Survey of Kenya National Atlas*, the Kitale landscape system receives between 1,000 mm and 1,500 mm in annual rainfall, while the Eldoret system receives 750–1,000 mm. The Kitale system is a broadly undulating

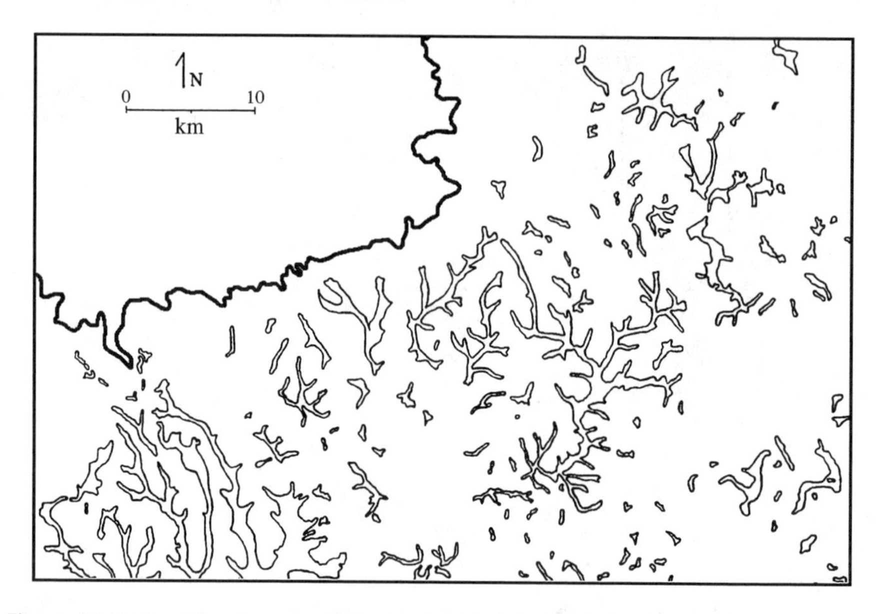

Figure 13.9 Edaphic sedge-grasslands (wetlands) of the Kitale landscape system, reflecting postvolcanic erosion of the area. Mt. Elgon occupies the northwest portion of the area depicted. The fully integrated drainage system in the southwest corner is part of the adjacent upper Bungoma landscape system (depicted in more detail in Raunet 1985, his fig. 4). The relatively unintegrated wetlands in the southeast corner are a projecting portion of the Eldoret landscape system to the east. Extracted from Trapnell et al. 1966.

landscape (see fig. 13.8A) at 1,550–2,050 m elevation, with a relief of about 60 m (Scott, Webster, and Lawrence, 1970–71). Stripped of its volcanic (Mt. Elgon tephra) cover over the past 12–15 million years (fig. 13.6), the earlier Kitale (basement complex) surface has been reexposed and dissected by widely spaced narrow V-shaped valleys with broader lower-valley bottomlands (fig. 13.9). In contrast, the Eldoret system is an almost level plateau (fig. 13.8B) at 1,950–2,300 m elevation, with broad gentle depressions and relatively infrequent, weakly entrenched streams (fig. 13.10), with a relief of only a few meters. In the Late Miocene, circa 12–13 Ma, phonolite flood lavas (fig. 13.6) were erupted out of the shallow Eastern Rift, covering an extensive area, including (creating) what is now the Eldoret system, where they buried and leveled the low-relief prevolcanic Miocene erosion surface (Scott, Webster, and Lawrance, 1970–71).

The dominant wetlands in both the Kitale and Eldoret landscapes are edaphic sedge grasslands (figs. 13.9 and 13.10), often on dambo ("vlei") soils (Trapnell et al., 1966). Across the currently more mature Kitale landscape, they are usually associated with drainage lines, less so over the Eldoret system. Perhaps the most striking initial observation that can be made about the plain geometry of these wetlands is that they are pan-like or, more often, irregularly linear, and usually about 1 km or less in breadth, while commonly 3–5 km or more in length. They maximize the boundary edge of the surrounding woodlands and the grassland-woodland ecotone.

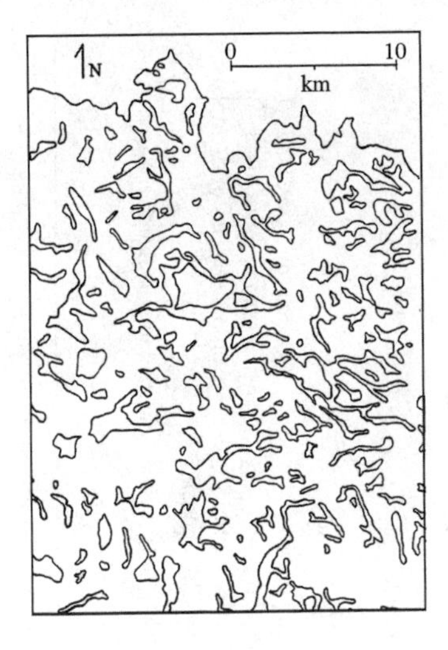

Figure 13.10 Edaphic sedge-grasslands (wetlands) of the Eldoret landscape system. Extracted from Trapnell et al. 1966.

They are individually much smaller as open areas than the open Serengeti plains, which cover distances of 15–25 km or more, but those volcanic ash edaphic grasslands are probably later Pleistocene in origin (Peters et al., n.d.).

Moving theoretically toward generality, the intensification of selection for ecophenotypic trophic traits, as climate dynamics forced demographic contraction and allowed expansion, centered on the Mt. Elgon core area with at least two consequences likely: (1) divergent innovative specializations occurred as a result of easy-to-use resource depletion; and (2) stabilizing selection occurred for successful specializations, refined across time in canalized ways. The Mt. Elgon area can be thought of as a microcosm in the broadscale (continental) scenario of hominin evolution and divergence. The identification of the Mt. Elgon area as a Plio-Pleistocene keystone landscape complex for both woodland and grassland mammals is also part of the endeavor to identify the main environments probably occupied by early hominins across Africa. The evolutionary roles of the environments represented by the fossil sites will not be known until this broad scale realism is added to our models. It may turn out that the fossil sites are not located in the landscapes that drove hominin evolution, either with regard to origins or divergence in the speciation process. (See Wiens, 2004, for a clear theoretical separation of the question of origins from the question of divergence in the process of speciation, and an argument for the importance of the ecological perspective.) The geography of origins may not be the same as that of evolutionary divergence (e.g., cf. figs. 13.3–13.6).

A few comments can be made on the carnivorous specializations of Oldowan *Homo* in the theoretical context of dambo landscapes. The fine-scale dambo-woodland mosaic maximizes the opportunity for scavenging from the remains of both leopard and lion kills (see Blumenschine and Pobiner, chapter 10, in this volume).

Hominin scavenging from lion kills is in part the business of landscape risk management. The broader the open landscape facet, the greater the distance to the safety

Table 13.2 Facets of the Kitale Landscape System, Western Kenya

Facet Number	Form	Soils, Materials, and Hydrology	Land Cover (Reconstructed)[a]
1	Plateaus and gentle slopes.		Woodland and drier semievergreen forest
	(a) Plateau remnants, level, even, with gently sloping margins. About 500 m across	(a) Dark reddish brown sandy clay loam up to 15 cm over reddish brown friable sandy clay with frequent MnO_2 and iron concretions plus quartz stones below 60 cm. Massive laterite occurs at about 125 cm	
	(b) Gentle and very gentle slopes frequently forming convex interfluves or flanks to 1(a). Even, up to 1–3 km wide	(b) Dark reddish brown sandy clay loam up to 15 cm over dark red to red friable clays over 170 cm deep	
2	Valley sides. Short, moderate to steep, straight, with concave upper margins, extending along nearly all river valleys, up to 100 m wide	Dark grey to brown sandy loam to sandy clay loam up to 40 cm, over yellow red friable sandy clay to sandy clay loam over 170 cm deep	Woodland and dry forest
3	Footslopes. Gentle to very gently sloping, even or slightly uneven, with concave upper margins and locally convex lower margins up to 500 m across and frequently with diffuse boundaries to facet 1(b)	Dark grey to brown sandy loam up to 30 cm over dark yellowish to strong brown mottled sandy loam to sandy clay loam over 170 cm deep. High seasonal water table	Woodland and/or forest
4	Small floodplains and dambos. Narrow level valley floors, usually including stream channel. Up to 200 m across, usually with sharp margins to facet 2	Dark grey sandy clay loam up to 30 cm over dark greyish brown to pale brown mottled sandy clay loam to loose coarse sand. High water table (at 120 cm). Seasonally flooded	Sedges and grassland. *Syzygium* forest along upper margins of floodplain
5	Extensive flats. Broad flats up to 3 km across and narrow elongated flats along main dainage lines	Dambo soils and peats over mud. Generally perennially flooded	Papyrus marsh and evergreen sedge-grassland

Sources: Modified from Scott et al. (1970–1971), with additional information from Trapnell et al. (1966), White (1983), Trapnell and Brunt (1987), and personal observations.
[a]Natural vegetation destroyed by modern land-use practices, cultivation, and grazing.

Table 13.3 Facets of the Eldoret Landscape System, Western Kenya

Facet Number	Form	Soils, Materials, and Hydrology	Land Cover (Reconstructed)[a]
1	Plateau. Extensive level to gently sloping areas; locally undulating	Dark brown clay loams up to 15 cm over strong brown to reddish brown friable clay with iron and MnO_2 concretions. Massive laterite occurs at depths from 30–90 cm, locally deeper where plateau undulating	Woodland and evergreen bushland
2	Clay depressions. (a) Gentle broad flat bottomed depressions with sharp or diffuse concave margins. Up to ca. 1 km wide (b) Moderately extensive low-lying land along stream and river margins, discontinuous; up to 300 m wide	Greyish brown mottled clay loams to clay 15–30 cm deep over grey clay over 170 cm deep. Subject to seasonal flooding	Sedgeland and evergreen grassland
3	Valley floors. (a) Broad, about 200–400 m wide, flat bottomed	Dambo soils and peats over mud. Permanently flooded drainage lines; seasonally flooded broad margins	Papyrus marsh and associated sedgelands plus evergreen grassland
	(b) River channel, narrow (5–15 m), sinuous	Seasonal to permanent flow	Nil

Sources: Modified from Scott et al. (1970–1971), with additional information from Trapnell et al. (1966), White (1983), and personal observations.
[a]Natural vegetation largely destroyed by modern land-use practices, cultivation, and grazing.

of trees, that is, the greater the risk of death from encounters with large carnivores (Blumenschine and Peters, 1998, develop some of the landscape correlates for this factor). The fine-scale dambo-woodland mosaic minimizes this risk and takes a good deal of the danger out of the problematic open-habitat adaptation concept applied to early hominins. Puzzling evidence for broad-open habitat (high risk) thorough butchery of large mammals (e.g., de Heinzelin et al., 1999) should be independently reassessed, as it goes against conservative theoretical models of early hominin capabilities (cf. Blumenschine and Peters, 1998). The apparent evidence offered may result from mistaken or incomplete analyses and be equivocal on reexamination. If not, then it has rather profound implications for early hominin offensive-defensive behavior and theoretical models will be gratefully revised. This might also open up the possibility not just of confrontational (interference) scavenging but of hunting larger game as well. Theoretical exploration of the kinds of hunting techniques that would be favored by dambo ecostructure have yet to be made systematically, but their linear segment offers a predictable flight corridor. Panicked game tends to follow lineation in landscape, where present. Here, linear structure is reinforced by the well-grazed edge along the dambo-woodland ecotone. The theoretical importance of this for the pursuer technique of the cheetah, or the "relay" technique of the wild dog, readily comes to mind.

Acknowledgments A special word of thanks to B. Akalo, R. Barnley, M. Bingham, A. Brunson, R.J. Clarke, S. Kamande, M.J. McFarlane, L. Oertel, and H.R. Pulliam; and to the Anthropology Department, University of Georgia, for continued research support as Professor Emeritus. Portions of this chapter (including background notes for bipedality) are from an unpublished 1999 paper I prepared for an unfortunately aborted JHE special issue on landscape archaeology. The comments of two anonymous reviewers helped with the final revision.

Notes

1. The higher vertebrate species richness of the Plio-Pleistocene suggests that (1) potential competition was much higher then; and/or (2) population densities per species were much lower; and/or (3) vegetation productivity was much higher; or (4) potential competition was realized in very high rates of local extinctions; or (5) other, currently unimagined possibilities were operating in super species-rich prehistoric ecosystems. Simple uniformitarian assumptions seem impoverished in the presence of even "time-averaged" prehistoric vertebrate checklists. We are moving into the unknown, and by the looks of it, the unknowable. Some potential theoretical relief is provided by the concepts of niche versatility with specialization and source-sink demographics (see text).
2. Unlike the nut-producing fruits of temperate zone woodlands, the nut-producing fruits of African tropical/subtropical woodlands have nutritious mesocarps in addition to edible oil-rich nut seeds (Peters, 1987a; see Peters, 1993 for additional notes).
3. In current usage, the unfortunate term "savanna" covers a range of anthropogenically degraded forms of wooded land in eastern and southern Africa. These are not the natural grasslands of Africa. See O'Brien and Peters (1999) for references to the latter.

References

Acres, B.D., Rains, A.B., King, R.B., Lawton, R.M., Mitchell, A.J.B., and Rackham, L.J., 1985. African dambos: Their distribution, characteristics and use. In: Thomas, M.F., and Goudie, A.S. (Eds.), *Dambos: Small Channelless Valleys in the Tropics—Characteristics, Formation, Utilisation.* Zeitschrift für Geomorphologie, Supplementband 52. Gebruder Borntraeger, Berlin, pp. 63–86.

Andrewartha, H.G., and Birch, L.C., 1984. *The Ecological Web: More on the Distribution and Relative Abundance of Animals.* University of Chicago Press, Chicago.

Baker, B.H., and Wohlenberg, J., 1971. Structure and evolution of the Kenya rift valley. *Nature* 229, 538–542.

Bishop, W.W., and Trendall, A.C., 1966–67. Erosion-surfaces, tectonics and volcanic activity in Uganda. *Q. J. Geol. Soc. Lond.* 122, 385–420.

Blumenschine, R.J., and Peters, C.R., 1998. Archaeological predictions for hominid land use in the paleo-Olduvai Basin, Tanzania, during lowermost Bed II times. *J. Hum. Evol.* 34, 565–607.

Brain, C.K., and Sillen, A., 1988. Evidence from the Swartkrans Cave for the earliest use of fire. *Nature* 336, 464–466.

Butzer, K.W., 1976. *Geomorphology from the Earth.* Harper & Row, New York.

Clark, J.D., and Harris, J.W.K., 1985. Fire and its roles in early hominid lifeways. *Afr. Archaeol. Rev.* 3, 3–27.

de Heinzelin, J., Clark, J.D., White, T., Hart, W., Renne, P., WoldeGabriel, G., Beyene, Y., and Vrba, E., 1999. Environment and behavior of 2.5-million-year-old Bouri hominids. *Science* 284, 625–629.

Demes, B., and Creel, N., 1988. Bite forces, diet and cranial morphology of fossil hominids. *J. Hum. Evol.* 17, 657–670.

Ebinger, C.J., Bechtel, T. D., Forsyth, D. W., and Bowin, C. O., 1989. Effective elastic plate thickness beneath the East African and Afar plateaus and dynamic compensation of the uplifts. *J. Geophys. Res.* B 94(3), 2883–2901.

Field, R., O'Brien, E.M., and Whittaker, R.J., 2005. Global models for predicting woody plant richness from climate: Development and evaluation. *Ecol.* 86, 2263–2277.

Hutchinson, G.E., 1957. Concluding remarks. *Cold Spring Harb. Symp. Quant. Biol.* 22, 415–427.

Iwu, M.M., 1993. *Handbook of African Medicinal Plants*. CRC Press, Baton Rouge, LA.

Kuchka, H.E., 2001. Method for theory: A prelude to human ecosystems. *J. Ecol. Anthropol.*, special issue, 5, 1–78.

Leopold, A.C., and Ardrey R., 1972. Toxic substances in plants and the food habits of early man. *Science* 176, 512–514.

Levins, R., 1966. The strategy of model building in population biology. *Am. Sci.* 54, 421–431.

Liebenberg, L., 1990. *The Art of Tracking: The Origin of Science*. David Philip, Claremont.

Lucas, P.W., and Peters, C.R., 2000. Function of postcanine tooth crown shape in mammals. In: Teaford, M.F., Smith, M.M., and Ferguson, M.W. (Eds.), *Development, Function and Evolution of Teeth*. Cambridge University Press, Cambridge, pp. 282–289.

Lucas, P.W., Peters, C.R., and Arrandale, S.R., 1994. Seed-breaking forces exerted by orangutans with their teeth in captivity and a new technique for estimating forces produced in the wild. *Am. J. Phys. Anthropol.* 94, 365–378.

Marzke, M.W., 1996. Evolution of the hand and bipedality. In: Lock, A., and Peters, C.R. (Eds.), *Handbook of Human Symbolic Evolution*. Oxford University Press, Oxford, pp. 126–154. Repr., Blackwell, Oxford, 1999.

McFarlane, M.J., 1995a. Dambo gullying in parts of Zimbabwe and Malawi: A reassessment of the causes. In: Owen, R., Verbeek, K., Jackson, J., and Steenhuis, T. (Eds.), *Dambo Farming in Zimbabwe: Water Management, Cropping and Soil Potentials for Smallholder Farming in Wetlands*. University of Zimbabwe Publications, Harare, pp. 105–116.

McFarlane, M.J., 1995b. *Pans and Dambos of Western Province, Zambia*. Department of Agriculture, Mongu.

O'Brien, E.M., 1989. Climate and woody plant species richness: Analyses based upon Southern Africa's native flora with extrapolations to Subsaharan Africa. D.Phil. diss., University of Oxford.

O'Brien, E.M., and Peters, C.R., 1991. Ecobotanical contexts for African hominids. In: Clark, J.D. (Ed.), *Cultural Beginnings: Approaches to Understanding Early Hominid Life-ways in the African Savanna*. Dr. Rudolf Habelt GMBH, Bonn, pp. 1–15.

O'Brien, E.M., and Peters, C.R., 1998. Wild fruit trees and shrubs of southern Africa: Geographic distribution of species richness. *Econ. Bot.* 52, 267–278.

O'Brien, E.M., and Peters, C.R., 1999. Landforms, climate, ecogeographic mosaics, and the potential for hominid diversity in Pliocene Africa. In: Bromage, T.G., and Schrenk, F. (Eds.), *African Biogeography, Climate Change, and Human Evolution*. Oxford University Press, Oxford, pp. 115–137.

Ollier, C.D., Lawrance, C.J., Webster, R., and Beckett, P.H.T., 1967. Land System Map of Uganda. Scale 1 : 1,000,000. United Kingdom: Director of Military Survey, Ministry of Defense.

Peters, C.R., 1979. Toward an ecological model of African Plio-Pleistocene hominid adaptations. *Am. Anthropol.* 81, 261–278.

Peters, C.R., 1981. Robust vs. gracile early hominid masticatory capabilities: The advantages of the megadonts. In: Mai, L.L., Shahklin, E., and Sussman, R.W. (Eds.), *The Perception of*

Evolution: Essays Honoring Joseph B. Birdsell. Anthropology UCLA Vol. 7. Department of Anthropology, University of California at Los Angeles, Los Angeles, pp. 161–181.

Peters, C.R., 1987a. Nut-like oil seeds: Food for monkeys, chimpanzees, humans, and probably ape-men. *Am. J. Phys. Anthropol.* 73, 333–363.

Peters, C.R., 1987b. *Ricinodendron rautanenii* (Euphorbiaceae): Zambezian wild food plant for all seasons. *Econ. Bot.* 41(4), 494–502.

Peters, C.R., 1988. Notes on the distribution and relative abundance of *Sclerocarya birrea* (A. Rich.) Hochst. (Anacardiaceae). *Monogr. Syst. Bot. MO Bot. Gard.* 25, 403–410.

Peters, C.R., 1990. African wild plants with rootstocks reported to be eaten raw: The monocolyledons, pt. I. *Mitt. Inst. Allg. Bot. Hamburg* 23, 935–952.

Peters, C.R., 1993. Shell strength and primate seed predation of nontoxic species in eastern and southern Africa. *Int. J. Primatol.* 14, 315–344.

Peters, C.R., 1994. African wild plants with rootstocks reported to be eaten raw: The monocotyledons, pt. II. In: Seyani, J.H., and Chikuni, A.C. (Eds.), *Proceedings of the XIIIth Plenary Meeting of AETFAT*. Zomba, Malawi, 2–11 April 1991, National Herbarium and Botanic Gardens of Malawi, Zomba, pp. 25–38.

Peters, C.R., 1996. African wild plants with rootstocks reported to be eaten raw: The monocotyledons, pt. III. In: van der Maesen, L.J.G., van der Burgt, X.M., and van Medenbach de Rooy, J.M. (Eds.), *Biodiversity of African Plants*. Kluwer, Dordrecht, pp. 665–677.

Peters, C.R., 1999. African wild plants with rootstocks reported to be eaten raw: The monocotyledons, pt. IV. In: Timberlake, J., and Kativu, S. (Eds.), *African Plants: Biodiversity, Taxonomy and Uses*. Royal Botanic Gardens, Kew, pp. 483–503.

Peters, C.R., and Blumenschine, R.J., 1995. Landscape perspectives on possible land use patterns for Early Pleistocene hominids in the Olduvai Basin, Tanzania. *J. Hum. Evol.* 29, 321–362.

Peters, C.R., and Blumenschine, R.J., 1996. Landscape perspectives on possible land use patterns for Early Pleistocene hominids in the Olduvai Basin, Tanzania. Pt. II: Expanding the landscape models. In: Magori, C.C., Saanane, C.B., and Schrenk, F. (Eds.), *Four Million Years of Hominid Evolution in Africa: Papers in Honour of Dr. Mary Douglas Leakey's Outstanding Contribution in Palaeoanthropology*. Darmstadter Beitrage zur Naturgeschichte, Kaupia 6, Darmstadt, pp. 175–221.

Peters, C.R., and Maguire, B., 1981. Wild plant foods of the Makapansgat area: A modern ecosystems analogue for *Australopithecus africanus* adaptations. *J. Hum. Evol.* 10, 565–583.

Peters, C.R., and O'Brien, E.M., 1981. The early hominid plant-food niche: Insights from an analysis of plant exploitation by *Homo*, *Pan*, and *Papio* in eastern and southern Africa. *Curr. Anthropol.* 22, 127–140.

Peters, C.R., and O'Brien, E.M., 1982. On early hominid plant-food niches: Reply. *Curr. Anthropol.* 23, 214–218.

Peters, C.R., and O'Brien, E.M., 1984. On hominid diet before fire. *Curr. Anthropol.* 25, 358–360.

Peters, C.R., and O'Brien, E.M., 1994. Potential hominid plant foods from woody species in semi-arid vs. sub-humid subtropical Africa. In: Chivers, D.J., and Langer, P. (Eds.), *The Digestive System in Mammals: Food, Form and Function*. Cambridge University Press, Cambridge, pp. 166–192.

Peters, C.R., and Vogel, J.C., 2005. Africa's wild C_4 plant foods and possible early hominid diets. *J. Hum. Evol.* 48, 219–236.

Peters, C.R., O'Brien, E.M., and Box, E.O., 1984. Plant types and seasonality of wild-plant foods, Tanzania to southwestern Africa: Resources for models of the natural environment. *J. Hum. Evol.* 13, 397–414.

Peters, C.R., O'Brien, E.M., and Drummond, R.B., 1992. *Edible Wild Plants of Subsaharan Africa*. Royal Botanic Gardens, Kew.

Peters, C.R., Blumenschine, R.J., Hay, R.L., Livingstone, D.A., Marean, C.W., Harrison, T., Armour-Chelu, M., Andrews, P., Bernor, R.L., Bonnefille, R., and Werdelin, L., n.d. Paleoecology of the Serengeti-Mara Ecosystem. In: Sinclair, A.R.E., et al. (Eds.),

Serengeti III: Biodiversity and Biocomplexity in a Human Influenced Ecosystem. University of Chicago Press, Chicago.

Pickett, S.T.A., Kolasa, J., and Jones, C.G., 1994. *Ecological Understanding: The Nature of Theory and the Theory of Nature*. Academic Press, London.

Pritchard, J. M., 1979. *Landform and Landscape in Africa*. Edward Arnold, London.

Pulliam, H.R., 1988. Sources, sinks, and population regulation. *Am. Nat.* 135, 652–661.

Raunet, M., 1985. Les bas-fonds en Afrique et à Madagascar: Géormorphologie-géochímíe-pédologie-hydrologie. In: Thomas, M.F., and Goudie, A.S. (Eds.), *Dambos: Small Channelless Valleys in the Tropics—Characteristics, Formation, Utilization*. Zeitschrift für Geomorphologie Supplementband 52. Gebrüder Borntraeger, Berlin, pp. 25–62.

Robinson, B.W., and Wilson, D.S., 1998. Optimal foraging, specialization, and a solution to Liem's paradox. *Am. Nat.* 151, 223–235.

Scott, R.M., Webster, R., and Lawrance, C.J., 1970–71. *A Land System Atlas of Western Kenya*. Military Engineering Experimental Establishment, Christchurch, England.

Scott, R.M., Webster, R., and Lawrance, C.J., 1972. *Land System Map of Western Kenya. Scale 1 : 500,000*. Director of Military Survey, Ministry of Defense, United Kingdom.

Stahl, A.B., 1984. Hominid dietary selection before fire. *Curr. Anthropol.* 25, 151–168.

Suzuki, A., 1969. An ecological study of chimpanzees in a savanna woodland. *Primates* 10, 103–148.

Trapnell, C.G., and Brunt, M.A., 1987. *Vegetation and Climate Maps of South-Western Kenya*. Land Resources Development Centre, Overseas Development Administration, Surbiton, Surrey.

Trapnell, C.G., Birch, W.R., Brunt, M.A., and Pratt, O. J., 1966. *Vegetation Map: Land Use System of South-Western Kenya*. Sheet 1, scale 1 : 250,000. British Directorate of Overseas Surveys.

Van Wyk, B.-E., Van Oudtshoorn, B., and Gericke, N., 1997. *Medicinal Plants of Southern Africa*. Briza, Pretoria.

Watt, J.M., and Breyer-Brandwijk, M.G., 1962. *The Medicinal and Poisonous Plants of Southern and Eastern Africa*. E& S Livingstone, Edinburgh.

Western. D., 1997. *In the Dust of Kilimanjaro*. Island Press, Washington DC.

White, F., 1983. *The Vegetation of Africa*. UNESCO, Paris.

Wiens, J.J., 2004. What is speciation and how should we study it? *Am. Nat.* 163, 914–923.

Youlatos, D., 2004. Multivariate analysis of organismal and habitat parameters in two neotropical primate communities. *Am. J. Phys. Anthropol.* 123, 181–194.

14

African Pliocene Paleoecology

Hominin Habitats, Resources, and Diets

KAYE E. REED
AMY L. RECTOR

Paleoecology is not only the reconstruction of habitat; it is also the study of the interactions of hominins with their environment. The paleoenvironment includes the climate, geomorphology, vegetation structure, and the faunal community, essentially the factors that make up an ecosystem. Climate includes abiotic parameters such as rainfall, temperature, and evapotranspiration. Climatic factors combined with soil nutrient properties determine the ultimate vegetation structure in an area, or broad-based habitat (e.g., forests or grasslands). The primary productivity of the plants in a given habitat is thus also dependent on these factors. The plant productivity within various habitats ultimately provides the resources for a variety of herbivorous animals who, in turn, supply carnivorous creatures. Geomorphology is important to climatic regimes such that each geographic region and/or locale is uniquely affected by these abiotic climatic strictures due to the influences of mountains (rain shadows), rivers (flooding), lakes (subterranean water), and so on. The fauna as a community is dependent on the habitat, as the vegetation must support various ecological niches for the animals adapted to particular trophic and substrate resources. Herbivores are more influenced by plants as trophic resources than carnivores. However, carnivores are often dependent on vegetation structure due to adaptive hunting practices and/or preferred prey. Seasonal differences caused by various climatic regimes and geomorphology influence plant productivity on a yearly basis. This means resource use of these plants must also affect the fauna. All of these factors, which are observable to some degree in extant habitats, are in the provenance of paleoecological reconstruction.

Understanding the paleoecology in which early hominins existed is important for providing the necessary contextual evidence to evaluate hominin foraging behavior and ultimately diet, especially when the hominins in question left no material culture. As other chapters in this book discuss reconstructing diet for hominins that have left material culture, we focus on habitat reconstruction for early hominins from the late Miocene until ~2.5 Ma., although we do mention some overall habitat patterns in later time periods. In this chapter, we discuss the relationship between climate, habitats, and mammal communities in Africa today as known quantities that provide analogs for reconstructing the same three tiers of relationships in the past. Of course, once the attempt is made to evaluate or reconstruct anything beyond what species particular fossil specimens are, we are in the realm of the unknown. As with all other researchers providing information for this book, we are interested in the ecological behavior of our ancestors and as such in combining as much data as we can to arrive at the best possible guess for foods consumed. In addition, we are interested in how dietary differences among species or populations of the same species may have led to evolutionary successes or failures. In this respect, we might be getting into the ultimately unknowable, but without ideas and subsequent hypotheses testing, science would cease to exist.

The Known

It is expected that African Pliocene habitats and extant African habitats under similar climatic regimes would have virtually the same vegetation structure. Disciplines studying the past often use parameters in the present as analogies, thereby creating templates on which to understand how patterns in the past may have been organized. Paleoecological reconstructions of habitats are no different. This does not necessarily mean that the various habitats in the past possessed the same plant species as those today; only that the overall structure would be quite similar. To recognize these Pliocene habitats, an understanding of extant African habitats must first provide the basis for paleoecological reconstructions.

Extant African Habitats

Climatic factors such as mean annual rainfall (MAR), temperature/sunlight, and evapotranspiration interact with other abiotic factors, such as latitude and soil properties, to produce the environments in which plants exist. While actual plant species differ depending on biogeographic factors and historical circumstances, the structure of habitats can be predicted based on differences among these abiotic factors (Archibold, 1995). In Africa, these broadly defined habitat structures include forests, woodlands, bushlands with thickets, shrublands, grasslands, wooded grasslands, deserts, and so on (White, 1983). Each habitat is distinct through vegetation structure, rather than the specific plant species within them. This does not mean that specific plant species cannot define various habitats, (e.g., the Miombo woodlands of southern Africa), but extant habitats need to be defined in a way that will be applicable to the fossil record.

Extant forests in Africa lie at one end of a habitat gradient that is dependent on sunlight, water, and soils. Forests require high MAR and/or subterranean groundwater, nutrient-rich soils, and long wet seasons if moisture is acquired from rainfall. In fact, in South Africa, rain accounts for almost 80 percent of the variation in plant-species richness, and this coincides with habitat type such that forests have the highest species diversity (O'Brien, 1998). Forests have tall, pillar-like trees that have interlaced crowns and multiple canopies (White, 1983). Deserts are at the opposite end of this habitat spectrum. Deserts exist with low MAR, few rainy months, and the least productive soils. Desert environments result in plants with high drought tolerance, such as euphorbias and cacti, but few trees.

In Africa, habitats that receive less MAR than in forests and more MAR than occurs in deserts fall under the rubric of "savanna" (White, 1983). The term *savanna* thus covers a multitude of habitats, such as woodland, bushland, shrubland, and grassland, and is therefore somewhat useless in discussions of early hominin habitats. Instead, more precise terms and habitat definitions can be applied. The structure of woodlands is different from forests in two major ways. First, the ground cover in woodlands consists of grasses, whereas in forests epiphytes and ferns abound (White, 1983). Second, no matter how dense the woodland, there are no canopy levels, and the crowns of the trees are not consistently interlaced. Density of woodlands (i.e., open to closed) is based on the percentage of ground covered by tree canopies if viewed from above (Pratt and Gwynne, 1977). Thus there are open woodlands, intermediate woodlands, and closed woodlands wherein tree coverage is least in open and greatest in closed woodlands. The MAR in woodlands ranges from 600 mm to about 1,000 mm, with greater tree cover resulting from higher rainfall (Bourliere and Hadley, 1983).

Bushlands are characterized as receiving from 250 mm to 600 mm MAR. Poorer soils often cause tree species to grow as bushes (between 3 and 9 m in height, with multiple stems rather than a trunk) and fewer grasses to grow as ground cover (White, 1983). Sept (1994) has reported, however, that bushlands have more fruiting and likely edible plant species than woodlands. Moving along the habitat gradient to drier realms, shrublands usually have the least amount of MAR combined with a lack of nutrients in soils, especially contrasted with woodlands and bushlands. These factors often produce scrub or dwarf versions of tree species (1 to 3 m in height), and/or few seasonal grass species. Shrublands therefore may also be referred to as *scrub*, or *dwarf* woodlands (White, 1983).

Finally, grasslands are obviously characterized by the presence of few trees. In Africa, there are essentially two types of grasslands. Edaphic grasslands are exemplified by floodplain grasslands and/or wetlands like the Kafue Flats in Zambia, or by the unique Serengeti Plains, which is grassland on volcanic soils (Archibold, 1995). These types of grasslands may receive abundant water in some months of the year either from rainfall or from flooding rivers. Secondary grasslands are derived and are caused by overgrazing, regular burning, or both. Much of Africa that is covered by secondary grassland today would revert to some type of woodland if they were not either consistently burned or overgrazed (Pratt and Gwynne, 1977). This may mean that the African Pliocene, and indeed much of the Pleistocene, had dominant habitats unlike the prevailing habitats of eastern and southern Africa today.

Habitats across the landscape occur in mosaic patterns because of changes in soil types, the presence of subterranean water, and other factors. This is important to consider when reconstructing past environments because river courses and lakes provide much of the mosaicism, and almost all of the East African hominin localities were accumulated under these depositional environments. Riverine forests often hug the banks of many major rivers in Africa no matter what the nature of the more distant, surrounding habitat. Consequently, there are riverine forests that can be delimited by grassland or shrubland habitats. Rivers also flood from either locally heavy rains or from distant montane rains. Whatever the source, over-bank flood and deltaic deposits provide an abundance of water for part of the year. In addition, these floods often deposit fertile sediments, thus changing the nature of the existing soils. These processes result in a mosaic of edaphic grasslands surrounded by woodlands, bushlands, and/or shrublands, such as occurs within the Okavango Delta. Lakes, such as Chamo and Abaya in Ethiopia's Rift Valley, provide extensive groundwater and can produce edaphic grasslands, thicker bushland, or even forests that are unexpected for the climate. That is, subterranean water has caused probable shrubland-like habitat to be one with more trees than expected for the MAR and soils. If habitat mosaicism can be reconstructed in the past, it will give a better idea of the types of broad-based ecosystems in which hominins interacted.

Plant Productivity and Seasonality

In addition to influencing the habitat structure, climate and latitude have a bearing on both primary productivity and seasonality. Primary productivity is energy produced through photosynthesis and is represented by biomass. Habitat structure is somewhat predictive of plant productivity because high productivity is a result of warm temperatures and high moisture. Therefore, the least productive habitats in Africa are deserts due to lack of rainfall, while rain forests are most productive. High-primary productivity is indicative of high-species diversity of plants, and subsequent high net productivity, that is, the energy that is available for the animals that ingest the plants. Communities with high-primary productivity support more herbivores (Ritchie and Oloff, 1999; Janis, Damuth, and Theodor, 2002). Tropical rain forests have an average net primary productivity of 9,000 kilocalories per m^2 per year (k/m^2yr), while savanna ecosystems (woodlands, bushlands, etc.) have an average of 3,000 k/m^2 yr (Archibold, 1995). Swamps and marshes have equivalent productivity to rain forests.

Higher latitudes usually experience extremes in seasons that include not only precipitation but wide shifts in temperatures. In the tropical and subtropical belt of Africa, however, the differences in seasons are dependent on when and how much rain is received instead of fluctuations in temperature. Seasonality, therefore, is reflected in the presence of wet and dry seasons. Tropical rain forest habitats in west and central Africa that receive high MAR are less seasonal when compared with tropical woodland habitats receiving less MAR. In southern Africa today, there is a unimodal rainfall pattern resulting, in general, in six months of rain and six dry months. In eastern Africa, the rainfall distribution tends to be bimodal with two wet and two dry seasons per year, none of which lasts six months. In general, the more open and arid the habitat, the greater the number of dry-season months per annum.

Local and regional geomorphology and monsoonal patterns also play a part in the number of wet- and dry-season months experienced by any habitat.

The numbers of wet- and dry-season months as well as plant productivity are important to consider in reconstructing both the paleohabitats and diets of early hominins. Primates that live in tropical rain forests often must switch to different foods in the dry or lean season if their primary resource is unavailable (Wrangham, Conklin-Brittain, and Hunt, 1998). If the differences in the physiognomy of plant species are magnified in habitats that have longer dry seasons, such as woodlands and bushlands, then hominins, and indeed other primates, that existed in these habitats may have also needed to switch to different and possibly less-nourishing food resources for the lean season. Understanding productivity is important for identifying differences between extant and past ecosystems (Janis, Damuth, and Theodor, 2002). Therefore, habitat reconstruction and consideration of these other factors will provide crucial contextual information with which to analyze hypotheses regarding hominin diets.

Extant African Mammal Communities

Many classes of animals other than Mammalia exist in extant and extinct ecosystems. When these animals are recovered, they can also contribute to paleoecological understanding. However, many Plio-Pleistocene paleoecological techniques focus on the use of mammals because of both their fairly large body sizes and their utility in reconstructing diets and locomotor propensities. We limit our discussion here to mammals but recognize that other animals can also provide additional information regarding community structure. As described, various vegetation structures provide different foods and other resources for mammals that live within them. Mammals that share resources in the same vegetation structure (i.e., the mammal community), have partitioned these resources such that their ecological adaptations to substrate and diet are fairly predictable based on the structure of the habitat in which they live (Reed, 1997, 1998, 2002, 2005). For example, the mammals in forests divide vertical (i.e., substrate) niches as well as dietary ones and therefore the community exhibits higher proportions of arboreal substrate use and frugivory than communities that live in any other habitat. Grasslands have no arboreal and few mammals that eat fruit but have high proportions of mammals that consume grass. The range of habitats, including closed woodlands and shrublands, are not as simply typified by specific mammalian adaptations, although they do have identifiable characteristics (Reed, 1997, 1998, 2002). Woodlands, for example, can be identified by percentages of arboreal mammals such that closed woodlands have greater proportions of these animals than do open woodlands. Proportions of frugivorous mammals in bushlands are often greater than in woodlands (Reed, 1997). Various types of mosaic habitats also have slight differences that allow identification through mammalian adaptations. For example, gallery forests within grasslands cause higher proportions of both arboreal and grazing mammals than expected in either woodland or bushland habitats. Sites with extensive lakes and/or river systems have higher proportions of aquatic animals than sites that do not.

The Unknown: Hominin Habitats

Hominin habitats and their ecology are, of course, unknown. Because we use modern analogs to compare with fragments of the past there is only the probability that the paleoecology of early hominins will be known. We review habitat reconstructions for late Miocene through early Pleistocene hominin fossil localities and address some anomalies that we have noticed between modern mammal communities and mammal communities based on fossil collections from various areas.

Materials and Methods

As we briefly discussed, climates and soils cause vegetation structure, which then supports various mammal communities. In habitat reconstruction, we must move backward from individual taxa, to the mammal communities, to the habitat, to the climate. Because Pliocene MAR, temperature, or evapotranspiration cannot be measured, a variety of methods have been developed for reconstructing past environments apart from analyses focusing on mammals. If plant fossils and/or pollen were abundant in each hominin locality, these would be ideal first-order references for reconstructing habitats. Piecing together remnants of the actual species that made up the vegetation would be one of the best methods of analysis. Unfortunately, hominin fossil localities often provide little direct evidence of the plants that previously existed. Studies of fossil leaves and pollen have supplied vegetation information when these items have been recovered (Cadman and Rayner, 1989, Retallack, 1992; Bonnefille et al., 2004; Jacobs and Herendeen, 2004), but some of these studies are not without problems. A second-order analysis (i.e., one removed from the actual plant species) is of soil carbonates for isotopic fractionation, which can give rough percentages of the types of plants that shed debris into and contributed to the formation of the paleosol (Kingston, Marino, and Hill 1994; Quade and Cerling, 1995; Wynn and Retallack, 2001; Retallack et al., 2002; Cerling, Wang, and Quade, 2003). The results of these studies can give percentages of C_4 plants (monocots, i.e., grasses) versus the percentages of C_3 plants (dicots, i.e., trees, bushes, and shrubs). However, one may not always be lucky to find paleosols; there are none in South African caves. Therefore, the most widely used techniques to reconstruct environments involve fossil mammals because they are found in great abundance at hominin fossil localities. Mammals have the potential to provide the best evidence of ancient habitats if we understand the interrelationship between animals and the plant structure in which they exist because of mammalian adaptations to a wide variety of foods and locomotor behaviors. In addition, understanding the mammalian community can shed light on possible ecological interactions faced by early hominins.

The study of a single taxon or taxonomic group and how their functional morphology is related to their ecology and thus, by extension, their habitat, is one method that has been used to suggest past habitats (Robinson, 1963; Vrba, 1974, 1975, 1980, 1988; Kay, 1975, 1978; Grine, 1981; Stern and Susman, 1983; Kappelman, 1988; Benefit and McCrossin, 1990; Ciochon, 1993; Spencer, 1995; Lewis, 1997; Elton, 2001, 2002). Other methods include using species diversity indices (e.g., Rose, 1981; Avery, 1982), faunal resemblance indices (e.g., Van Couvering and Van Couvering, 1976), analyses of abundances of various taxa through time by associating the abundances of various taxa with habitat preference of extant relatives (Bobe and Eck, 2001; Alemseged, 2003), and ecological structure analysis (e.g., Andrews, Lord, and Nesbitt-Evans, 1979; Andrews, 1989), wherein the adaptations of the mammals are analyzed instead of particular species.

For illustrative purposes, we use ecological diversity of the large mammal communities to reconstruct habitats of the fossil localities. All figures are based on comparisons of forty-three extant mammal communities from African habitats ranging from forests through grasslands (table 14.1). Each extant site is assigned a broad-based habitat structure based on observation or written accounts of the area (Reed, 1997, 1998). However, most of these localities are mosaic in nature and assignments to one overall habitat type are problematical. We try to associate ancient habitats with specific extant sites through the use of correspondence analysis irrespective of the assigned habitat type.

Mammal species from each extant site were assigned substrate and trophic adaptations (table 14.2) based on observations or examinations of stomach contents previously reported (Dorst and Dandelot, 1969; Hoffman and Stewart, 1972; Kruuk, 1972; Kingdon, 1974a; 1974b; 1977, 1979, 1982a, 1982b; Smithers, 1978; Delaney and Happold, 1979; Happold, 1989; Skinner and Smithers, 1990; Estes, 1991; Kitchener, 1991). Proportions of each adaptation within each community were then calculated for each extant site. Three different data sets were developed to simulate taphonomic issues that have likely affected fossil collections. First, none of the data sets rely on abundances of mammals, thereby minimizing confounding factors that may inflate individual species abundances such as a specific accumulating agent. Second, several taphonomic processes often create a bias against medium and small mammals in fossil collections. Therefore, one extant data set includes species from each extant locality that are larger than 500 g (fig. 14.1a; data used previously in Reed, 1997, 1998, 2002); a second extant data set uses species that are larger than 4 kg (fig. 14.1b). Finally, most African fossil localities are depauperate in carnivores and other rarer species such as those found in the orders Pholidota, Tubulidentata, Lagomorpha, Rodentia, Insectivora, and Hyracoidea. We removed these species from the final extant data set as well. Thus, the final data set of extant mammals from each locality reflects the orders that are recovered most often from fossil hominin localities: Artiodactyla, Perissodactyla, Proboscidea, and Primates. For this data set, we used only trophic adaptations and two taxonomic classifications (fig. 14.2). The discrimination power to identify habitat type of each data set diminishes with removal of data, and the only clear separation in the extant localities with the final data set is the distinction between rain forests and grasslands. However, the pattern holds with all the data sets and the positions of the fossil localities also tend to follow the same distributions.

For many of the South African fossil localities and Hadar, we measured the dentitions and postcrania (when available) of the mammals to identify species trophic and substrate adaptations through functional comparisons with extant taxa (i.e. primates, carnivores, etc.). Most ungulates are considered to use a terrestrial substrate except for hippos, which we consider aquatic. For other localities mentioned in East Africa, we rely on published information regarding trophic adaptations (e.g., Harris and Leakey, 2003a) and taxonomic uniformitarianism to assign adaptations to fossil

Table 14.1 Extant African localities

Locality and Abbreviation	Average Annual Rainfall	Biome	Number of Species > 500 g
Aberdares NP (Ab)	750	Montane heath and forest	31
Amboseli NP (Amb)	510	Shrubland/bushland	40
Bicaur National Park (Bi)	850	Woodland	46
Chobe NP (Cho)	650	Scrub woodland	47
Congo Basin (Co)	1,832	Rain forest	36
East of Cross (Ec)	1,550	Rain forest	24
East of Niger (En)	1,596	Rain forest	19
Gemsbok NP (Gem)	140	Shrubland	27
Golden Gate NP (Gg)	450	Grassland/woodland	20
Guinea Woodland (Gw)	1,000	Closed woodland	43
Hadar (Had)	250	Shrubland/desert	34
Hluhluwe NP (Hnp)	750	Woodland/floodplains	43
Kafue Flats (Kf)	821	Edaphic grasslands (floodplain)	30
Kafue NP (Kfn)	821	Edaphic grassland (Floodplain)/woodland	42
Kapama NP (Kap)	400	Woodland	40
Karoo (Kar)	350	Shrubland	41
Kidepo NP (Kid)	625	Bushland/grassland	39
Kilimanjaro (Kil)	1,050	Montane forest	16
Kruger NP (Oli)	675	Woodland/bushland	49
Lake Mweru (Lmw)	750	Bushland/woodland	29
Lake Nkuru (Lnk)	1,000	Forest/grassland	33
Linyanti Swamp (Lin)	650	Marsh/woodland	29
Liwonde NP (Liw)	750	Edaphic grassland (Floodplain)/woodland	29
Makakou (Mak)	1,800	Rain forest	28
Masai Mara (Mm)	1,000	Forest/grassland	55
Namib Desert (Nam)	<150	Desert	18
Natal Woodland (Nat)	875	Closed woodland	26
Nyika NP (Ny)	1,200	Forest/grassland	29
Okavango Delta (Ov)	600	Shrubland/woodland/edaphic grassland	44
Rukwa Valley (Ru)	700	Bushland/woodland	44
Rwenzori NP (Rw)	900	Closed woodland	37
Sahel Savanna (Sahel)	450	Shrubland/grasses	23
Serengeti Bushland (Sb)	803	Bushland	52
Serengeti NP (Sn)	750	Bushland/woodland/grassland	58
Serengeti Plains (Sp)	500	Grassland	15
SS Grasslands (Sg)	500	Grassland	26
SS Woodland (Sw)	650	Woodland/bushland/grassland	41
Sudan Woodland (Sud)	689	Woodland	37
Tarangire NP (Ta)	600	Scrub woodland	39
Tongwe NP (To)	1,012	Forest/woodland	29
Tsavo NP (Ts)	500	Bushland/woodland	49
West Lunga NP (Wl)	875	Bushland/woodland	34
West of Niger (Wn)	1,600	Rain forest	19

Table 14.2 Adaptations for Extant and Fossil Mammals

Adaptation and Abbreviation	Description	Example
Substrate		
Arboreal (A)		*Colobus guereza,* black and white colobus
Aquatic (AQ)		*Hippopotamus amphibious,* hippopotamus
Fossorial (F)		*Mellivora capensis,* honey badger
Terrestrial/arboreal (TA)		*Panthera pardus,* leopard
Terrestrial (T)		*Connochaetes taurinus,* wildebeest
Trophic		
Browser (B)	Dicot bushes, trees	*Tragelaphus strepsicerus,* kudu
Meat eater (C)	Flesh	*Panthera leo,* lion
Meat/bone eater (CB)	Flesh, bone	*Crocuta crocuta,* spotted hyena
Meat/invertebrates (CI)	Flesh, insects	*Genetta genetta,* small spotted genet
Fresh grazer (FG)	Floodplain/wetland grasses	*Kobus vardoni,* puku
Frugivore (FL)	Fruit with leaves or insects	*Chlorocebus aethiops,* vervet
Grazer (G)	Grass	*Alcelaphus bucelaphus,* hartebeest
Insectivore	Insects	*Orycteropus afer,* aardvark
Browser/grazer (MF)	Mixed grass and leaves	*Gazella granti,* Grant's gazelle
Omnivore (OM)	No preference	*Mellivora capensis,* honey badger
Roots, tubers (R)	USOs	*Hystrix africaustralis,* porcupine
Reduced Data Set		
Suidae (S)		
Primates (P)		

species. Taxonomic uniformitarianism means that fossil species are expected to have the same adaptations as closely related species in extant families or tribes.

There is one anomaly regarding extant trophic assignments. Some suids such as the bush pig (*Potamochoerus*) are considered omnivorous as they eat roots and tubers, fruits, carrion, and a variety of other foods. The warthog (*Phacochoerus*) is considered a grazer. However, Janis, Damuth, and Theodor (2002) consider all pigs omnivorous based on dental morphology and behavioral ecology suggests grass is only part of the *Phacochoerus* diet (Kingdon, 1979). We have assigned the trophic category of "grazing" to *Phacochoerus* but acknowledge that research on feeding behavior and comparative anatomy of suid dentition would be valuable, especially because of the number of fossil suids in the African Pliocene.

Climatic Patterns of the Miocene through the Pleistocene

Short-term climate changes are seasonal and in any time period are usually somewhat predictable on a year-to-year basis. Long-term climate change is caused by a variety of factors, but in Africa over the last 10 million years the major influence has been glacial cycling. The cooling caused by this cycling has caused overall drier and more open habitats to dominate in Africa today, contrasting the forests and decidu-

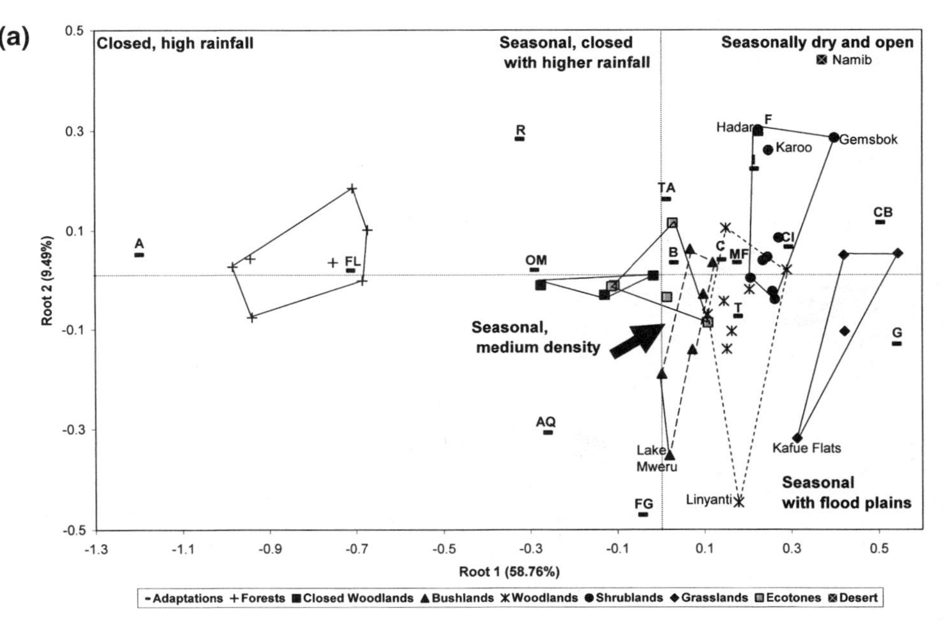

Figure 14.1 (*a*) Correspondence analysis of extant localities using mammals >500 g. The CA was performed on percentages of adaptations in each community. The CA accounts for 69% of the variation among the habitats. The sites are distributed from closed, wet (forests) on the left through open grassland habitats on the right. The y-axis separates dry (*top*) and wet (*bottom*) habitats based on lacustrine and floodplain systems. Selected sites labeled (see table 14.1 for more information on locality); adaptation abbreviations as in table 14.2. (*b*) Correspondence analysis of extant localities using mammals >4 kg. The CA was performed on percentages of adaptations in each community. The CA accounts for 67% of the variation among the habitats, but there is more variation on the y-axis. The sites are distributed from closed, wet (forests) on the left through open grassland habitats on the right. The y-axis separates dry (*top*) and wet (*bottom*) habitats based on lacustrine and floodplain systems. The medium density habitats are less well separated with this data set, although the closed woodlands and ecotones are better separated. Selected sites labeled (see table 14.1 for more information on locality); adaptation abbreviations as in table 14.2.

(continued)

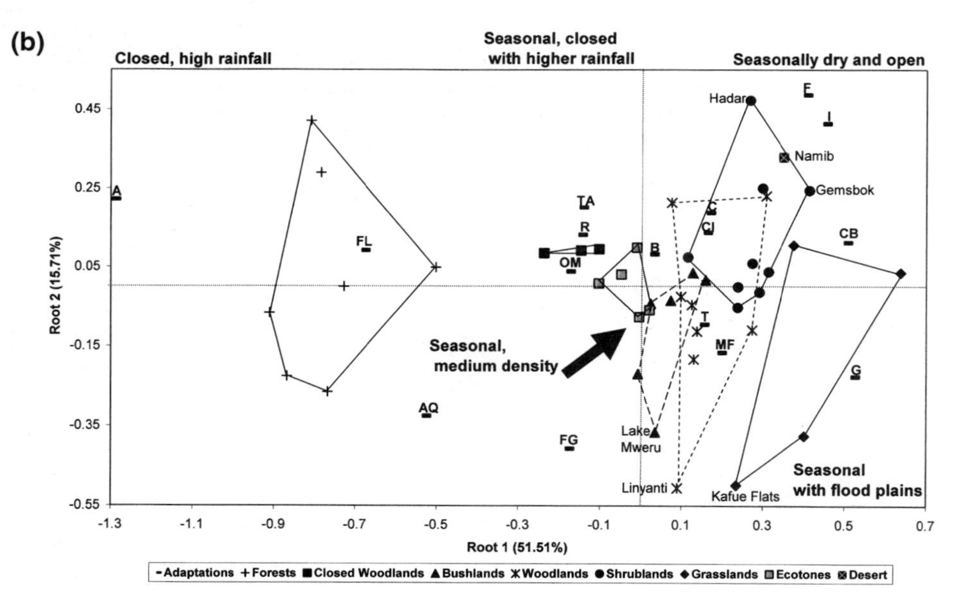

Figure 14.1 *(Continued)*

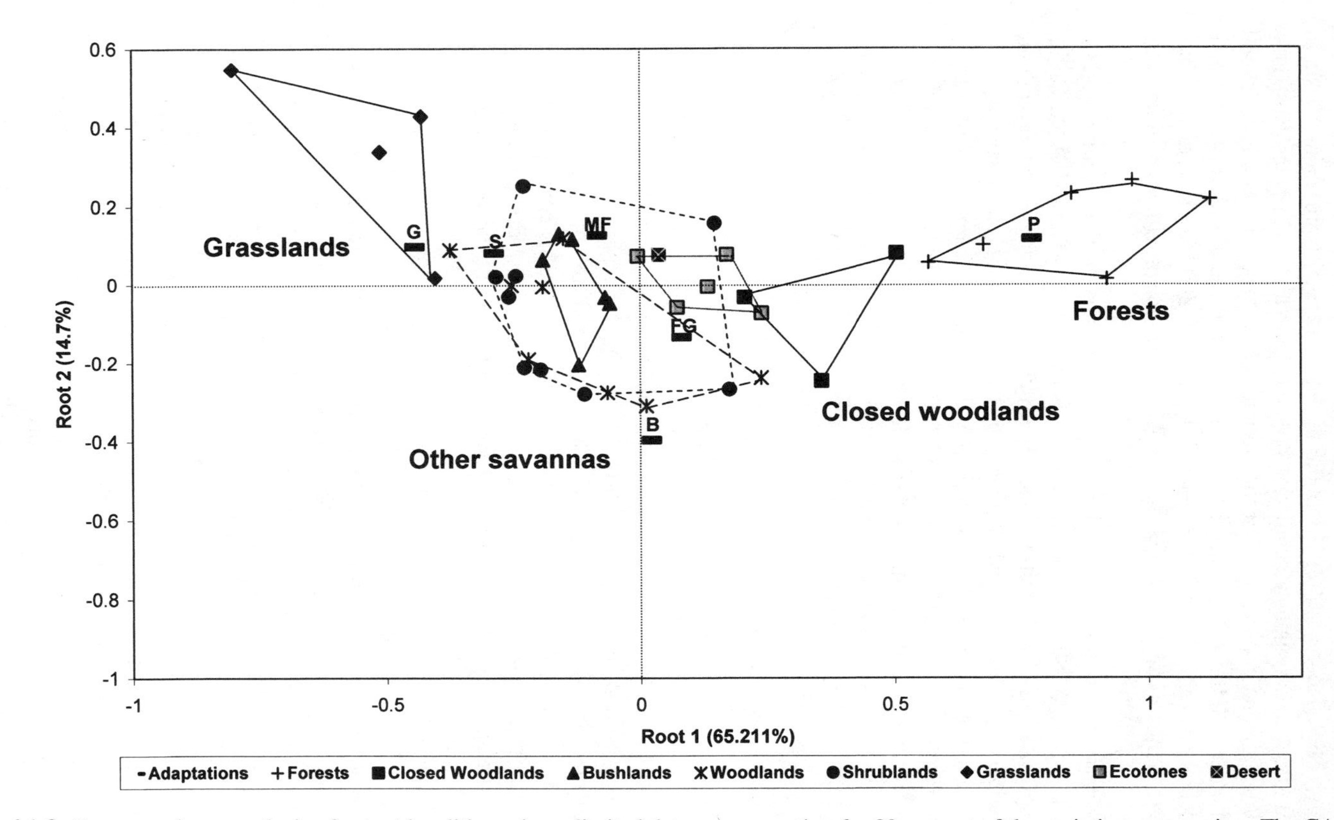

Figure 14.2 Correspondence analysis of extant localities using a limited data set accounting for 89 percent of the variation among sites. The CA was performed on raw numbers of browsers (B), grazers (G), mixed feeders (MF), fresh grass grazers (FG), suids (S), and primates (P). Grasslands, forests, and closed woodlands are identified, but the woodlands, bushlands and shrublands overlap.

ous woodlands that are reconstructed for the early to middle Miocene (Crowley and North, 1991). These forested and woodland habitats of the Miocene indicate higher MAR than seen today in the same regions of Africa (Foley, 1987; Andrews and Humphrey, 1999; Jacobs, 2002). After the middle Miocene, from about 10 Ma to about 7.0 Ma, the continuous glaciation at both the north and south poles caused the first aridification and cooling trend of the climate across Africa. However, at the end of the Miocene (~6.0–5.4 Ma) this drying trend was reversed as only the south pole maintained an ice cap. The early Pliocene (5.4 Ma until ~3.0 Ma) was consequently characterized by a wetter, warmer climate in Africa than the previous 5 million years (Marlow et al., 2000).

Another glacial cooling cycle evinced by low sea surface temperatures (SSTs) began at about 3.2 Ma and continued through about 2.2 Ma (Marlow et al., 2000). Accordingly, terrestrial temperatures and MAR must have been reduced through the same time period. deMenocal and Bloemendal (1995) note increased levels of eolian dust in sea cores at circa 2.8 Ma, which indicate terrestrial xerification. SSTs recorded from 2.1 to 1.9 Ma suggest fairly swift cooling of both sea and land temperatures, resulting in further terrestrial aridity by 1.8 Ma (Marlow et al., 2000). Each of these events is associated with discernible vegetation fluctuations and resulted in likely dietary changes of early hominins.

Hominin Habitats

Before the drying trend of the late middle Miocene, African habitats were characterized as more forested than any habitat associated with hominins of the Pliocene. Work by Jacobs (1999a, 1999b, 2002) on fossil plant leaves shows changing seasonality across 6 million years at Tugen Hills. At this site, the length of the dry season changed from about 0 to 4 months at 12.8 Ma to about 3 to 7 months at 9 Ma and then reverted to the original 0 to 4 months at circa 5.5 Ma. This corresponds to the climatic aridification patterns iterated earlier and indicates that dry seasons in parts of eastern Africa were likely less than four months long at the beginning of the Pliocene.

The climatic data would also suggest that the habitat of Toros-Menalla in Chad where *Sahelanthropus* has been recovered (Brunet et al., 2002) was more arid and perhaps open than the later Pliocene habitats. However, other fauna recovered with this taxon indicate the presence of water, which may have locally altered the habitat to be wetter and more closed (Vignaud et al., 2002). The earliest east African hominins (*Ardipithecus* spp.) from sites dating from 5.8 Ma to 4.3 Ma are suggested to have lived in fairly closed woodlands or possibly forested environments based on the mammals found in the same strata with the hominin fossils (WoldeGabriel et al., 1994; Haile-Selassie, 2001; WoldeGabriel et al., 2001). The Gona *Ardipithecus* sites, however, appear to be more varied and include some wooded grasslands (Semaw et al., 2005). We do not include these sites in our analyses because the fossil species found are far enough removed from extant groups that trophic assignments based on uniformitarianism might be problematic.

Hominins recovered from sites dated from about 4.2 Ma to 3.0 Ma include *Australopithecus anamensis, Australopithecus afarensis*, *Kenyanthropus platyops*, and possibly *Australopithecus africanus*. These hominins lived in regions that were of-

Table 14.3 Fossil Localities Used in Analyses

Formation	Member or Site and Abbreviation	Date in Ma from Base of Member	Reconstructed Habitat
Hadar	Danauli (DN)	2[a]	Edaphic grassland
	Makaamitalu (MA)	2.33	Edaphic grassland
	Kada Hadar 2 (K2)	3[a]	Woodland
	Kada Hadar 1 (K1)	3.18	Shrubland/edaphic grassland
	Denan Dora 3 (D3)	3.19[a]	Woodland
	Denan Dora 2 (D2)	3.2[a]	Woodland/edaphic grassland
	Denan Dora 1 (D1)	3.22	Woodland/lacustrine
	Sidi Hakoma 4 (S4)	3.25[a]	Closed woodland
	Sidi Hakoma 3 (S3)	3.3[a]	Woodland (Liwonde)
	Sidi Hakoma 2 (S2)	3.35[a]	Bushland (Tsavo)
	Sidi Hakoma 1 (S1)	3.4	Closed woodland (Ruwenzori)
	Hadar Basal (BA)	3.45	Woodland
Koobi Fora	Okote (OK)	1.6	Shrubland (Kapama)
	KBS (KBS)	1.88	Edaphic grassland (Okavango)
	Upper Burgi (UB)	2	Woodland (Chobe)
	Tulu Bor (TB)	3.4	Woodland (Hluhluwe)
	Lokochot (LKC)	3.6	Bushland
	Kanapoi (KAN)	4	Woodland
	Moiti (MO)	4.1	
	Kataboi (KBOI)	4.1	?Bushland
	Allia Bay (AB)	4.2	
Shungura	Shungura G (SG)	2.33	?Bushland
	Shungura F (SF)	2.36	?Woodland
	Shungura E (SE)	2.4	?Bushland/ecotone
	Shungura D (SD)	2.52	?Bushland/ecotone
	Shungura C (SC)	2.85	?Woodland/shrubland
	Shungura B (SB)	2.95	?Bushland/ecotone
South Africa	Swartkrans 3 (SK3)	1.6[a]	Grassland
	Swartkrans 2 (SK2)	1.7[a]	?Shrubland
	Swartkrans 1 (SK1)	1.8[a]	?Shrubland
	Sterkfontein 4 (ST4)	2.7[a]	?Shrubland
	Makapansgat 3 (M3)	3[a]	Bushland (Tsavo)
West Turkana	Nariokotome (NR)	1.3	Shrubland (Tarangire)
	Natoo (NT)	1.6	Grassland
	Kaito (KAI)	1.88	Grassland
	Kalachoro (KC)	2.35	Grassland/shrubland
	WT17000 (W17)	2.5	Woodland (Sudan savanna)
	Lokalalei (LOK)	2.52	Closed woodland/ecotone
	Upper Lomwkwi (UL)		Ecotone
	Middle Lomekwi (ML)		Ecotone
	Lower Lomekwi (LL)	3.4	Closed woodland
Olduvai	Olduvai IV	1[a]	Grassland
	Olduvai III	1.15	Grassland
	Olduvai II	1.17	Grassland
	Olduvai I	2.1	Grassland
Laetoli	L1	3.6	Woodland
	L7	3.6	?Woodland

[a]Dates are estimated; reconstructions based on limited data set.

ten mosaic and included, in various proportions, closed to open woodlands, bushlands, riverine forests, and seasonal floodplains (see table 14.3; Andrews, 1989; White et al., 1993; Bonnefille, 1995; Spencer, 1997; Reed 1997, 1998, 2002, 2005; Wynn, 2000; Bobe and Eck, 2001; Leakey et al., 2001; Alemseged, 2003; Harris and Leakey, 2003a). However, some localities such as Allia Bay (*A. anamensis*) and the Tulu Bor Member at Koobi Fora (*A. afarensis*) are suggested to have been fairly dry and open (Reed, 1997; Wynn, 2000). The *A. africanus* sites (Makapansgat and Sterkfontein) are also mosaic, but the Sterkfontein 4 deposit, from which the majority of *A. africanus* specimens come, reconstructs as drier than the other localities (fig. 14.3b).

As the climate in the earliest Pliocene was comparatively wet and warm, extended rainy seasons that supplied preferred food items to early hominins for long periods of each year are expected. Evidence for short numbers of dry-season months is extrapolated both from the habitat type and from other evidence. Nonmammalian evidence includes isotopic analyses of gastropods from the Sidi Hakoma Member of the Hadar Formation (3.4–3.22 Ma) that reveal dry seasons of approximately three months (Hailemichael, 1999). In addition, various vertisols at Koobi Fora and at Hadar also reveal slickensided clays that have had at least a three-month period without water in which to form (Wynn and Feibel, 1995; C. Campisano personal communication, 2005). It is possible that southern Africa had longer dry seasons than East Africa during the mid-Pliocene as there are six-month dry seasons today that differ from the current bimodal distribution of rainfall in East Africa.

Australopithecus afarensis, which has the most data on varying habitats at different sites (Laetoli, Tulu Bor, Maka, Hadar) and through time (Hadar), seemed to have existed in fairly different habitats across space and through time (table 14.3; figs. 14.3a and 14.4). In addition, the mammal community at Laetoli is quite different from any of those at Hadar because there have been no water-adapted species recovered from the Tanzanian site. The fact that *A. afarensis* existed in various habitats is interesting with regard to diet as the species was obviously not stenotopic and its diet must have been available in a variety of environments.

Sometime between 3.0 and 2.5 Ma two distinct lineages of hominins appeared, *Paranthropus* and *Homo*. Habitats such as open woodlands, bushlands, and shrublands with wetlands have been reconstructed for these hominins before circa 2.0 Ma (Reed, 1997; Spencer, 1997; Bobe and Eck, 2001). There appears to be a general trend toward more open habitats in this time period (fig. 14.4), although this depends on particular regions in which fossils are recovered. The Koobi Fora fossil localities appear drier overall than the West Turkana sites on the other side of the lake. The Shungura Formation has some interesting differences from either of the other Turkana Basin sites as the Shungura communities have high numbers of browsing bovid species and few suids. Olduvai Bed I from this time period appears extremely open and arid in our analysis, as do the Swartkrans sites from South Africa. However, although these two regions appear to be open and in the grassland realm, they are distant from each other on the graph, which indicates differences between the habitats.

Seasonality in the 3.0–2.0-Ma time period seems to be more in line with current seasonal regimes in eastern Africa. Whether this was unimodal or bimodal is probably an unknowable. However, these habitats had large river and lake systems that

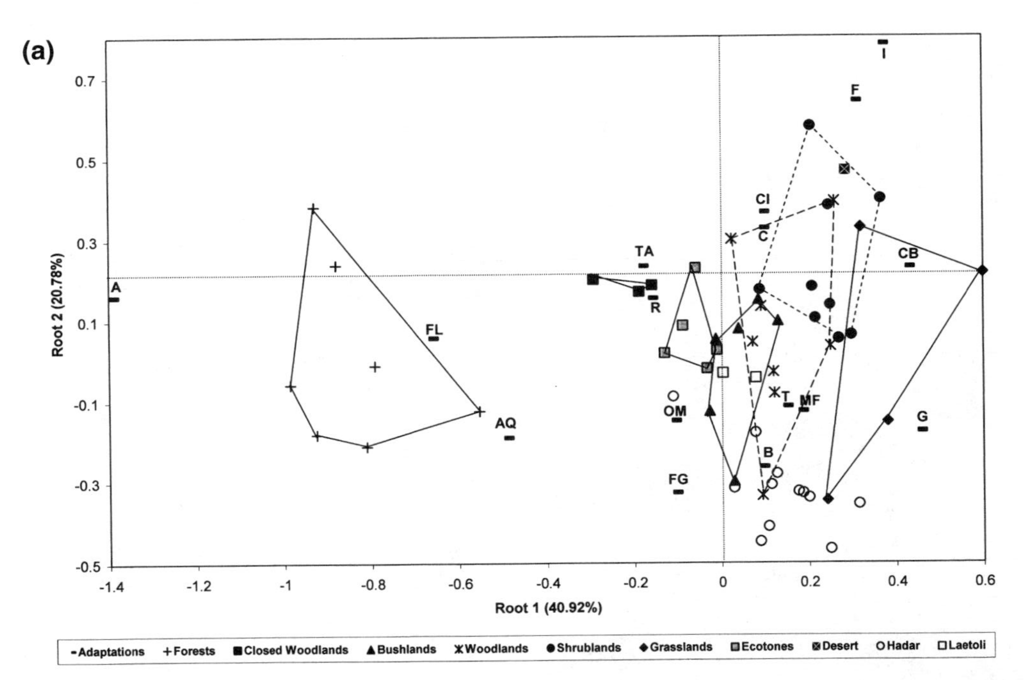

Figure 14.3 (*a*) Correspondence analysis of extant localities of mammals >4 kg with the fossil sites of Hadar and Laetoli. The CA accounts for 62% of the variation among the habitats, but only a few of the fossil members are positioned with extant sites. Compare the relationships of the adaptations in this figure to figure 14.1b as the browsing adaptation is much lower on the y-axis. The y-axis also accounts for much more of the variation among the sites. (*b*) Correspondence analysis of extant localities of mammals >4 kg with the fossil sites of South Africa. The CA accounts for 60% of the variation among the habitats, but only a few of the fossil members are positioned with extant sites. Compare the relationships of the adaptations in this figure to figure 14.1b as the browsing adaptation is much lower on the y-axis. The y-axis also accounts for much more of the variation among the sites.

(*continued*)

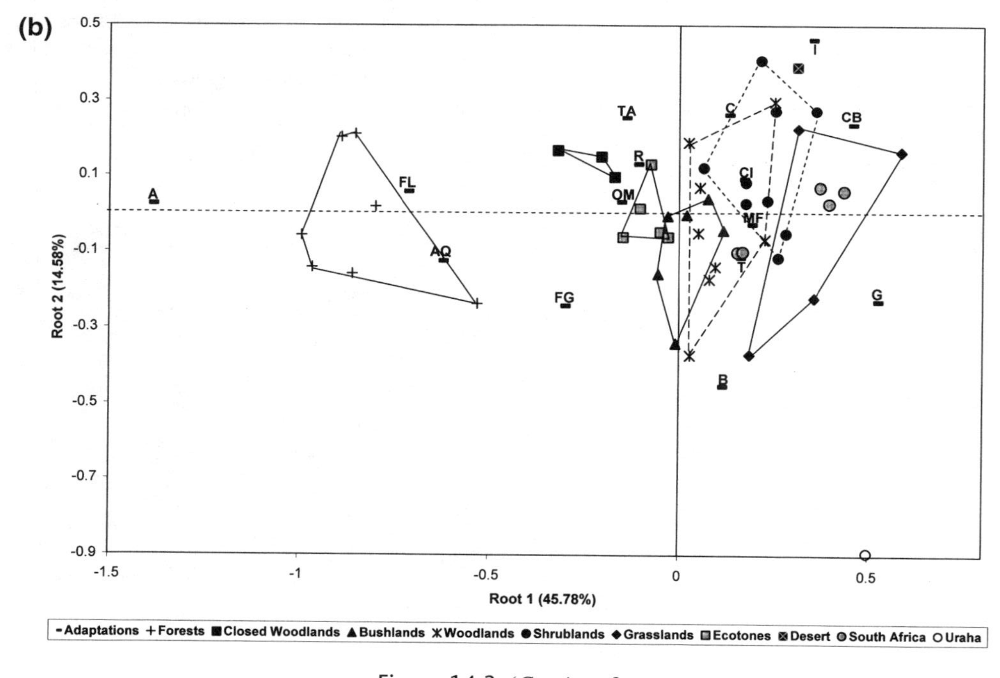

Figure 14.3 (*Continued*)

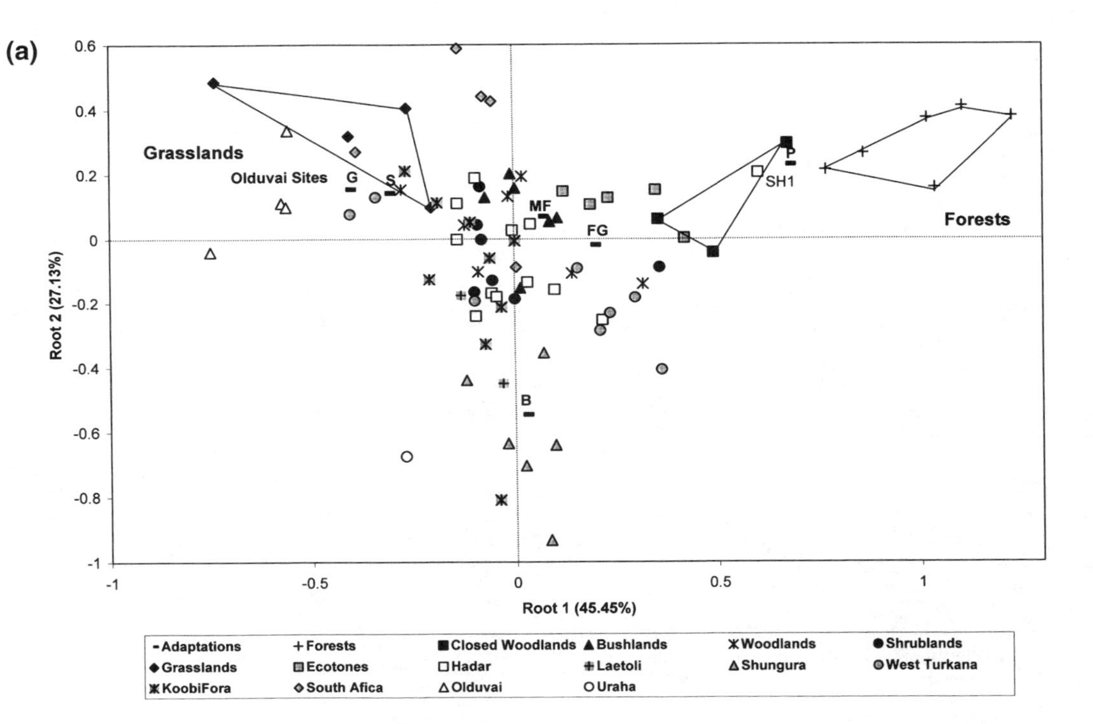

Figure 14.4 (*a*) Correspondence analysis of extant and all fossil localities using the limited data set. The CA accounts for 72% of the variation among the sites. The CA was performed on raw numbers of browsers, grazers, mixed feeders, fresh grass grazers, suids, and primates (abbreviations as in fig. 14.2). The fossil localities tend to range to the lower part of the graph on the y-axis, which accounts for 27% of the variation among the sites. Abbreviations for localities can be found in tables 14.1 and 14.3. (*b*) Correspondence analysis of extant and all fossil localities using the limited data set with the midsection expanded in order to see the localities better. The CA was performed on raw numbers of browsers, grazers, mixed feeders, fresh grass grazers, suids, and primates (abbreviations as in fig. 14.2). Abbreviations for localities can be found in tables 14.1 and 14.3.

(continued)

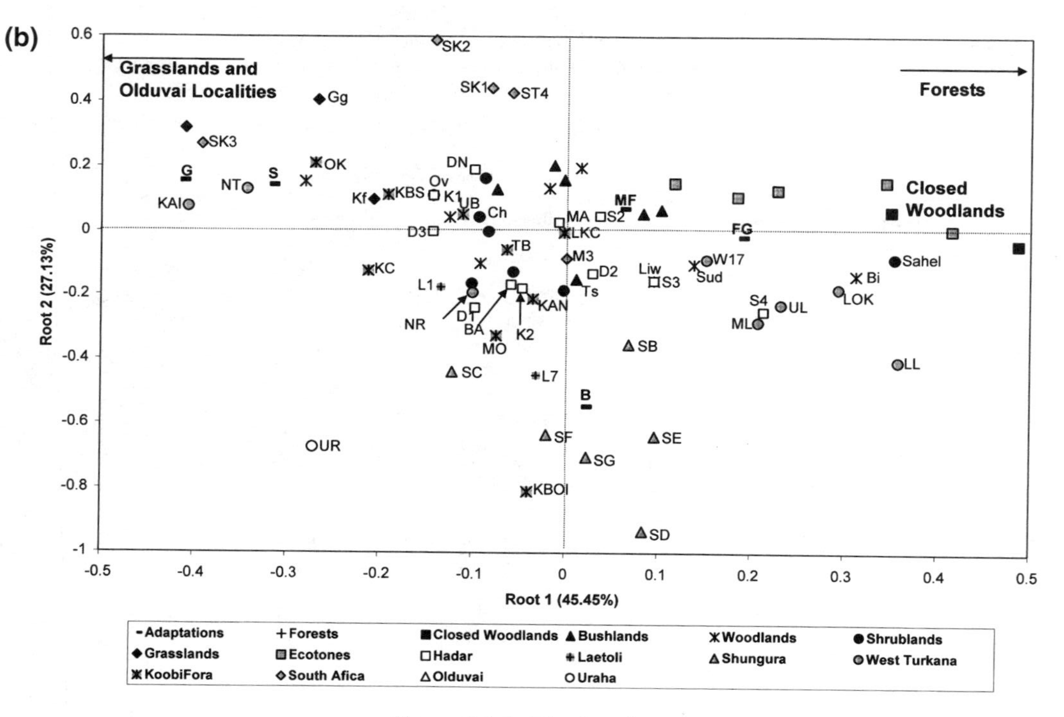

Figure 14.4 (*Continued*)

would have supplied water in dry seasons, perhaps to a greater extent than what we see today.

After about 2.0 Ma, more coverage of grasslands is indicated by carbon isotopes in soils and increased numbers of grazing mammals at about 1.8 Ma (Cerling, 1992; Reed, 1997; Spencer, 1997; Marlow et al., 2000). More open, arid, and thus seasonal environments typified the existence of hominins recovered from African sites from circa 2.0 to circa 1.0 Ma. Grasslands did not come to stay, however, and Cerling (1992) has shown two extensive grassland peaks at about 1.7 Ma and 1.2 Ma with subsequent returns to bushlands and woodlands. Sites in this time range from all of the localities are in the open and dry range of the graph (fig. 14.4), indicating a fairly broad change in climate and habitat. This is the climate in which *Homo erectus* appears and in which *Paranthropus* continued to exist.

The trend from the middle Miocene to the late Pleistocene was thus from wet, warm climates in east and southern Africa to decreased rainfall and increased seasonality caused by glaciation of one or both poles (Marlow et al., 2000). However, many differences have been noted in the rates of species turnover and habitat change at different localities, indicating that the trend was not uniform across time and space (fig. 14.4; Behrensmeyer et al., 1997; Bobe and Eck, 2001; Alemseged, 2003; Reed, Lockwood, and Arrowsmith, 2003).

In addition, there are several differences that bear mentioning between fossil and extant localities that may influence the results of these analyses. These may be caused by taphonomic problems, such as time averaging of fossil collections, or there may be disparities that represent different types of habitats that supported slightly different resource focuses for the animals that live in them. We discuss the possibilities in the next section.

The Unknowable

We like to think that at least some aspects of the following scenarios will eventually be placed in an unknown but "reconstructable" category. First, however, there are things that will never be completely known, such as all of the individual plant and mammal species in each fossil community. We will likely also never know the exact amount of MAR or number of wet- and dry-season months. That said, what are the differences between mammal communities in the past and the present? What can be said about associating habitats with hominin diets? Is there a relationship between habitat, diets, and patterns of speciation and extinction of these hominins?

Mammal Community Anomalies as Evidence for Habitat Differences

Suidae

Many of the fossil localities have more suid species than are present in any extant site. Irrespective of whether one classifies them trophically as grazers or omnivores, they cause the fossil sites to have different mammalian community structures than any extant locality. This suggests that the habitats might have been different in some way from any that exist today, and/or there was more plant productivity. As previously mentioned, *Phacochoerus* does ingest grasses and this likely makes up the majority of its diet. But *Phacochoerus* also has the ability to dig up rhizomes and roots with its hard, sharp-edged rostrum and focuses on these resources in the dry season (Kingdon, 1979).

Localities of the African Pliocene tend to have at least two and usually more fossil pigs while extant habitats usually only support two (*Phacochoerus* spp. and *Potamochoerus* spp.). Occasionally, *Hylochoerus*, the giant forest hog, will also be found in the vicinity of these other suids, but only in highly variable environments (e.g., ecotones). Fossil suids have been shown to have eaten C_4 plants (Harris and Leakey, 2003b) even if their dentition did not have the shearing crests that are hallmarks of grass and leaf diets (Janis, Damuth, and Theodor, 2002).

There are three possibilities for the meaning of finding many suids at hominin localities in the fossil record. First is the possibility that the suid lineages underwent an adaptive radiation between 4.5 and 1.4 Ma, and as such, there were an abundance of similar species but the competitive process had not yet eliminated the less well adapted. White and Suwa (2004) note that it is likely that *Notochoerus*, at least, had finely partitioned its dietary niches to allow two species at least to exist at the same time. Second, it is possible that the fossil deposits from which the highest numbers of pigs are recovered are time averaged and that the species actually did not overlap in time but appear to do so. Third, the habitats in which these pigs lived had different resources than habitats today. This is intriguing because it is difficult to imagine a habitat that had a different structure (i.e., that would show a different pattern in the fossil record). However, various processes may produce a similar pattern, and it may be that these ancient habitats are achieved through climatic factors that are not in operation today.

Browsers

Second, the earliest fossil localities in particular have higher percentages and numbers of browsing ungulates in their communities than many of the extant sites (fig. 14.4). An increase in the number of browsing species supported by a habitat may indicate increased plant productivity (Janis, Damuth, and Theodor, 2002). Climatological evidence may indicate higher productivity as well. According to Marlow and colleagues (2000), glaciation instigated further cooling and aridification beginning at about 3.2 Ma. Pearson and Palmer (2000) report one of the few discernible reductions of pCO_2 (partial pressure of carbon dioxide with respect to oceanic carbon content) in the Neogene at approximately 3.0 Ma. Plants are likely more productive under higher pressures of pCO_2 (Janis, Damuth, and Theodor, 2002). Both of these factors suggest that plant productivity globally declined around 3.2–3.0 Ma. In ancient Miocene habitats of North America, higher pCO_2 levels have been suggested to be responsible for increased plant productivity and thus high numbers of browsing species (Janis, Damuth, and Theodor, 2002). If these factors were at work in the Pliocene of circa 3.0 Ma, then we would expect the habitats to produce greater numbers of browsing mammals before about 3.0 Ma and fewer sometime later. In fact, numbers of browsing mammals appear to decrease at about 1.88 Ma, about the time of one of the grassland expansions described by Cerling (1992).

A second possibility for increased numbers of browsing mammals involves limited dry-season months. No African savanna habitat today has dry-season months limited to three or four per year such as that suggested for Hadar at 3.4 Ma (Hailemichael, 1999). Rainfall spread over longer wet seasons may cause plant species to increase leaf production at the expense of fruit production (Foley, 1987), thus resulting in more resources for browsing or leaf-eating mammals.

The importance of increased browsers and suids is unknown and may be unknowable. However, we propose that these differences between fossil and extant mammal communities, and thus habitats, should be further explored with regard to seasonality, productivity, and climatic differences that are possibly unknown today. As with any study of the past, we are limited to looking at the patterns the fossil record provides and then testing and conjecturing about the processes that may have caused the patterns. Ecological patterns may be caused by more than one process. These explorations may result in information useful to reconstructing hominin diets.

Hominin Habitats and Diet

Finally, can we relate these reconstructed habitats with hominin diets? The earliest hominins appear to exist in all types of savanna habitats from closed woodlands through shrublands. The plants in extant habitats of this type have a propensity to be covered in more dust and grit than are plants in extant forests and likely the early Pliocene sites had more dust and grit adhering to the plants than did mid-Miocene plants in forested habitats. Lack of rainfall in savanna habitats allows more dust to be produced in the dry season in comparison to rain forests. Although ungulates that eat grasses have more hypsodont teeth than browsers, Janis (1988) attributes the highest hypsodonty in various feeding categories to foraging closer to the ground.

Most early Pliocene hominins had similar dental morphology to one another although some, such as *Ardipithecus*, lacked characteristics shared by the other species. Primates, not being ungulates, likely had a different dental response to savanna dust. Teaford and Ungar (2000) review research that shows food items in *Australopithecus* spp. diets probably included malleable fruits, that is, smaller fruits that did not require incisors to peel or extensively prepare: the pop in the mouth variety of harder fruits. They also suggest that these items could have been abrasive to the teeth. Ergo, it is possible that early hominins were eating soft and small hard fruits, flowers, seeds, and so on, that simply required thicker enamel and larger teeth because they were foraged from bushes and trees closer to the ground in these dusty habitats.

However, if fruit production was slightly diminished because of more leafing in plants, early hominins may have had to shift to other resources in the short dry season. These resources might have included underground storage organs (USOs), which may have been quite plentiful if the number of suids in these fossil sites is an indication of the presence of roots and tubers. In addition, in times of water stress, plants concentrate nutrients in these USOs. Dry-season months increased as habitats became drier and there might have been more reliance by some populations of hominins on USOs. This situation would be reflected in hominin dentition as increased crushing and more contact with grit would require stronger and more robust dental morphology. This appears to be the case with *Paranthropus* dentition.

These scenarios can be explored further using combined methods of inquiry such as elemental isotope ratios, dental microwear, functional morphology, and further refinement of hominin habitats. Additionally, comparing hominins with mammals other than primates that may have been eating similar types of food and have existed at the same time would also be beneficial. As more research is accomplished on habitats, hominins, and diet, the answers to some of these questions may move from the unknown to the possible.

Acknowledgments We thank Peter Ungar for inviting our participation in this volume and Blaine Schubert for very useful comments on the manuscript.

References

Alemseged, Z., 2003. An integrated approach to taphonomy and faunal change in the Shungura Formation (Ethiopia) and its implication for hominid evolution. *J. Hum. Evol.* 44, 451–478.

Andrews, P., 1989. Paleoecology of Laetoli. *J. Hum. Evol.* 18, 173–181.

Andrews, P., and Humphrey, L., 1999. African Miocene environments and the transition to early hominins. In: Vrba, E.S., Denton, G.H., Partridge, T.C., and Burckle, L.C. (Eds.), *Paleoclimate and Evolution with Emphasis on Human Origins*. Yale University Press, New Haven, CT, pp. 282–300.

Andrews, P., Lord, J.M., and Nesbitt-Evans, E.M., 1979. Patterns of ecological diversity in fossil and modern mammalian faunas. *Biol. J. Linn. Soc.* 11, 177–205.

Archibold, W., 1995. *Ecology of World Vegetation*. Chapman & Hall, London.

Avery, D.M., 1982. Micromammals as paleoenvironmental indicators and an interpretation of the late Quaternary in the Southern Cape Province. *Ann. S. Afr. Mus.* 85, 183–374.

Behrensmeyer, A.K., Todd, N., Potts, R., and McBrinn, G.E., 1997. Late Pliocene faunal turnover in the Turkana Basin, Kenya and Ethiopia. *Science* 278, 1589–1594.

Benefit, B.R, and McCrossin, M.L., 1990. Diet, species diversity and distribution of African fossil baboons. *Kroeber Anthropol. Soc. Pap.* 71, 77–93.

Bobe, R., and Eck, G.G., 2001. Responses of African bovids to Pliocene climatic change. *Paleobiology* Suppl. 27, 1–47.

Bonnefille, R., 1995. A reassessment of the Plio-Pleistocene pollen record of East Africa. In: Vrba, E.S., Denton, G.H., Partridge, T.C., and Burckle, L.C. (Eds.), *Paleoclimate and Evolution with Emphasis on Human Origins*. Yale University Press, New Haven, CT, pp. 299–310.

Bonnefille, R., Potts, R., Chalie, F., Jolly, D., and Peyron, O., 2004. High-resolution vegetation and climate change associated with Pliocene *Australopithecus afarensis*. *Proc. Natl. Acad. Sci.* 101, 12125–12129.

Bourliere, F., and Hadley, M., 1983. Present day savannas: An overview. In: Bourliere, F. (Ed.), *Ecosystems of the World: Tropical Savannas*. UNESCO, Paris, pp. 1–17.

Brunet, M., Guy, F., Pilbeam, D., Mackaye, H.T., Likius, A., Ahounta, D., Beauvilain, A., Blondel, C., Bocherens, H., Boisserie, J.R., De Bonis, L., Coppens, Y., Dejax, J., Denys, C., Duringer, P., Eisenmann, V.R., Fanone, G., Fronty, P., Geraads, D., Lehmann, T., Lihoreau, F., Louchart, A., Mahamat, A., Merceron, G., Mouchelin, G., Otero, O., Campomanes, P.P., De Leon, M.P., Rage, J.C., Sapanet, M., Schuster, M., Sudre, J., Tassy, P., Valentin, X., Vignaud, P., Viriot, L., Zazzo, A., and Zollikofer, C., 2002. A new hominid from the Upper Miocene of Chad, central Africa. *Nature* 418, 145–151.

Cadman, A., and Rayner, R.J., 1989. Climatic change and the appearance of Australopithecus *africanus* in the Makapansgat sediments. *J. Hum. Evol.* 18, 107–113.

Cerling, T.E., 1992. Development of grasslands and savannas in East Africa during the Neogene. *Palaeogeogr. Palaeoclimatol. Palaeoecol.* 97, 241–247.

Cerling, T.E., Wang, Y., and Quade, J., 2003. Expansion of C4 ecosystems as an indicator of global ecological change in the Late Miocene. *Nature* 361, 344–345.

Ciochon, R.L., 1993. *Evolution of the Cercopithecoid Forelimb: Phylogenetic and Functional Implications from Morphometric Analyses. Geological Sciences.* Vol. 138. University of California Press, Berkeley.

Crowley, T.J., and North, G.R., 1991. *Paleoclimatology*. Oxford University Press, Oxford.

Delaney, M.J., and Happold, D.C., 1979. *Ecology of African Mammals*. Longman, London.

deMenocal, P.B., and Bloemendal, J., 1995. Plio-Pleistocene subtropical African climate variability and the paleoenvironment of hominid evolution: A combined data-model approach. In: Vrba, E.S., Denton, G.H., Partridge, T.C., and Burckle, L.C. (Eds.), *Paleoclimate and Evolution with Emphasis on Human Origins*. Yale University Press, New Haven, CT, pp. 262–288.

Dorst, J., and Dandelot, P., 1969. A Field Guide to the Larger Mammals of Africa. Collins, London.

Elton, S., 2001. Locomotor and habitat classifications of cercopithecoid postcranial material from Sterkfontein Member 4, Bolt's Farm Swartkrans Members 1 and 2. South Africa. *Palaeontol. Afr.* 37, 115–126.

Elton, S., 2002. A reappraisal of the locomotion and habitat preference of *Theropithecus oswaldi. Folia Primatol.* 73, 252–280.

Estes, R.D., 1991. *The Behavior Guide to African Mammals*. University of California Press, Berkeley.

Foley, R., 1987. *Another Unique Species. Longman Scientific and Technical*, Essex.

Grine, F.F., 1981. Trophic differences between gracile and robust australopithecines: A scanning electron-microscope analysis of occlusal events. *S. Afr. J. Sci.* 77, 203–230.

Hailemichael, M., 1999. The Pliocene environment of Hadar, Ethiopia: A Comparative isotopic study of paleosol carbonates and lacustrine mollusk shells of the Hadar Formation and of modern analog. PhD diss., Case Western Reserve University.

Haile-Selassie, Y., 2001. Late Miocene hominids from the Middle Awash, Ethiopia. *Nature* 412, 178–181.

Happold, D.C.D., 1989. *The Mammals of Nigeria*. Clarendon Press, Oxford.

Harris, J.M., and Leakey, M.G., 2003a. Kanapoi: Fauna and paleoenvironments. *Am. J. Phys. Anthropol.* Suppl. 36, 110.

Harris, J.M., and Leakey, M.G., 2003b. Lothagam Suidae. In: Leakey, M.G., and Harris, J.M. (Eds.), Lothagam: The Dawn of Humanity in Eastern Africa. Columbia University Press, New York, pp. 485–523.

Hoffman, R.R., and Stewart, D.R.M., 1972. Grazer or browser: A classification based on the stomach-structure and feeding habits of East African ruminants. *Mammalia* 36, 226–240.

Jacobs, B. F., 1999a. The use of leaf form to estimate Miocene rainfall variables in tropical Africa. XVI International Botanical Congress. Abstract 4570.

Jacobs, B.F., 1999b. Estimation of rainfall variables from leaf characters in tropical Africa. *Palaeogeogr. Palaeoclimatol. Palaeoecol.* 145, 231–250.

Jacobs, B.F., 2002. Estimation of low latitude paleoclimates using fossil angiosperm leaves: Examples from the Miocene Tugen Hills, Kenya. *Paleobiology* 28, 399–421.

Jacobs, B.F., and Herendeen, P.S., 2004. Eocene dry climate and woodland vegetation in tropical Africa reconstructed from fossil leaves from northern Tanzania. *Palaeogeogr. Palaeoclimatol. Palaeoecol.* 213, 115–123.

Janis, C.M., 1988. New ideas in ungulate phylogeny and evolution. *Trends Ecol. Evol.* 3, 291–297.

Janis, C.M., Damuth, J., and Theodor, J.M., 2002. The origins and evolution of the North American grassland biome: The story from the hoofed mammals. *Palaeogeogr. Palaeoclimatol. Palaeoecol.* 177, 183–198.

Kappelman, J., 1988. Morphology and locomotor adaptations of the bovid femur in relation to habitat. *J. Morphol.* 198, 119–130.

Kay, R.F., 1975. The functional adaptations of primate molar teeth. *Am. J. Phys. Anthropol.* 43, 195–216.

Kay, R.F., 1978. Molar structure and diet in extant Cercopithecidae. In: Joysey, K., and Butler, P. (Eds.), *Development, Function, and Evolution of Teeth*. Academic Press, London, pp. 309–339.

Kingdon, J., 1974a. *East African Mammals*. Vol. 1. University of Chicago Press, Chicago.

Kingdon, J., 1974b. *East African Mammals: Hares and Rodents*. Vol. IIB. University of Chicago Press, Chicago.

Kingdon, J., 1977. *East African Mammals: Carnivores*. Vol. IIIA. University of Chicago Press, Chicago.

Kingdon, J., 1979. *East African Mammals: Large Mammals*. Vol. IIIB. University of Chicago Press, Chicago.

Kingdon, J., 1982a. *East African Mammals: Bovids*. Vol. IIIC. University of Chicago Press, Chicago.

Kingdon, J., 1982b. *East African Mammals: Bovids*. Vol. IIID. University of Chicago Press, Chicago.

Kingston, J.D., Marino, B.D., and Hill, A., 1994. Isotopic evidence for Neogene hominid paleoenvironments in the Kenya Rift valley. *Science* 264, 955–959.

Kitchener, A., 1991. *The Natural History of the Wild Cats*. Comstock, Ithaca, NY.

Kruuk, H., 1972. *The Spotted Hyena: A Study of Predation and Social Behavior*. University of Chicago Press, Chicago.

Leakey, M.G., Spoor, F., Brown, F.H., Gathogo, P.N., Kiarie, C., Leakey, L.N., and McDougall, I., 2001. New hominin genus from eastern Africa shows diverse middle Pliocene lineages. *Nature* 410, 433–440.

Lewis, M.E., 1997. Carnivoran paleoguilds of Africa: Implications for hominid food procurement strategies. *J. Hum. Evol.* 32, 257–288.

Marlow, J.R., Lange, C.B., Wefer, G., and Rosell-Mele, A., 2000. Upwelling intensification as part of the Pliocene-Pleistocene climate transition. *Science* 290, 288–291.

O'Brien, E., 1998. Water-energy dynamics, climate, and prediction of woody plant species richness: an interim general model. *J. Biogeogr.* 25, 379–398.

Pearson, P.N., Palmer, M.R., 2000. Atmospheric carbon dioxide concentrations over the past 60 million years. *Nature* 406, 695–699.

Pratt, D.J., and Gwynne, M.D., 1977. *Rangeland Management and Ecology in East Africa*. Hooder & Stoughton, London.

Quade, J., and Cerling, T.E., 1995. Expansion of C4 grasses in the Late Miocene of northern Pakistan—Evidence from stable isotopes in paleosols. *Palaeogeogr. Palaeoclimatol. Palaeoecol.* 115, 91–116.

Reed, K.E., 1997. Early Hominid Evolution and Ecological Change through the African Plio-Pleistocene. *J. Hum. Evol.* 32, 289–322.

Reed, K.E., 1998. Using large mammal communities to examine ecological and taxonomic organization and predict vegetation in extant and extinct assemblages. *Paleobiology* 24, 384–408.

Reed, K.E., 2002. The use of paleocommunity and taphonomic studies in reconstructing primate behavior. In: Plavcan, M.J., Kay, R., van Schaik, C., and Jungers, W. L. (Eds.), *Reconstructing Primate Behavior in the Fossil Record*. Kluwer Academic/Plenum Press, New York, pp. 217–259.

Reed, K.E., 2005. Tropical and temperate seasonal changes on human evolution. In: Brockman, D.K., and van Schaik, C. P. (Eds.), *Seasonality in Primates: Studies of Living and Extinct Human and Non-Human Primates*. Cambridge University Press, Cambridge, chap. 17.

Reed, K.E., Lockwood, C.A., and Arrowsmith, J.R., 2003. Faunal comparison between the Middle Ledi and Hadar hominin sites, Ethiopia: Time, landscape, and depositional environment. *Am. J. Phys. Anthropol.* Suppl. 36, 176–177.

Retallack, G.J., 1992. Middle Miocene fossil plants from Fort Ternan (Kenya) and evolution of African grasslands. *Paleobiology* 18, 383–400.

Retallack, G.J., Wynn, J.G., Benefit, B.R., and McCrossin, M.L., 2002. Paleosols and paleoenvironments of the middle Miocene, Maboko Formation, Kenya. *J. Hum. Evol.* 42, 659–703.

Ritchie, M.E., and Olff, H., 1999. Spatial scaling laws yield a synthetic theory of biodiversity. *Nature* 400, 557–560.

Robinson, J.T., 1963. Adaptive radiation in the australopithecines and the origin of man. In: Howell, F.C., and Bourliere, F. (Eds.), *African Ecology and Human Evolution*. Aldine, Chicago, pp. 385–416.

Rose, K.D., 1981. Composition and species diversity in Paleocene and Eocene mammal assemblages: An empirical study. *J. Vertebr. Paleontol.* 1, 367–388.

Semaw, S., Simpson, S.W., Quade, J., Renne, P.R., Butler, R.F., McIntosh, W.C., Levin, N., Dominguez-Rodrigo, M., and Rogers, M.J., 2005. Paleoenvironments of the earliest stone toolmakers, Gona, Ethiopia. *Geol. Soc. Am. Bull.* 116, 1529–1544.

Sept, J.M., 1994. Beyond bones—archaeological sites, early hominid subsistence, and the costs and benefits of exploiting wild plant foods in east African riverine landscapes. *J. Hum. Evol.* 27, 295–320.

Skinner, J.D., and Smithers, R.H.N., 1990. *The Mammals of the Southern African Subregion*. University of Pretoria Press, Pretoria.

Smithers, R.H.N., 1978. *A Checklist of the Mammals of Botswana*. Trustees of the National Museum of Rhodesia, Salisbury.

Spencer, L.M., 1995. Morphological correlates of dietary resource partitioning in the African Bovidae. *J. Mammal.* 76, 448–471.

Spencer, L.M., 1997. Dietary adaptations of Plio-Pleistocene Bovidae: Implications for hominid habitat use. *J. Hum. Evol.* 32, 201–228.

Stern, J.T., and Susman, R.L., 1983. The locomotor anatomy of *Australopithecus afaransis*. *Am. J. Phys. Anthropol.* 60, 279–317.

Teaford, M.F., and Ungar, P.S., 2000. Diet and the evolution of the earliest human ancestors. *Proc. Natl. Acad. Sci.* 97, 13506–13511.

Van Couvering, J.A.H., and Van Couvering, J.A., 1976. Early miocene mammal Fossils from East Africa: Aspects of geology, faunistics, and paleoecology. In: Isaac, G.L., and McCown, E. R. (Eds.), *Human Origins: Louis Leakey and the East African Evidence*. Staples Press, Menlo Park, pp. 155–207.

Vignaud, P., Duringer, P., Mackaye, H.T., Likius, A., Blondel, C., Boisserie, J.R., de Bonis, L., Eisenmann, V., Etienne, M.E., Geraads, D., Guy, F., Lehmann, T., Lihoreau, F., Lopez-Martinez, N., Mourer-Chauvire, C., Otero, O., Rage, J.C., Schuster, M., Viriot, L., Zazzo, A., and Brunet, M., 2002. Geology and palaeontology of the Upper Miocene Toros-Menalla hominid locality, Chad. *Nature* 418, 152–155.

Vrba, E.S., 1974. Chronological and ecological implications of the fossil Bovidae at the Sterkfontein australopithecine site. *Nature* 250, 19–23.

Vrba, E.S., 1975. Some evidence of chronology and palaeocology of Sterkfontein, Swartkrans, and Kromdraai from the fossil Bovidae. *Nature* 254, 301–304.

Vrba, E.S., 1980. The significance of bovid remains as indicators of environment and prediction patterns. In: Behrensmeyer, A.K., and Hill, A. (Eds.), *Fossils in the making, Vertebrate Taphonomy and Paleoecology*. University of Chicago Press, Chicago, pp. 247–271.

Vrba, E.S., 1988. Late Pliocene climatic events and hominid evolution. In: Grine, F.E. (Ed.), *Evolutionary History of the "Robust" Australopithecines*. Aldine de Gruyter, New York, pp. 405–426.

White, F., 1983. *The Vegetation of Africa: A Descriptive Memoir to Accompany UNESCO/AETFAT/UNSO Vegetation Maps of Africa*. UNESCO, Paris.

White, T.D., Suwa, G., 2004. A new species of *Notochoerus* (Aartiodactyla, Suidae) from the Pliocene of Ethiopia. *J. Vertebr. Paleontol.* 24, 474–480.

White, T.D., Suwa, G., Hart, W.K., Walter, R.C., WoldeGabriel, G., de Heinzelin, J., Clark, J.D., Asfaw, B., and Vrba, E.S., 1993. New discoveries of *Australopithecus* at Maka in Ethiopia. *Nature* 366, 261–265.

WoldeGabriel, G., White, T.D., Suwa, G. Renne, P., de Heinzelin, J., Hart, W., and Heiken, G., 1994. Ecological and temporal placement of early Pliocene hominids at Aramis, Ethiopia. *Nature* 371, 330–333.

WoldeGabriel, G., Haile-Selassie, Y., Renne, P.R., Hart, W.K., Ambrose, S.H., Asfaw, B., Heiken, G., and White, T., 2001. Geology and palaeontology of the Late Miocene Middle Awash Valley, Afar Rift, Ethiopia. *Nature* 412, 175–178.

Wrangham, R.W., Conklin-Brittain, N.L., and Hunt, K.D., 1998. Dietary response of chimpanzees and cercopithecines to seasonal variation in fruit abundance. *Int. J. Primatol.* 19, 949–970.

Wynn, J.G., 2000. Paleosols, stable carbon isotopes, and paleoenvironmental interpretation of Kanapoi, Northern Kenya. *J. Hum. Evol.* 39, 411–432.

Wynn, J.G., and Feibel, C.S., 1995. Paleoclimatic implications of vertisols within the Koobi Fora Formation, Turkana Basin, Northern Kenya. *Univ. Utah J. Undergrad. Res.* 6, 32–42.

Wynn, J.G., Retallack, G.J., 2001. Paleoenvironmental reconstruction of middle Miocene paleosols bearing *Kenyapithecus* and *Victoriapithecus*, Nyakach Formation, southwestern Kenya. *J. Hum. Evol.* 40, 263–288.

15

Modeling the Significance of Paleoenvironmental Context for Early Hominin Diets

JEANNE SEPT

From a dietary perspective, did it matter which habitats early hominins lived in? We certainly assume it did, given our understanding of basic principles of evolutionary ecology, but to what extent can we demonstrate this? Hypotheses about early hominin diet have been central to our interpretations of human evolution since Robinson focused attention on dietary niches (Robinson, 1954, 1963) and challenged the field to develop ecological models of human ancestry. As described elsewhere in this volume, paleoanthropologists have made significant progress in recovering and interpreting different types of empirical data that can guide reconstructions of both ancient environments and hominin dietary adaptations. Yet, as richer and more detailed paleoecological data accumulate, the theoretical challenges also grow as we seek to compare more varied types of evidence with different sampling scales, taphonomic histories and interpretive frameworks.

This chapter reviews the role of modeling in developing our understanding of early hominin diet and subsistence behavior. It evaluates how combining data from paleoenvironments and contemporary analogs allows us to model both the ecological constraints on early hominin subsistence behavior and feeding opportunities—the variety, abundance, and distribution of foods hominins would have encountered in different ancient habitats. In particular, models of ancient "edible landscapes" can provide a critical framework for evaluating the archaeological and paleontological evidence for early hominin diets and help refine hypotheses about protohuman subsistence strategies to guide future research. Within the broad framework of evolutionary biology, various theories target processes and questions related to the evolution of human diet for analysis, and models can develop explicit assumptions and limited sets of parameters to explore the these questions in different contexts. In general, modeling is a critical methodological bridge between the known and the unknown, and a way of framing debate about what may remain unknowable.

Models of complex socioecological systems are both heuristic devices and essential tools, designed to help evaluate the state of scientific knowledge, weigh alternative explanations, and point to gaps in understanding (Levins, 1966; Costanza et al., 2001; Dunbar, 2002). All models make different trade-offs between the three dimensions of generality, realism, and precision and are always simplifications of a complex world, in effect, a balancing of the known against the unknown for purposes of defining the unknowable, at least in theoretical terms. A general model can represent a wide range of systems and thus is often low resolution or large scale. A realistic model accurately reflects the complexity of multivariate processes in a specific system, such as behavioral observations from an individual field site, with a correspondingly small scale. A precise model represents information in an exact, quantitative way that is empirically calibrated but often focused on limited properties of a system. Ultimately, models need to accommodate known data and are most useful as scientific tools when they are relatively simple, explicit, and test assumptions about the comparative significance of variables or multivariate thresholds in a complex system (Costanza and Ruth, 2001).

So, to what extent can models help us establish what we know about early hominin diet and help us extend the current limits of our knowledge? The paleoanthropological literature includes a variety of paleoecological "models" of early hominin diet and subsistence behavior that not only balance generality, precision and realism, but also vary in biogeographical and temporal scale. Models of hominin diet range in scale from narrow evaluations of a few morphological traits in a limited number of fossils or artifacts, to evolutionary arguments encompassing a broad range of systems. They range in focus from explicit to vague, and often include a mix of methods from computational to conceptual.

A sample of paleoanthropological models are discussed to illustrate the range of approaches and the relative strengths and weaknesses typical of such methodologies, following the classification scheme of Costanza and Ruth (2001). As some authors seem to use the terms "model," "framework," and "scenario" loosely and almost interchangeably, I have focused on more formal attempts to model early hominin diet in an environmental context for this selective review. Figure 15.1 illustrates the theoretical focus and scale of a sample of featured models in a framework that plots the model's relative balance of generality, realism and precision on a triangular graph and also positions the model in terms of its paleoenvironmental scale, ranging from the macroevolutionary models to models of microscale phenomena.

The Known

High-Generality Conceptual Models

The most common type of early hominin subsistence models address basic questions about the limits of early hominin behavioral systems in broad contexts. They focus on simple relationships of a few, critical variables, such as the effect of cli-

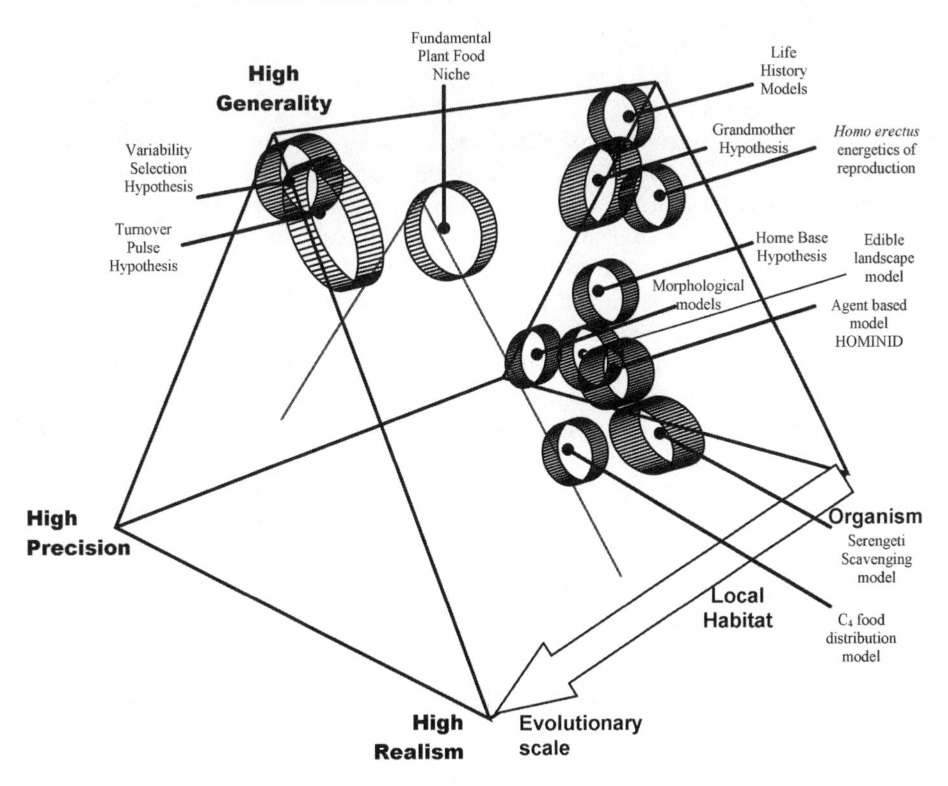

Figure 15.1 Four Dimensions of Paleoanthropological Modeling: all models balance generality, precision, and realism and also have a biogeographical application to a particular scale of time and space, ranging from large scale (macroevolutionary time, global space) to intermediate scales (microevolutionary time, regional landscapes) and small scales (short-term processes scaled within the lifetime and physiology of particular organisms). Examples of models discussed in the text are positioned in this matrix, represented as cylinders that approximate their relative focus.

matic conditions or life history variables on dietary parameters. Sometimes they incorporate quantitative relationships from linear equations derived from economics or game theory but often they are qualitative and sacrifice both realism and precision in favor of broad, evolutionary-scale generalizations.

A variety of high-generality models have framed debate about the effect of climate change on early hominin evolution and early hominin subsistence adaptations. For years, following in the footsteps of Raymond Dart (1925), advocates of the "savanna hypothesis" broadly suggested that hominin evolution and behavior in the late Miocene/Pliocene was driven by the widespread emergence of open, grassy habitats in Africa (Bartholomew and Birdsell, 1953; Jolly, 1970; Brain, 1981; Grine, 1986; Stanley, 1992; Owen-Smith, 1999). Spurred by recent advances in the analysis and synthesis of marine and terrestrial paleoenvironmental records (deMenocal, 1995, 2004; Reed, 1997; Potts, 1998), several authors have tried to develop models with testable hypotheses about how global climatic changes influenced the evolution of

African biotic communities. Elisabeth Vrba has argued in her "turnover pulse hypothesis" that climatic shifts that caused the expansion of African grassland habitats generated broad patterns of bovid speciation and extinction and that such significant habitat shifts were also correlated with hominin evolutionary patterns (Vrba, 1985, 1988, 1995a, 1995b). Potts (1996a, 1996b, 1998, 2004) has developed an alternative paleoenvironmental model of hominin evolution called the "variability selection hypothesis" that emphasizes adaptive responses to environmental instability, rather than habitat-specific adaptations, as the key selection factor. Efforts to compare the predictions of these two models against the faunal record in the Turkana basin (Bobe and Behrensmeyer, 2004) suggest that, while both models provide important insights into how shifting climatic conditions could be expected to have broadly affect patterns of human evolution, neither model effectively accounts for the variation in the paleontological record sampled in the Turkana basin. This may be due, in part, to the taphonomic and sampling problems inherent in comparing such surface assemblages (White, 1988, 1995) or to issues related to the tectonic history of the basin (Feibel, 1999). In any case, the scale of these models is so broad that they include only very generalized assumptions about what foods would have been available in such habitats and how climate change would have influenced hominin subsistence, as illustrated by statements such as the follow:

> The separate lineage leading to *Homo* spp. became dependent on the procurement of meat to bridge the seasonal bottleneck in food resources. (Owen-Smith 1999, p. 149)

> The highly specialized, megadont dentition of *P. boisei* may indicate an adaptive response to this shift toward open environments, with concomitant changes in dietary preferences. (Bobe and Behrensmeyer, 2004, p. 413)

> Instability in Pleistocene resource landscapes heightened the problem of sustaining soft-fruit frugivory. Since ripe fruits are a temporary and ephemeral resource even within a stable climatic setting, long-term and sometimes rapid variation during the Pleistocene in the timing, length, and intensity of wet and dry seasons not only led to repeated forest contraction and expansion, but would have dramatically altered the spatial and temporal pattern of fruiting, the phenological properties of fruit trees, and the size and distribution of food patches. (Potts, 2004, p. 222)

In this broad, evolutionary context, a number of authors have speculated about different types of keystone foods that might have been critical to early hominin survival, including tubers (Hatley and Kappelman, 1980; Vincent, 1985) and various seeds or nuts (Jolly, 1970; Peters, 1987; Schoeninger et al., 2001), scavenging opportunities (Schaller and Lowther, 1969), insects (McGrew, 2001), or the overall balance of likely plant/animal foods or cooked foods (Peters and O'Brien, 1981; Stahl, 1984; Conklin-Brittain, Wrangham, and Smith, 2002; Wrangham and Conklin-Brittain, 2003). These arguments are generally framed with reference to the dietary preferences of living primate or human populations and the relative nutritional quality of the food, such as its caloric return rate or high lipid value. But when framed as conceptual arguments in this way, food models tend to define the possible but not the probable. For example, when Peters and O'Brien (1981) compared the food lists of selected human and primate species, they sought to define a "fundamental plant food niche" for early hominins—a sort of master menu of African plant foods likely to have been considered edible by our ancestors but not "realized niches" of foods that were likely to have been available and eaten by different species in different habitats.

Conceptual models of early hominin subsistence have come a long way since Glynn Isaac first developed the food-sharing hypothesis to explain patterns of archaeological site data in the East African Rift (Isaac, 1978), but few have had comparable scientific influence or such a lasting impact. Isaac sought to explain the basic patterns of stones and bones at the earliest sites in bold, evolutionary terms, focusing on the established behavioral ecological principles of human foragers and primates and urging and inspiring his colleagues to delve deeper into both details of the archaeological data, processes of site formation (taphonomy), and systemic understanding of the evolution of subsistence behavior in related species (Isaac, 1984). While others were making basic analogical arguments about the value of understanding human foraging and nonhuman primate and carnivore behaviors as references for early hominin diet (Dunbar, 1976; Hatley and Kappelman, 1980; Binford, 1981; Harding and Teleki, 1981; Peters and O'Brien, 1981; Tooby and DeVore, 1987), Isaac's framework was unique in its formal attempt to organize and integrate the inferences from different fields in a way that would help archaeologists develop and test their interpretations of early archaeological sites (Sept, 1992a). Indeed, Isaac's model spawned a generation of archaeological and actualistic (e.g., ethnoarchaeological) research devoted to testing different hypotheses derived from his general food-sharing, home-base model. Avenues of research focused specifically on empirically evaluating the taphonomic history of specific stone and bone assemblages (Schick, 1987; Potts, 1988; Bunn, 1991) and theoretically evaluating dimensions of the problem of equifinality in site formation processes (Binford, 1981; Blumenschine, 1986a; O'Connell and Hawkes, 1988; Gifford-Gonzalez, 1991; Sept, 1992b). Ultimately, from this work, we can have confidence that at least one species of early hominin used tools to extract meat and marrow from animal carcasses and carried off pieces to eat away from the rest of the carcass at sites near rivers and lakes. However, how these meaty meals were typically acquired, the extent to which they were eaten or shared in relation to other types of food, and how frequently or dependably hunting or scavenging opportunities were accessed in different environmental contexts remain matters of significant debate, as discussed later in this chapter and in other contributions to this volume.

At the other end of the high-generality, conceptual spectrum, authors have developed hominin diet models with little reference to environmental context that focus, instead, on the physiology of the organism. In particular, a number of models explore the relationship between early hominin brain development, life history parameters, and broad dietary patterns, beginning with the relatively high-metabolic costs of brain tissue as a key developmental variable. Foley and Lee, for example, developed a conceptual model focusing on a few life history parameters "to model the effects on energy requirements of changing growth rates" (Foley and Lee, 1991, p. 228). The work by Aiello and Key follows this lead, modeling the effects of the increase in *Homo erectus* female body mass on reproductive strategy, based on aggregate daily energy expenditure figures for living primates. They focus on only a few properties of the system, carefully calibrated with baseline data on primate daily energy expenditure (Aiello and Key, 2002).

Complementary studies have suggested that the higher costs of growing and maintaining brain tissue were offset in *Homo* by a reduction in overall gut volume

(Aiello and Wheeler, 1995) and selection for nutrient-rich foods that could be easily absorbed (Milton, 1987, 1999, 2002). From a life history perspective, Leonard, Robinson, and colleagues used comparative generalizations on primate brain/body ratios and metabolic rates to argue that the emergence of larger-brained *Homo* would have necessitated a maternal dietary shift to foods of improved "stability and quality," which they think would have been animal foods (Leonard and Robertson, 1992, 1994; Leonard et al., 2003). Vasey and Walker argue from a comparative primate perspective that fossil morphology suggests the secondarily altricial condition of human neonates typical today was first established with early *Homo*; they suggest that this condition would have been dependent on a trophic-level shift and regular maternal access to nutritionally dense foods, such as animal protein and fat (Vasey and Walker, 2001), for support during gestation and extended lactation. Kennedy (2003, 2005) also argues that meat eating would have been a critical element to support the reproductive costs of early *Homo*, although in contrast to Vasey and Walker, she hypothesizes that meat eating encouraged a shift to early weaning in large-brained hominins; this key difference may be in large part due to her use of agricultural groups as the human "natural fertility populations" on which her model is based.

In an ambitious attempt to integrate climatic, dietary, and life history variables, Kristen Hawkes, James O'Connell, and their colleagues have set out to build "a comprehensive model of early hominin ecology and evolution" (Hawkes et al., 1998a, 1998b; O'Connell, Hawkes, and Blurton Jones, 1999, 2002). The Grandmother Hypothesis posits that *Homo ergaster* evolved in response to climate-driven changes in female-foraging and food-sharing practices. In brief, they argue that climate change decreased the availability of food sources easily accessible to children, such as soft fruits, making *Homo* more dependent on adult-access foods with high-processing costs (tubers, nuts, hard seeds), especially during the dry season. They argue that under such conditions there would have been selective advantages to behaviors such as tool use and social strategies such as food sharing. They suggest that selection pressures would also have favored longevity in postmenopausal females because, by analogy with Hadza and other human foragers today, females with reduced maternal burdens could have shared food with the group and raised inclusive fitness. Their model combines broad concepts from Charnov's (1993) formal, quantitative model of life history patterns, in the context of basic relationships derived from evolutionary ecology and optimal foraging theory (Krebs and Davies, 1984; Stephens and Krebs, 1986; Smith and Winterhalder, 1992). They have calibrated some aspects of their model with reference to short-term field data on contemporary time allocation and foraging strategies collected from the Hadza (Hawkes, O'Connell, and Blurton Jones, 1995; Hawkes, O'Connell, and Jones, 1997). However, data from other human and nonhuman primate groups have been used to challenge the generality of these arguments and the relevance of their application to early hominins (Peccei, 2001; Kennedy, 2003).

Overall, the development of these ideas about the relationship between life history parameters and subsistence strategies has effectively drawn attention to the limits of hominin subsistence systems in the context of dependence on different types of resources (particularly plant foods vs. animal foods), and it helps frame the challenges involved in the analysis of cost-benefit patterns as they relate to reproductive and provisioning strategies in models that involve either more precision or realism.

High-Precision Models

Although econometric models and small-scale ecological models often emphasize the quantitative correspondence between empirical data and the model, providing analytical precision (Costanza and Ruth, 2001), because of the limitations of the fossil and archaeological records, precise models of early hominin diet are quite limited in scope and focus on interpreting discrete aspects of known morphology, rather than our incomplete samples of ecology or behavioral artifacts. Because other contributions to this volume focus on the application of biomechanical and osteological/developmental models to the interpretation of hominin craniodental morphology, pathology, and toothwear, they will not be discussed here.

High-Realism Models

Realistic models seek to represent accurately the underlying processes of a specific system, rather than to match precise records of behavior or to be generally applicable. In the ecological literature, realistic models are very site specific, and often include high-spatial and temporal resolution and complex, dynamic processes, with the goal of evaluating threshold effects or analyzing the effect of changes to particular variable states within a system (Costanza and Ruth, 2001). A number of models of primate foraging can be seen as attempts to develop this type of realistic model. Altman's model of young baboon foraging in the Amboseli, Kenya (Altmann, 1998), is a classic of this genre, and a number of primate field studies have led to the development of optimality models to help evaluate different aspects of primate social behavior as well (Dunbar, 2002).

Ethnoarchaeological and actualistic studies can also attempt to develop relatively realistic, processual models of behavior that are specific to particular conditions or questions of material culture relevant to paleoanthropology (Bunn, 1991; Sept, 1992a; Marshall, 1994; O'Connell, 1995). One example explicitly evaluating subsistence opportunities in environmental contexts is the Serengeti model of scavenging opportunities, developed by Rob Blumenschine and his students (Blumenschine, 1986a, 1986b, 1988, 1989; Cavallo and Blumenschine, 1989; Blumenschine and Madrigal, 1993; Selvaggio, 1994). Blumenschine combined observations of carcass frequency and consumption sequences in different microhabitats, with taphonomic experiments and naturalistic observations of bone damage patterns, to develop a relatively realistic model of how and why scavenging opportunities vary in space and time in the contemporary Serengeti. He argues that scavenging opportunities are strongly mediated by carnivore competition and that the patterns of such competition vary predictably between riparian habitats and more open habitats in such African savanna landscapes. Two key questions emerge from this modeling effort. First, which aspects of this scavenging opportunity model can be generalized to other modern ecosystems? Second, to what extent is it useful to apply the predictions of this model directly to the interpretation of ancient faunal assemblages? Tappen has argued, for example, that the scavenging patterns studied by Blumenschine are region specific and that the Serengeti scavenging model should not be assumed to hold for other habitats. Only after the Serengeti scavenging model has been compared with results of scavenging research at other study sites should we have confidence in its general applicability (Tappen, 2001).

Dominguez-Rodrigo undertook such comparative scavenging studies in two savanna habitats in Kenya, the Masai Mara, a northern extension of the Serengeti ecosystem, and the Galana Ranch, a semiarid bushland, and observed the degree of dry-season competition between individual carnivores at carcasses (Dominguez-Rodrigo, 2001). Although his methods differed from Blumenschine's, and his results differ in detail, he argues that his research supports the Serengeti model in many fundamental respects. However, his statement that "clearly competition is very low in riparian woodlands, irrespective of the ecosystem" (Dominguez-Rodrigo, 2001, p. 87) seems premature because it is based on his observations of only two carcasses in Galana river woodlands, compared with fifteen in the Mara riverine samples; most of the carcasses he observed were in the Mara system (43/50), which is very similar to the Serengeti, so his total sample of carcasses from the bushland habitats seems too small to justify his claim that the Serengeti model can be generalized. However, Dominguez-Rodrigo's carcass sample size is significantly larger than the unsystematic sample of carcasses observed by Tappen (2001) in eastern Democratic Republic of the Congo, which were the empirical basis of her critique of the Serengeti model. Also, his earlier studies of landscape bone taphonomy in the same region (Dominguez-Rodrigo, 1996) included larger samples of bones from Galana riparian transects (twenty-four of the total fifty-nine MNI tallied); the preservation of skeletal parts in these assemblages (as reflected in MNE/MNI ratios, for example) supports the general idea of less bone destruction in closed woodlands, which is partially dependent on carnivore density and feeding competition. Overall, as comparable scavenging studies are undertaken, we will have greater confidence in which variables and relationships can form part of a more generalized model of scavenging. Until then, this case study is a good example of why we should proceed cautiously when applying principles derived from site-specific models as tools to interpret Plio-Pleistocene behavioral evidence, when the mammalian species composition, diversity, and trophic dynamics of the ancient ecosystems were distinctive and unlike any contemporary setting (Lewis, 1997; Dominguez-Rodrigo, 2001).

Intermediate Models of Moderate Realism and Precision

The most realistic models of early hominin diet are the ones that focus on conditions and evidence of specific ancient sites or assemblages, but obviously no such model can match the level of realistic detail possible when modeling living systems. As a result, many such models seek to determine accurately the relative magnitude and direction of aggregate trends but at the expense of realism and with limited precision and/or generality. Interesting examples of such intermediate or middle-range models are being developed for early hominin landscapes.

For example, in South Africa, recent applications of stable isotope analysis to reconstructions of early hominin diet have been made in the context of building isotopic models of ancient landscapes. Could sedges have been the main C_4 foods contributing to the relatively positive ^{13}C isotopic signatures of different South African hominin species, or was it more likely to have been grass seeds, or termites, or the meat from grazing herbivores? Evaluations of C_4 food distribution in the context of regional vegetation patterns have helped frame the debate (Peters and Vogel, 2005), but because an almost infinite number of dietary ways exist to create an average isotopic signature (the principle of equifinality), this generalized overview does not really narrow the field of dietary possibilities or establish probabilities. However, sampling the isotopic signatures of variable plant and animal distributions in specific modern habitats has helped focus models in ways that can lead to testable hypotheses. For example, strategic sampling of modern habitats by Matt Sponheimer, Julia Lee-Thorpe, and colleagues (Sponheimer et al., 2005) has revealed unanticipated variability in the isotopic patterning of sedges and termites across landscapes and allowed them to weight realistically the relative contributions that different foods could have made to the diets of *Australopithecus* and *Paranthropus* in different habitats. Their argument that significant omnivory was quite probable for both taxa is well supported because their model is context specific and quantitatively calibrated with reference to both modern and fossil data sets in ways that are replicable and contribute to both its precision and realism.

Peters and Blumenschine (1995, 1996) have developed a different type of landscape model, focused on a specific time period in the Olduvai Basin, and aimed to generate hypotheses testable with the archaeological record (Blumenschine and Peters, 1998). They have tried to model a distribution of key resources, or "affordances," across a landscape patterned after their paleogeographic reconstructions of lower Bed II, which spans a 50,000- to 90,000-year time period and use this to predict the relative, average land-use frequency of hominins in each "facet" of the landscape as it would be reflected in the distribution of artifacts and fossilized faunal remains. Some elements of their model are very realistic. The underlying paleogeographic basin reconstructions are based on years of sedimentological work in the Olduvai Gorge. Their models of probable scavenging opportunities are based on Blumenschine's Serengeti scavenging model, described earlier. Other elements of their model, such as plant food availability, are based on informal, subjective assessments, pending systematic field study (Copeland, 1999) and not calibrated methodologically, so their realistic accuracy is difficult to evaluate. While the goal of their basin-specific, integrative approach is laudable, their attempts to develop testable hypotheses from this model are undermined by their methodology, which cannot be replicated or scientifically evaluated, at least as currently published (Blumenschine and Peters, 1998, pp. 579–580). "Initial predictions about the facet-specific artifact and bone assemblages were made by the authors discussing the likely outcomes based upon our field experience and general knowledge of the wildlife/ecosystem literature. This procedure allowed us to narrow down the possibilities to the most likely outcomes. These outcomes were cross-checked through thought experiments involving imaginative and sometimes hands-on reenactments of hominin affordance interactions in and across facets." In other words, if their predictions do not match the empirical evidence, it is unclear whether that is because early hominins were behaving in unanticipated ways or because some or all parts of their landscape model are flawed. If the next generation of their model is calibrated more explicitly, then it will be easier to evaluate its performance.

I have also developed a landscape approach to understanding early hominin subsistence strategies, focused on plant foods (Sept, 1984, 1986, 1990, 1992a, 1994, 2001). My initial, formalized model of "the edible landscape" for early hominins (Sept, 2001) was based on measured samples of plant-food density in selected contemporary habitats. These were extrapolated using a geographical information system grid that created a model of the differential costs and benefits distributed seasonally across two landscapes. The model was developed from quantified field data and explicit assumptions about harvesting rates for different plant-food consumers following a simple set of foraging rules, including living humans, baboons, chimpanzees, two types of australopithecine, and early *Homo*. This approach had the advantage that values for individual variables (e.g., the costs of digging up tubers to a consumer with or without a digging stick) can be adjusted systematically, producing different model results that can then be compared or critiqued. It also had the advantage of creating simple, explicit predictions of foraging options in different zones and seasons for different consumers. For example, the model predicted the most significant seasonal differences in the plant foods available to robust australopithecines and early *Homo* in arid habitats, suggesting that during the "crunch seasons," when australopithecines could have exploited a wider variety of tough fruits, seeds, and roots in riparian habitats, early *Homo* would have faced less foraging competition by searching for higher-quality plant (and animal) foods across habitats. Weaknesses of this modeling approach included its plants-only focus and lack of a dynamic behavioral component to the model, which prevents a formal analysis of the implications of this model for processes of archaeological site formation, for example.

In terms of modeling methodology, as computing power has become cheap and widely distributed during the past few years, computational models have grown in popularity across all disciplines. Among anthropologists interested in modeling the complexities of human, protohuman, or nonhuman primate systems, agent-based models have emerged as useful analytical tools. Dunbar (2002) summarizes the goals of recent agent-based models in primatology, most of which focus on social attributes of primate populations, such as group size or cognitive processes. While there is a rich literature on the use of agent-based foraging models, to this point, few agent-based foraging models have been developed for hominins. Lake's work on Mesolithic foraging for hazelnuts is one example (Lake, 1999, 2001), where Lake was especially interested in the role of information exchange and relocations of a base camp. Reynolds modeled early hominin foraging in an abstract landscape with patches of resources, exploring different algorithms for group decision making (Reynolds, Whallon, and Goodhall, 2001). Costopoulos (2001) analyzed the effects of memory on foraging in a population of hominin agents. But none of these agent-based models have focused on modeling an explicit paleoenvironmental framework; instead, they have created generic environmental contexts within which early hominin cognitive and social parameters are simulated. Following the lead of my original "edible landscape" model (Sept 2001), Marco Jansen, Cameron Griffith, and I are developing an agent-based model to focus on understanding the effect of environmental context on early hominin foraging behaviors. Our model is based on carefully defined landscape parameters and simple decision rules for hominin foragers and simulates agents foraging and interacting on this landscape through time. It allows us to explore how resource availability and handling rates could be expected to affect the foraging patterns of different hominin taxa through time (Janssen, Sept, and Griffith, 2005). This approach should help us begin to model the adaptive significance of habitat for early hominin subsistence patterns. Quantification of the variables and relationships used to develop such landscape subsistence models can improve their utility because the sensitivity of the model to changes in variable states and the overall performance of the model can be formally evaluated and help us tease apart the differences between the known and the unknown.

The Unknown

Although the term "model" is sometimes construed to mean mere speculation in the absence of actual "facts" about hominin evolution, it should be evident from this brief review that all good models are explicitly derived from an empirical base but that they vary in scale and emphasis with the scope of the questions they are designed to explore. For example, generalized conceptual models focus on large-scale evolutionary patterns, but they can also be used to generate hypotheses that can be evaluated in the context of more precise or realistic models. But empirical evaluations of large-scale models can run up against limits in the structure of available data. For example, Bobe and Behrensmeyer (2004) used the paleontological record from a single basin as a context for evaluating competing predictions from the large-scale environmental models of Vrba and Potts but were unable to account for some variation in the data sets with either model. This may partly be a resolution issue. Costanza and Ruth (2001) describe how spatially explicit system models must struggle with the trade-off between resolution and predictability. Although increasing resolution provides more descriptive information about realistic patterns in the data, it also makes it more difficult to model the patterns accurately through time and space. This challenge is particularly tough when a model is trying to integrate variables that are measured at different sampling scales, with different methods, in space or in time. Bobe and Behrensmeyer acknowledge this challenge in their selective, probabilistic use of fossil specimens catalogued in the Turkana basin paleontological database. However, it is unclear from their analysis how their results would have differed if they had not manipulated data sets in this way, or had calibrated them at higher- or lower-temporal intervals; in other words, how sensitive is their analysis to the assumptions of their methodology?

As effective heuristic tools, models can reveal gaps in our empirical knowledge. They also remind us that we should be cautious in our eagerness to generalize from normative fossil or archaeological patterns. Wood and Strait, for example, question common assumptions made about the adaptive significance of the morphological differences between *Homo* and *Paranthropus* (Wood and Strait, 2004). Their critique reminds us that generalized models assuming simplistic contrasts between the diets and subsistence strategies of fossil taxa run the risk of neglecting ecological realism in our efforts to chart evolutionary trends and challenge us to make the assumptions of each component of a model as explicit as possible, so that they can be formally evaluated and adjusted to compare alternatives.

Sometimes we err by assuming we know more than we do. For example, some archaeological analyses that might appear to be "precise" to nonspecialists, such as the SEM analysis of cut marks on faunal remains from archaeological sites, are not. When identifying cut marks, or other types of surface damage to bones, each judgment about the cause of an individual mark on a bone can be made with reference to experimental standards. But, the overall interpretation of the associations of marks on an individual bone is often done subjectively, in the context of other bones in the faunal assemblage. Zooarchaeological models for such interpretations could benefit from a wider range of both experiments and ethnoarchaeological research, to make them more generalizable. Also, because each archaeological site has a unique taphonomic history, in addition to whatever behaviors contributed to the formation of the site, there is no way to match precisely quantitative results from different archaeological sites. This basic interpretive challenge lies at the heart of much of the zooarchaeological debate about whether early toolmakers acquired animal foods through hunting or active/passive scavenging, which remains unresolved (Dominguez-Rodrigo and Pickering, 2003).

Similarly, the challenges of trying to jump from realistic zooarchaeological models developed from individual field studies to generalizations can be illustrated by a cautionary tale—the zooarchaeological debate that developed between the two groups of anthropologists doing ethnoarchaeological research on the Hadza. Hadza populations living along Lake Eyasi in Tanzania first received anthropological attention through the studies of Woodburn presented at the influential Man the Hunter conference (Woodburn, 1968). Because they obtained significant proportions of their subsistence from foraging for wild foods, during the 1980s and 1990s, anthropologists studied the Hadza to learn about their use of tubers and other plant foods and their hunting and scavenging strategies (Vincent, 1985; Bunn, Bartram, and Kroll, 1988; O'Connell and Hawkes, 1988; O'Connell, Hawkes, and Blurton Jones, 1990; Hawkes, O'Connell, and Blurton Jones, 1991; Hawkes, 1993; Hawkes, O'Connell, and Jones, 1995; Bunn, 2001). But because the faunal exploitation studies were done during slightly different periods in different areas with different subgroups, the studies produced different results. Opportunities to work out how these differences could be understood in terms of a model of Hadza faunal exploitation were passed by in the authors' eagerness to claim that their results could be generalized to interpret definitively different aspects of the early archaeological record.

From a botanical perspective, so little is preserved of the behavioral interactions between hominins and plants that there is not much empirical data to generalize from. The occasional phytolith on a tool (Dominguez-Rodrigo et al., 2001) or the growing number of fossil specimens analyzed for isotopic signatures and/or tooth microwear (Teaford, Ungar, and Grine, 2002; Lee-Thorp et al., 2003; van der Merwe et al., 2003) are still overwhelmed by broad assumptions about early hominin plant food diet derived from analogies with living human and nonhuman primates and circumstantial evidence that ancient taxa lived, ate, and died in different types of habitats. Models can help us make progress on understanding the contributions of different types of foods to early hominin diets if they have a degree of realism that corresponds in scale to expected ranging behaviors. Models with a good balance of realism will help evaluate the significance of variability in the feeding opportunities that would have been available in specific habitats and help contextualize comparisons with archaeological, morphological, or chemical signatures of dietary choices that were made in those habitats.

The Unknowable

What can we never know from a modeling perspective on paleoanthropology? Because of the trade-offs between generality, realism, and precision, any individual model will always be constructed with large elements of "unknowable" built right in. By definition, a model is a simplification, a selection, a theoretical inquiry, and an iterative process. Models can be thought of as multilayered, dynamic, cognitive maps of the structure of our knowledge. They have value because they encourage us to define explicitly and redefine variables and to move from an analysis of individual variables or specific data sets toward an understanding of dependent relationships and complex systems. Without modeling we have questions to answer based on lists of hominin specimens with various morphological traits, found at fossil localities associated with lists of other fossil taxa and artifacts. With modeling we have the same empirical data, but our questions are framed in intersecting ways, like the challenge of a crossword puzzle, where answers to one question provide clues to the next. We face real taphonomic and sampling limits on our ability to recover empirical evidence of the past, but our abilities to interpret the evidence we do have are enhanced by modeling processes that help us generate expectations that can be tested against the paleoanthropological record.

Lewis Binford (2001) used a theater metaphor developed by Hutchinson (1965) to describe the process used to employ data from ethnography and environmental sciences in the interpretation of archaeological data. This metaphor can also characterize the theoretical challenges we face when trying to understand what is known, unknown, and unknowable about early hominin diet in paleoenvironmental context. In many ways, we have created a hominin puppet theater in the literature—a series of static environmental backdrops in front of which we portray different species of hominin, each one based on an underlying primate or human template, distinguished by a few, key morphological traits derived from our existing data sets. Sometimes those traits may seem overly exaggerated because of taphonomic or theoretical biases. All our puppets appear as patchworks of old and new, familiar and novel. The familiar traits are based on foundations of primate or human analogs, with unique features or novel traits derived from empirical evidence. Some hominin puppets have been carefully developed over the years, with their features created as layers of papier-mâché, while others look more like unfinished collages, each trait glued on as separate element. What are the theater backdrops for these hominin puppets? Some environmental sets are detailed and intimate, filled with foreground details of microfauna, phytoliths, soil chemistry, and other data that reflect relatively short-term, local indicators of ancient habitat that literally would have filled the soil under bipedal toes. Other settings are painted in broad strokes, portraying selected regional views that are unfocused in time and place but dramatize themes of forest and savanna. What scenes are played out on these stages? In a data-driven puppet show, there are limited interactions between the puppets or between the puppet and the setting. We can watch short paleontological scenes of death and burial. Sometimes we view acts of behavior—walking across a floodplain, dieing near the mouth of a cavern, flaking a river cobble, slicing an antelope forelimb, chewing on something tough and fibrous—glimpses of discrete behaviors portrayed in the context of an environmental setting. But socioecological interactions are often missing from these palimpsests, and time is fleeting or compressed. The relationships between the hominins and their environments seem either disconnected, superficial, or generalized. To add action to the scenes, and connect them into stories of life history or evolution, we must rely on our contemporary understanding of principles of behavioral ecology and our ability to develop models of species and habitats that are now extinct. Am I trying to characterize our science as mere puppetry? Of course not. But this metaphor may help explain the broad dimensions that shape "the unknown" and of the challenges we face modeling the significance of paleoenvironmental context for early hominin diet.

References

Aiello, L.C., and Key, C., 2002. Energetic consequences of being a *Homo erectus* female. *Am. J. Hum. Biol.* 14, 551–565.

Aiello, L.C., and Wheeler, P., 1995. The expensive-tissue hypothesis. *Curr. Anthropol.* 36, 199–222.

Altmann, S.A., 1998. *Foraging for Survival: Yearling Baboons in Africa*. University of Chicago Press, Chicago.

Bartholomew, G.A., and Birdsell, J.B., 1953. Ecology and the protohominins. *Am. Anthropol.* 55, 481–498.

Binford, L.R., 1981. *Bones: Ancient Men and Modern Myths*. Academic Press, New York.

Binford, L.R., 2001. Constructing Frames of Reference: An Analytical Method for Archaeological Theory Building Using Ethnographic and Environmental Data Sets. University of California Press, Berkeley.

Blumenschine, R.J., 1986a. Carcass consumption sequences and the archaeological distinction of scavenging and hunting. *J. Hum. Evol.* 15, 639–659.

Blumenschine, R.J., 1986b. *Early Hominid Scavenging Opportunities: Implications of Carcass Availability in the Serengeti and Ngorongoro Ecosystems*. British Archaeological Reports, Oxford.

Blumenschine, R.J., 1988. An experimental model of the timing of hominid and carnivore influence on archaeological bone assemblages. *J. Archaeol. Sci.* 15, 483–502.

Blumenschine, R.J., 1989. A landscape taphonomic model of the scale of prehistoric scavenging opportunities. *J. Hum. Evol.* 18, 345–371.

Blumenschine, R.J., and Madrigal, T.C., 1993. Variability in long bone marrow yields of East African ungulates and its zooarchaeological implications. *J. Archaeol. Sci.* 20, 555–587.

Blumenschine, R.J., and Peters, C.R., 1998. Archaeological predictions for hominid land use in the paleo-Olduvai basin, Tanzania, during lowermost Bed II times. *J. Hum. Evol.* 34, 565–607.

Bobe, R., and Behrensmeyer, A.K., 2004. The expansion of grassland ecosystems in Africa in relation to mammalian evolution and the origin of the genus Homo. *Palaeogeogr. Palaeoclimatol. Palaeoecol.* 207, 399–420.

Brain, C.K., 1981. The evolution of man in Africa: Was it the result of Cainozoic cooling? *Annex. Trans. Geol. Soc. S. Afr. J. Sci.* 84, 1–19.

Bunn, H.T., 1991. A taphonomic perspective in the archaeology of human origins. *Annu. Rev. Anthropol.* 20, 433–467.

Bunn, H.T., 2001. Hunting, power scavenging, and butchering by Hadza foragers and by Plio-Pleistocene *Homo*. In: Stanford, C.B., and Bunn, H.T. (Eds.), *Meat Eating and Human Evolution*. Oxford University Press, New York, pp. 199–218.

Bunn, H.T., Bartram, L.E., and Kroll, E.M., 1988. Variability in bone assemblage formation from Hadza hunting, scavenging and carcass processing. *J. Anthropol. Archaeol.* 7, 412–457.

Cavallo, J.A., and Blumenschine, R.J., 1989. Tree-stored leopard kills: Expanding the hominid scavenging niche. *J. Hum. Evol.* 18, 393–399.

Charnov, E.L., 1993. *Life History Invariants: Some Explorations of Symmetry in Evolutionary Ecology*. Oxford University Press, Oxford.

Conklin-Brittain, N.L., Wrangham, R.W., and Smith, C.C., 2002. A two-stage model of increased dietary quality in early hominid evolution: The role of fiber. In: Ungar, P.S., and Teaford, M.F. (Eds.), *Human Diet: Its Origin and Evolution*. Bergin & Garvey, Westport CT, pp. 61–76.

Copeland, S.R., 1999. Plant foods for hominins in east African habitats. *Crosscurrents* 11, 24–31.

Costanza, R., Low, B.S., Ostrom, E., and Wilson, J.A., 2001. Ecosystems and human systems: A framework for exploring the linkages. In: Constanza, R., Low, B.S., Ostrom, E., and Wilson, J. (Eds.), *Institutions, Ecosystems and Sustainability*. Lewis, Boca Raton, FL, pp. 3–20.

Costanza, R., and Ruth, M., 2001. Dynamic systems modeling. In: Constanza, R., Low, B.S., Ostrom, E., and Wilson, J. (Eds.), *Institutions, Ecosystems and Sustainability*. Lewis Publishers, Boca Raton, pp. 21–29.

Costopoulos, A., 2001. Evaluating the impact of increasing memory on agent behavior: Adaptive patterns in an agent based simulation of subsistence. *J. Artif. Soc. Social Simul.* 4.

Dart, R., 1925. *Australopithecus africanus*: The man ape of South Africa. *Nature* 115, 195–199.

deMenocal, P.B., 1995. Plio-Pleistocene African Climate. *Science* 270, 53–59.

deMenocal, P.B., 2004. African climate change and faunal evolution during the Pliocene-Pleistocene. *Earth Planet. Sci. Lett.* 220, 3–24.

Dominguez-Rodrigo, M., 1996. A landscape study of bone preservation in the Galana and Kulalu (Kenya) ecosystem. *Origini* 20, 17–38.

Dominguez-Rodrigo, M., 2001. A study of carnivore competition in riparian and open habitats of modern savannas and its implications for hominid behavioral modeling. *J. Hum. Evol.* 40, 77–98.

Dominguez-Rodrigo, M., and Pickering, T.R., 2003. Early hominid hunting and scavenging: A zooarchaeological review. *Evol. Anthropol.* 12, 275–282.

Dominguez-Rodrigo, M., Serrallonga, J., Juan-Tresserras, J., Alcala, L., and Luque, I., 2001. Woodworking activities by early humans: A plant residue analysis on Acheulian stone tools from Peninj (Tanzania). *J. Hum. Evol.* 40, 289–299.

Dunbar, R.I.M., 1976. Australopithecine diet based on a baboon analogy. *J. Hum. Evol.* 5, 161–167.

Dunbar, R.I.M., 2002. Modelling primate behavioral ecology. *Int. J. Primatol.* 23, 785–819.

Feibel, C., 1999. Basin evolution, sedimentary dynamics, and hominid habitats in East Africa. In: Bromage, T.G., and Schrenk, F. (Eds.), *African Biogeography, Climate Change, and Human Evolution*. Oxford University Press, Oxford, pp. 276–281.

Foley, R.A., and Lee, P.C., 1991. Ecology and energetics of encephalization in hominid evolution. *Philos. Trans. R. Soc. Lond.* B, 63–73.

Gifford-Gonzalez, D., 1991. Bones are not enough: analogues, knowledge, and interpretive strategies in zooarchaeology. *J. Anthropol. Archaeol.* 10, 215–254.

Grine, F.E., 1986. Ecological causality and the pattern of Plio-Pleistocene hominid evolution in Africa. *S. Afr. J. Sci.* 82, 87–89.

Harding, R.S.O., and Teleki, G. (Eds.), 1981. *Omnivorous Primates: Gathering and Hunting in Human Evolution*. Columbia University Press, New York.

Hatley, T., and Kappelman, J., 1980. Bears, pigs, and Plio-Pleistocene hominins: A case for the exploitation of belowground food resources. *Hum. Ecol.* 8, 371–387.

Hawkes, K., 1993. Why hunter-gatherers work: An ancient version of the problem of public goods. *Curr. Anthropol.* 34, 341–362.

Hawkes, K., O'Connell, J.F., and Blurton Jones, N., 1991. Hunting income patterns among the Hadza: Big game, common goods, foraging goals, and the evolution of the human diet. *Philos. Trans. R. Soc. Lond.* B 334, 243–251.

Hawkes, K., O'Connell, J.F., and Blurton Jones, N.G., 1995. Hadza children's foraging: Juvenile dependency, social arrangements and mobility among hunter-gatherers. *Curr. Anthropol.* 36, 688–700.

Hawkes, K., O'Connell, J.F., Blurton Jones, N.G., Charnov, E.L., Alvarez, H., 1998a. Grandmothering, menopause, and the evolution of human life histories. *Proc. Natl. Acad. Sci.* 95, 1336–1339.

Hawkes, K., O'Connell, J.F., Blurton Jones, N.G., Charnov, E.L., and Alvarez, H., 1998b. The grandmother hypothesis and human evolution. In: Cronk, L., Chagnon, N., and Irons, W. (Eds.), *Adaptation and Human Behavior: An Anthropological Perspective*. Aldine de Gruyter, Hawthorne NY, pp. 237–260.

Hawkes, K., O'Connell, J.F., and Jones, N.G.B., 1997. Hadza women's time allocation, offspring provisioning, and the evolution of long postmenopausal life spans. *Curr. Anthropol.* 38, 551–577.

Hutchinson, G.E., 1965. *The Ecological Theatre and the Evolutionary Play*. Yale University Press, New Haven, CT.

Isaac, G.L., 1978. The food-sharing behavior of protohuman hominins. *Sci. Am.* 311–325.

Isaac, G.L., 1984. The archaeology of human origins: Studies of the lower Pleistocene in East Africa 1971–1981. In: Wendorf, F. (Ed.), *Advances in World Archaeology*. Academic Press, New York, pp. 1–87.

Janssen, M.A., Sept, J., and Griffith, C., 2005. Foraging of *Homo ergaster* and *A. boisei* in East African environments. Available at the NAACSOS Conference 2005 Proceedings: http://www.casos.cs.cmu.edu/events/conferences/2005/conference_papers.php

Jolly, C.J., 1970. The seed-eaters: A new model of hominid differentiation based on a baboon analogy. *Man* 5, 5–26.

Kennedy, G.E., 2003. Palaeolithic grandmothers? Life history theory and early *Homo*. *J. R. Anthropol. Inst.*, n.s., 9, 549–572.

Kennedy, G.E., 2005. From the ape's dilemma to the weanling's dilemma: early weaning and its evolutionary context. *J. Hum. Evol.* 48, 123–145.

Krebs, J.R., and Davies, N.B. (Eds.), 1984. *Behavioral Ecology: An Evolutionary Approach*. Sinauer, Sunderland, MA.

Lake, M.W., 1999. MAGICAL Computer simulation of Mesolithic foraging. In: Kohler, T.A., and Gumerman, G.J. (Eds.), *Dynamics in Human and Primate Societies: Agent-Based Modeling of Social and Spatial Processes*. Oxford University Press, New York, pp. 107–143.

Lake, M.W., 2001. MAGICAL computer simulation of Mesolithic foraging in Islay. In: Mithen, S. (Ed.), *Hunter-Gatherer Landscape Archaeology, The Southern Hebrides Mesolithic Project 1988–1995*. Vol 2: Archaeological Fieldwork on Colonsay, Computer Modelling, Experimental Archaeology and Final Interpretations. McDonald Institute for Archaeological Research, Cambridge, pp. 465–495.

Lee-Thorp, J., Sponheimer, M., and van der Merwe, N.J., 2003. What do stable isotopes tell us about hominid dietary and ecological niches in the Pliocene? *Int. J. Osteoarchaeol.* 13, 104–113.

Leonard, W.R., and Robertson, M.L., 1992. Nutritional requirements and human evolution: A bioenergetics model. *Am. J. Hum. Biol.* 4, 179–195.

Leonard, W.R., and Robertson, M.L., 1994. Evolutionary perspectives on human nutrition: the influence of brain and body size on diet and metabolism. *Am. J. Hum. Biol.* 6, 77–88.

Leonard, W.R., Robertson, M.L., Snodgrass, J.J., and Kuzawa, C.W., 2003. Metabolic correlates of hominid brain evolution. *Comp. Biochem. Physiol.* A 136, 5–15.

Levins, R., 1966. The strategy of model building in population biology. *Am. Sci.* 54, 421–431.

Lewis, M.E., 1997. Carnivoran paleoguilds of Africa: Implications for hominid food procurement strategies. *J. Hum. Evol.* 32, 257–288.

Marshall, F., 1994. Food sharing and body part representation in Okiek faunal assemblages. *J. Archaeol. Sci.* 21, 65–77.

McGrew, W.C., 2001.The other faunivory: Primate insectivory and early human diet. In: Stanford, C.B., and Bunn, H.T. (Eds.), *Meat-Eating and Human Evolution*. Oxford University Press, New York, pp. 160–178.

Milton, K., 1987. Primate diets and gut morphology: implications for hominid evolution. In: Harris, M., and Ross, E. (Eds.), *Food and Evolution: Toward a Theory of Human Food Habits*. Temple University Press, pp. 93–115.

Milton, K., 1999. A hypothesis to explain the role of meat-eating in human evolution. *Evol. Anthropol.* 8, 11–21.

Milton, K., 2002. Hunter-gatherer diets: Wild foods signal relief from the diseases of affluence. In: Ungar, P.S., and Teaford, M.F. (Eds.), *Human Diet: Its Origin and Evolution*. Bergin & Garvey, Westport, CT, pp. 111–122.

O'Connell, J.F., 1995. Ethnoarchaeology needs a general theory of behavior. *J. Archaeol. Res.* 3, 205–255.

O'Connell, J.F., and Hawkes, K., 1988. Hadza hunting, butchering, and bone transport and their archaeological implications. *J. Anthropol. Res.* 44, 113–161.

O'Connell, J.F., Hawkes, K., and Blurton Jones, N., 1990. Reanalysis of large mammal body part transport among the Hadza. *J. Archaeol. Sci.* 17, 301–316.

O'Connell, J.F., Hawkes, K., and Blurton Jones, N.G., 1999. Grandmothering and the evolution of *Homo erectus*. *J. Hum. Evol.* 36, 461–485.

O'Connell, J., Hawkes, K., and Blurton Jones, N., 2002. Meat-eating, grandmothering, and the evolution of early human diets. In: Ungar, P.S., and Teaford, M.F. (Eds.), *Human Diet: Its Origin and Evolution*. Bergin & Garvey, Westport, CT, pp. 49–60.

O'Connell, J.F., Hawkes, K., Lupo, K.D., and Blurton Jones, N.G., 2002. Male strategies and Plio-Pleistocene archaeology. *J. Hum. Evol.* 43, 831–872.

Owen-Smith, N., 1999. Ecological links between African savanna environments, climate change and early hominid evolution. In: Bromage, T.G., and Schrenk, F. (Eds.), *African Biogeography, Climate Change and Human Evolution*. Oxford University Press, Oxford, pp. 138–149.

Peccei, J., 2001. A critique of the grandmother hypotheses, old and new. *Am. J. Hum. Biol.* 13, 434–452.

Peters, C.R., 1987. Nut-like oil seeds: Food for monkeys, chimpanzees, humans, and probably ape-men. *Am. J. Phys. Anthropol.* 73, 333–363.

Peters, C.R., and O'Brien, E.M., 1981. The early hominid plant-food niche: Insights from an analysis of plant exploitation by *Homo*, *Pan*, and *Papio* in eastern and southern Africa. *Curr. Anthropol.* 22, 127–140.

Peters, C.R., and Blumenschine, R.J., 1995. Landscape perspectives on possible land use patterns for Early Pleistocene hominins in the Olduvai Basin, Tanzania. *J. Hum. Evol.* 29, 321–362.

Peters, C.R., and Blumenschine, R.J., 1996. Landscape perspectives on possible land use patterns for early Pleistocene hominins in the Olduvai Basin. In: Magori, C.C., Saanane, C.B., and Schrenk, F. (Eds.), *Four Million Years of Hominid Evolution in Africa. Papers in honour of Dr. Mary Douglas Leakey'sOutstanding Contribution in Palaeoanthropology*. Arusha, Tanzania, Kaupia Darmstadter Beitrage zur Naturgeschichte, pp. 175–222.

Peters, C.R., and Vogel, J.C. 2005. Africa's wild C4 plant foods and possible early hominid diets. *J. Hum. Evol.* 48, 219–236.

Potts, R., 1988. *Early Hominid Activities at Olduvai*. Aldine de Gruyter, New York.

Potts, R., 1996a. Evolution and climatic variability. *Science* 273, 922–923.

Potts, R., 1996b. *Humanity's Descent: The Consequences of Ecological Instability*. Avon Books, New York.

Potts, R., 1998. Environmental hypotheses of hominid evolution. *Yearb. Phys. Anthropol.* 41, 93–136.

Potts, R., 2004. Paleoenvironmental basis of cognitive evolution in Great Apes. *Am. J. Primatol.* 62, 209–228.

Reed, K.E., 1997. Early hominid evolution and ecological change through the African Plio-Pleistocene. *J. Hum. Evol.* 32, 289–322.

Reynolds, R., Whallon, R., and Goodhall, S., 2001. Transmission of cultural traits by emulation: an agent-based model of group foraging behavior. *J. Memet.* 4, March 2001.

Robinson, J.T., 1954. The genera and species of the Australopithecinae. *Am. J. Phys. Anthropol.* 12, 181–200.

Robinson, J.T., 1963.Adaptive radiation in the Australopithecines and the origin of man. In: Howell, F.C., and Bourlière, F. (Eds.), *African Ecology and Human Evolution.* Aldine, Chicago, pp. 385–416.

Schaller, G.B., and Lowther, G.P., 1969. The relevance of carnivore behavior to the study of early hominins. *Southwest. J. Anthropol.* 25, 307–341.

Schick, K.D., 1987. Modeling the formation of Early Stone Age artifact concentrations. *J. Hum. Evol.* 16, 789–808.

Schoeninger, M.J., Bunn, H.T., Murray, S., Pickering, T., and Moore, J., 2001. Meat-eating by the fourth African ape. In: Stanford, C.B., Bunn, H.T. (Eds.), *Meat-Eating and Human Evolution.* Oxford University Press, New York, pp. 179–195.

Selvaggio, M.M., 1994. Carnivore tooth marks and stone tool butchery marks on scavenged bones: Archaeological implications. *J. Hum. Evol.* 27, 215–228.

Sept, J., 2001. Modeling the edible landscape. In: Stanford, C.B., and Bunn, H.T. (Eds.), *Meat Eating and Human Evolution.* Oxford University Press, New York, pp. 73–98.

Sept, J., 1986. Plant foods and early hominins at site FxJj50, Koobi Fora, Kenya. *J. Hum. Evol.* 15, 751–770.

Sept, J., 1984. *Plants and Early Hominids: A Study of Vegetation in Situations Comparable to Early Archaeological Site Locations.* University of California Press, Berkeley.

Sept, J., 1990. Vegetation studies in the Semliki Valley, Zaire as a guide to paleoanthropological research. Va. Mus. Nat. Hist. Mem. 1, 95–121.

Sept, J., 1992a. Archaeological evidence and ecological perspectives for reconstructing early hominid subsistence behavior. In: Schiffer, M.B. (Ed.), *Archaeological Method and Theory.* University of Arizona Press, Tucson, pp. 1–56.

Sept, J., 1992b. Was there no place like home? A new perspective on early hominid sites from the mapping of chimpanzee nests. *Curr. Anthropol.* 33, 187–207.

Sept, J., 1994. Beyond bones: Archaeological sites, early hominid subsistence, and the costs and benefits of exploiting wild plant foods in east African riverine landscapes. *J. Hum. Evol.* 27, 295–230.

Smith, E.A., and Winterhalder, B. (Eds.), 1992. *Evolutionary Ecology and Human Behavior.* Aldine de Gruyter, New York.

Sponheimer, M., Lee-Thorp, J., de Ruiter, D., Codron, D., Codron, J., Baugh, A.T., and Thackeray, F., 2005. Hominins, sedges, and termites: New carbon isotope data from the Sterkfontein valley and Kruger National Park. *J. Hum. Evol.* 48, 301–312.

Stahl, A., 1984. Hominid dietary selection before fire. *Curr. Anthropol.* 25, 151–168.

Stanley, S.M., 1992. An ecological theory for the origin of *Homo*. Paleobiology 18, 237–257.

Stephens, D.W., and Krebs, J.R., 1986. *Foraging Theory.* Princeton University Press, Princeton, NJ.

Tappen, M., 2001. Deconstructing the Serengeti. In: Stanford, C.B., and Bunn, H.T. (Eds.), *Meat-Eating and Human Evolution. Oxford University Press*, New York, pp. 13–32.

Teaford, M.F., Ungar, P.S., and Grine, F.E., 2002. Paleontological evidence for the diets of African Plio-Pleistocene hominins with special reference to early Homo. In: Ungar, P.S., and Teaford, M.F. (Eds.), *Human Diet: Its Origin and Evolution.* Bergin and Garvey, Westport, CT, pp. 143–166.

Tooby, J., and DeVore, I., 1987.The reconstruction of hominid behavioral evolution through strategic modeling. In: Kinzey, W.G. (Ed.), *The Evolution of Human Behavior: Primate Models.* State University of New York Press, Albany, pp. 183–238.

van der Merwe, N. J., Thackeray, J. F., Lee-Thorp, J., and Luyt, J., 2003. The carbon isotope ecology and diet of *Australopithecus africanus* at Sterkfontein, South Africa. *J. Hum. Evol.* 44, 581–597.

Vasey, N., and Walker, A., 2001. Neonate body size and hominid carnivory. In: Stanford, C.B., Bunn, H.T. (Eds.), *Meat-Eating and Human Evolution.* Oxford University Press, New York, pp. 332–349.

Vincent, A., 1985. Plant foods in savanna environments: A preliminary report of tubers eaten by the Hadza of Northern Tanzania. *World Archaeol.* 17, 1–4.

Vrba, E., 1988. Late Pliocene climatic events and hominid evolution. In: Grine, F.,E. (Ed.), *Evolutionary History of the Australopithecines.* Aldine de Gruyter, New York, pp. 405–426.

Vrba, E.S., 1985. Ecological and adaptive changes associated with early hominid evolution. In: Delson, E. (Ed.), *Ancestors: The Hard Evidence.* Alan R. Liss, New York, pp. 63–71.

Vrba, E.S., 1995a. On the connections between paleoclimate and evolution. In: Vrba, E.S., Denton, G.H., Partridge, T.C., and Burckle, L.H. (Eds.), *Paleoclimate and Evolution, with Emphasis on Human Origins.* Yale University Press, New Haven, CT, pp. 24–48.

Vrba, E.S., 1995b.The fossil record of African antelopes (Mammalia, Bovidae) in relation to human evolution and paleoclimate. In: Vrba, E.S., Denton, G.H., Partridge, T.C., and Burckle, L.H. (Eds.), *Paleoclimate and Evolution, with Emphasis on Human Origins.* Yale University Press, New Haven, CT, pp. 385–424.

White, T.D., 1988. The comparative biology of "robust" *Australopithecus*: Clues from context. In: Grine, F.E. (Ed.), *Evolutionary History of the "Robust" Australopithecines.* Aldine de Gruyter, New York, pp. 449–484.

White, T.D., 1995. African omnivores: Global climatic change and Plio-Pleistocene hominins and suids. In: Vrba, E.S., Denton, G.H., Partridge, T.C., and Burckle, L.H. (Eds.), *Paleoclimate and Evolution with Emphasis on Human Origins.* Yale University Press, New Haven, CT, pp. 369–384.

Wood, B., and Strait, D., 2004. Patterns of resource use in early *Homo* and *Paranthropus*. *J. Hum. Evol.* 46, 119–162.

Woodburn, J.C., 1968. An introduction to Hadza ecology. In: Lee, R.B., and DeVore, I. (Eds.), *Man the Hunter.* University of Chicago Press, Chicago, pp. 49–55.

Wrangham, R.W., and Conklin-Brittain, N.L., 2003. Cooking as a biological trait. *Comp. Biochem. Physiol.* A 136, 35–46.

16

The Cooking Enigma

RICHARD WRANGHAM

This chapter considers the role of cooking in the evolution of human diet. People in every culture know how to make fire, and everywhere they use it to improve their food (Tylor, 1878; Gott, 2002; Wrangham and Conklin-Brittain, 2003; fig. 16.1). But "no beast is a cook," as Boswell (1773) asserted. This difference between humans and other animals has long been appreciated, and some have even used it to define us: Boswell (1773) called humans the "cooking animal." But while cooking is indisputably unique to humans, its evolutionary significance is a matter of debate. There are two contrasting views.

The first, which is conventional wisdom, sees cooking as merely one of many extra-oral food-processing techniques (such as pounding, grinding, or drying) that can raise food quality. According to this view, cooking may be valuable in facilitating meal preparation but it is not important for understanding human adaptation. For example, it would not be considered to have led to fundamental changes in the human digestive system. In line with this idea, the ecological effect of human diet choice and foraging strategies is often discussed without considering the influence of cooking (e.g., Kaplan et al., 2000). Similarly, studies of the evolution of human feeding behavior often focus on dietary composition without considering extra-oral food-processing in general or cooking in particular (e.g., Eaton and Konner, 1985; Ungar and Teaford, 2002). The essential implication is that human biological evolution was not influenced in any major ways by the adoption of cooking and that evolutionists can therefore ignore it.

The radical alternative is that cooking is a core human adaptation that has importantly directed our evolution, or as Coon (1954) wrote, that cooking was "the decisive factor in leading man from a primarily animal existence into one that was more fully human." This perspective suggests that for humans, unlike other

Figure 16.1 A baboon on a Hadza fire. © Frank Marlowe.

species, cooked food is a need rather than an option. Accordingly, our reliance on cooking results from certain features of human biology that have evolved in response to the control of fire, such as our small guts, small teeth, and slow life histories (Wrangham and Conklin-Brittain, 2003). From a dietary perspective, it means that humans are distinguished as much by what we do with our food as by the food sources themselves (whether meat, roots or grasses, for example). In short, this view sees Boswell's characterization of humans as the "cooking animal" as not only biologically but also evolutionarily significant (Wrangham et al., 1999; Ulijaszek, 2002).

A few authors take an intermediate position. Notably, Brace (1995) has argued that cooking is an important option that has led to limited evolutionary effects, particularly a reduction in tooth size.

In this chapter, I present arguments relevant to resolving this debate.

Why Cooking Is Expected to Have Evolutionary Effects

The contrasting views on the role of cooking agree in at least one respect. Both acknowledge that cooking improves food. Some benefits vary across food types, such as reducing physical barriers, changing molecular structure, reducing toxin loads, and de-frosting (Stahl, 1984; Brace, 1995; Wrangham and Conklin-Brittain, 2003). Others appear to be consistent. For example, cooking leads to bursting of cells, making food molecules more available. It also tenderizes meat and softens plant foods, thereby making chewing easier. In addition, it reduces water content and increases the proportion of edible material (Wrangham and Conklin-Brittain, 2003).

Exactly how these benefits translate into fitness has not been well established. However, current data suggest that they may lead to significant energetic savings. Thus, the cost of digestion is a high proportion of total energy expenditure in all animals. In humans it has been measured at around 5%–15% of energy expenditure (Westerterp 2004). In other animals, the cost may be higher, for example, up to 43% of energy intake in snakes (Secor and Faulkner, 2002). But the cost of digestion varies not only between species but also with food quality. For instance, high-protein diets increase the cost of digestion by about 30% compared with high-fat diets (Westerterp-Plantenga et al., 1999). Relevant to cooking, large meals that are physically hard cost more (e.g., 50%–100% increase in cost of digestion in toads; Secor and Faulkner, 2002).

By softening food and reducing meal size, therefore, cooking can be expected to reduce the cost of digestion, for example, by accelerating the digestive process. One measure of the rate at which foods are digested is the glycemic index, which assesses the rate of appearance of glucose in the blood following ingestion. As expected, the glycemic index is indeed consistently increased by cooking (Brand et al., 1985; Bjorck, Liljeberg, and Ostman, 2000). Experiments are needed to test the hypothesis that proteins and lipids are also digested and absorbed more rapidly in cooked than raw foods. If so, cooked food may prove to offer consistent energy savings across all food types. Possible avenues for cost-saving include reduced energetic cost per gram of food, reduced time for the gut to be metabolically active, and a reduced size of gut needed to digest the food.

Evidence that cooking consistently improves food quality is suggestive in the context of evolution because even a small change in food quality can have very important effects. Among Galápagos finches, for example, a brief period of ecological constraint that causes a shift in diet can lead to the rapid evolution of larger or smaller beaks by natural selection (Boag and Grant, 1981; Grant and Grant, 2002). Such evolutionary changes in digestive anatomy then constrain future diet choices. Even minor changes in dietary adaptations, in their turn, are known to have widespread effects on various aspects of species biology.

Chimpanzees (*Pan troglodytes*) and gorillas (*Gorilla gorilla*) offer an instructive example of this process. These two species have closely similar diets. Both choose ripe fruits when they are available, being almost equally frugivorous (Tutin and Fernandez, 1985; Remis, 1997; Wrangham, Conklin-Brittain, and Hunt, 1998). When ripe fruits are scarce, both species also supplement their diets with fibrous foods such as piths and leaves. Despite this strong overall dietary similarity, there is one important difference that emerges from multiple field studies of the two species (gorillas, 11 sites; chimpanzees, 12 sites; Wrangham, 2005). In habitats with little or no fruit, gorillas can survive by eating fibrous foods for 100% of their feeding time (Doran et al., 2002). Chimpanzees never do so (Basabose, 2002). This contrast is attributable to differences in digestive adaptation between the two species, probably including both dental traits and features of gut anatomy and dynamics. Thus, gorilla molars have long shearing edges compared with those of chimpanzees, and gut passage rates in gorillas are longer than in chimpanzees, allowing more opportunity for fermentation of plant fiber (Milton, 1999; Remis et al., 2001; Remis and Dierenfeld, 2004).

The relative ability of these two apes to rely on the foliar component of the diet might at first glance appear to be a trivial matter. But many consequences appear to follow from it, even aside from digestive adaptations. There are differences in distributional range for example. Gorillas successfully occupy high-altitude forests without fruits, where they live at high density with excellent survival and reproduction (Robbins, Sicotte, and Stewart, 2001). Chimpanzees, by contrast, continue to be selective frugivores when living at high altitudes. As a result they are limited to habitats below 2,600 m altitude, unlike gorillas, and at high altitudes they live in small, scattered groups (Yamagiwa et al., 1996; Basabose, 2002).

Life history differences also appear to be influenced by the dietary shift. In contrast to expectations from their being larger than chimpanzees, gorillas have a shorter and faster life history pattern compared with chimpanzees (Tutin, 1994). For example gorillas mature earlier, have an earlier first birth (gorillas around nine years; chimpanzees around fourteen years), and have a shorter interbirth interval (gorillas 3.9 years; chimpanzees 5.0–6.2 years; Knott, 2001). While the reasons for the accelerated schedule of growth and reproduction in gorillas are debated (Knott, 2001), they conform precisely to life history differences that are found between folivorous and nonfolivorous primates in general (Leigh, 1994). The ability to digest a leaf diet may demand rapid development of digestive abilities and is thought to allow a sufficiently predictable food regime that it permits the evolution of rapid rates of growth and reproduction (Janson and van Schaik, 1993; Leigh, 1994; Knott, 2001).

Finally, there is a striking species difference in grouping patterns. During periods with temporary fruit scarcity chimpanzees experience intense scramble competition for fruits, such that they then tend to travel in small groups or alone (Wrangham, 2000). But when fruit is scarce in gorilla habitats, groups of gorillas respond by eating more terrestrial foliage, with little tendency for groups to fragment (Goldsmith, 1999). Because this foliage is distributed homogeneously compared with the widely separated fruit trees required by chimpanzees, it apparently allows gorillas to remain in more stable groups than chimpanzees. The difference in grouping patterns, in turn, is probably responsible for other important contrasts between the species, such as in sexual behavior, the degree of sexual dimorphism, and aggression (Wrangham, 1979; Yamagiwa, 1999).

The comparison of chimpanzees and gorillas thus illustrates how a relatively small change in diet (an ability to survive on foliage without fruits) implies substantial effects on biogeography, life history and social behavior. Because cooking is universal and has many effects on the diet, it can reasonably be expected to have effects at least as large. For example, it should increase the range of edible foods and therefore allow extension into new biogeographical zones. Other things being equal, it should also provide a more predictable food supply during periods of scarcity because it enables a range of otherwise inedible items to be used. It should have further effects by softening food. For example, it should lead to a greater ability of adults to provision infants, whose dentition is too immature to allow hard chewing, other than by giving milk. It should likewise cause a substantial drop in the amount of time that individuals spend chewing, with large consequences for the species activity budgets (Wrangham and Conklin-Brittain, 2003).

In addition to raising food quality, cooking also radically changes the nature of food distribution. Thus a species that cooks is obliged to assemble food items into a location (onto or next to a fire) that is fixed for at least the time that it takes to cook. Unlike the ordinary feeding pattern of any nonhuman ape, therefore, this means that a cooking population is exposed to intragroup competition over a valuable accumulated food pile of food. Among other animals, including primates, the distribution of food is considered to be a key variable that sculpts social relationships. For example, among chimpanzees by far the most valuable type of concentrated food supply is meat. A successful hunt therefore commonly leads to intense competition, including direct aggression and various complex forms of social manipulation (Goodall, 1986).

In short, the adoption of cooking is expected to be accompanied by a series of large influences on various important biological systems, such as foraging behavior, digestive strategy, infant development, geographical range, and the regulation of social competition. The fact that cooking is a human universal, therefore, ought to be intensely provocative for students of human evolution because it raises the question of whether the practice of cooking has indeed influenced these and other systems.

Why Cooking Appears Not to Have Had Important Evolutionary Effects

Cooking was once considered to be an important influence on human evolution. In 1871, Charles Darwin considered "the art of making fire" as "probably the greatest [discovery], excepting language, ever made by man." He specifically cited the process "by which hard and stringy roots can be rendered digestible, and poisonous roots or herbs innocuous" (Darwin, 1871, 1:132). Likewise Frazer (1930, p. 1) similarly suggested that "Of all human inventions the discovery of the method of kindling fire has probably been the most momentous and far-reaching," and he discussed the importance of cooking in this respect.

But in the second half of the twentieth century, such ideas largely disappeared. The main reason appears to have been the pattern that has crystallized in archaeological data. Data showed that fire was controlled in several sites in southern Europe during the Middle Paleolithic back to at least 250,000 years ago, and probably as early as 300,000–500,000 years ago (James, 1989; Straus, 1989; Gamble, 1993; Monnier et al. 1994; Brace, 1995). This evidence has been widely regarded as so much stronger than any indications of the control of fire in earlier times that the Middle Paleolithic is now conventionally interpreted as the first time that humans used fire. There is admittedly scattered evidence for earlier control of fire, but none of it is sufficiently convincing to persuade the skeptics (table 16.1). In addition, there are some well-known sites dated earlier than 500,000 years that show no evidence for control of fire. For example, the *Homo antecessor* site of Atapuerca, Spain, is dated at 800,000 years and has been examined sufficiently carefully to suggest that the hominids there did not use fire (Arsuaga et al., 1997; de Castro et al., 1997). To those who regard absence of evidence as evidence of absence, these facts suggest

Table 16.1 Suggestive Evidence of the Control of Fire Before 400,000 Years Ago

Site	Mya	Evidence	Reference
Gesher-Benet, Ya'aqov, Israel	0.8	Burnt seeds, wood, flint; hearthlike pattern	Goren-Inbar, 2004
Chesowanja, Baringo, Kenya	>1.4	Burnt clay + stone tools; low disturbance; 600°C temperature; hearthlike pattern	Gowlett et al., 1981; Gowlett, 1999
Middle Awash, Ethiopia	>1.42	Red patches, phytoliths	Clark and Harris, 1985
Gadeb, Ethiopia	0.7–1.4	Thermomagnetism	Clark and Kurashina, 1980; Clark and Harris, 1985
Swartkrans, South Africa	1.5	Burned bones, 600°C temperature	Brain, 1993
Olduvai (+ Turkana), Tanzania	1.6	"Pot-lid" burning of basalt/quartz tools	Ludwig, 2000
Koobi Fora, Turkana, Kenya	1.6	Red patches ($n = 20$); archaeomagnetic; thermoluminescence; palynology	Barbetti, 1986; Bellomo, 1991; Rowlett, 2000

that the Middle Paleolithic was the first time when humans controlled fire (e.g., Carbonell, 1999).

According to such evidence, therefore, *Homo* must have relied on raw food before the Middle Paleolithic. In that case, if the adoption of cooking strongly influenced human biology, a suite of changes should be visible in our ancestors' evolutionary anatomy around 300,000–500,000 years ago. In fact, however, the evolutionary changes in anatomy that are recorded around that time were trivial. In Europe *Homo neanderthalensis* evolved from *Homo heidelbergensis*, while in Africa *H. heidelbergensis* (*rhodesiensis*) continued with little change until the origin of *Homo sapiens* around 200,000 years ago. This evolutionary quiescence implies that if fire was indeed first controlled in the Middle Paleolithic, cooking had little impact on human evolution.

The Cooking Enigma

These facts constitute a profound puzzle. On the one hand, cooking is absent among animals, universal in humans, and rich in potent biological consequences. It is therefore expected to have a strong impact on evolutionary biology. On the other hand, archaeological data place the acquisition of cooking at a time when nothing dramatic was happening in human evolution. The cooking enigma, therefore, is how cooking became a human universal without having visible effects on our evolutionary biology.

Three kinds of solution have been suggested. I label them here by the time when they conclude that cooking became a human universal. The first two ("Late" and "Sneak") are both based on the assumption that cooking has done little or nothing to influence our biology.

The Late Solution

The Late solution suggests that cooking has been adopted too recently to have had time to influence our evolutionary biology. Milton (2002, p. 112) suggested this idea: "Relatively recent changes in certain features of the modern human diet (e.g., cooking of most foods . . .) may, in an evolutionary sense, have occurred so rapidly and so recently that human biology has not yet had time to adapt to them." Since speciation can occur in less than 25,000 years (Gould, 2002), the Late solution implies that if cooking has not had the time to affect our species' biology, it must have been adopted very recently indeed (e.g. less than 25,000 years ago). To be reconciled with the Middle Paleolithic evidence for the control of fire, this hypothesis would have to suggest that fire was controlled for a long period without leading to cooking. That solution is hard to imagine. Even wild chimpanzees take advantage of natural fires to eat foods that have been cooked by chance (Brewer, 1978).

The Sneak Solution

The Sneak solution accepts the idea that cooking was adopted during the Middle Paleolithic and therefore concludes that it did little to affect human evolution or biology beyond eventually causing a reduction in tooth size (beginning around 100,000 years ago, Brace, 1995). In other words, cooking "sneaked" into human culture with minimal effect. In its favor, the Sneak solution provides a logical interpretation of the Middle Paleolithic archaeological evidence. Furthermore, it is compatible with the idea that cooking has only trivial nutritional effects, which was suggested by Lévi-Strauss (1969) and has not been completely abandoned. Against it lies the challenge of explaining why the apparently important results of adopting cooking, including a large improvement in the diet and a major change in the way in which it was distributed, did little or nothing to influence the course of human evolution. A possible solution is that previous food-processing techniques (such as pounding) closely mimicked the effects of cooking (Wrangham, 2006). However, such ideas have not been extensively developed.

The Basal Solution

The Basal solution is the radical hypothesis that cooking was adopted around the origin of *Homo erectus* and was responsible for many of the features that characterize human evolutionary changes from australopithecines. It was proposed on the basis that many of the evolutionary changes that accompany hominization are easily explicable as responses to cooking, such as the reduced jaw and teeth, evidence of smaller gut, and yet higher energy expenditure (Wrangham et al., 1999). This solution faces the challenge of explaining why evidence of the control of fire is scarce before about 400,000 years ago. It must also be reconciled with the traditional idea that meat eating was the prime dietary mover of the evolution of the genus *Homo*.

I now consider evidence relevant to each of the three solutions.

The Known

It is known that all human populations cook their food and that cooking consistently increases the palatability of food. It is also known that a diet of raw plant food creates substantial energetic problems for humans under even the best conditions. The implication is that humans are adapted to eating cooked food.

First, under subsistence conditions, there is no evidence of any population of humans, or even any individual, having lived off raw wild foods for more than a few days at a time. The longest period that I have found was Helena Valero's report of living alone in the 1930s for some seven months in the forests of Venezuela and Brazil, after escaping from Yanomamö Indians (Valero and Biocca, 1970). She began her adventure carrying a firebrand wrapped in leaves. After a few days a heavy rain put out her only fire. Not daring to steal fire from the villages, she was close to starvation until she found an abandoned banana plantation. She survived by eating raw bananas. This exceptional case illustrates that there is nothing specifically impossible about living off raw food, but the fact that the food was domesticated means that it tells us nothing about the capacity of humans to survive off wild foods.

Likewise, people often eat particular items raw (including various fruits or roots, or choice animal products such as blood or the fat in the tail of a fat-tailed sheep). But such items are normally part of a diet that also includes cooked food. Diets that are restricted to raw items are rare in all societies. They appear to be recorded most often among religious extremists and warriors on the march, who benefit by not using fires but are delighted when they can revert to cooking (Fernández-Armesto, 2002).

Thus there appear to be no cases of long-term survival on raw food in the wild.

Second, even under the most favorable conditions people who attempt to restrict their diets to raw food do not thrive. The most extreme examples are members of modern raw-food movements, who tend to live under urban conditions in which activity levels are low and the diet consists of domesticated agricultural plants, with high-quality items available year-round. Much of the food that these devotees eat is actually processed: methods include sprouting, pressing, and even drying up to 60°C (i.e., the temperature at which enzymes are supposedly killed), which under some definition would be scored as cooking.

Even under such benign conditions, people experience low caloric intake. There have been few attempts to calculate caloric intake, but in one energy assessment of a "raw-foodist" diet (for 141 vegans), daily caloric intake was 1,460 kcal and 1,830 kcal for women and men, respectively (Donaldson, 2001). This is less than needed for modern humans or predicted for *H. erectus* (Aiello and Wells, 2002). In addition, female reproductive function is seriously impaired on a raw-foodist diet. Thus, approximately 50% of German women on a 100% raw-foodist diet were amenorrheic, and more could be assumed to be subfecund (Koebnick et al., 1999). In sum, the sparse current evidence suggests that raw-food diets produce inadequate energy for humans even under excellent conditions, at least when the diet is dominated by plant items. Therefore, the implication is that in the wild, they could not thrive.

Third, even the most committed raw-foodists find it difficult to keep to their régime because they are consistently hungry even when they eat as much as possible. Some reduce this problem by including small amounts of cooked food in their diets, for example, up to 30% of the diet (e.g., Koebnick et al., 1999).

These studies of raw-food diets are largely of vegetarians. They therefore do not resolve the question of whether humans can live off a raw diet that contains sufficient meat. However, they are still illuminating because they show that human digestion is incapable of the performance of a chimpanzee or gorilla.

Both chimpanzees and gorillas, by contrast to humans, would undoubtedly perform excellently on human raw-foodist diets, given that human raw-food diets offer superabundant access to foods with substantially lower-fiber concentrations than those in the diets of wild apes. We can therefore confidently conclude that digestion is adapted differently in humans from chimpanzees and that humans in subsistence society need cooked food under many and possibly all practical circumstances.

The Unknown

Humans are not known to be able to survive on raw food, which suggests that during our evolution, we became physiologically committed to eating foods of such high quality that in most circumstances they had to be cooked. However, important gaps in our knowledge concern whether humans can survive on a raw diet with sufficient meat, whether human guts are better adapted to raw meat or to cooked food, what effects cooking has on food quality, and the conditions under which cooking can be recognized archaeologically.

First, although raw plant food is evidently a poor diet for humans, a sufficient inclusion of raw meat might, in theory, create an energetically adequate diet. The diets of Arctic foragers included a substantial component of raw meat, for example. Experiments are therefore needed to assess the optimal balance of meat and plants for humans to maximize caloric intake on a raw diet.

The theoretical reasons why a raw-food diet (at least when dominated by plant items even of the highest quality) is expected to be difficult for humans have not yet been elaborated. However, two obvious possibilities are that the plant component is bulky and that it would take a long time for humans to chew enough raw wild meat to satisfy requirements (Wrangham and Conklin-Brittain, 2003). The problem of excessive bulk comes from the fact that humans not only have small guts in total, and small hindguts in particular, in comparison to great apes, but they also have fast rates of gut passage. Small fermenting volumes (such as hindguts) and fast-gut passage rates are generally associated with diets having low levels of plant fiber and high density of calories because they do not allow adequate time or volume for retention of digesta, or for fermentation of fiber to volatile fatty acids (Milton, 1999; Lambert, 2002; Remis and Dierenfeld, 2004). Raw plant diets of chimpanzees have fiber levels averaging 32% NDF in one study, far higher than the values in modern human diets (around 10%; Conklin-Brittain, Wrangham, and Hunt, 1998; Conklin-Brittain, Wrangham, and Smith, 2002).

Second, although the small and distinctive guts of humans appear well adapted to eating cooked food, the traditional conclusion from studies of gut anatomy and function is that the signals of adaptation to low-fiber foods of high caloric density reflect adaptation to a diet containing large amounts of high-quality raw food. For example, following Chivers and Hladik (1980), MacLarnon et al. (1986), Milton (1987), and others compared human gut proportions to other animals eating principally fauna, fruit, or insects. Milton (1987, p. 103) noted that human gut proportions were similar to those of *Cebus* and *Papio*, species that eat many insects and process their food with their hands. She concluded that the similarity represented "similar adaptive trends in gut morphology in response to diets made up of unusually high-quality dietary items that are capable of being digested and absorbed primarily in the small intestine."

The "unusually high-quality dietary items" to which humans are evidently adapted are normally considered to be meat and high-quality plant items such as fruits and seeds. But because researchers in this tradition have rarely considered whether cooked food provides a more reasonable explanation for the human digestive characteristics, the hypothesis that human guts are more closely adapted to cooked food than to raw meat remains to be tested. It should take into account that if the adaptation was to raw meat, plants had also to be eaten raw or not at all. This makes the fact that the human gut is poorly adapted to eating raw plant food puzzling.

There is therefore an obvious case for cooked food as the cause of the distinctive human guts. Human adaptation to a meat diet is certainly not as complete as in carnivores. Thus human teeth show little evidence of adaptation for carnivory, despite evidence of a slight shift toward long shearing edges in early hominids (Ungar, 2004). However, small teeth, as characterized by *Homo* compared to australopithecines, appear to be well adapted for eating cooked food (Lieberman et al., 2004; Lucas, 2004). Likewise, although the human digestive system does not appear to have been compared systematically with those of carnivores, it is clear that gut kinetics are radically different. Important, for example, although gut passage rates are similar between humans and dogs, dogs retain food in their stomachs for much longer (around four to twelve hours) than humans (around one to two hours; Ragir, Rosenberg, and Tierno, 2000). In addition, humans show little evidence of being able to survive purely on meat diets, even when cooked (Speth and Spielmann, 1983).

Third, the nutritional effects of cooking are not well understood. For example, it is often suggested that cooking might increase digestibility. But if digestibility means "the proportion (by dry weight) of a food item that can be digested," which is the standard usage, it is not necessarily an important variable. For example, the digestibility of a piece of pure meat appears to be close to 100% regardless of whether it is cooked.

But even if the digestibility of meat is not affected by cooking, we can expect its nutritional value of meat to be increased as a result of cooking's effect in making it tender. Tenderizing should lead to a shorter time for a given weight of meat to be chewed and/or for it to be subsequently digested. By accelerating the digestive process, it can likewise be expected to increase the rate of gastric emptying and therefore to allow a shorter intermeal interval between meals (Petring and Blake, 1993; Pera et al., 2002).

Such effects of cooking should affect both meat and plant foods. A major effect of cooking, accordingly, may be that by tenderizing or softening food, it shortens the digestive process and therefore both reduces the energetic costs of digestion and increases the rate at which total calories can be absorbed (if the individual maximizes the rate of ingestion). This prediction conforms to evidence that merely by eating a softer diet animals have such reduced energetic costs of digestion that they have significantly higher energy gain (Oka et al., 2003; Wrangham, 2006).

Fourth, it is too early to be confident that the Middle Paleolithic marked the first control of fire, given the persistence of evidence from earlier times (table 16.1). The recent claim for control of fire at 790,000 years at the Benot Ya'aqov site in Israel is supported by burned flints, seeds, and wood of six species found in hearthlike spatial patterns (Goren-Inbar et al., 2004). If further evidence for control of fire is found around this time but not earlier, it will support the notion that the first fire-using species was *H. heidelbergensis*, as Foley (2002) suggested. This would not solve the cooking enigma, however, because the anatomical changes from *H. erectus* to *H. heidelbergensis* were rather small.

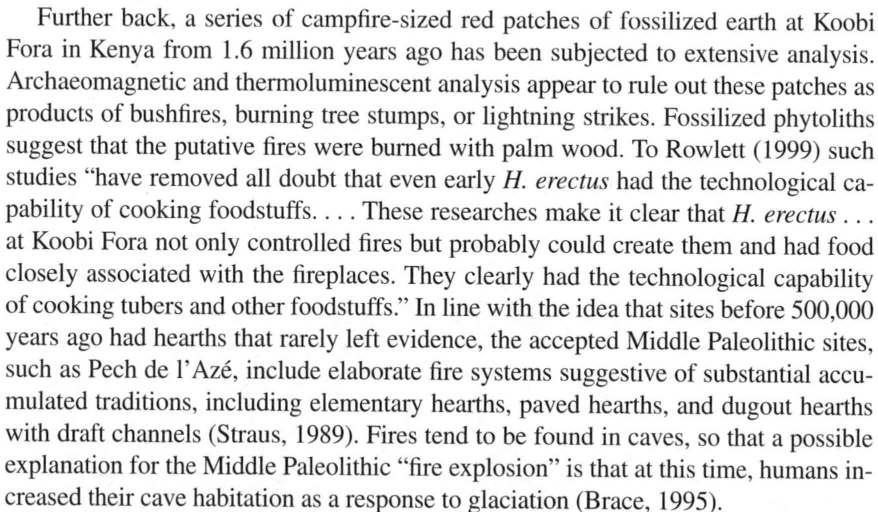

Further back, a series of campfire-sized red patches of fossilized earth at Koobi Fora in Kenya from 1.6 million years ago has been subjected to extensive analysis. Archaeomagnetic and thermoluminescent analysis appear to rule out these patches as products of bushfires, burning tree stumps, or lightning strikes. Fossilized phytoliths suggest that the putative fires were burned with palm wood. To Rowlett (1999) such studies "have removed all doubt that even early *H. erectus* had the technological capability of cooking foodstuffs. . . . These researches make it clear that *H. erectus* . . . at Koobi Fora not only controlled fires but probably could create them and had food closely associated with the fireplaces. They clearly had the technological capability of cooking tubers and other foodstuffs." In line with the idea that sites before 500,000 years ago had hearths that rarely left evidence, the accepted Middle Paleolithic sites, such as Pech de l'Azé, include elaborate fire systems suggestive of substantial accumulated traditions, including elementary hearths, paved hearths, and dugout hearths with draft channels (Straus, 1989). Fires tend to be found in caves, so that a possible explanation for the Middle Paleolithic "fire explosion" is that at this time, humans increased their cave habitation as a response to glaciation (Brace, 1995).

There are many other unknowns, of course; several of which are likely to be knowable eventually. We know nothing about the genetic differences between *Pan* and *Homo* relevant to adaptation to a cooked diet, or when the relevant genes spread in the human lineage. We have no developed theory of how the change in food distribution resulting from eating at cooking fires would have influenced the regulation of social competition, social norms, group size, or cognition. Also, scant thought has been given to how the more predictable and energy-rich diet offered by cooking would have affected life histories. It is intriguing to speculate that it would have allowed more investment in the immune system and therefore in longer lives.

The Unknowable

Current data give us little hope of knowing much about the detailed origins of cooking. For example, how was fire first controlled? How long was it used without being made? How long was it controlled without being used for cooking? And what methods of food preparation were used by humans before cooking? There are no obvious ways to answer these questions. As a result, even when the archaeological evidence can more clearly distinguish between possible and definite sites for the control of fire, we may still be uncertain whether a given population of fire users cooked their food.

Summary

Evidence from raw-foodists indicate that under subsistence conditions humans would not survive long term on the kinds of raw foods that are available in the wild. This implies that, as indicated by our digestive systems, humans are adapted to dietary items of unusually high quality compared with other species and that humans' high diet quality normally comes from the food being cooked. Therefore, given that humans currently depend on cooking for their high-quality food, a key question is how long we have done so. There are three types of solution, each with its own puzzle.

The Late solution suggests that cooking is recent, that is, probably less than 25,000 years ago. This has the merit of explaining why the advent of cooking did little to influence the course of human evolution. It faces the considerable difficulty, however, of explaining how humans could have exerted sophisticated control of fire for at least 250,000 years without using it to cook their food. The Late solution is therefore highly improbable.

The Sneak solution suggests that cooking has been practiced for at least 250,000 years without causing any dramatic changes in human body size, sexual dimorphism, tooth size, or gross morphology. This response to the archaeological evidence leaves unanswered the puzzle of how an adaptive change in diet with large apparent consequences for various biological systems occurred without leaving its mark on human anatomy. Although the Sneak solution is the standard answer to the cooking enigma, no serious attempts have been made to explain why cooking apparently had such small effects on human evolution.

The Basal solution suggests that cooking has been practiced since the origin of *H. erectus* and was responsible for many of the morphological changes associated with the evolution of *H. erectus*. This fits the many indications that early *Homo* had an unusually high-quality diet (such as small teeth and jaws and long-distance locomotion). It faces the challenge, however, of understanding why archaeological sites before 500,000 years ago show no evidence of the control of fire that is sufficient to convince skeptics.

References

Aiello, L.C., and Wells, J.C.K., 2002. Energetics and the evolution of the genus *Homo*. *Annu. Rev. Anthropol.* 31, 323–338.

Arsuaga, J.L., Martinez, I., García, A., Carretero, J.M., Lorenzo, C., García, N., and Ortega, A.I., 1997. Sima de los Huesos (Sierra de Atapuerca, Spain). The Site. *J. Hum. Evol.* 33, 109–127.

Barbetti, M., 1986. Traces of fire in the archaeological record before one million years ago. *J. Hum. Evol.* 15, 771–781.

Basabose, A.K., 2002. Diet composition of chimpanzees inhabiting the montane forest of Kahuzi, Democratic Republic of Congo. *Am. J. Primatol.* 58, 1–21.

Bellomo, R.V., 1991. Identifying traces of natural and humanly-controlled fire in the archaeological record: The role of actualistic studies. *Archaeol. Mont. Butte* 32, 75–93.

Bjorck, I., Liljeberg, H., and Ostman, E., 2000. Low glycaemic-index foods. *Br. J. Nutr.* 83, S149–S155.

Boag, P.T., and Grant, P.R., 1981. Intense natural selection in a population of Darwin's finches (*Geospizinae*) in the Galápagos. *Science* 214, 82–85.

Boswell, J., 1773. *Journal of a Tour to the Hebrides with Samuel Johnson, LL. D.* Oxford University Press, London.

Brace, C.L., 1995. *The Stages of Human Evolution*. Prentice-Hall, Englewood Cliffs, NJ, Prentice-Hall.

Brain, C.K., 1993. The occurrence of burnt bones at Swartkrans and their implications for the control of fire by early hominids. In: Brain, C.K. (Ed.), Swartkrans. *A Cave's Chronicle of Early Man*. Transvaal Museum Monograph No. 8, Transvaal, pp. 229–242.

Brand, J.C., Nicholson, P.L., Thorburn, A.W., and Truswell, A.S., 1985. Food processing and the glycemic index. *Am. J. Clin. Nutr.* 42, 1192–1196.

Brewer, S., 1978. *The Forest Dwellers*. Collins, London.

Carbonell, E., 1999. "Comment" on Wrangham et al. (1999). *Curr. Anthropol.* 40, 580–581.

Chivers, D.J., and Hladik, C.M., 1980. Morphology of the gastrointestinal tract in primates: Comparison with other mammals in relation to diet. *J. Morphol.* 166, 337–386.

Clark, J.D., and Harris, J.W.K., 1985. Fire and its role in early hominid lifeways. *Afr. Archaeol. Rev.* 3, 3–27.

Clark, J.D., and Kurashina, H., 1980. New Plio-Pleistocene archaeological occurrences from the Plain of Gadeb, Upper Webi Schebele basin, Ethiopia, and a statistical comparison of the Gadeb sites with other Early Stone Age assemblages. *L'Anthropologie* 18, 161–187.

Conklin-Brittain, N.L., Wrangham, R.W., and Hunt, K.D., 1998. Dietary response of chimpanzees and cercopithecines to seasonal variation in fruit abundance. II. Macronutrients. *Int. J. Primatol.* 19, 971–998.

Conklin-Brittain, N.L., Wrangham, R.W., and Smith, C.C., 2002. A two-stage model of increased dietary quality in early hominid evolution: The role of fiber. In: Ungar, P., and Teaford, M. (Eds.), *Human Diet: Its Origin and Evolution*. Bergin & Garvey, Westport, CT, pp. 61–76.

Coon, C.S., 1954. *The Story of Man: From the First Human to Primitive Culture and Beyond*. Knopf, New York.

Darwin, C., 1871. The Descent of Man and Selection in Relation to Sex. *Encyclopaedia Britannica*, Chicago.

de Castro, J.M.B., Arsuaga, J.L., Carbonell, E., Rosas, A., Martinez, I., and Mosquera, M., 1997. A hominid from the Lower Pleistocene of Atapuerca, Spain: Possible ancestor to Neanderthals and modern humans. *Science* 276, 1392–1395.

Donaldson, M.S., 2001. Food and nutrient intake of Hallelujah vegetarians. *Nutr. Food Sci.* 31, 293–303.

Doran, D.M., and McNeilage, A., 1998. Gorilla ecology and behavior. *Evol. Anthropol.* 6, 120–131.

Doran, D.M., McNeilage, A., Greer, D., Bocian, C., Mehlman, P., and Shah, N., 2002. Western lowland gorilla diet and resource availability: New evidence, cross-site comparisons, and reflections on indirect sampling methods. *Am. J. Primatol.* 58, 91–116.

Eaton, S.B., and Konner, M., 1985. Paleolithic nutrition: A consideration of its nature and current implications. *New Engl. J. Med.* 312, 283–289.

Fernández-Armesto, F., 2002. *Near a Thousand Tables: A History of Food*. Free Press, New York.

Foley, R., 2002. Adaptive radiations and dispersals in hominin evolutionary ecology. *Evol. Anthropol.* 11, 32–37.

Frazer, J.G., 1930. *Myths of the Origins of Fire*. Hacker Art Books, New York.

Gamble, C.S., 1993. *Timewalkers: The Prehistory of Global Civilization*. Harvard University Press, Cambridge, MA.

Goldsmith, M.L., 1999. Ecological constraints on the foraging effort of western gorillas (*Gorilla gorilla gorilla*) at Bai Hokou, Central African Republic. *Int. J. Primatol.* 20, 1–23.

Goodall, J., 1986. *The Chimpanzees of Gombe: Patterns of Behavior*. Harvard University Press, Cambridge, MA.

Goren-Inbar, N., Alperson, N., Kislev, M.E., Simchoni, O., Melamed, Y., Ben-Nun, A., and Werker, E., 2004. Evidence of hominin control of fire at Gesher Benot Ya'aqov, Israel. *Science* 304, 725–727.

Gott, B., 2002. Fire-making in Tasmania: Absence of evidence is not evidence of absence. *Curr. Anthropol.* 43, 649–656.

Gould, S.J., 2002. *The Structure of Evolutionary Theory*. Harvard University Press, Cambridge, MA.

Gowlett, J.A.J., 1999. Lower and Middle Pleistocene archaeology of the Baringo Basin. In: Andrews, P., and Banham, P. (Eds.), *Late Cenozoic Environments and Hominid Evolution: A tribute to Bill Bishop*. Geological Society, London, pp. 123–141.

Gowlett, J.A.J., Harris, J.W.K., Walton, D.A., and Wood, B.A., 1981. Early archaeological sites, further hominid remains, and traces of fire from Chesowanja Kenya. *Nature* 294, 125–129.

Grant, P.R., and Grant, B.R., 2002. Unpredictable evolution in a 30-year study of Darwin's finches. *Science* 296, 707–711.

James, S.R., 1989. Hominid use of fire in the Lower and Middle Pleistocene: a review of the evidence. *Curr. Anthropol.* 30, 1–26.

Janson, C.H., and van Schaik, C.P., 1993. Ecological risk aversion in juvenile primates: Slow and steady wins the race. In: Pereira, M., and Fairbanks, L. (Eds.), *Juvenile Primates: Life History, Development and Behavior*. Oxford University Press, New York, pp. 57–76.

Kaplan, H., Hill, K., Lancaster, J., and Hurtado, A.M., 2000. A theory of human life history evolution: Diet, intelligence, and longevity. *Evol. Anthropol.* 9, 156–185.

Knott, C., 2001. Female reproductive ecology of the apes: Implications for human evolution. In: Ellison, P. (Ed.), *Reproductive Ecology and Human Evolution*. Aldine, New York, pp. 429–463.

Koebnick, C., Strassner, C., Hoffmann, I., and Leitzmann, C., 1999. Consequences of a longterm raw food diet on body weight and menstruation: results of a questionnaire survey. *Ann. Nutr. Metab.* 43, 69–79.

Lambert, J., 2002. Digestive retention times in forest guenons (*Cercopithecus spp.*) with reference to chimpanzees (*Pan troglodytes*). *Int. J. Primatol.* 23, 1169–1185.

Leigh, S.R., 1994. Ontogenetic correlates of diet in anthropoid primates. *Am. J. Phys. Anthropol.* 94, 499–522.

Lévi-Strauss, C., 1969. *The Raw and the Cooked. Introduction to a Science of Mythology, I.* Harper & Row, New York.

Lieberman, D.E., Krovitz, G.E., Yates, F.W., Devlin, M., and St. Claire, M., 2004. Effects of food processing on masticatory strain and craniofacial growth in a retrognathic face. *J. Hum. Evol.* 46, 655–677.

Lucas, P., 2004. *How Teeth Work*. Cambridge University Press, Cambridge, MA.

Ludwig, B., 2000. New evidence for the possible use of controlled fire from ESA sites in the Olduvai and Turkana basins. *J. Hum. Evol.* 38, A17.

MacLarnon, A.M.D., Martin, R.D., Chivers, D.J., and Hladik, C.M., 1986. Some aspects of gastro-intestinal allometry in primates and other mammals. In: Sakka. M. (Ed.), *Definition et Origines de L'Homme*. CNRS, Paris, pp. 293–302.

Milton, K., 1987. Primate diets and gut morphology: Implications for hominid evolution. In: Harris, M., and Ross, E.B. (Eds.), *Food and Evolution: Towards a Theory of Human Food Habits*. Temple University Press, Philadelphia, PA, pp. 93–115.

Milton, K., 1999. A hypothesis to explain the role of meat-eating in human evolution. *Evol. Anthropol.* 8, 11–21.

Milton, K., 2002. Hunter-gatherer diets: Wild foods signal relief from diseases of affluence. In: Ungar, P.S., and Teaford, M.F. (Eds.), *Human Diet: Its Origin and Evolution*. Bergin & Garvey, Westport, CT, pp. 111–122.

Monnier, J.L., Hallegoue, B., Hinguant, S., Laurent, M., Auguste, P., Bahain, J.J., Falgueres, C., Geebhardt, A., Mergueria, D., Molines, N., Morzadec, H., and Yokoama, Y., 1994. A new regional group of the Lower Paleolithic in Brittany (France), recently dated by electron spin resonance. *C. R. Acad. Sci.* 319, 155–160.

Oka, K., Sakuarae, A., Fujise, T., Yoshimatsu, H., Sakata, T., and Nakata, M., 2003. Food texture differences affect energy metabolism in rats. *J. Dent. Res.* 82, 491–494.

Pera, P., Bucca, C., Borro, P., Bernocco, C., De Lillo, A., and Carossa, S., 2002. Influence of mastication on gastric emptying. *J. Dent. Res.* 81, 179–181.

Petring, O.U., Blake, D.W., 1993. Gastric emptying in adults: an overview related to anaesthesia. *Anaesth. Intensive Care* 21, 774–781.

Ragir, S., Rosenberg, M., and Tierno, P., 2000. Gut morphology and the avoidance of carrion among chimpanzees, baboons, and early hominids. *J. Anthropol. Res.* 56, 477–512.

Remis, M., 1997. Western lowland gorillas (*Gorilla gorilla gorilla*) as seasonal frugivores; use of variable resources. *Am. J. Primatol.* 43, 87–109.

Remis, M., Dierenfeld, E.S., Mowry, C.B., and Carroll, R.W., 2001. Nutritional aspects of western lowland gorilla (*Gorilla gorilla gorilla*) diet during seasons of fruit scarcity at Bai Hokou, Central African Republic. *Int. J. Primatol.* 22, 807–836.

Remis, M.J., and Dierenfeld, E.S., 2004. Digesta passage, digestibility and behavior in captive gorillas under two dietary regimes. *Int. J. Primatol.* 25, 825–846.

Robbins, M., Sicotte, P., and Stewart, K.J., 2001. *Mountain Gorillas: Three Decades of Research at Karisoke*. Cambridge University Press, New York.

Rowlett, R.M., 1999. "Comment" on Wrangham et al. (1999). *Curr. Anthropol.* 40, 584–585.

Rowlett, R.M., 2000. Fire control by *Homo erectus* in East Africa and Asia. *Acta Anthropol. Sin.* 19, 198–208.

Secor, S.M., and Faulkner, A.C., 2002. Effects of meal size, meal type, body temperature, and body size on the specific dynamic action of the marine toad, *Bufo marinus*. *Physiol. Biochem. Zool.* 75, 557–571.

Speth, J., and Spielmann, K.A., 1983. Energy source, protein metabolism, and hunter-gatherer subsistence strategies. *J. Anthropol. Archaeol.* 2, 1–31.

Stahl, A.B., 1984. Hominid diet before fire. *Curr. Anthropol.* 25, 151–168.

Straus, L.G., 1989. On early hominid use of fire. *Curr. Anthropol.* 30, 488–491.

Tutin, C.E.G., 1994. Reproductive success story: variability among chimpanzees and comparison with gorillas. In: Wrangham, R.W., McGrew, W.C., de Waal, F.B.M., and Heltne, P.G (Eds.), *Chimpanzee Cultures*. Harvard University Press, Cambridge, MA, pp. 181–193.

Tutin, C.E.G., and Fernandez, M., 1985. Foods consumed by sympatric populations of *Gorilla g. gorilla* and *Pan t. troglodytes* in Gabon: Some preliminary data. *Int. J. Primatol.* 6, 27–43.

Tylor, E.B., 1878. *Researches into the Early History of Mankind*. London.

Ulijaszek, S.J., 2002. Human eating behaviour in an evolutionary ecological context. *Proc. Nutr. Soc.* 61, 517–526.

Ungar, P., 2004. Dental topography and diets of Australopithecus afarensis and early Homo. *J. Hum. Evol.* 46, 605–622.

Ungar, P.S., Teaford, M.F., 2002. *Human Diet: Its Origin and Evolution*. Bergin & Garvey, Westport, CT.

Valero, H., and Biocca, E., 1970. *Yanoáma: The Narrative of a White Girl kidnapped by Amazonian Indians*. E. P. Dutton, New York.

Westerterp, K.R., 2004. Diet induced thermogenesis. *Nutr. Metabol.* 1, 5.

Westerterp-Plantenga, M.S., Rolland, V., Wilson, S.A., and Westerterp, K.R. Satiety related to 24 h diet-induced thermogenesis during high protein/carbohydrate vs high fat diets measured in a respiration chamber. *Eur. J. Clin. Nutr.* 53, 495–502.

Wrangham, R.W., 1979. On the evolution of ape social systems. *Soc. Sci. Inf.* 18, 335–368.

Wrangham, R.W., 2000. Why are male chimpanzees more gregarious than mothers? A scramble competition hypothesis. In: Kappeler, P.M. (Ed.), *Primate Males*. Cambridge University Press, Cambridge, pp. 248–258.

Wrangham, R.W., 2005. The delta hypothesis: Hominoid ecology and hominin origins. In: Lieberman, D.E., Smith, R.J., and Kelley, J. (Eds.), *Interpreting the Past: Essays on Human, Primate and Mammal Evolution in Honor of David Pilbeam*. Brill, Boston.

Wrangham, R.W., 2006. Food-softening and the problem of Middle Paleolithic cooking. *J. Anthropol. Archaeol.*, in press.

Wrangham, R.W., and Conklin-Brittain, N.L., 2003. The biological significance of cooking in human evolution. *Comp. Biochem. Physiol.* A 136, 35–46.

Wrangham, R.W., Conklin-Brittain, N.L., and Hunt, K.D., 1998. Dietary response of chimpanzees and cercopithecines to seasonal variation in fruit abundance. I. Antifeedants. *Int. J. Primatol.* 19, 949–970.

Wrangham, R.W., Jones, J.H., Laden, G., Pilbeam, D., and Conklin-Brittain, N.L., 1999. The raw and the stolen: Cooking and the ecology of human origins. *Curr. Anthropol.* 40, 567–594.

Yamagiwa, J., 1999. Socioecological factors influencing population structure of gorillas and chimpanzees. *Primates* 40, 87–104.

Yamagiwa, J., Maruhashi, T., Yumoto, T., and Mwanza, N., 1996. Dietary and ranging overlap in sympatric gorillas and chimpanzees in Kahuzi-Biega National Park, Zaire. In: McGrew, W.C., Marchant, L.F., and Nishida, T. (Eds.), *Great Ape Societies*. Cambridge University Press, Cambridge, pp. 82–98.

17

Seasonality, Fallback Strategies, and Natural Selection

A Chimpanzee and Cercopithecoid Model for Interpreting the Evolution of Hominin Diet

JOANNA E. LAMBERT

Introduction: The Known, Unknown, and Unknowable

> *Natural selection, acting in various guises at various levels, seems together with genetic drift to account for almost all features of organisms once the appropriate raw material has arisen by mutation and recombination.*
> —(Futuyma, 1979, p 438)

Within this evolutionary framework, we can reasonably assume that feeding-related features observed in extant primates should be, at least in theory, demonstrably the result of natural selection. It is the *demonstrable* aspect of this important assumption that those concerned with dietary adaptations in the present and in the evolutionary past must confront. In addition to information derived from morphology, it is a truism that scientists studying extant primates have the luxury of observing the function of diet-related anatomy via direct observation of a feeding animal, while those scientists evaluating fossil species do so in the absence of such data (Kay, 1984). Yet, both scientists confront a similar challenge, and in evolutionary terms yield comparably synchronic interpretations; the ghost of selection past haunts us all. Hence, our use of the powerful combination of comparative models, extant species analogs, and correlative evaluations as tools for interpreting changes in form and function over evolutionary time.

It is exceedingly doubtful that tropical habitats are, or ever have been, in any form of equilibrium (Maley, 1996; Newberry, Songwe, and Chuyong, 1998). Indeed, the most useful models for evaluating both plant and faunal adaptations are those that admit *dis*equilibrium and environmental change, often as a function of climate and rainfall (Newberry, Songwe, and Chuyong, 1998; Potts, 1998). From the perspective of a plant-eating animal, climate shifts and their impact on food resource availability and abundance vary in predictability. For example, longer-term (e.g., glaciation) or more extreme (e.g., monsoons) shifts are relatively unpredictable and can have important evolutionary consequences, particularly with regard to speciation and extinction (Foley, 1994; Vrba, 1995). Conversely, climate shifts can also be more predictable, especially in terms of rainfall seasonality. Indeed, while obfuscating myriad details of ecological specificity and variability, a shorthand reference to habitat type is commonly made in terms of habitats being either seasonal or nonseasonal as a function of broad patterns of latitude, land area, and, most importantly, rainfall (Rathcke and Lacey, 1985; van Schaik, Terborgh, and Wright, 1993; White, 1998a; references in Fleagle, Janson, and Reed, 1999). This dichotomy is realistically best viewed as being the two ends of a complex continuum of relative degrees and intensity of seasonality (e.g., strongly seasonal, less seasonal, etc.).

Evaluations of climate, seasonality, and concomitant shifts in food availability and abundance have been central to hypotheses regarding the origin and extinction of hominin species, as well as evaluations of hominin feeding-related adaptations (Foley, 1993; Vrba, 1995; Reed, 1997; Potts, 1998; Teaford and Ungar, 2000; Ungar, 2004). With regard to the latter, most models have historically evaluated form-function relationships in light of the frequency with which that animal exploits favored, preferred foods. Such categorization occurs with extant and extinct animals at both species (e.g., a species is a frugivore because it spends most of its time eating fruit; Chivers and Hladik, 1980) and community levels (e.g., percentage of total frugivory in a species community; Fleagle and Reed, 1996).

But foods are not created equally. Some are intrinsically more nutrient dense, while others less so, and either can differ in their chemical and mechanical defense. Extrinsic factors such as effect of season and habitat also influence a food's value to a consumer as they strongly influence overall and relative availability and abundance of that food (Oates, 1987; Janson and Chapman, 1999; Lambert, 2007). As such, species exhibit an array of behavioral and anatomical adaptations for exploiting foods that are of limiting importance during periods when other, either more nutrient dense, more abundant, or less-protected foods are scarce. This suggests differences in degree of natural selection pressure, and highlights an interpretation of anatomy that shifts the lens away from frequency to critical function (*sensu* Rosenberger and Kinzey, 1976; Rosenberger, 1992). That is, a traits' function is critical to the harvesting or ingestion of a particular resource (fallback foods); the crucial or "critical" aspect of this argument relates not to the overall utility of this feature for consuming a food type in high quantity or in high frequency, but that it instead has extreme utility under particular—limiting—environmental circumstances, during which time an animal must fall back on foods that are uncommonly consumed (Robinson and Wilson, 1998).

It is increasingly evident that understanding the use of fallback foods and the evolution of fallback strategies are key to understanding dietary adaptations, in both extant and extinct species, including those within our own lineage (Wrangham et al. 1999, Ungar, 2004; Laden and Wrangham, 2005; Ungar et al., 2006). However, researchers vary in their use of the term "fallback" and differentially focus on one or another of the variables important to optimal feeding and foraging, including inherent energy yield of food (nutrient density), search time, and handling time (MacArthur and Pianka, 1966). In theory, an animal should exhibit preference for the most profitable (in terms of energy yield) foods, that is, foods that are nutrient dense, easy to find, and easy to access. Foods that are either less nutrient dense, harder to find, or more difficult to access are presumably less profitable and thus less preferred. As such, "fallback" is used variously as an indicator of food preference, as an index of fallback food availability, either in absolute abundance or relative to preferred foods, or to simply refer to the absolute inherent quality and nutrient density of the food (Gautier-Hion and Michaloud, 1989; Wrangham et al., 1996; Tutin et al., 1997; Yamakoshi, 1998; Wrangham et al., 1999; Gursky, 2000; Furuichi, Hashimoto, and Tashiro, 2001; Fox et al., 2004; Lambert et al., 2004; Ungar, 2004). Variability in how this term is used also often relates to habitat differences in food abundance and availability, which strongly influence search time (Malenky and Wrangham, 1994; White, 1998b; Furuichi, Hashimoto, and Tashiro, 2001; Laden and Wrangham, 2005).

My discussion is not novel in its evaluation of the importance of fallback foods. Rather, my purpose here is to refine our understanding of the types of fallback foods animals rely on during periods of potentially more intense selective pressure, and how these differences can relate to a species' overall fallback strategy and evolution of feeding related traits. In this discussion, I place fallback *foods* into two broad categories that ultimately relate to different fallback *strategies,* including (i) fallback foods of lower nutritional density (lower inherent energy yield). These foods typically, although not always, comprise nonreproductive plant anatomy such as bark, leaves, petioles. Such foods often have the advantage of being more abundant (less search time); however, they inherently require more processing on the part of the animal (more handling time). It is in the use of these fallback foods where we tend to see anatomical adaptation (e.g., thicker dental enamel, longer molar shearing crests, longer digestive retention times); extant Cercopithecoidea and *Gorilla* spp. are good examples of this strategy.

Alternatively, (ii) fallback foods can be of higher nutritional density (greater inherent energy yield). These foods are commonly, although not exclusively, reproductive plant parts such as fruit and seeds. The advantage of these foods is that they can facilitate the maintenance of a high-energy-yield diet throughout the year. However, they often carry with them the disadvantages of being mechanically protected (longer handling time; e.g., palm nuts, termite mounds) and difficult to find (more search time). It is these foods where we may observe behavioral innovation and tool use for coping with seasonal shortages—*Pan troglodytes* and *Cebus* spp. are examples of such a strategy.

It should be noted that neither strategy precludes the other, so that, for example, species that use tools to access a high energy, but mechanically protected, food may also have thick dental enamel and a robust masticatory apparatus (e.g., *Cebus* spp.). This classification facilitates evaluation of differences between sites because food abundance and availability are heavily influenced by habitat and season; resource availability strongly influence search time, which, in turn, directly affects a food's profitability (MacArthur and Pianka, 1966). With this framework in mind, my goals in this chapter are as follows:

1. To discuss how species relying on a plant-based diet in habitats that vary in their seasonality must evolve strategies to cope with periods of preferred food scarcity (*known*), and that this can result in the adoption of morphological and/or behavioral adaptations that facilitate the consumption of fallback foods which differ in chemical, mechanical, nutritional, and ecological properties (*known and knowable*).
2. To argue that the evolution of such strategies is influenced by the likelihood of an environment undergoing periods when food availability shifts (critical periods), and that environments differ in this regard as a function of latitude, altitude, rainfall and area—with some habitats being more inherently vulnerable (*known*).
3. To evaluate how Cercopithecoidea and African apes (*Gorilla* and *Pan* spp.) have adapted to consuming fallback foods. I suggest that cercopithecoids and gorillas have evolved a set of anatomical and physiological adaptations for facilitating a fallback strategy that relies on more abundant but lower-quality foods. Alternatively, *P. troglodytes* has evolved behavioral means by which to maintain key, higher-quality foods during such shortages. *Pan paniscus*, however, is distributed throughout relatively uniform, less-seasonal habitat, and a restricted geographical range; while we observe behavioral mechanisms in this species for reducing stress and intragroup competition over ripe fruit, critical periods are less extreme for this species than in monkeys, chimpanzees, and gorillas, and selection for adaptations for fallback foods relaxed.

Anatomical and Physiological Solutions for Falling Back on Lower-Quality Foods

Although the myth of "stable forests" persists (Janson and Chapman, 1999), primates have evolved in inhospitable habitats, both in terms of availability of preferred resources, and the not unrelated facts that plants have evolved myriad chemical and mechanical mechanisms to protect themselves from plant predators. As such, there is ample evidence that primates are extremely selective feeders and consume only a fraction of the plant species available to them in a habitat (Oates, 1977, 1987; Glander, 1978; Milton, 1984). Selective-feeding behavior is also exhibited in terms of the times of day a primate will consume a given plant species (quantity/quality of toxins and antifeedants in a plant ebb and flow throughout the day), as well as which portions of the plant (e.g., leaf tips or petioles but not the entire leaf), and the total quantity of a particular plant food that is consumed (Glander, 1978; Struhsaker, 1978; Oates, 1987). In addition to behavioral adaptations, like all herbivorous animals, primates have also evolved anatomical and physiological solutions for dealing with plant defenses and fiber.

Periods of preferred food scarcity differ in duration and intensity as a function of habitat type and seasonality. During such periods, species have the option of switching to foods that are more abundant but may require more dental and digestive processing because they are either less nutrient dense, higher in secondary metabolites and/or fiber, or more mechanically defended. Alternatively, species can fall back on foods that are either difficult to find or that may be of less-nutrient density than most preferred items but do not influence the overall quality of diet because the profitability of that food is relative to other foods in an already high-quality diet. Examples of

Figure 17.1 *Lophocebus albigena* consuming *Ficus* fruit in Kibale National Park, Uganda. Photo courtesy of Alain Houle.

both strategies are found among primates. For example, Wrangham, Conklin-Brittain, and Hunt (1998) and Conklin-Brittain, Wrangham, and Hunt (1998) have demonstrated the different ways in which cercopithecines respond to periods of seasonal scarcity relative to sympatric *P. troglodytes* in Kibale National Park, Uganda. While the monkeys tended to fall back on foods that were higher in digestion inhibitors and toxins, in this study, the chimpanzee fallback food (pith) did not significantly negatively influence their overall high-quality diet in terms of fiber or secondary metabolites. While pith of different plant species differs, they are overall higher than leaves in more readily digested fiber fractions (hemicellulose; Wrangham et al., 1991; Conklin-Brittain, Wrangham, and Smith, 2002). In addition, these authors found that chimpanzees maintained more ripe fruit throughout the year than did the cercopithecines and did not consume more fiber. Since *P. troglodytes* is roughly an order of magnitude larger than the monkeys, these were unexpected results based on body size predictions.

Primates that fall back on foods that differ from preferred or more commonly consumed foods in their chemical, mechanical, nutritional, and ecological characteristics are expected to evolve solutions (anatomical, physiological, behavioral) to the particular challenges presented by those fallback foods. For example, *Lophocebus albigena* (fig. 17.1) has among the thickest dental enamel in the order Primates, and thick enamel is generally argued to have been selected for (at some point in the evolutionary past) because of its function in facilitating the frequent consumption of hard foods (Kay, 1981; Dumont, 1995; Teaford, Maas, and Simons, 1996). Lambert et al. (2004) evaluated the hardness of foods consumed by *L. albigena* and sympatric *Cercopithecus ascanius* during a dry El Niño season in the Kibale National Park, Uganda (1997). During the study period, ripe fruit and other preferred, softer, and more commonly consumed foods were unavailable to these two species. The study animals instead consumed fallback foods, which were not only different in form and type, but also hardness. We found no difference in dietary hardness when the puncture resistance of all fruit (over 1991–1997) consumed by the two species was compared. However, when the hardness of only fallback foods (1997) was evaluated, *L. albigena* exploited a diet with greater resistant to puncture and crushing than *C. ascanius*. This difference was largely explained by the higher percentage of bark and seeds consumed by the *L. albigena* during this period. The diet of these two species can have high overlap, and *L. albigena* can be highly frugivorous; however, *L. albigena* fall back on very hard foods, while the guenon increases leaf intake. Bark is a highly lignified and particularly hard resource, and these results corroborate the hypothesis that it is not so much what is consumed most commonly (i.e., soft, fleshy fruit) that selects for enamel thickness but the hardness of foods that are consumed during critical periods, when other more preferred foods are not available.

Like many primates, including chimpanzees, lowland gorillas overall show a preference for ripe, soft fruit (Remis, 1997; Remis et al., 2001). Nishihara (1995), for example, has found that western lowland gorillas (*Gorilla gorilla gorilla*) spend 63% of their feeding time on fruit. And, in a recent review that included data from all long-term western lowland gorilla sites, Rogers et al. (2004) documented that these apes will maintain fruit in their diet throughout the year; the authors indeed call *G. gorilla gorilla* "fruit pursuers, with strong preferences for particular and often rare fruit species, for which they will incur significant foraging costs" (p 175). However, as noted by Ungar (2004), the dietary differences between gorillas and chimpanzees become most apparent in analyses of their fallback foods, such that during periods of preferred fruit scarcity, gorillas will tend to fall back on tougher, more fibrous foods than those consumed by chimpanzees. Rogers et al. (1994), for example, argue that the bark of *Milicia excelsa* is an important fallback resource for *G. gorilla gorilla* in Lope Reserve, Gabon, during seasons of low fruit availability. Ungar (2004) argues that it is the key differences in the mechanical properties of the fallback foods that can account for the longer shearing crests and steeper cusps in gorillas and flatter molar surfaces in chimpanzees.

Gorillas are argued to be able to consume their lower-quality fallback foods as a consequence of the aforementioned dental adaptations, as well as a high number of cellulose digesting ciliates and overall body size, which facilitates the slower digestive processing required for fermenting fiber (Demment and van Soest, 1983; Collet et al., 1984; Remis et al., 2001).

But, cercopithecoids do not have the body size advantage of gorillas. In addition to the case study of thick dental enamel in *Lophocebus*, are there other anatomical and/or physiological solutions that can be identified? The degree to which fiber (nonsoluble structural carbohydrates) of plant cell walls can be used as an energy source depends in part on the length of time that these components are retained in the fermenting chamber(s) of the gastrointestinal tract. Longer digestive retention times result in higher levels of fermentation, and many plant parts require considerable fermentation before they are useful as an energy source (Milton, 1981, 1984; 1986; 1993; Lambert, 1998; Remis, 2000). It has long been presumed that smaller mammals have (both absolutely and relatively) faster food passage rates than larger species, which limits their capacity to ferment fibrous plant components (Parra,

1978; Kay, 1985; Cork and Foley, 1991; Kay and Davies, 1994; Van Soest, 1994). Yet, as it turns out, both cercopithecoid subfamilies have relatively longer digestive retention times than the much larger African apes (Lambert, 2002a). This is to be expected among the Colobinae, who have specialized, sacculated, and alkaline stomachs for consuming a diet high in fiber. On average, their mean digestive retention times range from roughly 40 to 60 h (refs in Lambert, 1998; Caton, 1999). Neither cercopithecines nor hominoids exhibit such derived stomach anatomy (Chivers and Hladik, 1980; Milton, 1987). Yet, in an analysis regressing digestive retention times as a function of body size, Lambert (1998) found that the cercopithecines in the analysis were significantly further above the regression line than any other primate taxon. Indeed, despite being on average an order of magnitude smaller than African apes, all tested cercopithecines to date exhibit mean digestive retention times averaging 31 h (*P. troglodytes*: 31.5–48 h; *Gorilla gorilla*: 36.5–61.9 h) (Milton and Demment, 1988; Maisels, 1993; Lambert, 1998, 2002a; Remis, 2000). These digestive results have important implications for understanding how monkeys can consume either similar or greater levels of fiber than larger-bodied apes and rely on higher-fiber fallback foods during critical periods (Lambert, 2002a).

In addition to fiber, cercopithecoid fallback foods are often higher in toxins than those consumed by *Pan* (Wrangham, Conklin-Brittain, and Hunt, 1998) and must be detoxified either by microbial activity in a specialized stomach or microsomal enzymes activated in the liver (Freeland and Janzen, 1974). Although we know extremely little about microbial detoxification in colobines, the potential of the specialized stomach with a diverse and dense microbial community to act as a detoxification chamber has been demonstrated in ruminating ungulates (Keeler et al., 1978; Waterman and Kool, 1994). All noncolobine primates rely not on bacterial activity, but on microsomal enzymes (Freeland and Janzen, 1974), the production of which scales allometrically. Walker (1978) has demonstrated that rates of enzymatic activity scale negatively with mammal body size (Walker, 1978; Freeland, 1991). Freeland (1991) thus suggests that smaller mammals are at an advantage for detoxifying plant secondary metabolites and that the larger the mammal, the greater the preference for foods with low amounts of toxic plant metabolites. Cercopithecines, on average, are smaller than apes, which may facilitate their consumption of chemically defended plants not available to apes, and only available to colobines as a consequence of their specialized stomach.

Thus, cercopithecoids deal with the higher fiber and defenses of their fallback foods via dental and digestive adaptations. These adaptations for consuming less-nutrient dense foods as a fallback strategy may well have evolved at some time in the mid- to late Miocene. Most explanations for the evolutionary success of Cercopithecoidea (over apes) suggest that they adopted a more efficient strategy in competing for increasingly rare resources (Andrews, 1981; Temerin and Cant, 1983). The actual mechanisms of how these monkeys accomplished this have received less attention, although I have argued elsewhere that a combination of long retention times and digestive flexibility, a capacity to detoxify plant compounds (either by bacteria in the case of colobines, or microsomal enzymes in the case of cercopithecines), uriposia, and cheek pouches (which in cercopithecines facilitates fast harvesting) gave monkeys a competitive edge over ape counterparts in allowing them access to lower-quality food resources. These anatomical and physiological mechanisms facilitate access to a set of fallback foods not consumed by apes that are more abundantly distributed in a habitat and also less vulnerable (say, than fruiting phenology) to climate perturbations (Lambert, 1997, 2000, 2002a, 2002b, 2005; Lambert et. al., 2004; Lambert and Whitham, 2001). Smaller body size (along with its associated faster life history strategies and shorter generation times) is also correlated with greater potential for speciosity (Cowlishaw and Dunbar, 2000).

Maintaining Higher-Quality Foods during Critical Periods

Rather than switching to alternative, lower-quality foods, other species exhibit adaptations for coping with environmental fluctuations and scarcity of preferred food types by evolving behavioral mechanisms to maintain particular foods in the diet throughout the year regardless of their availability. *Pan troglodytes* stand out as an excellent example of such a strategy (fig. 17.2). Although their colon is proportionately larger than what is seen in *Homo sapiens*, chimpanzees have digestive retention times that are comparable to those of humans and, contrary to what might be predicted for their body size, consume an unexpectedly high-quality diet (Milton, 1999). Behavioral adaptations (e.g., fission-fusion social structure) that facilitate the maintenance of ripe fruit in the diet throughout the year, in addition to tool use, which can facilitate the consumption of foods that are high in nutrient density (but often mechanically protected), are behaviors that are key to their fallback strategy. For example, several authors (Chapman, White, and Wrangham, 1994; Wrangham et al., 1996) have demonstrated the correlation between fruit availability and the fissioning-fusing of chimpanzee feeding-party size; essentially as fruit becomes

Figure 17.2 *Pan troglodytes* consuming *Ficus* fruit in Kibale National Park, Uganda. Photo courtesy of Alain Houle.

scarce, feeding parties become smaller, and animals will increase day range to maintain fruit in the diet. In this way, the total yield of a food consumed during seasonal scarcity is influenced by the cost of searching time but not that of handling and processing time.

So, given that a foraging primates' diet comprises some number of foods, some of which are more profitable than others, when does it make sense for an animal to add the next most profitable item, with profit in this case influenced by handling time, rather than search time? It is under these circumstances where we observe the use of feeding-related tool use that facilitate consumption of resources that are mechanically protected (which increases handling time) but that have the potential to yield high amounts of energy (e.g., termite mounds, hard nuts; McGrew, 1992). Chimpanzees will use a variety of tools regardless of overall food availability (Whiten et al., 1999), but of interest here is what the selective advantage for this behavior was in the first place. Procuring requisite nutrients and calories is critical to the survival of every animal, with more successful foragers assumed to have greater fitness. At some point, costs incurred from increasing day range, regardless of party-size fissioning, will outweigh the benefits from consuming that food. The balance of this equation will vary by habitat, season, and rainfall and hence potential for selective pressure. For example, the chimpanzee (*P. troglodytes*) community at Bossou, Guinea, lives in a seasonal habitat characterized by low rainfall, low resource availability, and high seasonality of fruit availability (Yamkoshi, 1998). This forest is also isolated and discontiguous from other forest fragments; this essentially means that the chimpanzees are unable to expand their day and home range in search of readily accessed high-quality foods. Chimpanzees in this forest employ a number of feeding-related tools, including hammer and anvil, pestle pounding, and ant dipping. Most important, tool use in these chimpanzees facilitates the consumption of two important, nutrient-dense fallback resources: oil palm nuts and palm pith. Both resources are mechanically protected and cannot be exploited without the use of tools (Yamkoshi, 1998). During periods of extreme resource scarcity, nut cracking and pestle pounding for pith increase dramatically, providing critically needed calories from lipids and carbohydrates.

Such observations suggest that fission-fusion and foraging-related tool use can serve critical functions and are advantageous during critical periods when more profitable and preferred foods are not available. In the case of tool use, cognitive function is a necessary precursor, suggesting that in these cases, the critical, limiting habitats in which chimpanzees found themselves resulted in a situation that promoted innovation of these behaviors. Indeed, while he did not discuss critical function or fallback foods, Potts (2004) has recently proposed a "fruit-habitat hypothesis" that suggests the evolution of the relatively very large brains of great apes and their concomitant cognitive capacity, can be explained by a "causal connection between ape ancestral diets, habitats, and environmental history" (p. 224). In short, he argues that as preferred fruit resources became increasingly rare as a function of forest reductions and climatic shifts in the African Miocene, ancestral apes were under extreme selective pressure for evolving cognitive means (e.g., complex mental representational ability) to deal with food source uncertainty.

Although not directly related to diet and feeding, the recent observation (Pruetz, 2005) of cave use by western populations of *P. troglodytes* living in the hottest, most extreme habitat of its range, fits this same pattern of behavioral innovation under extreme environmental circumstances. Chimpanzees living in very dry areas will also dig drinking holes, or wells (Matsuzawa, 2002). These behaviors represent behavioral solutions to intense selective pressure that is more likely to occur in seasonal habitats. These examples pose a critical question: Can we use our understanding of fallback strategies in apes and Cercopithecoidea to explain the evolution of feeding related adaptations in *Pan* spp.—and by analogy and extrapolation—early hominins?

The Role of Critical Function and Environmental Shifts in *Pan* spp. Adaptations

I turn now to the oft-cited differences between our two closest living relatives: *P. troglodytes* and *Pan paniscus*. Despite glossing over myriad details and decades of research, it is evident with regard to the discussion of feeding-related adaptations and the environment, these two species differ in their: (1) total species distribution, (2) total range in their use of habitat types, and (3) proclivity for tool use and composition of toolkit.

Habitats differ in their patterning of rainfall and its impact on seasonal availability of foods; this not only has extreme effects on habitat carrying capacity but also, as argued, on the degree of potential for periods of intense selection as a consequence of environmental fluctuations. Indeed, Potts (1998) has evaluated the impact of variable selection in widely fluctuating circumstances on hominin evolution. He suggests that the environmental circumstances in which hominins evolved were particularly inconsistent, with episodic and extensive change in vegetation, water, and other resources. Habitats are on a continuum, as are species distribution with regard to habitat types, with habitats at the periphery of a species' distribution tending to be more challenging from the animal's perspective; it is indeed at the edges of species distributions where we see most evolution (Cowlishaw and Dunbar, 2000). Of the two *Pan* species, *P. troglodytes* is by far the more widely distributed—in discontinuous populations from East to West Africa, north of the Congo River. This wide distribution across much of equatorial Africa encompasses a diversity of habitat types, including less seasonal wet lowland forest; relatively more seasonal habitats, such as moist montane forest, savanna, and gallery forest; and highly seasonal, arid, and open woodland (fig. 17.3).

Pan paniscus, conversely, lives in a much smaller region south of the Congo River, with a distribution that is delineated on all sides by rivers. Increasing total species distribution is not an option for this species as its movement is impeded by the presence of large, impassable rivers (Eriksson et al., 2004). This region is characterized as being relatively undisturbed, lowland, climax-moist forest and is commonly commented on as being markedly less seasonal that the habitats used by *P. troglodytes* (Malenky, 1990; Foley, 1993; White, 1998b; Doran et al., 2002). Overall, the bonobo is argued to live in habitats where food resources are spatially and temporally more abundant than those of common chimpanzees (Malenky and Wrangham, 1994; Thompson, 2003). For example, in a comparison of *P. troglodytes* in Kibale National Park, Uganda, and *P. paniscus* in Lomako, Democratic Republic of Congo, White

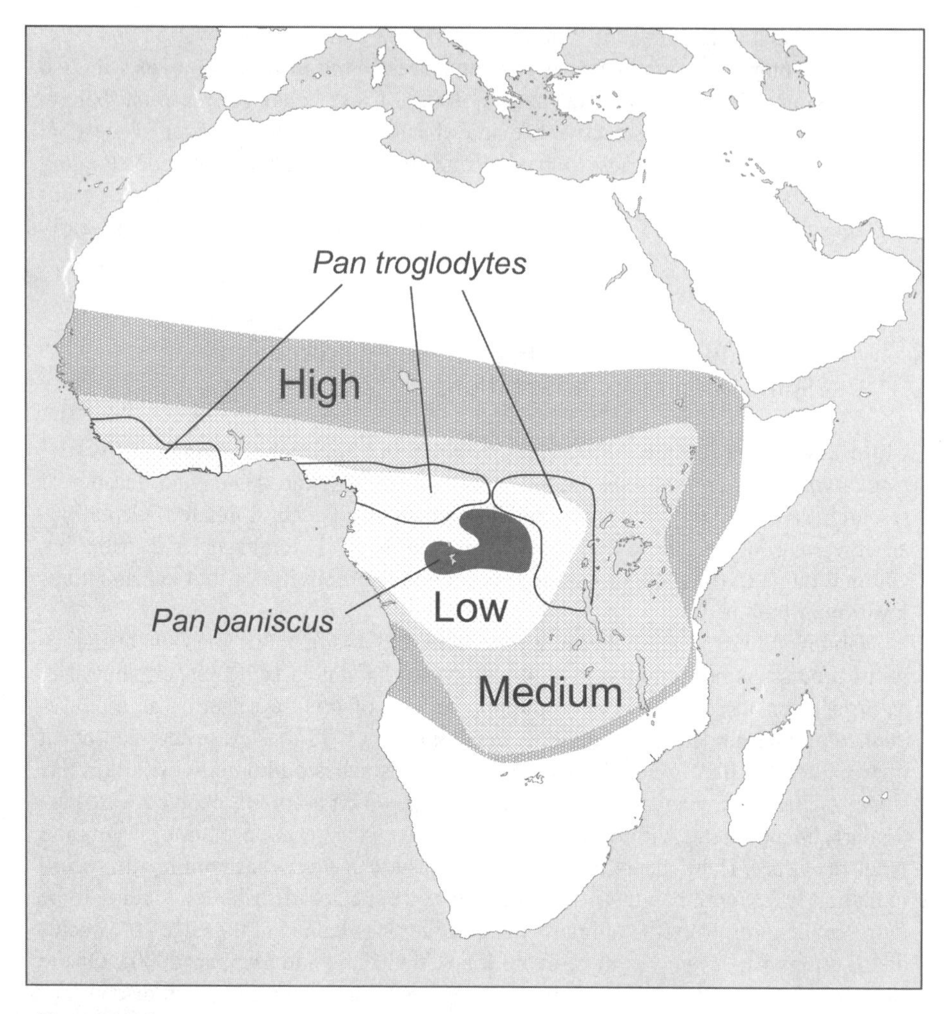

Figure 17.3. Map of Africa, indicating: (i) patterns (low, medium, high) of rainfall seasonality and (ii) geographic distribution of *Pan paniscus* and *Pan troglodytes.*

(1998b) suggests "there is more variation in food and distribution for *P. troglodytes* than is experienced is by *P. paniscus*, suggesting that it is the level of variation in food availability rather than the absolute amount that is the critical difference selecting for differences in social organization in these two species" (p 38).

Pan troglodytes can be found in habitats at the other extreme of the seasonality continuum: in savanna woodland habitats in both Eastern and Western regions of the continent. These habitats differ in many respects from the moist forests of the Congo Basin. Indeed, relative to the forests in Democratic Republic of Congo, savanna woodland is drier with more seasonal rainfall patterns, has xeric-adapted plant species with hard-shelled seed or nuts and underground storage organs, and has more patchily distributed preferred resources as a function of greater seasonality (Moore, 1996).

The common chimpanzee is also well known for its diverse toolkit, while feeding-related tool use in wild populations of *P. paniscus* has not been reported (Boesch, Hohmann, and Marchant, 2002; McGrew, 1992). Are these facts related? Moreover, can we use this information to interpret patterns in early hominins and the stimuli for tool use? We have entered the world of the unknown, but are still—potentially—within the realm of the knowable. Jim Moore (1996, p. 275) writes: "many a great ape grant proposal has waxed on about insights that the proposed research would provide into the behavioral ecology of our early ancestors, and the term 'model' is ubiquitous in the resulting literature." It is difficult to not fall into this trap, but my primary point here is not so much using these two *Pan* species as a model of what may or may not have happened some millions of years ago but, instead, to consider the evolution of particular traits as a function of periods of food scarcity when animals need to rely on less-profitable resources. Specifically, I am referring to interpreting the evolution of behaviors as a response to critical periods (more seasonal habitats) and, which, in the case of *P. troglodytes*, resulted in a fallback strategy that facilitated the maintenance of foods that are nutrient dense, despite being mechanically defended.

The most common explanation of speciation imputes a disruption of gene flow by geographical isolation of a previously panmictic population into two or more disparate population, thus allowing for the accumulation of change and divergence into separate species (Futuyama, 1979; Eriksson et al., 2004). One of the most striking examples of this is the distribution of *Pan paniscus* and *Pan troglodytes*. My model, evolutionary scenario, analog—whatever one wants to name it suggests the following:

1. Some ancestral proto-*Pan* population was isolated by changes in the course of the Congo River—this is not a new argument and is well supported by genetic and biogeographic evidence (Bradley and Vigilent, 2002; Thompson, 2002, 2003).
2. Early *P. troglodytes* was able to increase its range through equatorial Africa, while *P. paniscus* was not because of the numerous large impassable rivers—again these are not new arguments and are well-supported aspects of *Pan* spp. biogeography (Thompson, 2002, 2003; Eriksson et al., 2004).
3. As the distribution of *P. troglodytes* expanded across equatorial Africa, this species encountered increasingly more seasonal and unpredictable environments, and, with increasing seasonality, came increased potential for disruptive critical periods of intense selective pressure. Innovative behaviors in early *P. troglodytes* were selected for because of their utility during these critical times for maintaining a diet that was relatively high in quality.

It is part 3 of this argument that is potentially of interest to paleoanthropologists, as it arrives at the question of why *P. troglodytes* uses tools, while *P. paniscus* does not. Ancestral and extant populations of *P. paniscus* occupy a restricted, relatively less seasonal habitat, suggesting that this species is less likely to be presented with critical periods resulting in either fallback strategy; although, given the inherently more labile nature of behavior over, say, evolution of dental traits, it might be expected that on those occasions where food may be limiting, behavioral solutions might arise. In the case of *P. troglodytes*, concomitant to range expansion came an increasing likelihood of critical periods. But, why not adopt a fallback strategy that relies on more abundant albeit lower-quality foods, which is seen in Cercopithecoidea? Indeed, the body size of *P. troglodytes* would predict such a strategy, and yet the opposite is true: Despite their smaller size, Cercopithecoidea are generalist, eclectic feeders with an array of anatomical and physiological adaptations (dental, digestive, enzymatic, etc.) for consuming high-fiber plant foods replete with toxins that are not tolerated by sympatric chimpanzees. This suggests that as *P. troglodytes* expanded its ranges into increasingly seasonal habitat, the trophic space that included consumption of lower-quality food was occupied by monkeys who were simply better at it and had been since at least the end of the Miocene. No room at the inn. Gorillas, with their absolutely larger body size, proportionately larger and more ciliated gut, and higher molar-shearing blades and cusps, could use higher-fiber foods than common chimpanzees. Thus, selection on *Pan troglodytes* resulted in another strategy, namely, behavioral strategies for shifting feeding-party size and using feeding-related tools.

It is of note that the genus *Cebus* is known both for its tool use in the wild and has a relatively unspecialized digestive strategy. Relative to similarly sized cercopithecoids, capuchins have fast digestive times, which limits their ability to ferment fiber. For example, *Cebus apella*, weighing only 3.5 kg, has a transit time (TT) of 3.5 h, while *Cercopithecus pogonias* and *Cercopithecus ascanius* have species average weights of 3.75 kg and 3.6 kg, respectively, and have TT of 16.6 and 19.7 h (Milton, 1984; Maisels, 1993; Lambert, 1998, 2002a). At the same time, capuchins have been demonstrated to use tools to facilitate access to resources during periods when preferred foods were not available. Moura and Lee (2004) argue that "energy bottlenecks" create contexts for capuchins (*C. apella)* to derive benefits from tool technology. In the Caatinga dry forest of northeastern Brazil, capuchins have been observed to commonly use tools and do so during the extended dry season of this region. During such times, the resources that are available without tool use are not sufficient for nutritional requirements and the capuchins forage terrestrially. Several tools and tool-facilitated behaviors have been found in four habituated capuchin groups foraging in these areas, including digging for tubers with stones, cracking open seeds and branches with stones, breaking tubers with stones, and using stones as hammers in combination with wooden anvils to crack seeds. These monkeys consume forty-one plant species as food; tool use increases the use of at least three of these species. The researchers argue that *Cebus* foraging for embedded and mechanically defended resources in habitats that experience energy bottlenecks is facilitated by innovative tool use (Moura and Lee, 2004). A similar argument is made for *C. apella* in another dry region of Brazil, where capuchins commonly use hammer-and-anvil technology to crack nuts during the dry season when preferred resources are scarce (Fragaszy et al., 2004).

This chimpanzee model is directly testable by observing the behavior of *P. paniscus* in its most southern limit of its extant distribution, which is described as "transitional grassland habitat" (Thompson, 2003). This southern region is characterized by different soils and a marked seasonal distribution of rainfall. *Pan paniscus* are known to move through and feed in these areas. At least three grassland species have been consumed by *P. paniscus*, and there are behavioral correlates of being in a different habitat type, too, including no vocalizations (Thompson, 2003). It is exactly in this inherently more seasonal and less-predictable habitat where we might observe innovative solutions related to feeding on foods during crunch times.

Conclusions and Implications: Using What We Know

Pan paniscus has classically received less attention in reconstructions of early hominin behavior for exactly the reason that I employ it here: it is distributed in much less seasonal habitats than those are associated with early hominin sites (Moore, 1996; Hunt and McGrew, 2002). Foley (1993) argued that hominin origins are closely linked to the development of increasingly seasonal conditions; he also argues that differences among African ape species were driven by the degree to which their ancestral and extant environments are seasonal and differences in the selection for traits for dealing with food scarcity in more seasonal habitats. He purposely ruled out *P. paniscus* in his evaluation because this species is found in less-seasonal habitats. He instead applied data from *P. troglodytes* day range, time spent feeding, and feeding party size to an understanding of early hominin behavior and adaptations.

I suggest that the *P. troglodytes–P. paniscus* comparison, along with knowledge of cercopithecoid and gorilla feeding, can lend insight into the evolution of behavioral solutions to facilitate a fallback strategy on relatively high-quality foods. By analogy from this comparison, as early hominins expanded into increasingly seasonal habitats, the potential for critical periods of selection increased as well. We cannot observe behavior in the past (the unknown) directly, but we can document fine details of climate via ever increasingly sophisticated archaeological and geological methods, and we can use models derived from nonhuman primate models, which can illuminate the timing and rate of natural selection pressure on feeding adaptations. Moreover, we can evaluate distribution patterns and the degree to which habitats are marginal. *Pan troglodytes* occupy habitat that can have upward of five to seven dry (rainfall <100 mm) months/annum; *Australopithecus afarensis* has been found in habitats that probably had even more dry months (Foley, 1993; see Reed and Rector, chapter 14, in this volume). This suggests extreme potential for critical periods, which may select either for adaptations to lower-quality foods or adaptations for innovative behaviors—such as flexibly sized feeding parties and tool use—that maintain a higher-quality diet or some combination of both. As suggested earlier, it is not necessarily an either-or situation, and neither strategy precludes the other. *Cebus* spp., for example, have both thick dental enamel for exploiting hard foods, as well as tools to access high-quality, but protected fallback foods.

With regard to hominin diet and adaptations, Scott et al. (2005) have recently argued that there is greater dietary overlap and variability in *Paranthropus* and *Australopithecus* than previously thought. They indeed suggest "early hominin diet differences might relate more to microhabitat, seasonality, or fallback food choice than to oversimplified, dichotomous food preferences" (Scott et al., 2005, p. 694). I agree, but I would add another degree of complexity: there are differences among both fallback *foods* and, more importantly, overall fallback *strategies*. On the basis of the above framework, we can envisage a continuum of fallback strategies: at one

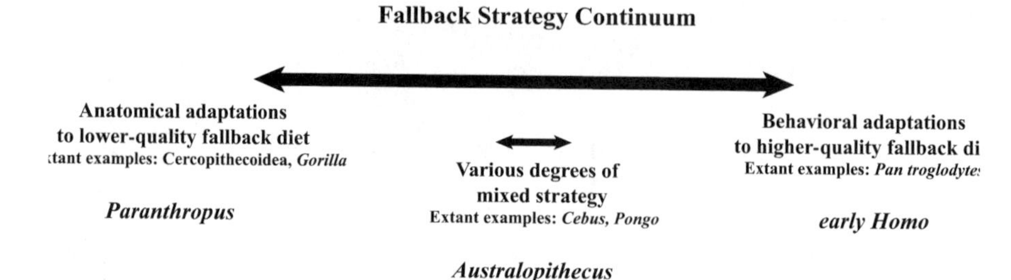

Figure 17.4 Graphic depicting the continuum of fallback strategies observed in extant primates, with implications for interpreting hominin diet-related adaptations.

end, we see the evolution of *anatomy* to exploit *lower-quality* (less nutritionally dense) foods; at the other end, we see evolution of *behavioral* mechanisms for maintaining *higher-quality* foods, with various degrees of a mixed strategy in between (fig. 17.4). Only in relatively aseasonal habitats like that of *P. paniscus* would we see relaxed selection for fallback strategies. We know that the evolution of hominin species is closely related to increasing seasonality, so some fallback strategy at a point along this continuum is expected. In the case of a comparative *Paranthropus, Australopithecus*, early *Homo* evaluation, we might expect, based on anatomy alone, for *Paranthropus* to be at one end of the continuum, early *Homo* at the other, and *Australopithecus* somewhere between the two.

Acknowledgments I owe many thanks to Peter Ungar for inviting me to contribute to this volume, his careful editing, and advice during critical periods. I also gratefully acknowledge and thank Karen Strier for discussion, input, and insight on the manuscript. Also, my thanks to Henry Bunn, John Hawks, Liza Moscovice, Karen Strier, Peter Ungar, and the University of Wisconsin Primate Ecology Lab Group for discussions that helped in the development of the ideas presented here. I am very grateful to Alain Houle who kindly provided the photographic images of *Pan* and *Lophocebus*. Finally, I would like to thank Jerry Jacka, who provided me with assistance on the graphic images, and emotional support for this and all of my endeavors.

References

Andrews, P., 1981. Species diversity and diet in monkeys and apes during the Miocene. In: Stringer, C.B. (Ed.), *Aspects of Human Evolution*. Taylor & Francis, London, pp. 25–26.

Boesch, C., Hohmann, G., and Marchant, L.F. (Eds.), 2002. *Behavioural Diversity in Chimpanzees and Bonobos*. Cambridge University Press, Cambridge.

Bradley, B.J., and Vigilent, L., 2002. The evolutionary genetics and molecular ecology of chimpanzees and bonobos. In: Boesch, C., Hohmann, G., and Marchant, L.F. (Eds.), *Behavioural Diversity in Chimpanzees and Bonobos*. Cambridge University Press, Cambridge, pp. 259–276.

Caton, J., 1999. Digestive strategy of the Asian colobine genus *Trachypithecus*. Primates 40, 311–325.

Chapman, C.A., White, F.J., and Wrangham, R.W., 1994, Party size in chimpanzees and bonobos: A reevaluation of theory based on two similarly forested sites. In: Wrangham,

R.W., McGrew, W.C, deWaal, F.B., and Heltne, P.G. (Eds.), *Chimpanzee Cultures*. Harvard University Press, Cambridge, pp. 41–58.

Chivers, D.J., and Hladik, C.M., 1980. Morphology of the gastrointestinal tract in primates: Comparisons with other mammals in relation to diet. *J. Morphol.* 116, 337–386.

Collet, J., Bourreau, E., Cooper, R.W., and Tutin, G.E.G., 1984. Experimental demonstration of cellulose digestion by *Troglodytella gorillae*, an intestinal ciliate of lowland gorillas. *Int. J. Primatol.* 5, 328.

Conklin-Brittain, N.L., Wrangham, R.W., and Hunt, R.D., 1998. Dietary response of chimpanzees and cercopithecines to seasonal variation in fruit abundance. II. Macronutrients. *Int. J. Primatol.* 19, 971–998.

Conklin-Brittain, N.L., Wrangham, R.W., and Smith, C.G., 2002. A two-staged model of increased dietary quality in early hominid evolution: the role of fiber. In: Ungar, P.S., and Teaford, M.F. (Eds.), *Human Diet: Its Origin and Evolution*. Bergin & Garvey Press, Westport, CT, pp. 61–76.

Cowlishaw, G., and Dunbar, R., 2002. *Primate Conservation Biology*. University of Chicago Press, Chicago.

Cork, S.J., and Foley, W.J., 1991. Digestive and metabolic strategies of arboreal folivores in-relation to chemical defenses in temperate and tropical forests. In: Palo, R.T., and Robbins, C.T. (Eds.), *Plant Defenses Against Mammalian Herbivory*. CRC Press, Boca Raton, FL, pp. 133–166.

Demment, M.W., and van Soest, P.J., 1983. *Body Size, Digestive Capacity, and Feeding Strategies of Herbivores*. Winrock International Livestock Research Training Center, Morrilton, AR.

Doran, D.M., McNeilage, A., Greer, D., Bocian, C., Mehlman, P., and Shah, N., 2002. Western lowland gorilla diet and resource availability: New evidence, cross-site comparisons, and reflections on indirect sampling methods. *Am. J. Primatol.* 58, 3, 91–116.

Dumont, E.R., 1995. Enamel thickness and dietary adaptation among extant primates and chiropterans. *J. Mammal.* 76, 1127–1136.

Eriksson, J., Hohmann, G., Boesch, C., and Vigilant, L., 2004. Rivers influence the population genetic structure of bonobos (*Pan paniscus*). *Mol. Ecol.* 13, 3425–3435.

Fleagle, J.G., and Reed, K.E., 1996. Comparing primate communities: a multivariate approach. *J. Hum. Evol.* 30, 489–510.

Fleagle, J.G., Janson, C., and Reed, K.E. (Eds.), 1999. *Primate Communities*. Cambridge University Press, Cambridge.

Foley, R.A., 1993. The influence of seasonality on hominid evolution. In: Ulijaszek, S.J., and Strickland, S.S. (Eds.), *Seasonality and Human Ecology: 35th Symposium Volume of the Society for the Study of Human Biology*. Cambridge University Press, Cambridge, pp. 17–37.

Foley, R.A., 1994. Speciation, extinction and climatic change in hominid evolution. *J. Hum. Evol.* 26, 275–289.

Fox, E.A., van Schaik, C.P., Sitompul, A., and Wright, D.N., 2005. Intra- and interpopulational differences in organgutan (*Pongo pygmaeus*) activity and diet: Implications for the invention of tool use. *Am. J. Anthropol.* 125, 162–174.

Fragaszy, D., Izar, P., Visalbergh, E., Ottoni, E.B., and Oliveira de, M.G., 2004. Wild capuchin monkeys (*Cebus libidinosus*) use anvils and stone pounding tools. *Am. J. Primatol.* 64, 359–366.

Freeland, W.J., 1991. Plant secondary metabolites: Biochemical coevolution with herbivores. In: Palo, R.T., and Robbins, C.T. (Eds.), *Plant Defenses against Mammalian Herbivory*. CRC Press, Boca Raton, FL, pp. 61–82.

Freeland, W.J., and Janzen, D.H., 1974. Strategies in herbivory by mammals: The role of plant secondary compounds. *Am. Nat.* 108, 269–289.

Furuichi, T., Hashimoto, C., and Tashiro, Y., 2001. Fruit availability and habitat use by chimpazees in the Kalinzu Forest, Uganda: Examination of fallback foods. *Int. J. Primatol.* 22, 929–945.

Futuyma, D.J., 1979. *Evolutionary Biology*. Sinauer, Sunderland, MA.

Gautier-Hion, A., and Michaloud, G., 1989. Are figs always keystone resources for tropical frugivorous vertebrates? A test in Gabon. *Ecology* 70, 1826–1833.

Glander, K.E., 1978. Howling monkey feeding behavior and plant secondary compounds: A study of strategies. In: Montgomery, G.G. (Ed.), *The Ecology of Arboreal Folivores*. Smithsonian Press, Washington DC, pp. 231–241.

Gursky, S., 2000. Effect of seasonality on the behavior of an insectivorous primate, *Tarsius spectrum*. *Int. J. Primatol.* 21, 477–495.

Hunt, K.D., and McGrew, W.C., 2002. Chimpanzees in the dry habitats of Assirik, Senegal and Semliki Wildlife Reserve, Uganda. In: Boesch, C., Hohmann, G., and Marchant, L.F. (Eds.), *Behavioural Diversity in Chimpanzees and Bonobos*. Cambridge University Press, Cambridge, pp. 35–51.

Janson, C.H., and Chapman, C.A., 1999. Resources and primate community structure. In: Fleagle, J.G., Janson, C., and Reed, K.E. (Eds.), *Primate Communities*. Cambridge University Press, Cambridge, pp. 237–260.

Kay, R.F., 1981. The nut-crackers: A new theory of the adaptations of the Ramapithecidae. *Am. J. Phys. Anthropol.* 55, 141–151.

Kay, R.F., 1984. On the use of anatomical features to infer foraging behavior in extinct primates. In: Rodman, P.S., and Cant, J.G.H. (Eds.), *Adaptations for Foraging in Nonhuman Primates: Contributions to an Organismal Biology of Prosimians, Monkeys, and Apes*. Columbia University Press, New York, pp. 21–53.

Kay, R.N.B., and Davies, A.G., 1994. Digestive physiology. In: Davis, A.G., and Oates, J.F. (Eds.), *Colobine Monkeys: Their Ecology, Behavior and Evolution*. Cambridge University Press, Cambridge, pp. 229–259.

Kay, R.N.B., 1985. Comparative studies of food propulsion in ruminants. In: Ooms, L.A.A., Degryse, A.D., and van Miert A.S.J.A.M. (Eds.), *Physiological and Pharmacological Aspects of the Reticulo-Rumen*. Martinus Nijhoff, Dordrecht, pp. 155–170.

Keeler, R.F., Kampen van, K.A., and James, L.F., 1978. *Effects of Poisonous Plants on Livestock*. Academic Press, New York.

Laden, G., and Wrangham, R.W., 2005. The rise of the hominids as an adaptive shift in fallback foods: Plant underground storage organs (USOs) and australopith origins. *J. Hum. Evol.* 49, 482–498.

Lambert, J.E., 1997. Digestive strategies, fruit processing, and seed dispersal in the chimpanzees (*Pan troglodytes*) and redtail monkeys (*Cercopithecus ascanius*) of Kibale National Park, Uganda. PhD diss., University of Illinois, Urbana–Champaign.

Lambert, J.E., 1998. Primate digestion: Interactions among anatomy, physiology, and feeding ecology. *Evol. Anthropol.* 7, 1, 8–20.

Lambert, J.E., 2000. Urine drinking in wild *Cercopithecus ascanius*: Evidence of nitrogen balancing? *Afr. J. Ecol.* 389, 4, 360–363.

Lambert, J.E., 2002a. Digestive retention times in forest guenons with reference to chimpanzees. *Int. J. Primatol.* 26(6), 1169–1185.

Lambert, J.E., 2002b. Resource switching in guenons: a community analysis of dietary flexibility. In: Glenn, M., and Cords, M. (Eds.), *The Guenons: Diversity and Adaptation in African Monkeys*. Kluwer Academic Press, New York, pp. 303–317.

Lambert, J.E., 2005. Competition, predation and the evolution of the cercopithecine cheek pouch: The case of *Cercopithecus* and *Lophocebus*. *Am. J. Phys. Anthropol.* 126(2), 183–192.

Lambert, J.E., 2007. Primate nutritional ecology: Feeding biology and diet at ecological and evolutionary scales. In: Campbell, C., Fuentes, A., MacKinnon, K.C., Panger, M., and Bearder, S. (Eds.), *Primates in Perspective*. Oxford University Press, pp. 482–495.

Lambert, J.E., Chapman, C.A., Wrangham, R.W., and Conklin-Brittain, N.L., 2004. The hardness of mangabey and guenon foods: Implications for the critical function of enamel thickness in exploiting fallback foods. *Am. J. Phys. Anthropol.* 125(4), 363–368.

Lambert, J.E., and Whitham, J., 2001. Cheek pouch use in *Papio cynocephalus*. *Folia Primatol.* 72, 89–91.

MacArthur, R., Pianka, E., 1966. On optimal use of a patchy environment. *Am. Nat.* 100, 603–609.

Maisels. F., 1993. Gut passage rate in guenons and mangabeys: Another indicator of a flexible dietary niche? *Folia Primatol.* 61, 35–37.

Malenky, R.K., 1990. Ecological factors affecting food choice and social organization in *Pan paniscus*. PhD diss., University of Michigan.

Malenky, R.K., and Wrangham, R.W., 1994. A quantitative comparison of terrestrial herbaceous vegetation and its consumption by *Pan paniscus* in the Lomako Forest, Zaire, and *Pan troglodytes* in the Kibale Forest, Uganda. *Am. J. Primatol.* 32, 1–12.

Maley, J., 1996. The African rain forest——Main characteristics of changes in vegetation and climate from Upper Cretaceous to Quaternary. *Proc. R. Soc. Edinburgh* B 104, 31–73.

Matsuzawa, T., 2002. Behavioural flexibility. In: Boesch, C., Hohmann, G., and Marchant, L.F. (Eds.), *Behavioural Diversity in Chimpanzees and Bonobos*. Cambridge University Press, Cambridge, pp. 11–13.

McGrew, W.C., 1992. *Chimpanzee Material Culture: Implications for Human Evolution.* Cambridge University Press, Cambridge.

Milton, K., 1981. Food choice and digestive strategies of two sympatric primate species. *Am. Nat.* 117, 476–495.

Milton, K., 1984. The role of food processing factors in primate food choice. In: Rodman, P.S., and Cant, J.G.H. (Eds.), *Adaptations for Foraging in Nonhuman Primates: Contributions to an Organismal Biology of Prosimians, Monkeys, and Apes.* Columbia University Press, New York, pp. 249–279.

Milton, K., 1986. Digestive physiology in primates. *News Physiol. Sci.* 1, 76–79.

Milton, K., 1987. Primate diets and gut morphology: implications for hominid evolution. In: Harris, M., and Ross, E.B. (Eds.), *Food and Evolution: Toward a Theory of Human Food Habits*. Temple University Press, Philadelphia, pp. 93–115.

Milton, K., 1993. Diet and primate evolution. *Sci. Am.* August, 86–93.

Milton, K., 1999. A hypothesis to explain the role of meat-eating in human evolution. *Evol. Anthropol.* 8, 11–21.

Milton, K., and Demment, M.W., 1988. Digestion and passage kinetics of chimpanzees fed high and low fiber diets and comparison with human data. *J. Nutr.* 118, 1082–1088.

Moore, J., 1996. Savanna chimpanzees, referential models and the last common ancestor. In: McGrew, W.C., Marchant, L.F., and Nishida, T. (Eds.), *Great Ape Societies*. Cambridge University Press, Cambridge, pp. 275–292.

Moura, A.C., de, and Lee, P.C., 2004. Capuchin stone tool use in Caatinga dry forest. *Science* 306, 1909.

Newberry, D.M., Songwe, N.C., and Chuyong, G.B., 1998. Phenology and dynamics of an African rainforest at Korup, Cameroon. In: Newberry, D.M., Prins, H.H.T., and Brown, N. (Eds.), *Dynamics of Tropical Communities*. Blackwell Science Publishers, Oxford. pp. 267–308.

Nishihara, T., 1995. Feeding ecology of western lowland gorillas in the Nouabale-Ndoki National Park, Congo. *Primates* 36, 151–168.

Oates, J.F., 1977. The guereza and its food. In: Clutton-Brock, T.H. (Ed.), *Primate Ecology*. Academic Press, London, pp. 275–321.

Oates, J.F., 1987. Food distribution and foraging behavior. In: Smuts, B.B., Cheyney, D.L., Seyfarth, R.M., Wrangham, R.W., and Struhsaker, T.T. (Eds.), *Primate Societies*. University of Chicago Press, Chicago, pp. 197–209.

Parra, R., 1978. Comparison of foregut and hindgut fermentation in herbivores. In: Montgomery, G.G. (Ed.), *The Ecology of Arboreal Folivores*. Smithsonian Institution Press, Washington DC, pp. 205–229.

Potts, R., 1998. Variability selection in hominid evolution. *Evol. Anthropol.* 7, 81–96.

Potts, R., 2004. Paleoenvironmental basis of cognitive evolution in great apes. *Am. J. Primatol.* 62, 209–228.

Pruetz, J.D., 2005. Cave use by wild savanna chimpanzees (*Pan troglodytes verus*) in Senegal: Behavioral adaptation to heat stress? *Am. J. Phys. Anthropol.* Suppl. 40, 168.

Rathcke, B., and Lacey, E.P., 1985. Phenological patterns of terrestrial plants. *Annu. Rev. Ecol. Syst.* 16, 179–214.

Reed, K.E., 1997. Early hominid evolution and ecological change through the African Plio-Pleistocene. *J. Hum. Evol.* 32, 289–322.

Remis, M.J., 1997. Western lowland gorillas (*Gorilla gorilla gorilla*) as seasonal frugivores: Use of variable resources. *Am. J. Primatol.* 43, 87–109.

Remis, M.J., 2000. Initial studies on the contributions of body size and gastrointestinal times to dietary flexibility among gorillas. *Am. J. Phys. Anthropol.* 112, 171–180.

Remis, M.J., Dierenfeld, E.S., Mowry, C.B., and Carroll, R.W., 2001. Nutritional aspects of western lowland gorilla diet during seasons of fruit scarcity at Bai Hokou, Central African Republic. *Int. J. Primatol.* 22, 807–836.

Robinson, B.W. and Wilson, D.S., 1998. Optimal foraging, specialization, and a solution to Liem's Paradox. *Am. Nat.* 151, 223–235.

Rogers, E., Tutin, C., Parnell, R., Voysey, B., and Fernandez, M., 1994. Seasonal feeding on bark by gorillas: An unexpected keystone food? Abstracts from the XIVTH Congress of the International Primatological Society. IPS, Strasbourg p. 154.

Rogers, E.M., Aberneth, Y. K., Bermejo, M., Cipolletta, Doran, D., McFarland, K., Nishihara, T., Remis, M., and Tutin, C.E.G., 2004. Western gorilla diet: A synthesis from six sites. *Am. J. Primatol.* 64, 173–192.

Rosenberger, A.L., 1992. Evolution of feeding niches in New World monkeys. *Am. J. Phys. Anthropol.* 88, 525–562.

Rosenberger, A.L, and Kinzey, W., 1976. Functional patterns of molar occlusion in platyrrhine primates. *Am. J. Phys. Anthropol.* 45, 281–298.

Scott, R.S., Ungar, P.S., Bergstrom, T.S., Brown, C.A., Grine, E.E., Teaford, M.F., and Walker, A., 2005. Dental microwear texture analysis shows within-species diet variability in fossil hominins. *Nature* 436, 4, 693–695.

Struhsaker, T.T., 1978. Food habits of five monkey species in the Kibale Forest, Uganda. In: Chivers, D.J., and Herbert, J. (Eds.), *Recent Advances in Primatology*. Vol. 2: *Conservation*. Academic Press, London, pp. 87–94.

Teaford, M.F., and Ungar P.S., 2000. Diet and the evolution of the earliest human ancestors. *Proc. Natl. Acad. Sci.* 97, 13506–13511.

Teaford, M., Maas, M., and Simons, E.L., 1996. Dental microwear and microstructure in early Oligocene primates from the Fayum, Egypt: Implications for diet. *Am. J. Phys. Anthropol.* 101, 527–544.

Temerin, L.A., and Cant, J.G.H., 1983. The evolutionary divergence of Old World monkeys and apes. *Am. Nat.* 12(2), 355–351.

Thompson, J.A.M., 2002. Bonobos of the Lukuru Wildlife Research Project. In: Boesch, C., Hohmann, G., and Marchant, L.F. (Eds.), *Behavioural Diversity in Chimpanzees and Bonobos*. Cambridge University Press, Cambridge, pp. 61–70.

Thompson, J.A.M., 2003. A model of the biogeographical journey from *Proto-Pan* to *Pan paniscus*. *Primates* 44, 191–197.

Tutin, C.E.G., Hamm, R.M., White, L.J.T., and Harrison, M.J.S., 1997. The primate community of Lope Reserve, Gabon: Diets, responses to fruit scarcity, and effects on biomass. *Am. J. Primatol.* 42, 1–24.

Ungar, P.S., 2004. Dental topography and diets of *Australopithecus afarensis* and early Homo. *J. Hum. Evol.* 46, 605–622.

Ungar, P.S., Grine, F.E., Teaford, M.F., and el Zaatari, S., 2006. Dental microwear and diets of African early Homo. *J. Hum. Evol.* 50(1), 78–95.

van Schaik, C.P., Terborgh, J.W., and Wright, S.J., 1993. The phenology of tropical forests: Adaptive significance and consequences for primary consumers. *Annu. Rev. Ecol. Syst.* 24, 353–377.

Van Soest, P.J., 1994. *Nutritional Ecology of the Ruminant*. Cornell University Press, New York.

Vrba, E.S., 1995. On the connections between paleoclimate and evolution. In: Vrba, E.S., Denton, G.H., Partridge, T.C., and Burkle, L.H. (Eds.), *Paleoclimate and Evolution, with Emphasis on Human Origins*. Yale University Press, New Haven, CT, pp. 24–48.

Walker, C.H., 1978. Species differences in microsomal monooxygenase activity and their relationship to biological half-lives. *Drug Metab. Rev.* 7, 295–310.

Waterman, P.G., and Kool, K, 1994. Colobine food selection and plant chemistry. In: Davies, A.G., and Oates, J.F., (Eds.), *Colobine Monkeys: Their ecology, behavior and evolution.* Cambridge University Press, Cambridge, pp. 251–284.

White, F.J., 1998a. The importance of seasonality in primatology. *Int. J. Primatol.* 19, 925–927.

White, F.J., 1998b. Seasonality and socioecology: The importance of variation in fruit abundance to bonobo sociality. *Int. J. Primatol.* 19, 925–927.

Whiten, A, Goodall, J., McGrew, W.C., Nishida, T., Reynolds, C., Sugiyama, Y., Tutin, C.E.G., Wrangham, R.W., and Boesch, C., 1999. Cultures in chimpanzees. *Nature* 399, 6826–6836.

Wrangham, R.W., Chapman, C.A., Clark-Arcadi, A.P., and Isabirye-Basuta, G., 1996. Social ecology of Kanyawara chimpanzees: Implications for understanding the costs of great apes groups. In: McGrew, W.C., Marchant, L.F., and Nishida, T. (Eds.), *Great Ape Societies.* Cambridge University Press, Cambridge, pp. 45–57.

Wrangham, R.W., Conklin, N.L., Chapman, C.A., and Hunt, K.D., 1991. The significance of fibrous foods for Kibale Forest chimpanzees. *Philos. Trans. R. Soc. Lond.* B 334, 171–178.

Wrangham, R.W., Conklin-Brittain, N.L., and Hunt, K.D., 1998. Dietary response of chimpanzees and cercopithecines to seasonal variation in fruit abundance. I. Antifeedants. *Int. J. Primatol.* 19, 949–970.

Wrangham, R.W., Jones, J.H., Laden, G., Pilbeam, D., and Conklin-Brittain, N.L., 1999. The raw and the stolen: Cooking and the ecology of human origins. *Curr. Anthropol.* 40, 567–594.

Yamakoshi, G., 1998. Dietary response to fruit scarcity of wild chimpanzees at Bossou, Guniea: Possible implications for ecological importance of tool use. *Am. J. Phys. Anthropol.* 106, 283–295.

18

Energetic Models of Human Nutritional Evolution

WILLIAM R. LEONARD
MARCIA L. ROBERTSON
J. JOSH SNODGRASS

Over the past decade, biological anthropologists have increasingly begun to rely on energetic models for understanding patterns and trends in hominin evolution (e.g., Leonard and Robertson, 1994, 1997a; Aiello and Wheeler, 1995; Leonard, 2002). The study of energetics is important to evolutionary research for several reasons. First, food energy represents a critical interface between organisms and their environment. The search for food energy, its consumption, and ultimately its allocation for biological processes are all critical aspects of an organism's ecology (McNab, 2002). In addition, the energy dynamic between organisms and their environments—energy expenditure in relation to energy consumed—has important adaptive consequences for both survival and reproduction. Energy thus provides a useful currency for measuring fitness. Indeed, the two components of Darwinian fitness—survival and reproduction—are reflected in the way that total energy budgets for animals are typically divided (see fig. 18.1). "Maintenance" energy expenditure represents the costs of keeping an animal alive on a day-to-day basis. This includes resting (or basal) energy expenditure and energy expenditure associated with normal daily activities of work, play, and recreation. "Productive" energy costs, on the other hand, are those associated with growth from infancy into adulthood and the production of offspring for the next generation. For mammals like ourselves, productive energy costs include the increased energy costs of a mother during pregnancy and lactation.

Clearly, the type of environment that an organism lives in will strongly shape the relative allocation of energy to these different components. However, from an evolutionary perspective, the goal for all organisms is the same: to allocate sufficient energy to "production" to ensure their genes are passed on to future generations. Consequently, by looking at the ways that animals go about acquiring and then allo-

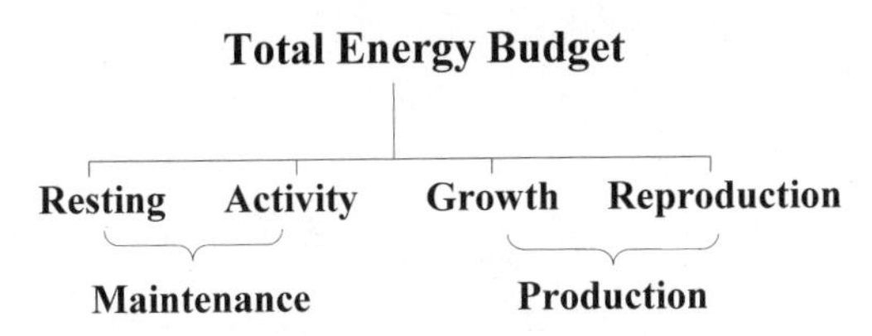

Figure 18.1 Components of an animal's total energy budget. Maintenance energy costs are those associated with keeping the animal alive on a daily basis. Productive energy costs are those required for growth and reproduction.

cating food energy, we can better understand how natural selection produces important patterns of evolutionary change. This approach is particularly useful in studying human evolution because it appears that key transitions in our evolutionary past likely had important implications for energy allocation. In this chapter, we examine the use of energetic models to help understand two of these "turning points" in human evolution: (1) the evolution of bipedal (upright) movement and (2) the rapid evolution of brain size with the emergence of the genus *Homo.*

Energetics and Bipedality

The Known

The potential energetic benefits of hominin bipedal locomotion have been long debated among biological anthropologists and other evolutionary biologists (Taylor and Rowntree, 1973; Rodman and McHenry, 1980; Steudel, 1994, 1996; Leonard and Robertson, 1995, 1997b, 2001; Steudel-Numbers, 2001). To address whether bipedality was a more "economical" form of movement, scholars have examined the energy costs of human locomotion relative to those of other mammalian species. Locomotor energy costs are measured using standard methods of indirect calorimetry in which respiratory gases (oxygen and carbon dioxide) and ventilation (breathing) rates are measured while the subject is moving at a known speed (see McLean and Tobin, 1987; McArdle et al., 2001). Energy costs of human movement have been well studied in the fields of nutrition and exercise science (e.g., Consolazio, Johnson, and Pecora, 1963; Margaria et al, 1963; Morgan and Craib, 1992), whereas those of other animal species have been examined by comparative and evolutionary physiologists (e.g., Taylor, Schmidt-Nielsen, and Raab, 1970; Taylor, Heglund, and Maloiy, 1982; Tucker, 1970).

Both within and between species, the strongest predictors of locomotor energy costs are body weight (mass) and speed. For most species, weight-specific energy costs increase as a simple linear function of speed (Taylor, Heglund, and Maloiy, 1982). However, for humans, this linear relationship is evident only at running

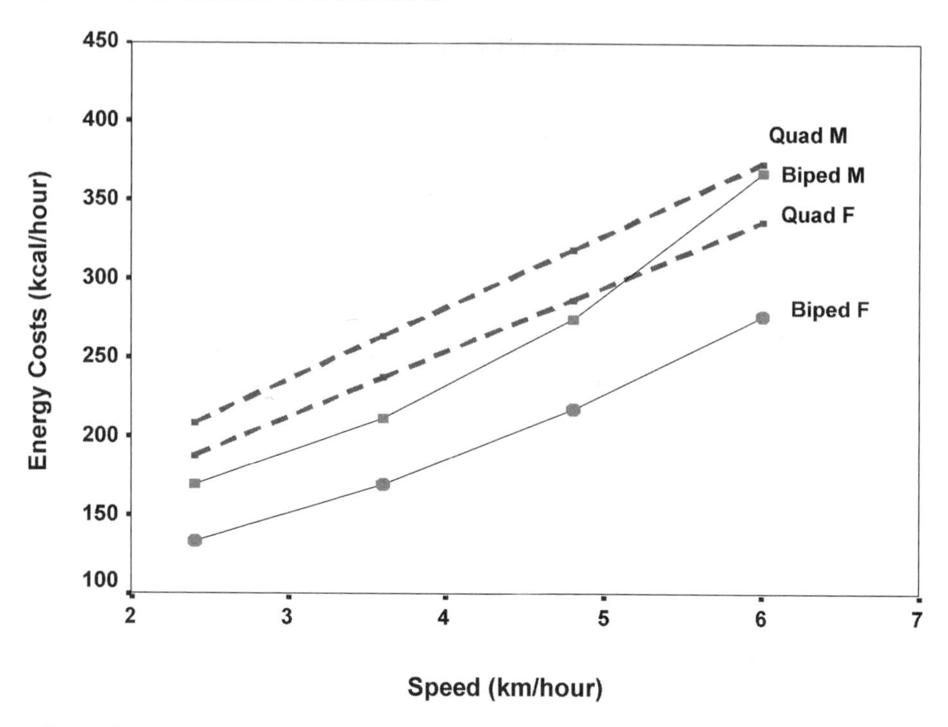

Figure 18.2 Energy costs (kcal/hour) of bipedal versus generalized quadrupedal locomotion for males and females moving at speeds between 2.4 and 6.0 km/hour. Values are average costs for females of 60 kg and males of 70 kg. For both males and females, bipedal energy costs are less than quadrupedal costs over the entire range of speeds. Modified from Leonard and Robertson (1997b).

speeds (of about 6 to 8 km/hour or more); at walking speeds, the relationship is curvilinear (Margaria et al., 1963; Menier and Pugh, 1968). Comparative analyses indicate that at walking speeds (between 2.4 and 6. 0 km/hour), humans expend significantly *less* energy than most quadrupeds (Rodman and McHenry, 1980; Leonard and Robertson, 1995, 1997b). At running speeds, however, bipedality is less economical, as human expend significantly more energy than quadrupeds (Taylor, Heglund, and Maloiy, 1982).

The relative economy of human bipedal walking is evident in figure 18.2, which shows the energy costs of locomotion (in kcal/hour) for adult men and women compared with those of typical (average) mammalian quadrupeds of the same weight over a range of typical walking speeds. Note that the human energy costs are lower than those of the typical quadrupeds over all these speeds, with the relative energy savings of walking bipedally being greatest at slower rates. Over the entire range of walking speeds, human males expend 13% less energy their quadrupedal counterparts, whereas human females expend 25% less (Leonard and Robertson, 1997b).

In contrast to humans, the available data on modern apes indicates that they have relatively high energy costs for locomotion. Taylor and Rowntree (1973) and Taylor, Heglund, and Maloiy (1982) found that chimpanzees expend 35% to 40% more energy for quadrupedal locomotion than an "average" mammalian quadruped of the same size. The differences in relative energy costs of movement for humans and apes appear to reflect differences in the environment in which they evolved and adapted. The great apes evolved and continue to exist in forested environments. Conversely, much of early hominin evolution involved living in more open mixed woodland and grassland environments (Reed, 1997). As a consequence, hominin forms living in more open environments would have moved over much greater areas in pursuit of food, whereas apes typically do not have to move over large distances during a course of the day.

These differences in typical movement patterns between humans and apes are evident when we compare the day ranges of modern apes to those of human hunter-gatherers. Modern apes move only about 1.8 km/day, whereas human foragers move an average of 13.1 km/day (Leonard and Robertson, 1997b). Differences in day range have important implications for the total energy costs of locomotion. Because apes move only short distances each day, the potential energetic differences between moving "efficiently" or "inefficiently" are very small. However, for animals that must move over longer distances, moving more efficiently has the potential to save substantial energy from maintenance.

This point is evident in table 18.1, which shows the results of a model estimating the energy costs of moving bipedally and quadrupedally for an animal the size of an early australopithecine (*Australopithecus afarensis*) with day ranges of 2.0, 5.0, and 13.0 km/day. With movement over larger day ranges, the net energetic benefit of moving bipedally increases from 20–25 kcal/day up to 120–150 kcal/day. For the smaller day ranges, the energetic savings accounts for about 1% to 3% of estimated resting metabolic rate (RMR, kcal/day); however, for day ranges similar to those of modern human foragers, the savings is 12% to 16% of RMR, a substantial portion of the daily energy budget. Thus, for movement over larger areas, the greater economy of bipedal movement offers the potential for substantial reductions in maintenance energy demands, allowing for the energy savings to be allocated to reproduction. Selection for energetically efficient locomotion is therefore likely to be more intense

Table 18.1 Estimated Weight, Resting Metabolic Rate, and Energy Costs of Bipedal and Quadrupedal Movement for *Australopithecus afarensis* Moving Over Day Ranges of 2, 5, and 13 Kilometers

				Energy Costs of Movement			
Sex	Weight (kg)	RMR (kcal/day)	Range (km)	Biped (kcal)	Quad. (kcal)	Difference (kcal)	RMR (%)
Male	45	1,216	2	86	109	23	1.8
Female	29	874	2	59	81	22	2.4
Male	45	1,216	5	216	272	56	4.6
Female	29	874	5	148	201	53	6.1
Male	45	1,216	13	565	712	147	12.1
Female	29	874	13	389	528	139	15.9

Sources: Body weight estimates are from McHenry and Coffing (2000); RMR and energy costs for bipedal movement estimated from FAO/WHO/UNU (1985); quadrupedal energy costs are estimated from Taylor et al. (1982).

among animals that must move over larger areas because they have the most to gain by being more efficient.

The Unknown

What remains unclear about the energetics of human bipedal locomotion is whether our earliest hominin ancestors had a sufficiently humanlike gait to reap the full benefits of this locomotor strategy. This uncertainty is reflected in the ongoing debate over the effectiveness of early australopithecine bipedality. Lovejoy (1988, 2005) and Latimer and Lovejoy (1989, 1990a, 1990b), for example, have argued that the skeletal anatomy of Lucy and other members of *A. afarensis* is consistent with a fully functional biped that was committed to terrestrial movement. In contrast, Stern and Susman (1983); Susman, Stern, and Junger (1984); and Jungers (1982, 1988) have argued that *A. afarensis* remained partially arboreal and did not walk as effectively as modern humans. Steudel (1994, 1996) has further suggested that the apelike body proportions of the earliest australopithecines (i.e., long arms and short legs) likely would have reduced their locomotor efficiency relative to modern human bipedality. Stuedel-Numbers and Tilkins (2004) have shown that among humans, relatively longer legs are associated with greater locomotor economy.

Recent paleoenvironmental analyses have also suggested that the first hominins may have emerged in environments that were more heavily forested than previously thought (Woldegabriel et al., 1994; Potts, 1998). This raises the question about how widely our earliest hominin ancestors moved. If day ranges for the earliest australopithecines were similar to those of modern chimpanzees, then the initial benefits of upright movement would have been more modest. If true, this would suggest that the full energetic benefits of human bipedal locomotion were not gained during a limited period of time but were accrued over a larger span of evolutionary time. The archaeological evidence suggests that with the drying of the African landscape between 4 and 1.5 million years ago, hominins were moving over larger areas and adapting to more open grassland environments (Reed, 1997; Potts, 1998). It is over this time that we also see the evolution of humanlike body size and proportions that become fully realized with the advent of *Homo erectus* at 1.8 million years ago (McHenry and Coffing, 2000).

Further understanding of the energetics of hominin locomotion will also depend on the collection of more data from living apes and other large-bodied primates. The available information on locomotor energetics in modern apes is limited to data collected on two small chimpanzees by Taylor and Rowntree (1973). Additional information on a broader sample of apes is necessary to better understand how variation in body size, proportions, and movement patterns may shape differences in energy demands.

Energetics and Hominin Brain Evolution

The Known

Energetic models have also provided important insights into hominin brain evolution. What is remarkable about the large human brain is its high metabolic cost. The

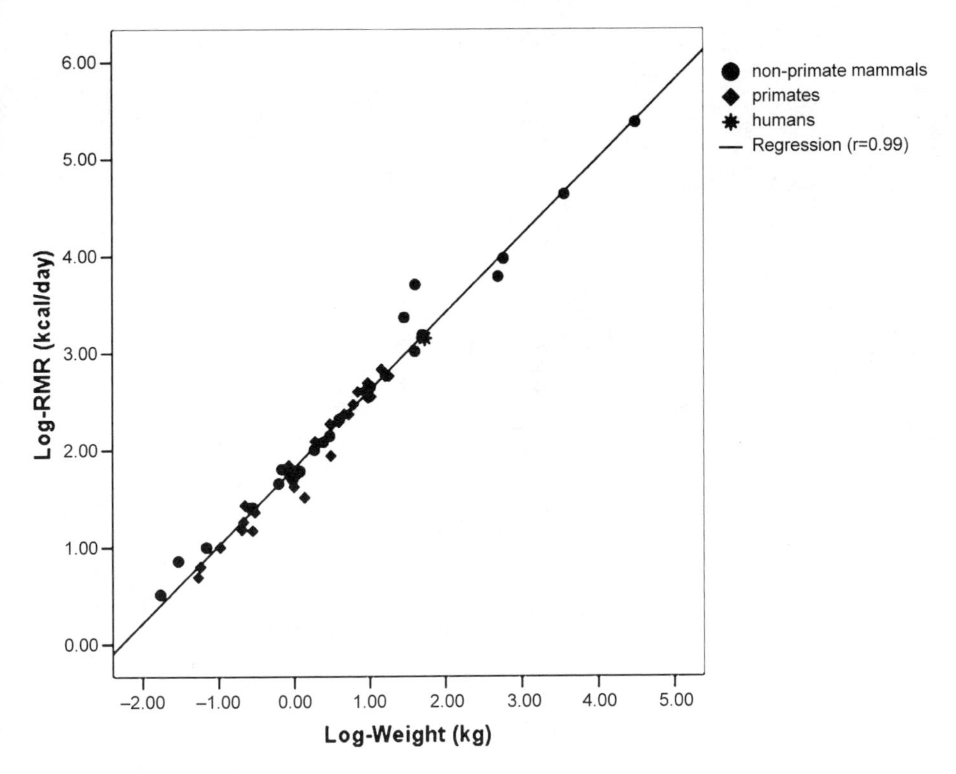

Figure 18.3 Log-log plot of resting metabolic rate (RMR; kcal/day) versus body weight (kg) for fifty-two mammalian species (twenty-one nonprimate mammals, thirty primates, and humans). Humans conform to the general mammalian scaling relationship, as described by Kleiber (1961). Data are from Leonard et al. (2003) and Snodgrass et al. (1999).

energy requirements of brain tissue are about 29 kcal/100grams/day, roughly sixteen times that of skeletal muscle tissue (Kety, 1957; Holliday, 1986). This means that for a 70-kg adult human (with a brain weight of about 1,400 grams), approximately 400 kcal per day are allocated to brain metabolism. Although humans have much larger brains than most other mammals, the total energy demands for our body—our resting energy requirements—are no greater than those of a comparably sized mammal (Kleiber, 1961; Leonard and Robertson, 1992). This point is evident in figure 18.3, which shows the relationship between RMR (kcal/day) and body weight (kg) in nonprimate mammals, primates, and humans. Humans conform to the general mammalian scaling relationship between RMR and body weight (the "Kleiber relationship"), and as a consequence, we allocate a much larger share of our daily energy budget for brain metabolism. In humans, brain metabolism accounts for 20% to 25% of RMR, as compared to 8% to 10% in other primate species, and 3% to 5% in nonprimate mammals (Leonard and Robertson, 1994).

Key aspects of human nutritional biology appear to be associated with the high-energy demands of our large brains. In particular, humans consume diets that are more energy and nutrient dense than other primates of similar size. Recent analyses

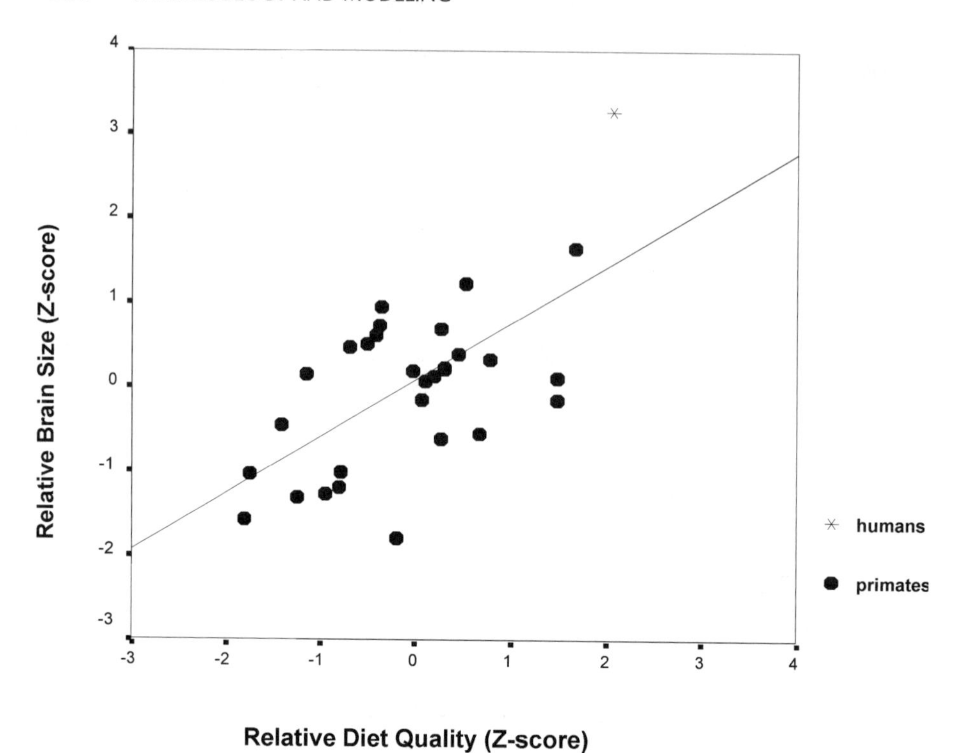

Figure 18.4 Plot of relative brain size versus relative diet quality for thirty-one primate species (including humans). Primates with higher quality diets for their size have relatively larger brain size ($r = 0.63$; $P < 0.001$). Humans represent the positive extremes for both measures, having relatively large brains and a substantially higher quality diet than expected for their size. Adapted from Leonard et al. (2003).

by Cordain et al. (2000) have shown that modern human foraging populations typically derive 45% to 65% of their dietary energy intake from animal foods. In contrast, modern great apes obtain much of their diet from low-quality plant foods. Gorillas derive more than 80% of their diet from fibrous foods such as leaves and bark (Richard, 1985). Even among chimpanzees, the most predatory of the great apes, only about 5% of their calories are derived from animal foods, including insects (Teleki, 1981; Stanford, 1996). Meat and other animal foods are much more dense in calories and nutrients than most of the plant foods typically eaten by large-bodied primates. This higher-quality diet means that humans need to eat less volume of food to get the energy and nutrients we require.

Comparative analyses support the link between brain size and dietary quality. Figure 18.4 shows relative brain size versus dietary quality (an index based on the relative proportions of leaves, fruit, and animal foods in the diet) for 31 different primate species (from Leonard et al., 2003). There is a significant positive relationship ($r = 0.63$; $P < 0.001$) between the amount of energy allocated to the brain and the caloric and nutrient density of the diet. Across all primates, bigger brains require better-quality diets, and humans are the extreme example of this, having the largest relative brain size and the highest-quality diet. This relationship implies that the evolution of larger hominin brains would have necessitated the adoption of a sufficiently high-quality diet to support the elevated energy demands.

The morphology of the human gut also appears to reflect our high-quality diet. Most large-bodied primates have expanded large intestines (colons), an adaptation to fibrous, low-quality diets (Milton, 1987). Fermentation of plant fiber in these enlarged colons allows for extraction of additional energy in the form of volatile fatty acids (Milton and Demment, 1988; Milton, 1993). Humans, however, have relatively enlarged small intestines and a reduced colon. This morphology is more similar to a carnivore and reflects an adaptation to an easily digested, nutrient-rich diet (Sussman, 1987; Martin, 1989).

The comparative dietary and energetic data thus suggest that the dramatic expansion of brain size over the course of human evolution likely would have required the consumption of a diet that was more dense in energy and nutrients than is typically the case for most large primates. This *does not* imply that dietary change was the driving force behind major brain expansion during human evolution. Rather, the available evidence indicates that a sufficiently high-quality diet was probably a necessary condition for supporting the metabolic demands associated with evolving larger hominin brains.

Evidence from the human fossil record shows that the first major burst of evolutionary change in hominin brain size occurs about 2.0 to 1.7 million years ago, associated with the emergence and evolution of early members of the genus *Homo*. Data on evolutionary changes in hominin brain size (cm^3), estimated body weight (kg), and posterior tooth area (mm^2) are presented in table 18.2. The australopithecines showed only modest brain size evolution from about 430 to 530 cm^3 over more than 2 million years (from about 4 to 1.5 million years ago). However, with the evolution

Table 18.2 Geological Ages, Brain Size, Estimated Male and Female Body Weights, and Postcanine Tooth Surface Areas for Selected Hominin Species

Species	Geological Age (Mya)	Brain Size (cm^3)	Body Weight Male (kg)	Body Weight Female (kg)	Postcanine Tooth Surface Area (mm^2)
Australopithecus afarensis	3.9–3.0	438	45	29	460
A. africanus	3.0–2.4	452	41	30	516
A. boisei	2.3–1.4	521	49	34	756
A. robustus	1.9–1.4	530	40	32	588
Homo habilis (*sensu strictu*)	1.9–1.6	612	37	32	478
H. erectus (early)	1.8–1.5	863	66	54	377
H. erectus (late)	0.5–0.3	980	60	55	390
H. sapiens	0.4–0.0	1,350	58	49	334

Sources: All data from McHenry and Coffing (2000), except for *Homo erectus*. Early *H. erectus* brain size is the average of African specimens as presented in McHenry (1994a), Indonesian specimens from Antón and Swisher (2001) and Georgian specimens from Gabunia et al. (2000, 2001). Data for late *H. erectus* are from McHenry (1994a).

of the genus *Homo* there is rapid change, with brain sizes of over 600 cm^3 in *Homo habilis* (at 1.9–1.6 Mya) and 800–900 cm^3 in early members of *H. erectus* (at 1.8–1.5 Mya). Although the relative brain size of *H. erectus* is smaller than the average for modern humans, it is outside of the range seen among other living primate species (Leonard and Robertson, 1994).

Additionally, changes in the skeletal and dental anatomy of *H. erectus* relative to the australopithecines indicate that these forms were consuming different foods. As shown in table 18.2, with the evolution of the australopithecines the surface area of the postcanine teeth increased dramatically from 460 mm^2 in *A. afarensis* to 756 mm^2 in *Australopithecus boisei.* However, with the emergence of early *Homo* at approximately 2 million years ago, we find rapid reductions in the size of the posterior teeth. Postcanine tooth surface area is 478 in *H. habilis* and 377 mm^2 in early *H. erectus. H. erectus* also shows substantial reductions in craniofacial and mandibular robusticity relative to the australopithecines (Wolpoff, 1999). Yet, despite having smaller teeth and jaws, *H. erectus* was a much bigger animal than the australopithecines, being humanlike in its stature, body mass, and body proportions (McHenry, 1992, 1994b; Ruff and Walker, 1993; Ruff, Trinkaus, and Holliday, 1997; McHenry and Coffing, 2000). Together these features indicate that early *H. erectus* was consuming a richer, more calorically dense diet with less low-quality fibrous plant material.

The Unknown

The comparative evidence clearly indicates that humans allocate more energy to brain metabolism than other primate species. Moreover, it appears that a higher-quality, more nutritionally dense diet is critical for "fueling" our large brains. However, a number of issues remain unresolved with regard to our understanding of the energetics of hominin brain evolution. Two of the most important unresolved issues are (1) how energy metabolism was altered during the course of human evolution to support the costs of larger brain size and (2) what kinds of dietary/nutritional changes likely occurred to promote the initial increases in brain size that we find with the evolution of early *Homo*.

Analyses of human and primate body composition have suggested possible answers to the first question. Aiello (1997) and Aiello and Wheeler (1995) have argued that the increased energy demands of the human brain were accommodated by the reduction in size of the gastrointestinal tract. Since the intestines are similar to the brain in having very high energy demands (so-called expensive tissues), the reduction in size of the large intestines of humans, relative to other primates, is thought to provide the necessary energy savings required to support elevated brain metabolism. Aiello and Wheeler (1995) have shown that among a sample of eighteen primate species (including humans), increased brain size was associated with reduced gut size. However, recent analyses by Snodgrass, Leonard, and Roberson (1999) have failed to demonstrate significant differences in gastrointestinal size between primates and non-primate mammals that are predicted from the "expensive tissue hypothesis."

Leonard and colleagues (2003) and Kuzawa (1998) have suggested that differences in muscle and fat mass between humans and other primates may also account

Table 18.3 Body Weight, Brain Weight, Percent Body Fat, Resting Metabolic Rate, and Percent of RMR Allocated to Brain Metabolism for Humans from Birth to Adulthood

Age	Body Weight (kg)	Brain Weight (g)	Body Fat (%)	RMR (kcal/day)	BrMet (%)
Newborn	3.5	475	16	161	87
3 months	5.5	650	22	300	64
18 months	11.0	1,045	25	590	53
5 years	19.0	1,235	15	830	44
10 years	31.0	1,350	15	1,160	34
Adult male	70.0	1,400	11	1,800	23
Adult female	50.0	1,360	20	1,480	27

Sources: All data are from Holliday (1986), except for percent body fat data for children 18 months and younger, which are from Dewey et al. (1993).

for variation in the budgeting of metabolic energy. Relative to other primates and other mammals, humans have lower levels of muscle mass and higher levels of body fatness (Leonard et al., 2003). The relatively high levels of body fatness (adiposity) in humans have two important metabolic implications for brain metabolism. First, because fat has lower-energy requirements than muscle tissue, replacing muscle mass with fat mass results in energy "savings" that can be allocated to the brain. Additionally, fat provides a ready source of stored energy that can be drawn on during periods of limited food availability. Consequently, the higher levels of body fat in humans may also help to support larger brain size by providing stored energy to buffer against environmental fluctuations in nutritional resources.

The importance of body fat is particularly notable in human infants, which have both high brain-to-body weight ratios and high levels of body fatness. Table 18.3 shows age-related changes in body weight (kg), brain weight (g), fatness, RMR (kcal/day), and percent of RMR allocated to the brain for humans from birth to adulthood. We see that in infants, brain metabolism accounts for upward of 60% of RMR. Human infants are also considerably fatter than those of other mammalian species (Kuzawa, 1998). Body fatness in human infants is about 15% to 16% at birth and continues to increase to 25%–26% during the first 12 to 18 months of postnatal growth. Fatness then declines to about 15% by early childhood (Dewey et al., 1993). Thus, during early human growth and development, it appears that body fatness is highest during the periods of the greatest metabolic demand of the brain.

It is likely that fundamental changes in body composition (i.e., the relative sizes of different organ systems) during the course of hominin evolution allowed for the expansion of brain size without substantial increases in the total energy demands for the body. At present, we do not know which alterations were the most critical for accommodating brain expansion. Variation in body composition both within and between primates species is still not well understood. Among humans, our knowledge of variation in body composition is based largely on data from populations of industrialized world. Consequently, more and better data on interspecific and ontogenetic variation in primate and human body composition are necessary to further resolve these issues. In addition, new imaging techniques such as positron emission tomography

(PET scans) offer the potential to directly explore variation in organ-specific blood flow and energy demands in humans and other primates.

An additional unanswered question is what types of dietary changes allowed for the evolution of brain size in early *Homo*. The skeletal evidence (larger body size, reduced postcanine tooth size, and reduced craniofacial robusticity) indicates that these hominins were eating foods that were more nutritionally dense and required less chewing than those consumed by the australopithecines. The most widely held view is that the diet of early *Homo* included more animal foods. The environment at the Plio-Peistocene boundary (2.0–1.8 Mya) was continuing to become drier, creating more arid grasslands (Vrba, 1995; Reed, 1997; Owen-Smith, 1999). These changes in the African landscape made animal foods more abundant and thus, an increasingly attractive food resource (Behrensmeyer et al., 1997). Greater relative abundance of animal resources would have offered an opportunity for hominins with sufficient capability to exploit those resources. The archaeological record provides evidence that this occurred with *H. erectus* the development of the first rudimentary hunting and gathering economy in which game animals became a significant part of the diet and resources were shared within foraging groups (Potts, 1988; Harris and Capaldo, 1993; Roche et al., 1999). These changes in diet and foraging behavior would not have turned our hominin ancestors into carnivores; however, the addition of even modest amounts of meat to the diet combined with the sharing of resources that is typical of hunter-gatherer groups would have significantly increased the quality and stability of hominin diets.

Greater consumption of animal foods also may have promoted brain evolution in early *Homo* by providing higher amounts of two polyunsaturated fatty acids that are critically important to brain growth, docosahexaenoic acid (DHA) and arachidonic acid (AA); Crawford et al., 1999; Cordain et al., 2001). The composition of all mammalian brain tissue is similar with respect to these two fatty acids (Crawford et al., 1999). Consequently, higher levels of encephalization are associated with greater requirements of DHA and AA. Because mammals also appear to have limited capacity to synthesize these fatty acids from dietary precursors, it has been suggested that dietary sources of DHA and AA may have been limiting nutrients that constrained the evolution of larger brain size in many mammalian lineages (Crawford, 1992; Crawford et al., 1999).

Cordain and colleagues (2001) have shown that wild-plant foods available on the African savanna (e.g., tubers, nuts) contain, at most, only trace amounts of AA and DHA, whereas muscle tissue and organ meat of wild African ruminants provide moderate to high levels of these key fatty acids. Brain tissue is a rich source of both AA and DHA, whereas liver and muscle tissues are good sources of AA and moderate sources of DHA. Other good sources of AA and DHA are freshwater fish and shellfish (Broadhurst, Cunnane, and Crawford, 1998; Crawford et al., 1999; Cordain et al., 2001); however, there is little archaeological evidence for the systematic use of aquatic resources until later in human evolution (Klein, 1999).

Wrangham and colleagues (1999) and Wrangham and Conklin-Brittain (2003) have suggested an alternative strategy for increasing diet quality in early *Homo*—cooking. They note that cooking not only makes plant foods softer, it also increases the energy content of those foods, particularly starchy tubers such as potatoes and manioc. In their raw form, the starch in tubers is not absorbed in the small intestine and is passed through the body (Tagliabue et al., 1995; Englyst and Englyst, 2005). However, when heated, the starch granules swell and are disrupted from the cell walls. This process, known as gelatinization, makes the starch much more accessible to breakdown by digestive enzymes (García-Alonso and Goñi, 2000). Thus, cooking increases the nutritional quality of tubers by making more of the carbohydrate energy available for biological processes.

Although cooking is clearly an important innovation that would have improved diet quality, there is considerable debate about when hominins began to use fire systematically for cooking (Pennisi, 1999). At present, there is limited evidence for early (>1.5 million years ago) controlled use of fire by hominins (Brain and Sillen, 1988; Bellomo, 1994). The more widely held view is that the use of fire and cooking did not occur until later in human evolution, at 200,000–250,000 years ago (Straus, 1989; Weiner et al., 1998).

In addition, Cordain and colleagues (2001) have noted that even after being cooked, wild tubers still have lower-energy content than most animal foods and lack both DHA and AA. Consequently, there remain major questions about whether cooking was an important force for promoting rapid brain evolution with the emergence of early *Homo*.

The Unknowable

Biological anthropologists have long relied on the study of anatomical and morphological variation in modern species to provide a comparative context for understanding patterns and trends in the human fossil record. The more recent development of energetic models employs a similar approach in drawing on the study of physiological variation in contemporary species to shed light on ecological and behavioral trends in our hominin ancestors. By exploring the ecological correlates of variation in energy metabolism among living primates and other mammals, we are able to gain insights into the nutritional and metabolic implications of important changes in brain size, body size, and locomotor strategies during the course of human evolution. Because variation in energy allocation has direct adaptive consequences, energy represents a particularly useful currency for studying human evolutionary processes.

Nonetheless, in developing and applying energetic models, we must recognize the limits in what these models can tell us. Although we can create a model that will estimate the locomotor energy costs for a 29 kg australopithecine, we will never know how many calories "Lucy" or her contemporaries spent to move around or how far they moved on a typical day. Likewise, while we can estimate the increased energy costs associated with the larger brain and body sizes of *H. erectus* relative to the australopithecines, we will never know with accuracy how much energy our hominin ancestors were consuming or where those calories were derived from.

Critics of energetic modeling have noted the potential for "false quantification" and for drawing inferences with inappropriate levels of precision (e.g., Smith, 1996). However, this critique largely misinterprets the principal objectives of energetic models and ecological models in general. The goal of these approaches is not to quantify precisely some energetic or physiological parameter (e.g., calories consumed or expended) but rather to use the information on ecological variation in extant species to explore the energetic consequences of key changes in hominin evolution (e.g., changes in body size, proportions, day ranges, diet). For example, rather than measuring how many calories Lucy spent on a foraging trip, we seek to explore how changes in body size, foraging range, and locomotor efficiency may interact to produce meaningful variation in daily energy demands.

Obviously, the predictions of any model are only as good as the data and assumptions that go into that model. The fossil and archaeological records have limitations that will continue to constrain our insights. However, there are many insights still to be gained by more broadly studying the ecological correlates of nutritional and metabolic variation in modern humans, primates, and other mammals. A richer understanding of these patterns of modern physiological variation will provide us with a stronger and more rigorous framework for interpreting the fossil record and refining our understanding of the evolution of our species and its distinctive nutritional needs.

References

Aiello, L.C., 1997. Brains and guts in human evolution: The expensive tissue hypothesis. *Braz. J. Genet.* 20, 141–148.

Aiello, L.C., and Wheeler, P., 1995. The expensive-tissue hypothesis: The brain and the digestive system in human and primate evolution. *Curr. Anthropol.* 36, 199–221.

Antón, S.C., and Swisher, C.C., III, 2001. Evolution of cranial capacity in Asian *Homo erectus*. In: Indriati, E. (Ed.), *A Scientific Life: Papers in Honor of Dr. T. Jacob*. Bigraf Publishing, Yogyakarta, Indonesia, pp. 25–39.

Antón, S.C., and Swisher, C.C., III, 2004. Early dispersals of *Homo* from Africa. *Annu. Rev. Anthropol.* 33, 271–296.

Behrensmeyer, K., Todd, N.E., Potts, R., and McBrinn, G.E., 1997. Late Pliocene faunal turnover in the Turkana basin, Kenya and Ethiopia. *Science* 278, 1589–1594.

Bellomo, R.V., 1994. Methods of determining early hominid behavioral activities associated with the controlled use of fire at FxJj 20 Main, Koobi Fora. *J. Hum. Evol.* 27, 173–195.

Blumenschine, R.J., Cavallo, J.A., and Capaldo, S.D., 1994. Competition for carcasses and early hominid behavioral ecology: A case study and conceptual framework. *J. Hum. Evol.* 27, 197–213.

Brain, C.K., and Sillen, A., 1988. Evidence from the Swartkrans cave for the ealiest use of fire. *Nature* 336, 464–466.

Broadhurst, C.L., Cunnane, S.C., and Crawford, M.A., 1998. Rift Valley lake fish and shellfish provided brain-specific nutrition for early *Homo*. *Br. J. Nutr.* 79, 3–21.

Consolazio, C.F., Johnson, R.E., and Pecora, L.J., 1963. *Physiological Measurements of Metabolic Functions in Man*. McGraw-Hill, New York.

Cordain, L., Brand Miller, J., Eaton, S.B., Mann, N., Holt, S.H.A., and Speth, J.D., 2000. Plant to animal subsistence ratios and macronutrient energy estimations in world wide hunter-gatherer diets. *Am. J. Clin. Nutr.* 71, 682–692.

Cordain, L., Watkins, B.A., and Mann, N.J., 2001. Fatty acid composition and energy density of foods available to African hominids. *World Rev. Nutr. Diet.* 90, 144–161.

Crawford, M.A., 1992. The role of dietary fatty acids in biology: Their place in the evolution of the human brain. *Nutr. Rev.* 50, 3–11.

Crawford, M.A., Bloom, M., Broadhurst, C.L., Schmidt, W.F., Cunnane, S.C., Galli, C., Gehbremeskel, K., Linseisen, F., Lloyd-Smith, J., and Parkington, J., 1999. Evidence for unique function of docosahexaenoic acid during the evolution of the modern human brain. *Lipids* 34, S39–S47.

Dewey, K.G., Heinig, M.J., Nommsen, L.A., Peerson, J.M., and Lonnerdal, B., 1993. Breast-fed infants are leaner than formula-fed infants at 1 y of age: The Darling Study. *Am. J. Clin. Nutr.* 52, 140–145.

Englyst, K.N., and Englyst, H.N., 2005. Carbohydrate bioavailability. *Br. J. Nutr.* 94, 1–11.
FAO/WHO/UNU (Food and Agriculture Organization/World Health Organization/United Nations University), 1985. *Energy and Protein Requirements*. WHO Technical Support Series No.724. World Health Organization, Geneva.
Gabunia, L., Vekua, A., Lordkipanidze, D., Swisher, C.C., Ferring, R., Justus, A., Nioradze, M., Tvalchrelidze, M., Antón, S.C., Bosinski, G., Joris, O., de Lumley, M.-A., Majsuradze, G., and Mouskhelishivili, A., 2000. Earliest Pleistocene cranial remains from Dmanisi, Republic of Georgia: taxonomy, geological setting, and age. *Science* 288, 1019–1025.
Gabunia, L., Antón, S.C., Lordkipanidze, D., Vekua, A., Justus, A., and Swisher, C.C., III., 2001. Dmanisi and dispersal. *Evol. Anthropol.* 10, 158–170.
García-Alonso, A., and Goñi, I., 2000. Effect of processing on potato starch: In vitro availability and glycemic index. *Nahrung* 44, 19–22.
Harris, J.W.K., and Capaldo, S., 1993. The earliest stone tools: Their implications for an understanding of the activities and behavior of lat Pliocene hominids. In: Berthelet, A., and Chavaillon, J. (Eds.), *The Use of Tools by Human and Nonhuman Primates*. Oxford *Science* Publications, Oxford, pp. 196–220.
Holliday, M.A., 1986. Body composition and energy needs during growth. In: Falkner, F., and Tanner, J.M. (Eds.), *Human Growth: A Comprehensive Treatise*. Vol. 2, 2nd ed. Plenum Press, New York, pp. 101–117.
Jungers, W.L., 1982. Lucy's limbs: Skeletal allometry and locomotion in *Australopithecus afarensis*. *Nature* 297, 676–678.
Jungers, W.L., 1988. Lucy's length: Stature reconstruction in *Australopithecus afarensis* (A.L.288-1) with implications for other small-bodied hominids. *Am. J. Phys. Anthropol.* 76, 227–231.
Kety, S.S., 1957. The general metabolism of the brain *in vivo*. In: Richter, D. (Ed.), *Metabolism of the Central Nervous System*. Pergamon, New York, pp. 221–237.
Kleiber, M., 1961. *The Fire of Life*. Wiley, New York.
Klein, R.G., 1999. *The Human Career: Human Biological and Cultural Origins*. 2nd ed. University of Chicago Press, Chicago.
Kuzawa, C.W., 1998. Adipose tissue in human infancy and childhood: An evolutionary perspective. *Yearb. Phys. Anthropol.* 41, 177–209.
Latimer, B., and Lovejoy, C.O., 1989. The calcaneus of *Australopithecus afarensis* and its implications for the evolution of bipedality. *Am. J. Phys. Anthropol.* 78, 369–386.
Latimer, B., and Lovejoy, C.O., 1990a. Hallucal tarsometotarsal joint in the *Australopithecus afarensis*. *Am. J. Phys. Anthropol.* 82,125–133.
Latimer, B., and Lovejoy, C.O., 1990b. Metatarsophalangeal joints of *Australopithecus afarensis*. *Am. J. Phys. Anthropol.* 83, 13–23.
Leonard, W.R., 2002. Food for thought: Dietary change was a driving force in human evolution. *Sci. Am.* 287(6), 106–115.
Leonard, W.R., and Robertson, M.L., 1992. Nutritional requirements and human evolution: A bioenergetics model. *Am. J. Hum. Biol.* 4, 179–195.
Leonard, W.R., and Robertson, M.L., 1994. Evolutionary perspectives on human nutrition: The influence of brain and body size on diet and metabolism. *Am. J. Hum. Biol.* 6, 77–88.
Leonard, W.R., and Robertson, M.L., 1995. Energetic efficiency and human bipedality. *Am. J. Phys. Anthropol.* 97, 335–338.
Leonard, W.R., and Robertson, M.L., 1997a. Comparative primate energetics and hominid evolution. *Am. J. Phys. Anthropol.* 102, 265–281.
Leonard, W.R., and Robertson, M.L., 1997b. Rethinking the energetics of bipedality. *Curr. Anthropol.* 38, 304–309.
Leonard, W.R., and Robertson, M.L., 2001. Locomotor economy and the origin of bipedality: Reply to Steudel-Numbers. *Am. J. Phys. Anthropol.* 116, 174–176.
Leonard, W.R., Robertson, M.L., Snodgrass, J.J., and Kuzawa, C.W., 2003. Metabolic correlates of hominid brain evolution. *Comp. Biochem. Physiol.* A 135, 5–15.
Lovejoy, C.O., 1988. Evolution of human walking. *Sci. Am.* 259(5), 118–125.

Lovejoy, C.O., 2005. The natural history of human gait and posture. Part 1. Spine and pelvis. *Gait Posture* 21, 95–112.

Margaria, R., Cerretelli, P., Aghemo, P., and Sassi, G., 1963. Energy cost of running. *J. Appl. Physiol.* 18, 367–370.

Martin, R.D., *1989. Primate Origins and Evolution: A Phylogenetic Reconstruction*. Princeton University Press, Princeton, NJ.

McArdle, W.D., Katch, F.I., and Katch, V.L., 2001. *Exercise Physiology: Energy, Nutrition and Human Performance*. 5th ed. Lipincott, Williams, & Wilkins, Philadelphia.

McHenry, H.M., 1992. Body size and proportions in early hominids. *Am. J. Phys. Anthropol.* 87, 407–431.

McHenry, H.M., 1994a. Behavioral ecological implications of early hominid body size. *J. Hum. Evol.* 27, 77–87.

McHenry, H.M., 1994b. Tempo and mode in human evolution. *Proc. Natl. Acad. Sci.* 91, 6780–6786.

McHenry, H.M., Coffing, K., 2000. *Australopithecus* to *Homo*: Transformations in body and mind. *Annu. Rev. Anthropol.* 29,125–146.

McLean, J.A., and Tobin, G., 1987. *Animal and Human Calorimetry*. Cambridge University Press, Cambridge.

McNab, B.K., 2002. *The Physiological Ecology of Vertebrates: A View from Energetics*. Cornell University Press, Ithaca, NY.

Menier, D.R., and Pugh, L.G.C.E., 1968. The relation of oxygen intake and velocity of walking and running in competition walkers. *J. Physiol.* 197, 717–721.

Milton, K., 1987. Primate diets and gut morphology: Implications for hominid evolution. In: Harris, M., and Ross, E.B. (Eds.), *Food and Evolution: Toward a Theory of Human Food Habits*. Temple University Press, Philadelphia, PA, pp. 93–115.

Milton, K., 1993. Diet and primate evolution. Sci. Am. 269(2), 86–93.

Milton, K., and Demment, M.W., 1988. Digestion and passage kinetics of chimpanzees fed high and low-fiber diets and comparison with human data. *J. Nutr.* 118, 1082–1088.

Morgan, D.W., and Craib, M., 1992. Physiological aspects of running economy. *Med. Sci. Sport. Exerc.* 24, 456–461.

Owen-Smith, N., 1999. Ecological links between African savanna environments, climate change and early hominid evolution. In: Bromage, T.G., and Schrenk, F. (Eds.), *African Biogeography, Climate Change, and Human Evolution.* Oxford University Press, New York, pp. 138–149.

Pennisi, E., 1999. Did cooked tubers spur the evolution of big brains? *Science* 283, 2004–2005.

Potts, R., 1988. *Early Hominid Activities at Olduvai*. Aldine, New York.

Potts, R., 1998. Environmental hypotheses of hominin evolution. *Yearb. Phys. Anthropol.* 41, 93–136.

Reed, K., 1997. Early hominid evolution and ecological change through the African Plio-Pleistocene. *J. Hum. Evol.* 32, 289–322.

Richard, A.F., 1985. *Primates in Nature*. W.H. Freeman, New York.

Roche, H., Delagnes, A., Brugal, J.P., Feibel, C., Kibunjia, M., Mourre, V., and Texier, J.-P., 1999. Early hominid stone tool production and technical skill 2.34 Myr ago in West Turkana, Kenya. *Nature* 399, 57–60.

Rodman, P.S., and McHenry, H.M., 1980. Bioenergetics and the origin of hominid bipedalism. *Am. J. Phys. Anthropol.* 52, 103–106.

Ruff, C.B., Trinkaus, E., and Holliday, T.W., 1997. Body mass and encephalization in Pleistocene *Homo*. *Nature* 387, 173–176.

Ruff, C.B., and Walker, A., 1993. Body size and body shape. In: Walker, A., and Leakey, R. (Eds.), *The Nariokotome* Homo erectus *Skeleton*. Harvard University Press: Cambridge, MA, pp. 234–265.

Smith, R.J., 1996. Biology and body size in human evolution: Statistical inference misapplied. *Curr. Anthropol.* 37, 451–481.

Snodgrass, J.J., Leonard, W.R., and Robertson, M.L., 1999. Interspecific variation in body composition and its influence on metabolic variation in primates and other mammals. *Am. J. Phys. Anthropol.* Suppl. 28, 255.

Stanford, C.B, 1996. The hunting ecology of wild chimpanzees: Implications for the evolutionary ecology of Pliocene hominids. *Am. Anthropol.* 98, 96–113.

Stern, J.T., Susman, R.L., 1983. The locomotor anatomy of *Australopithecus afarensis*. *Am. J. Phys. Anthropol.* 60, 279–317.

Steudel, K.L., 1994. Locomotor energetics and hominid evolution. *Evol. Anthropol.* 3, 42–48.

Steudel, K.L., 1996. Morphology, bipedal gait, and the energetics of hominid locomotion. *Am. J. Phys. Anthropol.* 99, 345–355.

Steudel-Numbers, K.L., 2001. Role of locomotor economy in the origin of bipedal posture and gait. *Am. J. Phys. Anthropol.* 116, 171–173.

Steudel-Numbers, K.L., and Tilkins, M.J, 2004. The effect of lower limb length on the energetic cost of locomotion: implications for fossil hominins. *J. Hum. Evol.* 47, 95–109.

Straus, L.G., 1989. On early hominid use of fire. *Curr. Anthropol.* 30, 488–491.

Susman, R.L., Stern, J.T., and Junger, W.L., 1984. Arboreality and bipedality in the Hadar hominids. *Folia Primatol.* 43, 113–156.

Sussman, R.W., 1987. Species-specific dietary patterns in primates and human dietary adaptations. In: Kinzey, W.G. (Ed.), *The Evolution of Human Behavior: Primate Models*. SUNY Press: Albany, NY, pp. 131–179.

Tagliabue. A., Raben, A., Heijnen, M.L., Duerenberg, P., Pasquali, E., and Astrup, A., 1995. The effect of raw potato starch on energy expenditure and substrate oxidation. *Am. J. Clin. Nutr.* 61, 1070–1075.

Taylor, C.R., Heglund, N.C., and Maloiy, G.M.O., 1982. Energetics and mechanics of terrestrial locomotion. I. Metabolic energy consumption as a function of speed and body size in birds and mammals. *J. Exp. Biol.* 97, 1–21.

Taylor, C.R., and Rowntree, V.J., 1973. Running on two or four legs: Which consumes more energy? *Science* 179, 186–187.

Taylor, C.R., Schmidt-Nielsen, K., and Raab, J.L., 1970. Scaling of energetic costs to body size in mammals. *Am. J. Physiol.* 219, 1104–1107.

Teleki, G., 1981. The omnivorous diet and eclectic feeding habits of the chimpanzees of Gombe National Park. In: Harding, R.S.O., and Teleki, G. (Eds.), *Omnivorous Primates*. Columbia University Press, New York, pp. 303–343.

Tucker, V.A., 1970. Energetic cost of locomotion in animals. *Comp. Biochem. Physiol.* 34, 841–846.

Vrba, E.S., 1995. The fossil record of African antelopes relative to human evolution. In: Vrba, E.S., Denton, G.H., Partridge, T.C., and Burkle, L.H. (Eds.), *Paleoclimate and Evolution, with Emphasis on Human Origins*. Yale University Press, New Haven, pp. 385–424.

Weiner, S., Qunqu, X., Goldberg, P., Liu, J., and Bar-Yosef, O., 1998. Evidence for the use of fire at Zhoukoudian, China. *Science* 281, 251–253.

Woldegabriel, G., White, T.D., Suwa, G., Renne, P., deHeinzelin, J., Hart, W.K., and Heiken, G., 1994. Ecological and temporal placement of early Pliocene hominids at Aramis, Ethiopia. *Nature* 371, 330–333.

Wolpoff, M.H., 1999. *Paleoanthropology.* 2nd ed. McGraw-Hill, Boston.

Wrangham, R.W., and Conklin-Brittain, N.L., 2003. Cooking as a biological trait. *Comp. Biochem. Physiol.* A 136, 35–46.

Wrangham, R.W., Jones, J.H., Laden, G., Pilbeam, D., and Conklin-Brittain, N.L., 1999. The raw and the stolen: Cooking and the ecology of human origins. *Curr. Anthropol.* 40, 567–594.

PART V

IMPLICATIONS OF STUDIES OF EARLY HOMININ DIETS

19

Implications of Plio-Pleistocene Hominin Diets for Modern Humans

LOREN CORDAIN

Within the anthropology community, there has been a long and sustained interest in the diets and eating patterns of Plio and Pleistocene hominins, primarily because these nutritional practices provide a glimpse into their varied and distinctive lifeways, activities, and cultural patterns. In contrast, until relatively recent times, the nutritional and medical communities have largely ignored anthropological dietary information for a variety of reasons, not the least of which is its perceived lack of immediate relevance to their respective disciplines. Beginning in the mid 1980s, a series of key publications in mainstream medical and nutrition journals (Eaton and Konner, 1985; Eaton, Konner, and Shostak, 1988; Eaton and Nelson, 1991; Eaton, 1992) triggered an increased awareness of the relevance of ancestral diets to the health and well being of modern peoples. Because of that insight as well as others gleaned from a variety of medical branches of learning, an entirely new academic discipline was born, dubbed "evolutionary medicine" (Williams and Neese, 1991), or sometimes "Darwinian medicine" (Williams and Nesse, 1991) The primary tenet of evolutionary medicine is that the profound changes in the environment (e.g., in diet and other lifestyle conditions), which began with the introduction of agriculture and animal husbandry approximately 10,000 years ago, occurred too recently on an evolutionary timescale for natural selection to adjust the human genome (Eaton and Konner, 1985; Williams and Neese, 1991). In conjunction with this discordance between our ancient (millions of years ago), genetically determined biology and the nutritional, cultural, and activity patterns of contemporary western populations, many of the so-called diseases of civilization have emerged (Eaton, Konner, and Shostak, 1988; Williams and Neese, 1991).

With regard to diet and health, food staples and food-processing procedures introduced during the Neolithic and Industrial era have fundamentally altered seven

crucial nutritional characteristics of ancestral hominin diets: (1) glycemic load, (2) fatty acid composition, (3) macronutrient composition, (4) micronutrient density, (5) acid/base loads, (6) sodium/potassium ratio, and (7) fiber content. Each of these nutritional factors either alone or combined with some, or all, of the remaining factors underlie the pathogenesis of a wide variety of chronic diseases and maladies that almost universally afflict people living in western, industrialized societies. Increasingly, dietary interventions and clinical trials have demonstrated the therapeutic potential of contemporary diets, which emulate one or more of the seven universal characteristics common to preagricultural diets.

Plio-Pleistocene Hominin Diets: The Known

An Omnivorous Diet: The Evidence

Figure 19.1 demonstrates that since the evolutionary emergence of hominins twenty or more species may have existed (Wood, 2002). Similar to historically studied hunter-gatherers (Cordain et al., 2000; Cordain et al., 2002b), there would have been no single, universal diet consumed by all extinct hominin species. Rather, diets would have varied by geographic locale, climate, and specific ecologic niche. However, a number of universal dietary characteristics and trends within preagricultural hominins have emerged, which have important health ramifications for contemporary humans.

Since the evolutionary split between hominins and pongids approximately 7 million years ago, the available evidence shows that all species of hominins ate an omniv-

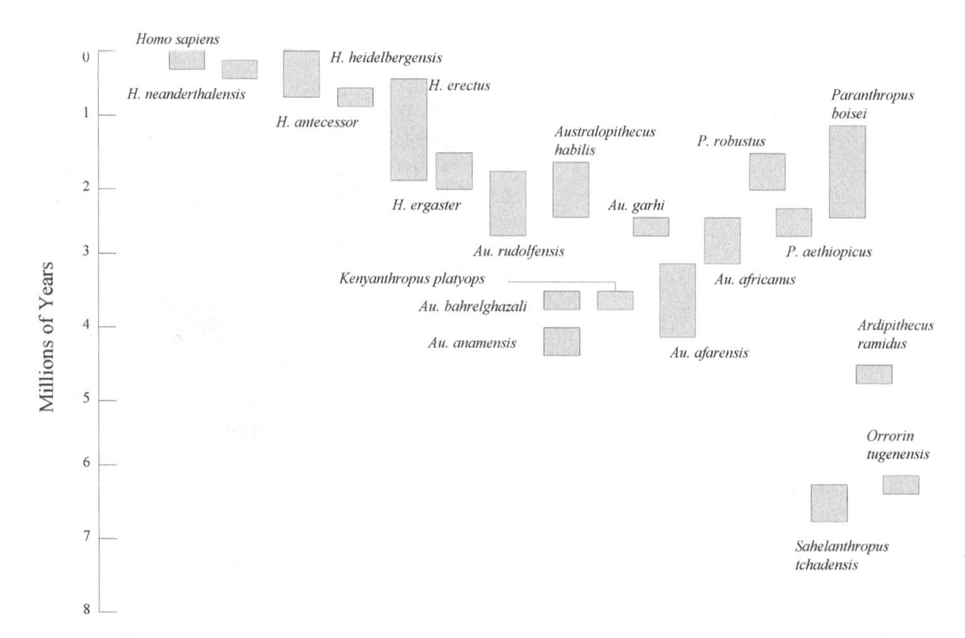

Figure 19.1 The hominin fossil record. Species are indicated with the dates of the earliest and latest fossil record. Adapted from Wood (2002).

orous diet composed of minimally processed, wild-plant, and animal foods. In support of this view is the omnivorous nature of chimpanzees, the closest living pongid link to hominins. Although chimpanzees (*Pan paniscus* and *Pan troglodytes*), our genetically closest nonhuman relatives, primarily consume a frugivorous diet, they still eat a substantial amount of meat obtained throughout the year from hunting and scavenging (Teleki, 1973; Stanford, 1996; Schoeninger, Moore, and Sept, 1999). Observational studies of wild chimpanzees demonstrate that during the dry season meat intake is about 65 g per day for adults (Stanford, 1996). Accordingly, it is likely that the very earliest hominins would have been capable of obtaining animal food through hunting and scavenging in a manner similar to chimpanzees.

Carbon isotope data also support the notion that early hominins were omnivorous. By about 3 million years ago (Mya), *Australopithecus africanus* obtained a significant portion of food from C_4 sources (grasses, particularly seeds and rhizomes; sedges; invertebrates, including locusts and termites; grazing mammals; and perhaps even insectivores and carnivores; van der Merwe et al., 2003). Other fossils of early African hominins, including *Australopithecus robustus* and *Homo ergaster*, maintain carbon isotope signatures characteristic of omnivores (Lee-Thorp, Thackeray, and van der Merwe, 2000; Sponheimer and Lee-Thorp, 2003). The finding of C_4 in *A. robustus* fossils refutes the earlier view that this hominin was vegetarian (Lee-Thorp, Thackeray, and van der Merwe, 2000).

Secular Increase in Animal Food: The Evidence

Beginning approximately 2.6 Mya, the hominin species that eventually led to *Homo* began to include more animal food in their diet. A number of lines of evidence support this viewpoint. First, Oldowan lithic technology appears in the fossil record 2.6 Mya (Semaw et al., 2003), and there is clear cut evidence to show that these tools were used to butcher and disarticulate animal carcasses (Bunn and Kroll, 1986; de Heinzelin et al., 1999). Stone tool cut marks on the bones of prey animals and evidence for marrow extraction appear concurrently in the fossil record with the development of Oldowan lithic technology by at least 2.5 Mya (de Heinzelin et al., 1999). It is not entirely clear which specific early hominin species or group of species manufactured and used these earliest of stone tools; however, *Australopithecus garhi* may have been a likely candidate (Asfaw et al., 1999; de Heinzelin et al., 1999).

The development of stone tools and the increased dietary reliance on animal foods allowed early African hominins to colonize northern latitudes outside of Africa where plant foods would have been seasonally restricted. Early *Homo* skeletal remains and Oldowan lithic technology appear at the Dmanisi site in the Republic of Georgia (40° N) by 1.75 Mya (Vekua et al., 2002), and more recently Oldowan tools dating to 1.66 Mya have been discovered at the Majuangou site in North China (40° N; Zhu et al., 2004). Both of these tool-producing hominins would likely have consumed considerably more animal food than prelithic hominins living in more temperate African climates.

In addition to the fossil evidence suggesting a trend for increased animal food consumption, hominins may have experienced a number of genetic adaptations to animal-based diets early on in our genus's evolution analogous to those of obligate carnivores such as felines. Carnivorous diets reduce evolutionary selective pressures that act to maintain certain anatomical and physiological characteristics needed to process and metabolize high amounts of plant foods. In this regard, hominins, like felines, have experienced a reduction in gut size and metabolic activity along with a concurrent expansion of brain size and metabolic activity as they included more energetically dense animal food into their diets (Leonard and Robertson, 1994; Aiello and Wheeler, 1995; Cordain, Watkins, and Mann, 2001). Further, similar to obligate carnivores (Pawlosky, Barnes, and Salem, 1994), humans maintain an inefficient ability to chain elongate and desaturate 18 carbon fatty acids to their product 20 and 22 carbon fatty acids (Emken et al., 1992). Since 20 and 22 carbon fatty acids are essential cellular lipids, then evolutionary reductions in desaturase and elongase activity in hominins indicate that preformed dietary 20 and 22 carbon fatty acids (found only in animal foods) were increasingly incorporated in lieu of their endogenously synthesized counterparts derived from 18 carbon plant fatty acids. Finally, our species has a limited ability to synthesize the biologically important amino acid, taurine, from precursor amino acids (Sturman et al., 1975; Chesney et al., 1998), and vegetarian diets in humans result in lowered plasma and urinary concentrations of taurine (Laidlaw et al., 1988). Like felines (Knopf et al., 1978; MacDonald, Rogers, and Morris, 1984), the need to endogenously synthesize taurine may have been evolutionarily reduced in humans because exogenous dietary sources of preformed taurine (found only in animal food) had relaxed the selective pressure formerly requiring the need to synthesize this conditionally essential amino acid.

Plio-Pleistocene Hominin Diets: The Uncertain

The Uncertain: How Much Plant Food, How Much Animal Food?

There is little evidence to the contrary that animal foods have always played a significant role in the diets of all hominin species. Increased reliance on animal foods not only allowed for enhanced encephalization and its concomitant behavioral sophistication (Leonard and Robertson, 1994; Aiello and Wheeler, 1995; Cordain, Watkins, and Mann, , 2001), but this dietary practice also permitted colonization of the world outside of Africa. An unresolved issue surrounding Plio-Pleistocene diets is the relative amounts of plant and animal foods that were typically consumed.

Before the advent of Oldowan lithic technology about 2.6 Mya, quantitative estimates of hominin energy intake from animal food sources are unclear, other than they were likely similar to, or greater than, estimated values (4%–8.5% total energy) for chimpanzees (Sussman, 1978; Stanford, 1996). Although all available data point to increasing animal food consumption following the arrival of lithic technology, the precise contribution of either animal or plant food energy to Plio-Pleistocene hominin diets is not known. Obviously, then as now, no single (animal/plant) subsistence ratio would have been necessarily representative of all populations or species of hominins. However, there are a number of lines of evidence which suggest more than half (>50%) of the average daily energy intake for most Paleolithic hominin species and populations of species was obtained from animal foods.

Richards, Pettitt, and colleagues (2000) have examined stable isotopes ($\delta^{13}C$ and $\delta^{15}N$) in two Neanderthal specimens (~28,000–29,000 years BP) from Vindija Cave in northern Croatia and contrasted these isotopic signatures to those in fossils of herbivorous and carnivorous mammals from the same ecosystem. The analysis demonstrated that Neanderthals, similar to wolves and arctic foxes, behaved as top-level carnivores, obtaining all of their protein from animal sources (Richards, Pettitt, et al., 2000). A similar analysis was made of five Upper Paleolithic *Homo sapiens* specimens dated to the Upper Paleolithic (~11,700–12,380 years BP) from Gough's and Sun Hole Caves in Britain (Richards, Hedges, et al. 2000). The data indicated these hunter-gatherers were consuming animal protein year-round at a higher trophic level than the artic fox.

Both studies by Richards, Hedges, and colleagues (2000) and Richards, Pettitt, and colleagues (2000) could be criticized as not being representative of typical hominin diets, as these two species lived in climates and ecosystems that fostered an abundance of large, huntable mammals, which were preyed on preferentially. Additional clues to the typical plant-to-animal subsistence ratio in Paleolithic hominin diets can be found in the foraging practices of historically studied hunter-gatherers.

Our analysis (fig. 19.2) of the *Ethnographic Atlas* data (Gray, 1999) showed that the dominant foods in the majority of historically studied hunter-gatherer diets were derived from animal food sources (Cordain et al., 2000). Most (73%) of the world's hunters-gatherers obtained >50% of their subsistence from hunted and fished animal foods, whereas only 14% of worldwide hunter gatherers obtained >50% of their subsistence from gathered plant foods. For all 229 hunter-gatherer societies, the median

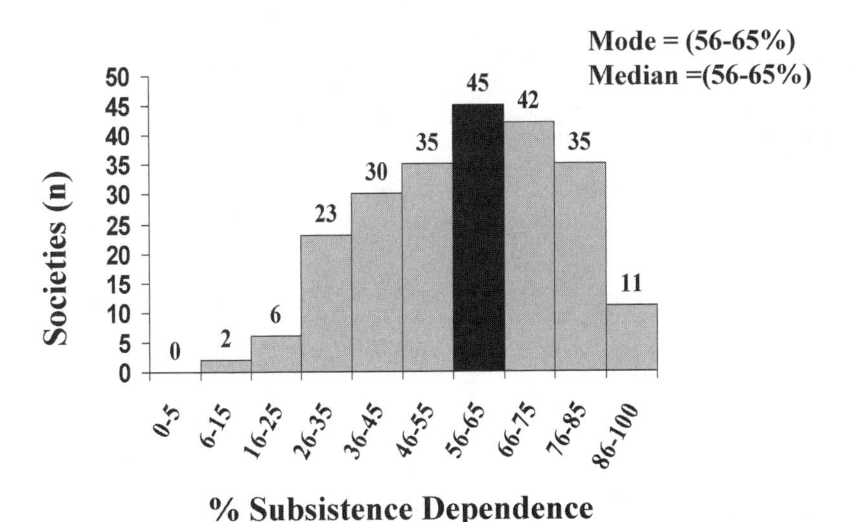

Figure 19.2 Frequency distribution of subsistence dependence upon total (fished + hunted) animal foods in worldwide hunter-gatherer societies ($n = 229$). Adapted from Cordain et al. (2000).

subsistence dependence on animal foods was 56% to 65%. In contrast, the median subsistence dependence on gathered plant foods was 26% to 35% (Cordain et al., 2000).

The major limitation of ethnographic data is that the preponderance of it is subjective in nature, and the assigned scores for the five basic subsistence economies in the *Ethnographic Atlas* are not precise, but rather are approximations (Hayden, 1981). Fortunately, more exact, quantitative dietary studies were carried out on a small percentage of the world's hunter gatherer societies. Table 19.1 lists these studies and shows the plant-to-animal subsistence ratios by energy. The average score for animal food subsistence is 65%, while that for plant-food subsistence is 35%. These values are similar to our analysis of the entire ($n = 229$) sample of hunter-gatherer societies listed in the *Ethnographic Atlas* in which the mean score for animal food subsistence was 68% and that for plant food was 32%. When the two polar hunter-gatherer populations, who have no choice but to eat animal food because of the inaccessibility of plant foods, are excluded from table 19.1, the mean score for animal subsistence is 59% and that for plant-food subsistence is 41%. These animal-to-plant subsistence values fall within the same respective class intervals (56%–65% for animal food; 26%–35% for plant food) as those we estimated from the ethnographic data when the confounding influence of latitude was eliminated (Cordain et al., 2000). Consequently, there is remarkably close agreement between the quantitative data in table 19.1 and the ethnographic data that animal food comprised more than half of the energy in historically studied hunter-gatherer diets.

Table 19.1 Quantitatively Determined Proportions of Plant and Animal Food in Hunter-Gatherer Diets

Population	Location	Latitude	% Animal Food	% Plant Food	Reference
Aborigines (Arhem Land)	Australia	12°S	77	23	McArthur, 1960
Ache	Paraguay	25°S	78	22	Hill et al., 1984
Anbarra	Australia	12°S	75	25	Meehan, 1982
Efe	Africa	2°N	44	56	Dietz et al., 1989
Eskimo	Greenland	69°N	96	4	Sinclair, 1953; Krogh and Krogh, 1914
Gwi	Africa	23°S	26	74	Silberbauer, 1981; Tanaka, 1980
Hadza	Africa	3°S	48	52	Blurton Jones et al., 1997; Hawkes et al., 1989
Hiwi	Venezuela	6°N	75	25	Hurtado and Hill, 1986, 1990
!Kung	Africa	20°S	33	67	Lee, 1968
!Kung	Africa	20°S	68	32	Yellen, 1977
Nukak	Columbia	2°N	41	59	Politis, 1996
Nunamiut	Alaska	68°N	99	1	Binford, 1978
Onge	Andaman Islands	12°N	79	21	Rao et al., 1989; Bose, 1964

Source: Adapted from Kaplan et al. (2000).

Differences between Postagricultural and Plio-Pleistocene Diets: The Known

In contrasting pre- and postagricultural diets, it is important to consider not only the nutrient qualities and categories of foods that were consumed by preagricultural hominins but also to recognize the categories of foods and their nutrient qualities that could not have been regularly consumed before the development of agriculture, industrialization, and advanced technology. Table 19.2 lists the food categories that would have generally been unavailable to preagricultural hominins. Tables 19.3–19.5 show when these novel Neolithic and Industrial era food introductions occurred.

Although dairy products, cereals, refined sugars, refined vegetable oils, and alcohol make up 72.1% of the total daily energy consumed by all people in the United States, these types of foods would have contributed little or no energy in the archetypal preagricultural hominin diet (Cordain et al., 2000). Furthermore, mixtures of foods shown in table 19.2 encompass the ubiquitous processed foods (e.g., cookies,

Table 19.2 Food and Food Types Found in Western Diets Generally Unavailable to Preagricultural Hominins

Food or Food Group	% Total Energy in U.S. Diet
Dairy products	
Whole milk	1.6
Low-fat milks	2.1
Cheese	3.2
Butter	1.1
Other	2.6
Total	10.6
Cereal grains	
Whole grains	3.5
Refined grains	20.4
Total	23.9
Refined sugars	
Sucrose	8.0
High-fructose corn syrup	7.8
Glucose	2.6
Syrups	0.1
Other	0.1
Total	18.6
Refined vegetable oils	
Salad, cooking oils	8.8
Shortening	6.6
Margarine	2.2
Total	17.6
Alcohol	1.4
Total energy	72.1
Added daily salt (NaCl)	9.6 g

Sources: Data adapted from Gerrior and Bente (2002), United States Department of Agriculture (1997), United States Department of Agriculture (2002).

Table 19.3 Neolithic Food Introductions

Food	Date	Reference
Domesticated Meats		
Sheep	~11,000 BP	Hiendleder et al., 2002
Goats	~10,000 BP	Luikart et al., 2001
Cows	~10,000 BP	Loftus et al., 1999
Pigs	~9,000 BP	Giuffra et al., 2000
Chickens	~8,000 BP	Fumihito et al., 1996
Dairy products	6,100–5,500 BP	Copley et al., 2003
Cereal grains		
Wheat	10,000–11,000 BP	Salamini et al., 2003
Barley	~10,000 BP	Badr et al., 2000
Rice	~10,000 BP	Vitte et al., 2004
Maize	~9,000 BP	Matsuoka et al., 2002
Wine	7,100–7,400 BP	McGovern et al., 1996
Salt	5,600–6,200 BP	Weller, 2002
Vegetable oils	5,000–6,000 BP	O'Keefe, 2000; Liphschitz et al., 1991
Beer	~4,000 B.C.	Rudolph et al., 1992

cake, bakery foods, breakfast cereals, bagels, rolls, muffins, crackers, chips, snack foods, pizza, soft drinks, candy, ice cream, condiments, salad dressings, etc.) that dominate the typical U.S. diet.

Health Ramifications of Neolithic and Industrial Era Foods

The novel foods (dairy products, cereals, refined cereals, refined sugars, refined vegetable oils, fatty meats, salt and combinations of these foods) introduced as staples during the Neolithic and Industrial eras fundamentally changed a number of major nutritional characteristics of ancestral hominin diets and ultimately had far-reaching effects on health and well being. As these foods gradually displaced the minimally

Table 19.4 Historical and Industrial Era Food-Type Introductions

Food	Date	Reference
Refined sugar (sucrose)	500 B.C.	Galloway, 2000
Distilled alcoholic beverages	800–1300 AD	Comer, 2000
Refined sugar (widely available)	1800 AD	Ziegler, 1967
Fatty, feedlot-produced meats	~1860 AD	Whitaker, 1975
Refined grains (widely available)	~1880 AD	Storck and Teague, 1952
Hydrogenated vegetable fats	1897 AD	Emken, 1984
Vegetable oils (widely available)	~1910 AD	Gerrior and Bente, 2002
High-fructose corn syrup	~1970s AD	Hanover and White, 1993

Table 19.5 Assorted Processed Food Introductions

Food	Date	Food	Date
Saltine crackers	1876	Processed cheese	1915
Pillsbury white flour	1881	Orange Crush soda	1916
Coca-Cola	1886	Moon Pie	1917
Log Cabin syrup	1888	Baby Ruth candy bar	1920
Fig Newtons	1891	Wonder Bread	1921
Triscuits	1895	Milky Way candy bar	1923
Tootsie Rolls	1896	Wheaties cereal	1924
Graham crackers	1898	Kool Aid	1927
Wesson oil	1899	Rice Krispies cereal	1928
Chiclets chewing gum	1900	Birds Eye frozen vegetables	1930
Hershey's chocolate bar	1900	Hostess Twinkies	1930
Karo corn syrup	1902	Frito Corn chips	1932
Pepsi	1902	Kraft Macaroni & Cheese	1937
Canned tuna fish	1903	Kit Kat candy bar	1937
Campbell's pork and beans	1904	M&M's candy	1941
Peanut butter	1904	Cheerios cereal	1945
Kellogg's corn flakes	1906	Frozen orange juice	1946
Hershey's kisses	1907	Instant mashed potatoes	1946
Quaker puffed wheat 1	1909	Cheetos	1948
Crisco	1911	Kellogg's Frosted Flakes	1952
Mazola corn oil	1911	Kentucky Fried Chicken	1952
Hamburger buns	1912	M&M's peanut candy	1954
Hellmann's mayonnaise	1912	Fruit Loops cereal	1963
Lifesavers	1912	Doritos	1966
Oreo cookies	1913	Pringles chips	1969

Source: Adapted from http://www.geocities.com/foodedge/timeline.htm

processed wild-plant and animal foods in hunter-gatherer diets, they adversely affected the following dietary parameters: (1) glycemic load, (2) fatty acid composition, (3) macronutrient composition, (4) micronutrient density, (5) acid/base load, (6) sodium/potassium ratio, and (7) fiber content.

The Glycemic Load

The glycemic index, originally developed in 1981, is a relative comparison of the blood-glucose-raising potential of various foods or combination of foods based on equal amounts of carbohydrate in the food (Jenkins et al., 1981). In 1997, the concept of glycemic load (glycemic index × the carbohydrate content per serving size) was introduced to assess blood-glucose-raising potential of a food based on both the quality and quantity of dietary carbohydrate (Liu and Willett, 2002). Refined grain and sugar products nearly always maintain much higher glycemic loads than unprocessed fruits and vegetables (Foster-Powell, Holt, and Brand-Miller, 2002). Unrefined wild-plant foods, like those available to contemporary hunter-gatherers, typically exhibit low glycemic indices (Thorburn, Brand, and Truswell, 1987).

Within the past two decades, substantial information has accumulated showing that long-term consumption of high glycemic load carbohydrates can adversely affect metabolism and health (Liu and Willett, 2002; Ludwig, 2002; Cordain, Eades, and Eades, 2003). Specifically, chronic hyperglycemia and hyperinsulinemia, induced by high glycemic load carbohydrates may elicit a number of hormonal and physiological changes that promote insulin resistance (Liu and Willett, 2002; Ludwig, 2002; Cordain, Eades, and Eades, 2003). Diseases of insulin resistance are frequently referred to as "diseases of civilization" (Eaton, Konner, and Shostak, 1988; Reaven, 1995) and include obesity, coronary heart disease, type 2 diabetes, hypertension, and dyslipidemia (elevated serum triacylglycerols; small-dense, low-density lipoprotein cholesterol; and reduced high-density lipoprotein cholesterol). It is likely that the metabolic syndrome may extend to other chronic illnesses and conditions that are widely prevalent in western societies, including myopia, acne, gout, polycystic ovary syndrome, epithelial cell cancers (breast, colon, and prostate), male vertex balding, skin tags, and acanthosis nigricans (Cordain, Eades, and Eades, 2003). Diseases of insulin resistance are rare or absent in hunter-gatherer and other less-westernized societies living and eating in their traditional manner (Schaeffer, 1971; Trowell, 1980; Eaton, Konner, and Shostak, 1988).

In addition to high glycemic load carbohydrates, other elements of Neolithic and Industrial era foods may contribute to the insulin resistance underlying metabolic syndrome diseases. Milk, yogurt, and ice cream, despite having relatively low glycemic loads are highly insulinotropic, with insulin indices comparable to white bread (Ostman, Liljeberg Elmsthal, and Bjorck, 2001). Fructose is a major constituent in high fructose corn syrup (table 19.2) and maintains a low glycemic index of 23 but paradoxically may worsen insulin sensitivity (Reiser et al., 1989) and cause acute insulin resistance in humans (Dirlewanger et al., 2000).

In the typical U.S. diet, high glycemic load sugars now supply 18.6% of total energy, whereas high glycemic load, refined cereal grains supply 20.4% of energy (table 19.2). Hence, at least 39% of the total energy in the typical U.S. diet is supplied by foods that may promote insulin resistance. Although high glycemic load sugars and grains now represent a dominant element of the modern urban diet, these foods were rarely or never consumed by average citizens as recently as 200 years ago (tables 19.3–19.5).

The Fatty Acid Composition

Chemically, fats are defined as acylglycerols, compounds in which a fatty acid molecule (acyl group) is linked to a glycerol molecule by an ester bond. Almost all dietary and storage fats are triacylglycerols, compounds in which three fatty acid molecules are bound to a single glycerol molecule. Fatty acids fall into one of three major categories: (1) saturated fatty acids (SFA), (2) monounsaturated fatty acids (MUFA), and (3) polyunsaturated fatty acids (PUFA). Additionally, essential PUFA occur in two biologically important families, the n-6 PUFA and the n-3 PUFA. Substantial evidence now indicates that for preventing the risk of chronic disease the absolute amount of dietary fat is less important than the type of fat (Institute of Medicine of the National Academies, 2003). Beneficial, health-promoting fats are MUFA and some PUFA, whereas most SFA and *trans* fatty acids are detrimental when consumed in excessive quantities (Institute of Medicine of the National Academies, 2003). Further, the balance of dietary n-6 and n-3 PUFA is integral in preventing the risk of chronic disease and promoting health (Kris-Etherton, Harris, and Appel, 2002; Simopoulos, 2002).

The western diet frequently contains excessive saturated and *trans* fatty acids and has too little n-3 PUFA relative to its n-6 PUFA (Simopoulos, 2002; Institute of Medicine of the National Academies, 2003). High dietary intakes of SFA and *trans* fatty acids increase the risk of cardiovascular disease by elevating blood levels of total and LDL cholesterol (97, 100–102). N-3 PUFA may reduce the risk for cardiovascular disease via a number of mechanisms, including reductions in ventricular arrhythmias, blood clotting, serum triacylglycerol concentrations, growth of atherosclerotic plaques, and blood pressure (Kris-Etherton et al., 2002).

The six major sources of SFA in the U.S. diet are fatty meats, baked goods, cheese, milk, margarine, and butter (Subar et al., 1998). Five of these six foods would not have been components of hominin diets before the advent of animal husbandry or the Industrial Revolution. Because of the inherently lean nature of wild-animal tissues throughout most of the year and the dominance of MUFA + PUFA in their muscle and organ tissues (Cordain et al., 2002a), high dietary levels of SFA on a year-round basis would have been infrequently encountered in preagricultural diets (O'Keefe and Cordain, 2004).

The advent of the oil seed processing industry at the beginning of the twentieth century significantly raised the total intake of vegetable fat (Gerrior and Bente, 2002), which directly increased the dietary level of n-6 PUFA at the expense of a lowered n-3 PUFA because of the inherently higher concentrations of n-6 PUFA and lower concentrations of n-3 PUFA in most vegetable oils (Simopoulos, 2002). The trend toward a higher n-6/n-3 PUFA ratio was exacerbated as meat from grain-fed cattle and livestock became the norm in the U.S. diet over the past one hundred years (Whitaker, 1975; Cordain et al., 2002a). In the current U.S. diet, the ratio of n-6/n-3 PUFA has risen to 10 : 1 (Kris-Etherton et al., 2000), whereas in wild-animal-food dominated, hunter-gatherer diets it has been estimated between 2 : 1 and 3 : 1 (Cordain et al., 2002a).

The invention of the hydrogenation process in 1897 (Emken, 1984) allowed vegetable oils to become solidified and marketed as shortening or margarine and as foods containing hydrogenated vegetable oils. The hydrogenation process introduced a novel *trans* fatty acid (*trans* elaidic acid) into the human diet, which elevates blood cholesterol levels and leads to an increased risk of cardiovascular disease (Ascherio et al., 1994). *Trans* fatty acids in the U.S. diet now are estimated to constitute 7.4% of the total fatty acid intake (Allison et al., 1999).

The Macronutrient Composition

In the present U.S. diet, the percentage of total food energy derived from the three major macronutrients is as follows: carbohydrate (51.8%), fat (32.8%), and protein (15.4%). Current advice for reducing the risk of cardiovascular and other chronic diseases is to limit fat intake to 30% of total energy, to maintain protein at 15% of total energy and to increase complex carbohydrates to 55%–60% of total energy (Krauss, et al., 2000). Both the current U.S. macronutrient intakes and suggested healthful levels differ considerably from average levels obtained from ethnographic (Cordain et al., 2000) and quantitative studies (Cordain et al., 2002b) of hunter-gatherers in which dietary protein is characteristically elevated (19%–35% energy) at the expense of carbohydrate (22%–40% energy; Cordain et al., 2000).

An increasing body of evidence indicates high-protein diets may improve blood lipid profiles (Cordain et al., 2002b) and thereby lessen the risk for cardiovascular disease. A four-week dietary intervention of hypertensive subjects has demonstrated that a high-protein diet (25% energy) was effective in significantly lowering blood pressure (Burke et al., 2001). Further, high-protein diets have been shown to improve metabolic control in type 2 diabetes patients (O'Dea, 1984; O'Dea, et al., 1989).

Because protein has more than three times the thermic effect of either fat or carbohydrate (Crovetti et al., 1998) and because it has a greater satiety value than fat or carbohydrate (Crovetti, et al., 1998), increased dietary protein may represent an effective weight-loss strategy for the overweight or obese. Recent clinical trials have demonstrated that calorie-restricted high-protein diets are more effective than calorie-restricted high-carbohydrate diets in promoting (Skov et al., 1999) and maintaining (Westerterp-Plantenga et al., 2004) weight loss in overweight subjects, while producing less hunger and more satisfaction (Johnston, Tjonn, and Swan, 2004).

The Micronutrient Density

Refined sugars are essentially devoid of any vitamin or mineral. Accordingly the consumption of refined sugar or foods containing refined sugar reduces the total vitamin, mineral (micronutrient) density of the diet by displacing more nutrient dense foods. A similar situation exists for refined vegetable oils, except that they contain two fat-soluble vitamins (vitamin E and vitamin K). Because vegetable oils and refined sugars contribute at least 36.2% of the energy in a typical U.S. diet (table 19.2), the widespread consumption of these substances—or food made with them—has considerable potential to influence the risk of vitamin and mineral deficiencies.

At least half the U.S. population fails to meet the recommended daily allowances (RDA) for vitamin B6, vitamin A, magnesium, calcium, and zinc, and 33% of the population does not meet the RDA for folate (United States Department of Agriculture, Agricultural Research Service, 1997). Adequate dietary intake of both folate and vitamin B6 prevent the accumulation of homocysteine in the bloodstream. Elevated blood concentrations of homocysteine represent an independent risk factor for development of cardiovascular disease, stroke, and deep-vein thrombosis (Wald, Law, and Morris, 2002).

Displacement of fruits, vegetables, lean meats, and seafood by whole grains and milk lowers the overall dietary micronutrient density because whole grains and milk maintain lower-nutrient densities than these other foods (Cordain, 2002). Additionally, wild-plant foods, known to be consumed by hunter-gatherers, generally maintain higher micronutrient concentrations than their domesticated counterparts (Brand-Miller and Holt, 1998), as does the muscle meat of wild animals (First Data Bank, 2000). Consequently, the Neolithic introduction of dairy foods and cereal grains as staples would have caused the average micronutrient content of the diet to decline. This situation worsened as cereal milling techniques developed in the Industrial era allowed for the production of bread flour devoid of the more nutrient dense bran and germ (Storck and Teague, 1952). The displacement of more nutrient-dense items (fruits, vegetables, lean meats, and seafood) by less dense foods (refined sugars, grains, vegetable oils, and dairy products) and the subsequent decline in dietary vitamin and mineral density has far-reaching health implications; consequences that not only promote the development of vitamin deficiency diseases but also numerous infectious and chronic diseases (Cordain, 1999).

Acid/Base Balance

After digestion, absorption, and metabolism, nearly all foods release either acid or bicarbonate (base) into the systemic circulation (Frassetto et al., 1998). Table 19.6 shows that fish, meat, poultry, eggs, cheese, milk, and cereal grains are net acid producing, whereas fresh fruits, vegetables, tubers, roots, and nuts are net-base producing. Legumes yield near-zero mean acid values, reflecting an overlapping distribution from slightly net acid producing to slightly net-base producing. Not shown in table 19.6 are energy-dense, nutrient-poor foods, such as separated fats and refined sugars, that contribute neither to the acid nor base load. Additionally, salt (NaCl) is net acid producing because of the chloride ion (Frassetto et al., 1998).

The typical western diet yields a net acid load estimated to be + 50 meq/day (Lemann, 1999). As a result, normal adults consuming the standard U.S. diet sustain a chronic, low-grade pathogenic metabolic acidosis that worsens with age as kidney function declines (Frassetto, Morris, and Sebastian, 1996). Virtually all preagricultural

Table 19.6 Potential Net Acid (or Base) Loads of 17 Food Groups

	n	Net Acid Load (meq/418 kJ)	Net Acid Load (meq/10,460 kJ)	Potassium (meq/418 kJ)	Protein (g/418 kJ)	Protein g/100 meq Potassium
Acid-producing foods						
Fish	8	+14.6	+398	8.1	16.8	207
Meat	3	+12.4	+342	7.6	18.4	242
Poultry	2	+7.8	+227	4.7	13.4	287
Egg	1	+7.3	+215	2.4	8.3	339
Shellfish	3	+7.3	+215	18.4	18.0	159
Cheese	9	+3.3	+115	0.8	7.1	982
Milk	4	+1.3	+64	6.4	5.7	90
Cereal grains	7	+1.1	+60	2.6	3.2	153
Near-neutral foods						
Legumes	6	−0.4	+24	12.6	10.6	100
Base-Producing Foods						
Nut	6	−1.1	+6	3.8	2.5	86
Fresh fruit	11	−5.2	−98	9.4	1.6	16
Tuber	2	−5.4	−102	11.8	2.2	18
Mushroom	1	−11.2	−247	62.3	25.7	41
Root	5	−17.1	−395	34.3	6.8	21
Vegetable fruit	1	−17.5	−404	35.5	5.6	15
Leafy greens	6	−23.4	−553	43.5	10.0	24
Plant stalks	1	−24.9	−590	54.8	4.6	8

Note: Acid load calculations were made using previously described procedures (Sebastian et al., 2002). Positive (+) and negative (−) values represent acid-producing and base-producing equivalents, respectively.

diets were net-base yielding because of the absence of cereals and energy-dense, nutrient-poor foods—foods that were introduced during the Neolithic and Industrial eras and which displaced base-yielding fruits and vegetables (Sebastian et al., 2002). Consequently, a net-base-producing diet was the norm throughout most of hominin evolution (Sebastian et al., 2002). The known health benefits of a net-base-yielding diet include preventing and treating osteoporosis (Sebastian et al., 1994), age-related muscle wasting (Frassetto, Morris, and Sebastian, 1997), calcium kidney stones (Pak et al., 1985), hypertension (Morris et al., 1999), exercise-induced asthma (Mickleborough et al., 2001), as well as slowing the progression of age- and disease-related chronic renal insufficiency (Alpern and Sakhaee, 1997).

Sodium/Potassium Ratio

The average sodium content (3,271 mg/day) of the typical U.S. diet is substantially higher than its potassium content (2, 620 mg/day; United States Department of Agriculture, Agricultural Research Service, 1997). Three dietary factors are primarily responsible for the dietary ratio of Na/K, which is greater than 1.0. First, 90% of the sodium in western diets comes from manufactured salt (NaCl); hence, the sodium content of naturally occurring foods in the average U.S. diet (~330 mg) is quite low. Second, vegetable oils and refined sugars, which are essentially devoid of potassium, make up 36% of the total food energy. The inclusion of these two foods into the diet displaces other foods with higher potassium concentrations and thereby reduces the total dietary potassium. Third, the displacement of vegetables and fruits by whole grains and milk products may further reduce the potassium intake because potassium concentrations in vegetables are four and twelve times higher than in milk and whole grains, respectively, whereas in fruits the potassium concentration is two and five times higher than in milk and whole grains (First Data Bank, 2000). Taken together, the addition of manufactured salt to the food supply and the displacement of traditional potassium-rich foods by foods introduced during the Neolithic and Industrial periods caused a 400% decline in the potassium intake while simultaneously initiating a 400% increase in sodium ingestion (Frassetto et al., 2001).

The inversion of potassium and sodium concentrations in hominin diets had no evolutionary precedent and now plays an integral role in eliciting and contributing to numerous diseases of civilization. Diets low in potassium and high in sodium may partially or directly underlie or exacerbate a variety of maladies and chronic illnesses including: hypertension (Morris et al., 1999), stroke (Antonios and MacGregor, 1996), kidney stones (Pak et al., 1985), osteoporosis (Devine et al., 1995), gastrointestinal tract cancers (Tuyns, 1988), asthma (Carey, Locke, and Cookson, 1993), exercise-induced asthma (Mickleborough et al., 2001), insomnia (Miller, 1945), airsickness (Lindseth and Lindseth, 1995), high-altitude sickness (Porcelli and Gugelchuk, 1995), and Menière's syndrome (ear ringing; Thai-Van, Bounaix, and Fraysse, 2001).

The Fiber Content

The fiber content (15.1 g/day; United States Department of Agriculture, Agricultural Research Service, 1997) of the typical U.S. diet is considerably lower than recommended values (25–30 g; Krauss et al., 2000). Refined sugars, vegetable oils, dairy products, and alcohol are devoid of fiber and compose an average of 48.2% of the energy in the typical U.S. diet (table 19.2). Further, fiber-depleted, refined grains represent 85% of the grains consumed in the United States (table 19.2), and because refined grains contain 400% less fiber than whole grains (by energy), they further dilute the total dietary fiber intake. Fresh fruits typically contain twice the fiber of whole grains, and nonstarchy vegetables contain almost eight times more fiber than whole grains on an energy basis (First Data Bank, 2000). Fruits and vegetables known to be consumed by hunter-gatherers also maintain considerably more fiber than their domestic counterparts (Brand-Miller and Holt, 1998). Contemporary diets devoid of cereal grains, dairy products, refined oils, and sugars and processed foods have been shown to contain significantly more fiber (42.5 g/day) than either current or recommended values (Cordain, 2002).

Once again, the displacement of fiber-rich plant foods by novel dietary staples, introduced during the Neolithic and Industrial periods, was instrumental in changing the diets our species had traditionally consumed—a diet that would have almost always been high in fiber. Soluble fibers (those found primarily in fruits and vegetables) modestly reduce LDL and total cholesterol concentrations beyond those achieved by a diet low in saturated fat, and fiber, by slowing gastric emptying, may reduce appetite and help to control caloric intake (Anderson, Smith, and Gustafson, 1994). Diets low in dietary fiber may underlie or exacerbate constipation, appendicitis, hemorrhoids, deep-vein thrombosis, varicose veins, diverticulitis, hiatal hernia, and gastroesophageal reflux (Trowell, 1985).

References

Aiello, L.C., and Wheeler, P., 1995. The expensive tissue hypothesis. *Curr. Anthropol.* 36, 199–222.

Allison, D.B., Egan, S.K., Barraj, L.M., Caughman, C., Infante, M., and Heimbach, J.T., 1999. Estimated intakes of trans fatty and other fatty acids in the U.S, population. *J. Am. Diet Assoc.* 99, 166–174.

Alpern, R.J., and Sakhaee, S., 1997. The clinical spectrum of chronic metabolic acidosis: Homeostatic mechanisms produce significant morbidity. *Am. J. Kidney Dis.* 29, 291–302.

Anderson, J.W., Smith, B.M., and Gustafson, N.J., 1994. Health benefits and practical aspects of high-fiber diets. *Am. J. Clin. Nutr.* 59, 1242S–1247S.

Antonios, T.F., and MacGregor, G.A., 1996. Salt—more adverse effects. *Lancet* 348, 250–251.

Ascherio, A., Hennekens, C.H., Buring, J.E., Master, C., Stampfer, M.J., and Willett, W.C., 1994. Trans-fatty acids intake and risk of myocardial infarction. *Circulation* 89, 94–101.

Asfaw, B., White, T., Lovejoy, O., Latimer, B., Simpson, S., and Suwa, G., 1999. *Australopithecus garhi*: A new species of early hominid from Ethiopia. *Science* 284, 629–635.

Badr, A., Muller, K., Schafer-Pregl, R., El Rabey, H., Effgen, S., Ibrahim, H.H., Pozzi, C., Rohde, W., and Salamini, F., 2000. On the origin and domestication history of Barley (*Hordeum vulgare*). *Mol. Biol. Evol.* 17, 499–510.

Binford, L.R., 1978. Food processing and consumption. In: Binford, L.R., (Ed.), *Nunamiut Ethnoarchaeology*. Academic Press, New York, pp. 135–167.

Blurton Jones, N.G., Hawkes, K., and O'Connell, J., 1997. Why do Hadza children forage? In: Segal, N.L., Weisfeld, G.E., and Weisfield, C.C., (Eds.), *Uniting Psychology and Biology: Integrative Perspectives on Human Development*. American Psychological Association, New York, pp. 297–331.

Bose, S., 1964. Economy of the Onge of Little Andaman. *Man India* 44, 298–310.

Brand-Miller, J.C., and Holt, S.H., 1998. Australian Aboriginal plant foods: A consideration of their nutritional composition and health implications. *Nutr. Res. Rev.* 11, 5–23.

Bunn, H.T., and Kroll, E.M., 1986. Systematic butchery by Plio-Pleistocene hominids at Olduvai Gorge, Tanzania. *Curr. Anthropol.* 20, 365–398.

Burke, V., Hodgson, J.M., Beilin, L.J., Giangiulioi, N., Rogers, P., and Puddey, I.B., 2001. Dietary protein and soluble fiber reduce ambulatory blood pressure in treated hypertensives. *Hypertension* 38, 821–826.

Carey, O.J., Locke, C., and Cookson, J.B., 1993. Effect of alterations of dietary sodium on the severity of asthma in men. *Thorax* 48, 714–718.

Chesney, R.W., Helms, R.A., Christensen, M., Budreau, A.M., Han, X., and Sturman, J.A., 1998. The role of taurine in infant nutrition. *Adv. Exp. Med. Biol.* 442, 463–476.

Comer, J., 2000. Distilled beverages. In: Kiple K.F., and Ornelas, K.C., (Eds.), *The Cambridge World History of Food.* Vol. 1. Cambridge University Press, Cambridge, pp. 653–664.

Copley, M.S., Berstan, R., Dudd, S.N., Docherty, G., Mukherjee, A.J., Straker, V., Payne, S., and Evershed, R.P., 2003. Direct chemical evidence for widespread dairying in prehistoric Britain. *Proc. Natl. Acad. Sci.* 100, 1524–1529.

Cordain, L., 1999. Cereal grains: Humanity's double edged sword. *World Rev. Nutr. Diet* 84, 19–73.

Cordain, L., 2002. The nutritional characteristics of a contemporary diet based upon Paleolithic food groups. *J. Am. Nutraceut. Assoc.* 5, 15–24.

Cordain, L., Brand-Miller, J., Eaton, S.B., Mann, N., Holt, S.H.A., and Speth, J.D., 2000. Plant to animal subsistence ratios and macronutrient energy estimations in world wide hunter-gatherer diets. *Am. J. Clin. Nutr.* 71, 682–692.

Cordain, L., Eades, M.R., and Eades, M.D., 2003. Hyperinsulinemic diseases of civilization: more than just syndrome X. *Comp. Biochem. Physiol.* A 136, 95–112.

Cordain, L., Eaton, S.B., Brand-Miller, J., Mann, N., and Hill, K., 2002b. The paradoxical nature of hunter-gatherer diets: Meat based, yet non-atherogenic. *Eur. J. Clin. Nutr.* Suppl. no. 1, 56, S42–S52.

Cordain, L., Watkins, B.A., Florant, G.L., Kelher, M., Rogers, L., and Li, Y., 2002a. Fatty acid analysis of wild ruminant tissues: Evolutionary implications for reducing diet-related chronic disease. *Eur. J. Clin. Nutr.* 56, 181–191.

Cordain, L., Watkins, B.A., and Mann, N.J., 2001. Fatty acid composition and energy density of foods available to African hominids. Evolutionary implications for human brain development. *World Rev. Nutr. Diet* 90:144–161.

Crovetti, R., Porrini, M., Santangelo, A., and Testolin, G., 1998. The influence of thermic effect of food on satiety. *Eur. J. Clin. Nutr.* 52, 482–488.

de Heinzelin, J., Clark, J.D., White, T., Hart, W., Renne, P., WoldeGabriel, G., Beyene, Y., and Vrba, E., 1999. Environment and behavior of 2.5-million-year-old Bouri hominids. *Science* 284, 625–629.

Devine, A., Criddle, R.A., Dick, I.M., Kerr, D.A., and Prince, R.L., 1995. A longitudinal study of the effect of sodium and calcium intakes on regional bone density in postmenopausal women. *Am. J. Clin. Nutr.* 62, 740–745.

Dietz, W.H., Marino, B., Peacock, N.R., and Bailey, R.C., 1989. Nutritional status of Efe pygmies and Lese horticulturalists. *Am. J. Phys. Anthropol.* 78, 509–518.

Dirlewanger, M., Schneiter, P., Jequier, E., and Tappy, L., 2000. Effects of fructose on hepatic glucose metabolism in humans. *Am. J. Physiol. Endocrinol. Metab.* 279, E907–E911.

Eaton, S.B., 1992. Humans, lipids and evolution. *Lipids* 27, 814–820.

Eaton, S.B., and Konner, M., 1985. Paleolithic nutrition: A consideration of its natureand current implications. *New Engl. J. Med.* 312, 283–289.

Eaton, S.B., Konner, M., and Shostak, M., 1988. Stone agers in the fast lane: Chronic degenerative diseases in evolutionary perspective. Am. J. Med. 84, 739–749.

Eaton, S.B., Nelson, D.A., 1991. Calcium in evolutionary perspective. *Am. J. Clin. Nutr.* Suppl. 54(1), 281S–287S.

Emken, E.A., 1984. Nutrition and biochemistry of trans and positional fatty acid isomers in hydrogenated oils. *Annu. Rev. Nutr.* 4, 339–376.

Emken, R.A., Adlof, R.O., Rohwedder, W.K., and Gulley, R.M., 1992. Comparison of linolenic and linoleic acid metabolism in man: Influence of dietary linoleic acid. In:

Sinclair, A., and Gibson, R. (Eds.), *Essential Fatty Acids and Eicosanoids*. Invited Papers from the Third International Conference. AOCS Press, Champaign IL, pp. 23–25.

First Data Bank, 2000. Nutritionist V nutrition software, version 2.3. San Bruno, CA.

Foster-Powell, K., Holt, S.H., and Brand-Miller, J.C., 2002. International table of glycemic index and glycemic load values: 2002. *Am. J. Clin. Nutr.* 76, 5–56.

Frassetto, L., Morris, R.C., and Sebastian, A., 1996. Effect of age on blood acid-base composition in adult humans: Role of age-related renal functional decline. *Am. J. Physiol.* 271, 1114–1122.

Frassetto, L., Morris, R.C., and Sebastian, A., 1997. Potassium bicarbonate reduces urinary nitrogen excretion in postmenopausal women. *J. Clin. Endocrinol. Metab.* 82, 254–259.

Frassetto, L., Morris, R.C., Sellmeyer, D.E., Todd, K., and Sebastian, A., 2001. Diet, evolution and aging: The pathophysiologic effects of the post-agricultural inversion of the potassium-to-sodium and base-to-chloride ratios in the human diet. *Eur. J. Nutr.* 40, 200–213.

Frassetto, L.A., Todd, K.M., Morris, R.C., and Sebastian, A., 1998. Estimation of net endogenous noncarbonic acid production in humans from diet potassium and protein contents. *Am. J. Clin. Nutr.* 68, 576–583.

Fumihito, A., Miyake, T., Takada, M., Shingu, R., Endo, T., Gojobori, T., Kondo, N., and Ohno S., 1996. Monophyletic origin and unique dispersal patterns of domestic fowls. *Proc. Natl. Acad. Sci.* 93, 6792–6795.

Galloway, J.H., 2000. Sugar. In: Kiple, K.F., and Ornelas, K.C. (Eds.), *The Cambridge World History of Food*. Vol. 1. Cambridge University Press, Cambridge, pp. 437–449.

Gerrior, S., and Bente, L., 2002. *Nutrient Content of the U.S. Food Supply, 1909–99: A Summary Report*. U.S. Department of Agriculture, Center for Nutrition Policy and Promotion. Home Economics Report No. 55.

Giuffra, E., Kijas, J.M., Amarger, V., Carlborg, O., Jeon, J.T., and Andersson, L., 2000. The origin of the domestic pig: Independent domestication and subsequent introgression. *Genetics* 154, 1785–1791.

Gray, J.P., 1999. A corrected ethnographic atlas. *World Cult. J.* 10, 24–85.

Hanover, L.M., and White, J.S., 1993. Manufacturing, composition, and applications of fructose. *Am. J. Clin. Nutr.* Suppl. 58, 724S–732S.

Hawkes, K., O'Connell, J.F., and Blurton Jones, N., 1989. Hardworking Hadza grandmothers. In: Standen, V., and Foley, R.A., (Eds.), *Comparative Socio-ecology of Humans and Other Mammals*. Basil Blackwell, London, pp. 341–366.

Hayden, B., 1981. Subsistence and ecological adaptations of modern hunter/gatherers. In: RSO Harding, R.S.O., and Teleki, G. (Eds.), *Omnivorous Primates*. Columbia University Press, New York, pp. 344–421.

Hiendleder, S., Kaupe, B., Wassmuth, R., and Janke, A., 2002. Molecular analysis of wild and domestic sheep questions current nomenclature and provides evidence for domestication from two different subspecies. *Proc. R. Soc. Lond.* B 269, 893–904.

Hill, K., Hawkes, K., Hurtado, M., and Kaplan, H., 1984. Seasonal variance in the diet of Ache hunter-gatherers in Eastern Paraguay. *Hum. Ecol.* 12, 101–135.

Hurtado, A.M., and Hill, K., 1986. Early dry season subsistence ecology of the Cuiva (Hiwi) foragers of Venezuela. *Hum. Ecol.* 15, 163–187.

Hurdado, A.M., and Hill, K., 1990. Seasonality in a foraging society: variation in diet, work effort, fertility, and the sexual division of labor among the Hiwi of Venezuela. *J. Anthropol. Res.* 46, 293–345.

Institute of Medicine of the National Academies, 2003. Dietary fats: Total fat and fatty acids. In: *Dietary Reference Intakes for Energy, Carbohydrate, Fiber, Fat, Fatty Acids, Cholesterol, Protein, and Amino Acids (Macronutrients)*. National Academy Press, Washington DC, pp., 335–432.

Jenkins, D.J., Wolever, T.M., Taylor, R.H., Barker, H., Fielden, H., Baldwin, J.M., Bowling, A.C., Newman, H.C., Jenkins, A.L., and Goff, D.V., 1981. Glycemic index of foods: A physiological basis for carbohydrate exchange. *Am. J. Clin. Nutr.* 34, 362–366.

Johnston, C.S., Tjonn, S.L., and Swan, P.D., 2004. High-protein, low-fat diets are effective for weight loss and favorably alter biomarkers in healthy adults. *J. Nutr.* 134, 586–591.

Kaplan, H., Hill, K., Lancaster, J., and Hurtado, A.M., 2000. A theory of human life history evolution: diet, intelligence, and longevity. *Evol. Anthropol.* 9, 156–185.

Knopf, K., Sturman, J.A., Armstrong, M., and Hayes, K.C., 1978. Taurine: An essential nutrient for the cat. *J. Nutr.* 108, 773–778.

Krauss, R.M., Eckel, R.H., Howard, B., Appel, L.J., Daniels, S.R., Deckelbaum, R.J., Erdman, J.W., Kris-Etherton, P., Goldberg, I.J., Kotchen, T.A., Lichtenstein, A.H., Mitch, W.E., Mullis, R., Robinson, K., Wylie-Rosett, J., Sachiko S.J., Suttie, J., Tribble, D.L., and Bazzarre, T.L., 2000. AHA dietary guidelines: Revision 2000: A statement for healthcare professionals from the Nutrition Committee of the American Heart Association. *Circulation* 102, 2284–2299.

Kris-Etherton, P.M., Harris, and W.S., Appel, L.J., 2002. Fish consumption, fish oil, omega-3 fatty acids, and cardiovascular disease. *Circulation* 106, 2747–2757.

Kris-Etherton, P.M., Taylor, D.S., Yu-Poth, S., Huth, P., Moriarty, K., Fishell, V., Hargrove, R.L., Zhao, G., and Etherton, T.D., 2000. Polyunsaturated fatty acids in the food chain in the United States. *Am. J. Clin. Nutr.* Suppl. no. 1, 71, 179S–188S.

Krogh, A., and Krogh, M., 1913. A study of the diet and metabolism of Eskimos undertaken in 1908 on an expedition to Greenland. *Medd. Gronl.* 51, 1–52.

Laidlaw, S.A., Shultz, T.D., Cecchino, J.T., and Kopple, J.D., 1988. Plasma and urine taurine levels in vegans. *Am. J. Clin. Nutr.* 47, 660–663.

Lee, R.B., 1968. What hunters do for a living, or how to make out on scarce resources. In: Lee, R.B., and DeVore, I., (Eds.), *Man the Hunter*. Aldine, Chicago, pp. 30–48.

Lee-Thorp, J., Thackeray, J.F., and van der Merwe, N., 2000. The hunters and the hunted revisited. *J. Hum. Evol.* 39, 565–576.

Lemann, J., 1999. Relationship between urinary calcium and net acid excretion as determined by dietary protein and potassium: A review. *Nephron* Suppl. no. 1, 81, 18–25.

Leonard, W.R., and Robertson, M.L., 1994. Evolutionary perspectives on human nutrition: The influence of brain and body size on diet and metabolism. *Am. J. Hum. Biol.* 6, 77–88.

Lindseth, G., and Lindseth, P.D., 1995. The relationship of diet to airsickness. *Aviat. Space Environ. Med.* 66, 537–541.

Liphschitz, N., Gophna, R., Hartman, M., and Biger, G., 1991. The beginning of Olive (*Olea europaea* L.) cuttivation in the old world: A reassessment. *J. Archaeol. Sci.* 18, 441, 453.

Liu, S., and Willett, W.C., 2002. Dietary glycemic load and atherothrombotic risk. *Curr. Atheroscler. Rep.* 4, 454–461.

Loftus, R.T., Ertugrul, O., Harba, A.H., El-Barody, M.A., MacHugh, D.E., Park, S.D., and Bradley, D.G., 1999. A microsatellite survey of cattle from a centre of origin: The Near East. *Mol. Ecol.* 8, 2015–2022.

Ludwig, D.S., 2002. The glycemic index: Physiological mechanisms relating obesity, diabetes, and cardiovascular disease. *J. Am. Med. Assoc.* 287, 2414–2243.

Luikart, G., Gielly, L., Excoffier, L., Vigne, J., Bouvet, J., and Taberlet, P., 2001. Multiple maternal origins and weak phylogeographic structure in domestic goats. *Proc. Natl. Acad. Sci.* 98, 5927–5932.

MacDonald, M.L., Rogers, Q.R., and Morris, J.G., 1984. Nutrition of the domestic cat, a mammalian carnivore. *Annu. Rev. Nutr.* 4, 521–562.

Matsuoka, Y., Vigouroux, Y., Goodman, M.M., Sanchez, G. J., Buckler, E., and Doebley, J., 2002. A single domestication for maize shown by multilocus microsatellite genotyping. *Proc. Natl. Acad. Sci.* 99:6080–6084.

McArthur, M., 1960. Food consumption and dietary levels of groups of aborigines living on naturally occurring foods. In: Mountford, C.P., (Ed.), *Records of the American-Australian Scientific Expedition to Arnhem Land*. Melbourne University Press, Melbourne, pp. 90–135.

McGovern, P.E., Voigt, M.M., Glusker, D.L., and Exner, L.J., 1996. Neolithic resinated wine. *Nature* 381, 480–481.

Meehan, B., 1982. *Shell Bed to Shell Midden*. Australian Institute of Aboriginal Studies, Canberra.

Mickleborough, T.D., Gotshall, R.W., Kluka, E.M., Miller, C.W., and Cordain, L., 2001. Dietary chloride as a possible determinant of the severity of exercise-induced asthma. *Eur. J. Appl. Physiol.* 85, 450–456.

Miller, M.M., 1945. Low sodium chloride intake in the treatment of insomnia and tension states. *J. Am. Med. Assoc.* 129, 262–266.

Morris, R.C., Sebastian, A., Forman, A., Tanaka, M., and Schmidlin, O., 1999. Normotensive salt sensitivity: effects of race and dietary potassium. *Hypertension* 33, 18–23.

O'Dea, K., 1984. Marked improvement in carbohydrate and lipid metabolism in diabetic Australian Aborigines after temporary reversion to traditional lifestyle. *Diabetes* 33, 596–603.

O'Dea, K., Traianedes, K., Ireland, P., Niall, M., Sadler, J., Hopper, J., and De Luise, M., 1989. The effects of diet differing in fat, carbohydrate, and fiber on carbohydrate and lipid metabolism in type II diabetes. *J. Am. Diet. Assoc.* 89, 1076–1086.

O'Keefe, S.F., 2000. An overview of oils and fats, with a special emphasis on olive oil. In: Kiple, K.F., and Ornelas, K.C., (Eds.), *The Cambridge World History of Food.* Vol. 1. Cambridge University Press, Cambridge, pp. 375–397.

O'Keefe, J.H., and Cordain, L., 2004. Cardiovascular disease resulting from a diet and lifestyle at odds with our Paleolithic genome: How to become a 21st-century hunter-gatherer. *Mayo Clin. Proc.* 79, 101–108.

Ostman, E.M., Liljeberg Elmstahl, H.G., and Bjorck, I.M., 2001. Inconsistency between glycemic and insulinemic responses to regular and fermented milk products. *Am. J. Clin. Nutr.* 74, 96–100.

Pak, C.Y., Fuller, C., Sakhaee, K., Preminger, G.M., and Britton, F., 1985. Long-term treatment of calcium nephrolithiasis with potassium citrate. *J. Urol.* 134, 11–19.

Pawlosky, R., Barnes, A., and Salem, N., 1994. Essential fatty acid metabolism in the feline: Relationship between liver and brain production of long-chain polyunsaturated fatty acids. *J. Lipid Res.* 35, 2032–2040.

Politis, G., 1996. *Nukak.* Instituto Amazonico de Investigaciones Cientificas-SINCHI, Columbia.

Porcelli, M.J., and Gugelchuk, G.M., 1995. A trek to the top: A review of acute mountain sickness. *J. Am. Osteopath. Assoc.* 95, 718–720.

Rao, D.H., Brahman, G.V., and Rao, N.D., 1989. Health and nutritional status of the Onge of Little Andaman Island. *J. Ind. Anthropol. Soc.* 24:69–78.

Reaven, G.M., 1995. Pathophysiology of insulin resistance in human disease. *Physiol. Rev.* 75, 473–86.

Reiser, S., Powell, A.S., Scholfield, D.J., Panda, P., Fields, M., and Canary, J.J., 1989. Day-long glucose, insulin, and fructose responses of hyperinsulinemic and non-hyperinsulinemic men adapted to diets containing either fructose or high-amylose cornstarch. *Am. J. Clin. Nutr.* 50, 1008–10014.

Richards, M.P., Hedges, R.E.M., Jacobi, R., Current, A., and Stringer, C., 2000. Focus: Gough's Cave and Sun Hole Cave human stable isotope values indicate a high animal protein diet in the British Upper Palaeolithic. *J. Archaeol. Sci.* 27, 1–3.

Richards, M.P., Pettitt, P,B., Trinkaus, E., Smith, F.H., Paunovic, M., and Karavanic, I., 2000. Neanderthal diet at Vindija and Neanderthal predation: The evidence from stable isotopes. *Proc. Natl. Acad. Sci.* 97, 7663–7666.

Rudolph, M.H., McGovern, P.E., and Badler, V.R., 1992. Chemical evidence for ancient beer. *Nature* 360, 24.

Salamini, F., Ozkan, H., Brandolini, A., Schafer-Pregl, R., and Martin, W., 2003. Genetics and geography of wild cereal domestication in the near east. *Nat. Rev. Genet.* 3:429–441.

Schaeffer, O., 1971. When the Eskimo comes to town. *Nutr. Today* 6, 8–16.

Schoeninger, M.J., Moore, J., and Sept, J.M., 1999. Subsistence strategies of two "savanna" chimpanzee populations: The stable isotope evidence. *Am. J. Primatol.* 49, 297–314.

Sebastian, A., Frassetto, L.A., Sellmeyer, D.E., Merriam, R.L., and Morris, R.C., 2002. Estimation of the net acid load of the diet of ancestral preagricultural *Homo sapiens* and their hominid ancestors. *Am. J. Clin. Nutr.* 76, 1308–1316.

Sebastian, A., Harris, S.T., Ottaway, J.H., Todd, K.M., and Morris, R.C., 1994. Improved mineral balance and skeletal metabolism in post-menopausal women treated with potassium bicarbonate. *New Engl. J. Med.* 330, 1776–1781.

Semaw, S., Rogers, M.J., Quade, J., Renne, P.R., Butler, R.F., Dominguez-Rodrigo, M., Stout, D., Hart, W.S., Pickering, T., and Simpson, S.W., 2003. 2.6-Million-year-old stone tools and associated bones from OGS-6 and OGS-7, Gona, Afar, Ethiopia. *J. Hum. Evol.* 45, 169–177.

Silberbauer, G., 1981. *Hunter and Habitat in the Central Kalahari Desert.* Cambridge University Press, Cambridge.

Simopoulos, A.P., 2002. Omega-3 fatty acids in inflammation and autoimmune disease. *J. Am. Coll. Nutr.* 21, 495–505.

Sinclair, H.M., 1953. The diet of Canadian Indians and Eskimos. *Proc. Nutr. Soc.* 12, 69–82.

Skov, A.R., Toubro, S., Ronn, B., Holm, L., and Astrup, A., 1999. Randomized trial on protein vs. carbohydrate in ad libitum fat reduced diet for the treatment of obesity. *Int. J. Obes. Relat. Metab. Disord.* 23, 528–536.

Sponheimer, M., and Lee-Thorp, J.A., 2003. Differential resource utilization by extant great apes and australopithecines: Towards solving the C4 conundrum. *Comp. Biochem. Physiol.* A 136, 27–34.

Stanford, C.B., 1996. The hunting ecology of wild chimpanzees: Implications for the evolutionary ecology of Pliocene hominids. *Am. Anthropol.* 98, 96–113.

Storck, J., and Teague, W.D., 1952. *Flour for Man's Bread, a History of Milling.* University of Minnesota Press, Minneapolis, Minnesota.

Sturman, J.A., Hepner, G.W., Hofmann, A.F., and Thomas, P.J., 1975. Metabolism of [35S] taurine in man. *J. Nutr.* 105, 1206–1214.

Subar, A.F., Krebs-Smith, S.M., Cook, A., and Kahle, L.L., 1998. Dietary sources of nutrients among US adults, 1989 to 1991. *J. Am. Diet Assoc.* 98, 537–547.

Sussman, R.W., 1978. Foraging patterns of nonhuman primates and the nature of food preferences in man. *Fed. Proc.* 37, 55–60.

Tanaka, J., 1980. *The San, Hunter-gatherers of the Kalahari: A Study in Ecological Anthropology.* Tokyo University Press, Tokyo.

Teleki, G., 1973. The omnivorous chimpanzee. *Sci. Am.* 228, 33–42.

Thai-Van, H., Bounaix, M.J., and Fraysse, B., 2001. Meniere's disease: Pathophysiology and treatment. *Drugs* 61, 1089–1102.

Thorburn, A.W., Brand, J.C., and Truswell, A.S., 1987. Slowly digested and absorbed carbohydrate in traditional bushfoods: A protective factor against diabetes? *Am. J. Clin. Nutr.* 45, 98–106.

Trowell, H., 1985. Dietary fiber: a paradigm. In: Trowell, H., Burkitt, D., Heaton, K., and Doll, R., (Eds.*), Dietary Fibre, Fibre-depleted Foods and Disease*. Academic Press, New York, 1–20.

Trowell, H.C., 1980. From normotension to hypertension in Kenyans and Ugandans 1928–1978. *East. Afr. Med. J.* 57, 167–173.

Tuyns, A.J., 1988. Salt and gastrointestinal cancer. *Nutr. Cancer* 11, 229–232.

United States Department of Agriculture, Agricultural Research Service, 1997. Data tables: Results from USDA's 1994–96 Continuing Survey of Food Intakes by Individuals and 1994-96 Diet and Health Knowledge Survey, [Online]. ARS Food Surveys Research Group, Available (under "Releases"): http://www.barc.usda.gov/bhnrc/foodsurvey/home.htm (accessed May 11, 2004).

United States Department of Agriculture, Economic Research Service, 2002. Food Consumption (per capita) data system, sugars/sweeteners, Washington DC, http://www.ers.usda.gov/Data/foodconsumption/datasystem.asp (accessed May 11, 2004).

van der Merwe, N.J., Thackeray, J.F., Lee-Thorp, J.A., Luyt, J., 2003. The carbon isotope ecology and diet of *Australopithecus africanus* at Sterkfontein, South Africa. *J. Hum. Evol.* 44, 581–597.

Vekua, A., Lordkipanidze, D., Rightmire, G.P., Agusti, J., Ferring, R., Maisuradze, G., Mouskhelishvili, A., Nioradze, M., De Leon, M.P., Tappen, M., Tvalchrelidze, M., and Zollikofer, C., 2002. A new skull of early Homo from Dmanisi, Georgia. *Science* 297:85–89.

Vitte, C., Ishii, T., Lamy, F., Brar, D., and Panaud, O., 2004. Genomic paleontology provides evidence for two distinct origins of Asian rice (*Oryza sativa.*). *Mol. Genet. Genom.* 272, 504–511.

Wald, D.S., Law, M., Morris, J.K., 2002. Homocysteine and cardiovascular disease: Evidence oncausality from a meta-analysis. *BMJ* 325, 1202–1208.

Weller, O., 2002. The earliest salt exploitation in Europe: A salt mountain in the Spanish Neolithic. *Antiquity* 76, 317–318.

Westerterp-Plantenga, M.S., Lejeune, M.P., Nijs, I., van Ooijen, M., and Kovacs, E.M., 2004. High protein intake sustains weight maintenance after body weight loss in humans. *Int. J. Obes. Relat. Metab. Disord.* 28, 57–64.

Whitaker, J.W., 1975. *Feedlot Empire: Beef Cattle Feeding in Illinois and Iowa, 1840–1900.* Iowa State University Press, Ames.

Williams, G.C., and Nesse, R.M., 1991. The dawn of Darwinian medicine. *Q. Rev. Biol.* 66, 1–22.

Wood, B., 2002. Palaeoanthropology: Hominid revelations from Chad. *Nature* 418, 133–135.

Yellen, J., 1977. *Archaeological Approaches to the Present: Models for Reconstructing the Past.* Academic Press, New York.

Zhu, R.X., Potts, R., Xie, F., Hoffman, K.A., Deng, C.L., Shi, C.D., Pan, Y.X., Wang, H.Q., Shi, R.P., Wang, Y.C., Shi, G.H., and Wu, N.Q., 2004. New evidence on the earliest human presence at high northern latitudes in northeast Asia. *Nature* 431, 559–562.

Ziegler, E., 1967. Secular changes in the stature of adults and the secular trend of modern sugar consumption. *Z Kinderheilkd* 99, 146–166.

20

Preagricultural Diets and Evolutionary Health Promotion

S. BOYD EATON

Preagricultural Diets and Evolutionary Health Promotion

During the past thirty years, health promotion advice, especially that involving nutrition, has been based primarily on epidemiological research findings. While not a paradigm in the Kuhnian sense (Kuhn, 1996), epidemiology has been by far the dominant force in the field of disease prevention. Grant funding, academic publications, official recommendations, conferences, and popular press accounts related to personal health all reflect an underlying faith that the epidemiological method will ultimately prevail. However, to "prevail" in health promotion terms means reducing the incidence and/or prevalence of targeted disease conditions and, viewed from this bottom-line perspective, epidemiology's record has been unimpressive. Essential hypertension (Hajjar and Kotchen, 2003), diabetes mellitus (Mokdad et al., 2001), age-related fractures (Surgeon General, 2004), obesity (Flegal, Carroll, and Ogden, 2002), depression (Murphy et al., 2000), melanoma (Bevona and Sober, 2002), breast cancer (Clarke et al., 2002), and asthma (Mannino et al., 2002) all occur as, or more, frequently now than they did thirty years ago. The decline in myocardial infarction mortality has ceased (Willett, 2002), while the rate of congestive heart failure (now chiefly a consequence of atherosclerotic heart disease) has skyrocketed (Gheorghiade and Bonow, 1998).

Our investment in epidemiological research is, in one sense, a kind of experiment and a proper respect for scientific method requires that when enormous sums of money, hundreds (if not thousands) of painstaking investigations and untold hours of human effort have produced contradictory (Ioannidis, 2005), inconclusive results (as disease incidence increases), it is appropriate to at least consider a different approach. In Kuhnian terms, it may be time for a paradigm shift.

Evolutionary health promotion, informed by an understanding of the ancestral human lifestyle, offers a potential alternative that could make a vital contribution. This candidate paradigm arises from the discordance hypothesis that focuses on dissonance between our genetic heritage and our contemporary lifestyle (Eaton et al., 2002).

The human genome was selected in adaptation to Stone Age living circumstances, culminating in the appearance of behaviorally modern humans, between 100,000 and 50,000 years ago (kya) (Henshilwood et al., 2002). Before that time, human ancestors, like all other organisms, adapted to changing environmental conditions mainly by genetic evolution. However, for humans during the past 50 kya, adaptation has increasingly involved cultural modification. Since the appearance of agriculture, about 10 kya, there have been few generally recognized genetic changes (e.g., hemoglobinopathies, adult lactose tolerance, etc.). To an overwhelming degree, our genome and its epigenetic regulatory mechanisms remain adapted for a Paleolithic lifestyle.

While our genes have been relatively constant, our culture has changed to an almost indescribable extent. This has produced discordance or mismatch between our genes and our lives, and this discordance promotes the chronic degenerative diseases that are responsible for most morbidity, mortality, and health expenditure in developed, westernized nations.

Differences in reproductive experience, physical activity, psychosocial factors, microbial interactions, and toxin-allergen exposure all play important roles in evolutionary discordance theory. However, nutritional changes are those most pertinent to this volume's focus. Also, an understanding of their nature can perhaps alleviate the academic and popular dissention about nutrition, which confuses and disillusions today's health-conscious public.

Nutritional Comparisons

Assessing the nutrition of early behaviorally modern humans requires assessment of data from diverse sources. Radioisotopic analyses of human skeletal remains and evaluation of archaeological finds (bones of animals and fish consumed, botanical remains, implements, paintings, etc.) are essential elements, as are studies of recent hunter-gatherer (HG) subsistence patterns. Such peoples were the best, if imperfect, surrogates for Paleolithic foragers; therefore, proximate analyses of the game, aquatic resources, and uncultivated plant foods they used have provided vital data.

Given the indirect, incomplete nature of the evidence, retrojective precision is necessarily limited and the figures cited imply more exactitude than is currently justified. Still, they represent the latest stage in an ongoing investigative exercise, which has consistently revealed important differences between ancestral and contemporary human nutrition. New data will refine the estimates but will almost surely fail to erase the striking contrasts already evident.

The retrojections presented in the following section pertain to ancestors in East Africa, roughly 60 kya, about the time and place from which the modern human diaspora originated.

Lipids

Total Fat Intake

Hunter-gatherers, unlike affluent westerners, consume the total edible available from game animals, not just that in muscle meat, hence the estimate for total ancestral fat intake is about 35% (Cordain et al., 2000). Fat intake for Paleolithic HGs varied drastically with latitude: the 35% estimate is for Stone Agers in northeast Africa, the region currently thought most pertinent to establishment of the contemporary human genome. A total fat intake of 35% is at the upper limit of the acceptable range proposed recently by the Institute of Medicine (IOM), 20%–35% (IOM, 2002) and above the 30% upper limit set by numerous other authoritative organizations.

Saturated Fats

Even though fat provided a substantial proportion of total energy for Stone Agers, the contribution of saturated fat was lower than in the average western diet. American adults obtain 11%–12% of total dietary energy from saturated fat (IOM, 2002). Partially acculturated Greenland Eskimos obtain 8.4% (Bang and Dyerberg, 1980), and the estimate for Paleolithic humans is lower still, perhaps 7.5% (Eaton, Eaton, and Konner, 1997; Cordain et al., 2000). While its composition varies seasonally, the fat of game animals (including that from muscle meat, brain, organs, bone marrow, and storage depots) tends to have more mono- and polyunsaturated fatty acid and less saturated fatty acid than is found in supermarket meat (Cordain, Watkins, et al., 2002).

Early recommendations for saturated fat intake were less than 10% of total energy. Thc IOM now suggests neither a specific recommended dietary allowance nor a tolerable upper intake level for saturated fats, but notes that any incremental increase in saturated fatty acid intake raises coronary heart disease risk (IOM, 2002).

Trans Fatty Acids

Ancestral humans did obtain a minimal amount of trans fatty acids from mother's milk during infancy and from the flesh of certain herbivores, but the total from these sources was a small fraction of American consumption, which approaches 2% of total caloric intake. In addition, most trans fatty acid from ruminants is converted, after absorption, into conjugated linoleic acid isomers, which appear to have antineoplastic and antiatherogenic properties (IOM, 2002).

Polyunsaturated Fats

In the United States, total polyunsaturated fatty acids (PUFA) consumption is estimated to average about 15 g/d with ω6 polyunsaturated fatty acid (ω6 PUFA) intake about ten times that of ω3 polyunsaturated fatty acids (ω3 PUFA; IOM, 2002). The best available estimate of ancestral human intake suggests that total PUFA intake was nearly twice present levels, due almost entirely to more ω3 PUFA so that the ω6 : ω3 ratio was closer to unity, perhaps 2 : 1 (Eaton et al., 1998; Cordain et al., 2000). Current recommendations are much different, about 8 : 1 (IOM, 2002).

Most contemporary ω6 and ω3 PUFA intake occurs as eighteen carbon ω6 linoleic acid (LA; C18 : 2 ω6) and eighteen carbon ω3 α-linoleic acid (LNA; C18 : 3 ω3). These can be converted slowly to longer chain derivatives: arachidonic acid (AA; C20 : 4, ω6), eicosapentaenoic acid (EPA; C20 : 5, ω3), and docosahexaenoic acid (DHA; C22 : 6, ω3). The latter appear to be more important structurally as membranous constituents and biochemically as eicosanoid precursors and biological response modifiers. It is likely that ancestral humans had much higher total intakes of these longer chain PUFA, and that the ratio of ω3 forms (EPA and DHA) to the chief ω6 constituent (AA) was substantially greater than at present, an AA : EPA + DHA ratio of from 1.5 to 2.0 (Eaton et al., 1998; Cordain et al., 2000). The IOM recommends that EPA and DHA intake not exceed 10% of total ω3 PUFA intake. This accords well with retrojected Stone Age experience but total ancestral EPA + DHA intake would still have been much higher than at present because of higher total ω3 PUFA intake.

Dietary Cholesterol

Recent HG cholesterol intake was generally greater (~480 mg/d; Eaton and Konner, 1985; Eaton, Eaton, and Conner, 1997; Cordain et al., 2000) than that of Americans (~260 mg/d; IOM, 2002) because of high consumption of animal flesh: whether from fat or lean animals, all cell membranes contain cholesterol. Nevertheless, mean total serum cholesterol levels for five groups of HGs (from three continents) averaged 123 mg% (Eaton, Konner, and Shostak, 1988) versus 200–210 mg% for Americans. The LDL/HDL fractionation of HG serum cholesterol has not been recorded (to the authors' knowledge), but a total serum value of 123 mg% roughly corresponds to the recently suggested LDL target of 70 mg% for high-risk coronary disease patients. Furthermore, the HG cholesterol value falls within the range found among free-living nonhuman primates (Eaton, 1992). That HG cholesterol intake and serum levels seem inversely related to American experience adds to the growing evidence that dietary cholesterol is not a major independent driver of serum cholesterol (Sinclair, 1992; Cordain, Brand-Miller, et al., 2002a).

Protein

Recently studied HG protein consumption varies with latitude, but in equatorial savanna populations, those most like the core group of ancestral humans, protein provides about 30% of daily energy intake (Eaton and Konner, 1985; Eaton et al., 1987; Cordain et al., 2000). This corresponds to consumption of just over 3 g/kg/d (for a 70 kg individual consuming 3,000 kcal/d). Such intake greatly exceeds the recommended daily requirement (RDA) most recently advanced by the IOM, 0.8 g/kg/d (IOM, 2002) but is well within the range observed for free-living higher primates, 1.6–5.9 g/kg/d (Eaton, 1992). Despite its RDA, the IOM states that, "for adults, protein intake may range from 10 to 35% of energy intake to ensure a nutritionally adequate diet" (IOM, 2002). The IOM's comprehensive literature survey failed to establish a definite (or even strongly suggestive) linkage between high-protein intake and osteoporosis, nephrolithiasis, renal failure, coronary artery disease, obesity, or cancer.

Carbohydrate

Contemporary Americans obtain about 50% of their daily energy from carbohydrate, of which about 15% is contributed by added sugar (sugars added to foods during processing, preparation, or consumption). In regions where behaviorly modern humans probably originated, ancestral carbohydrate consumption is thought to have made up about 35% of total energy intake (Cordain et al., 2000), of which perhaps 2%–3%, on average, came from honey, the closest Paleolithic parallel for added sugar (Eaton, Eaton, and Konner, 1997; Cordain et al., 2005). Cereal grains (85% refined, 15% whole) are the largest single carbohydrate source for current humans with dairy products another significant contributor (Cordain et al., 2005). Our ancestors had no dairy products, after weaning, and rarely used cereal grains, so nearly all their carbohydrate came from fruits and vegetables, which generally have more desirable glycemic responses than do cereal grain and dairy products. Fruits and vegetables now provide only 23% of total carbohydrate intake.

The IOM proposes that an acceptable range of carbohydrate intake is from 45%–65% of total energy intake (for adults) and that added sugars provide no more than 25% of total energy (IOM, 2002). If added sugar were almost wholly eliminated, the resulting carbohydrate contribution to daily energy would approximate Paleolithic experience quantitatively. For qualitative equivalence, nearly all carbohydrate would have to be derived from fruits and vegetables rather than dairy or cereal grain products.

Fiber

Proximate analyses of uncultivated vegetables and fruits consumed by recent HGs show that they are substantially more fibrous (13.3 g dietary fiber/100 g) than are those now commercially available (4.2 g/100 g), which have been modified by millennia of selective agriculture practice (Eaton, 1990). Hence the high intake of wild-plant foods by ancestral humans necessarily provided a great deal of fiber.

Calculation of ancestral dietary fiber intake, based on a 50 : 50 animal vegetable subsistence ratio (as opposed to the 1985 35 : 65 estimate), suggests an average total fiber intake of over 100 g/d. The IOM now suggests an adequate intake of total fiber from food ranges from 25 g/d for adult women to 38 g/d for adult men, values a bit more than twice current median intake for Americans (IOM, 2002). The IOM found insufficient evidence to set a tolerable upper intake level for dietary fiber. High-fiber intake may adversely affect mineral bioavailability, especially when phytate is present. Phytic acid is a prominent constituent of many cereal grains but is minimally, if at all, present in most uncultivated fruits and vegetables. Consequently, the high-fiber content of ancestral diets would probably have had less impact on mineral bioavailability than would a similarly high intake of fiber from contemporary sources (Eaton, 1990; Eaton, Eaton, and Konner, 1997).

By definition, dietary fiber is not digested by mammalian enzymes and passes relatively intact into the large intestine where a proportion is fermented by gut microflora. Fruits and vegetables contain fiber more completely fermentable than that found in cereals, a distinction which appears to influence the physiological and health effects related to fiber intake. The overall fermentability of ancestral dietary fiber would have much exceeded that typically found in today's fiber-containing foods.

Energy

Extrapolation from the estimated daily energy intake of foragers studied in the past century suggests that the taller and comparably active humans of 50,000 years ago probably consumed about 2,900 kcal/d (averaged for men and women) (Cordain, Gotshall, and Eaton et al., 1998; Eaton and Eaton, 2003). The best available figures suggest that mean energy intake for Americans is much less—about 2,093 kcal/d (males 2,511, females 1,674; IOM, 2002). Nevertheless, recently studied HGs have body mass indices averaging 21.6 (Jenike, 2001), well within the accepted normal range (18.5–25 kg/m^2; IOM, 2002). The American average is 26.5. Forager skin-fold thicknesses are typically half or less than those of age-matched contemporary North Americans (Eaton, Konner, and Shostak, 1988). Further, skeletal remains show that Paleolithic humans developed muscularity similar to that of today's superior athletes, substantially greater than that of typical males and females (Eaton and Eaton, 2003). These findings suggest our ancestors existed within a high-energy throughput metabolic environment characterized by both greater caloric intake and greater caloric expenditure than is now the case.

Most contemporary recommendations for energy intake revolve around consumption commensurate with energy expenditure so as to achieve energy balance (assuming preexisting desirable body weight). That Americans are becoming ever more obese while consuming far less food energy than did ancestral humans is convincing evidence against the value of hypocaloric dieting and for increasing physical activity in our lives.

Acid-Base Considerations

Ancestral diets were net base yielding because of their vegetable and fruit content; they tended to drive systemic pH toward alkalinity. Conversely, the cereal grains and dairy foods that make up such a large proportion of contemporary diets are net acid yielding and tend to drive pH toward acidity (Sebastian et al., 2002). Homeostatic mechanisms ordinarily maintain pH at about 7.4, but over prolonged periods (decades), the corrective metabolic measures necessary to offset persistent acid-yielding diets have deleterious effects, including urinary calcium loss (to balance hydrogen ion excretion), accelerated skeletal calcium depletion, calcific urolithiasis, age-related muscle wasting, and progressive renal function deterioration (Frassetto, Morris, and Sebastian, 2004).

Micronutrients

Ancestral diets generally provided more vitamins and minerals than are obtained by typical Americans, whether in absolute terms or relative to energy intake (Eaton and Konner, 1985; Eaton, Eaton, and Konner, 1997). The ratio between Paleolithic and current intake varied with the specific micronutrient being considered, but typically ranged from 1.5 to 8 times more for Stone Agers (Eaton and Cordain, 1997; Eaton, Eaton, and Konner, 1997; Eaton and Eaton, 2000). The RDA for certain micronutrients has been increased in the past few years; for example, the requirement for vitamin C was raised from 60 mg/d to 75–90 mg/d (for women and men, respectively). Also acceptable upper limits for micronutrients were established in 2000. Except for iron, all retrojected ancestral intakes fall within these limits. For example, the newly established acceptable upper limit for vitamin C intake is 2,000 mg, well above the estimated Paleolithic average of 513 mg/d.

Sodium is a glaring exception to the general rule that Paleolithic vitamin and mineral intake exceeded that at present. Extrinsic sodium is added to that which occurs naturally in foods during processing, preparation, and at the table; only 10% of our sodium intake is intrinsic to the foods we eat (Food and Nutrition Board, 1989; Eaton and Cordain, 1997). For ancestral humans, potassium intake substantially exceeded that of sodium; the K/Na ratio was probably around 5 : 1 in northeast Africa. Now, however, sodium intake exceeds that of potassium. Availability of cheap commercial sodium explains much of this radical electrolyte reversal, but replacement of potassium-rich fruits and vegetables by potassium-poor vegetable oils, refined sugars, dairy products, and cereals is an additional important factor.

Research Possibilities

This volume's focus on the known, unknown, and unknowable is directly applicable to establishment of a health promotion research agenda.

We know:

1. That conventional health promotion efforts have not produced desired results (*vide ante*).
2. That chronic disease incidence and prevalence in a population is determined by the interaction between the group's genetic makeup and its lifestyle.
3. That there is no guiding paradigm for health promotion research planning, selection (resource allocation), and interpretation.
4. That humans whose lifestyle more closely approaches ancestral experience have lower rates of chronic degenerative disease (Eaton et al., 1988, 2002).

We can never know, absent time machines, the exact biomarker (HDL and total cholesterol levels, blood pressures, glucose responsiveness, body composition, etc.) status of ancestral humans. Their demographic statistics, morbidity and mortality figures, and subsistence patterns can never be directly ascertained. Our retrojections of such factors must be based on the indirect, incomplete data sources to which we have access.

The Unknown Presents Many Promising Areas for Investigation

It is generally accepted that current humans differ little, genetically, from ancestral Africans of 60,000 years ago, but further evidence bearing on this relationship would be welcome. Individuals who have lived as HGs are rapidly disappearing. We need to access diet and health-related information as quickly as feasible from those who remain. Recent HGs (and, by extension, Paleolithic humans) prize dietary fat, consuming about as much, in total, as do typical Americans. Their dietary cholesterol intake would have exceeded present intake, but their consumption of serum cholesterol–raising fats was lower, in line with or slightly less than current recommendations. Their low levels of serum cholesterol should be considered the intriguing result of a natural experiment.

Because adult Stone Agers ate only high-fiber uncultivated fruits and vegetables and high-protein wild game, their fiber and/or protein intake (usually both) would necessarily have far exceeded current consumption. While the effects of a high protein diet have been investigated to a modest extent (Sinclair, 1992; Cordain, Eaton, et al., 2002a; Layman et al., 2003; Gannon and Nuttall, 2004; McAuley et al., 2005) the physiological and anatomic implications of what seems to us an extremely high fiber diet from childhood on have been studied little, if at all.

Cereal grain products have been staples for millennia. Their political, economic, and demographic necessity is unquestioned but are they required nutritionally? By displacing a substantial proportion of the foods our ancestors used, have they degraded dietary quality (e.g., decreased phytochemical intake; Eaton and Cordain, 1997)? Could they have intrinsic negative characteristics (Cordain, 1999)?

Reconstructions of ancestral nutrition indicate a micronutrient intake much above contemporary recommendations. This fuels the argument, made by some, that optimal levels of dietary vitamins and minerals exceed currently accepted minimum requirements. We suspect that ancestral intake of phytochemicals and antioxidants, like that of micronutrients in general, was greater in the Paleolithic than at present. To be more confident about this probability, we need better data on the phytochemical/antioxidant content of wild foods.

Sodium intake now exceeds that of potassium, a striking reversal of prior human experience (and that of all free-living terrestrial mammals). The physiological effects of this electrolytic inversion deserve intense scrutiny. Can commercial considerations have affected our evaluation of this critical alteration?

A related area of potentially similar importance concerns the impact of diet on the body's acid-base balance. That contemporary foods tend to drive systemic pH toward acidity, whereas those of Paleolithic humans had alkalinizing properties is a fundamental difference with potentially major implications (Frassetto, Morris, and Sebastian, 2004).

The foraging existence necessitated high-energy throughput. Hunter-gatherer body composition and mass indices were almost always in the highly desirable range despite caloric intake well above the American average. Their example suggests that effective prevention of obesity involves increasing physical activity more than decreasing caloric intake (Eaton et al., 1988; Eaton and Eaton, 2003).

We anticipate that recommendations informed by ancestral dietary (and general lifestyle) experience should produce better health promotion results than have current recommendations. This has not yet been tested.

Discussion

Thoughtful people interested in their own health typically pay careful attention to research findings and official recommendations about disease prevention. However, those who have been keeping track have surely noted that successive discoveries and advisories are frequently inconsistent and too often contradictory (Ioannidis, 2005). As the *New York Times* observed, advice to the public about healthy living seems to change with each research article (*New York Times*, 1998). Not surprisingly, inconsistency has been coupled with ineffectiveness: many targeted conditions, including, for instance, breast cancer (Clarke et al., 2002), type 2 diabetes (Mokdad et al., 2001), depression (Murphy et al., 2000), and hypertension (Hajjar and Kotchen, 2003) occur as, or more, frequently now than they did twenty years ago.

Thomas Kuhn's classic treatise, *The Structure of Scientific Revolutions* (Kuhn, 1996), argues that to be effective, mature sciences must operate under the aegis of a governing paradigm, a worldview that serves to integrate and coordinate an investigative discipline's endeavors. To date, disease prevention research has had no such guiding framework, and the consequent random epidemiological studies have produced conflicting, disorderly results, which have tended to confuse and discourage rather than to enlighten and motivate. Further, Joel Mokyr's *The Gifts of Athena* (Mokyr, 2002) argues that an understandable, persuasive paradigm is essential for public acceptance of, and action in line with, authoritative recommendations, no matter how sound the basic research on which they are based.

The lifestyle of ancestral humans, that for which our genome was originally selected, could be considered a paradigmatic baseline. Deviation from the essentials of that experience appears to underlie the pathophysiology of chronic disease propagation and, conversely, behavior, which tends to match the Stone Age lifestyle model, is likely to forestall development of chronic illness while positively enhancing health (Eaton et al., 2002).

The nutritional considerations presented in this chapter provide an example of how an approach to disease prevention based on the evolutionary-discordance paradigm might operate. The first requirement would be to determine as accurately as possible the pertinent lifestyle characteristics of ancestral humans. Next, the health effects of, deviation from, and reversion toward these original biobehavioral parameters would require meticulous scientific investigation. Thereafter, those factors whose importance is supported by careful research would be promulgated as recommendations for healthy living. Finally, the fundamental rationale, research findings, and actual recommendations would be integrated into an understandable, persuasive framework—that the essentials of our ancestral lifestyle constitute a guide for healthy living in the present.

The theoretical benefits of disease prevention are great, but to date, this potential has been unrealized and the health-conscious public deserves a fresh approach. A program of research and recommendations based on the evolutionary discordance paradigm might be a logical way to address this need.

References

Bang, H.O., and Dyerberg, J., 1980. Lipid metabolism and ischemic heart disease in Greenland Eskimos. *Adv. Nutr. Res.* 3, 1–22.

Bevona, C., Sober, A.J., 2002. Melanoma incidence trends. *Dermatol. Clin.* 20, 589–595.

Clarke, C.A., Glaser, S.L. West, D.W., Ereman, R.R., Erdman, C.A., Barlow, J.M., and Wrensch, M.R., 2002. Breast cancer incidence and mortality trends in an affluent population: Marin County, California, USA, 1990–1999. *Breast Cancer Res.* 4, R 13.

Cordain, L., 1999. Cereal grains: Humanity's double-edged sword. *World Rev. Nutr.* Diet. 84, 19–73.

Cordain, L., Brand-Miller, J., Eaton, S.B., Mann, N., Holt, S.H.A., and Speth, J.D., 2000. Plant-animal subsistence ratios and macro-nutrient energy estimations in worldwide hunter-gatherer diets. *Am. J. Clin. Nutr.* 71, 682–692.

Cordain, L., Eaton, S.B., Brand-Miller, J., Mann, N., and Hill, K., 2002. The paradoxical nature of hunter-gatherer diets: Meat-based, yet non-atherogenic. *Eur. J. Clin. Nutr.* Suppl. no. 1, 56, 542–552.

Cordain, L., Eaton, S.B., Sebastian, A., Mann, N., Lindeberg, S., Watkins, B.A., O'Keefe, J.H., and Brand-Miller, J., 2005. Origins and evolution of the Western diet: health implications for the 21st century. *Am. J. Clin. Nutr.* 81, 341–354.

Cordain, L., Gotshall, R.W., and Eaton S.B., 1998. Physical activity, energy expenditure and fitness: An evolutionary perspective. *Int. J. Sports Med.* 19, 328–335.

Cordain, L., Watkins, B.A., Florant, G.L., Kehler, M., Rogers, L., and Li, Y., 2002b. Fatty acid analysis of wild ruminant tissues: Evolutionary implications for reducing diet-related chronic disease. *Eur. J. Clin. Nutr.* 56, 181–191.

Eaton, S.B., 1990. Fibre intake in prehistoric times. In: Leeds, A.R. (Ed.), *Dietary Fibre Perspectives. Reviews and Bibliography*. 2nd ed. John Libby, London, pp. 27–40.

Eaton, S.B., 1992. Humans, lipids and evolution. *Lipids* 27, 814–820.

Eaton, S.B., Cordain, L., 1997. Evolutionary aspects of diet: old genes, new fuels. *World Rev. Nutr. Diet.* 81, 26–37.

Eaton, S.B. III, and Eaton, S.B., 2000. Consumption of trace elements and minerals by preagricultural humans. In: Bogden, J.D., and Klevay, L.M. (Eds.), *Clinical Nutrition of the Essential Trace Elements and Minerals*. Humana Press, Totowa, NJ., pp. 37–47.

Eaton, S.B., and Eaton, S.B., III, and Konner, M.J., 1997. Paleolithic nutrition revisited: A twelve year retrospective and implications. *Eur. J. Clin. Nutr.* 51, 207–216.

Eaton, S.B., and Eaton, S.B., III., 2003. An evolutionary perspective on human physical activity: Implications for health. *Comp. Biochem. Physiol* A 136, 153–159.

Eaton, S.B., Eaton, S.B. III, Sinclair, A.J., Cordain, L., and Mann, N.J., 1998. Dietary intake of long-chain polyunsaturated fatty acids during the Paleolithic. *World Rev. Nutr. Diet.* 83, 12–23.

Eaton, S.B., and Konner, M., 1985. Paleolithic nutrition: A consideration of its nature and current implications. *New Engl. J. Med.* 312, 283–289.

Eaton, S.B., Konner, M., and Shostak, M., 1988. Stone Agers in the fast lane: Chronic degenerative diseases in evolutionary perspective. *Am. J. Med.* 84, 739–749.

Eaton, S.B., Strassman, B.I., Nesse, R.M., Neel, J.V., Ewald, P.W., Williams, G.C., Weder, A.B., Eaton, S.B. III, Lindeberg, S., Konner, M.J., Mysterud, I., and Cordain, L., 2002. Evolutionary health promotion. *Prev. Med.* 34, 109–118.

Flegal, K.M., Carroll, M.P., and Ogden, C.L., 2002. Prevalence and trends in obesity among U.S. adults, 1999–2000. *J. Am. Med. Assoc.* 288, 1723–1727.

Food and Nutrition Board, National Research Council, 1989. *Recommended Dietary Allowances*. 10th ed. National Academy Press, Washington, DC, pp. 13, 55.

Frassetto, L., Morris, C., and Sebastian, A., 2004. Effects of diet acid load on bone health. In: Burkhardt, P., Dawson-Hughes, B., and Heaney, R.P. (Eds.), *Nutritional Aspects of Osteoporosis*. 2nd ed. Elsevier Press, Amsterdam, pp. 273–295.

Gannon, M.C., and Nuttall, F.Q., 2004. Effect of a high-protein, low-carbohydrate diet on blood glucose control in people with type-2 diabetes. *Diabetes* 53, 2375–2382.

Gheorghiade, M., and Bonow, R.O., 1998. Chronic heart failure in the United States: A manifestation of coronary artery disease. *Circulation* 97, 282–289.

Hajjar, I., and Kotchen, T.A., 2003. Trends in prevalence, awareness, treatment, and control of hypertension in the United States 1988–2000. *J. Am. Med. Assoc.* 290, 199–206.

Henshilwood, C.S., d'Errico, F., Yates, R., Jacobs, Z., Tribolo, C., Duller, G.A.T., Mercier, N., Sealy, J.C., Valladas, H., Watts, I., and Wintle, A.G., 2002. Emergence of modern human behavior: Middle stone age engravings from South Africa. *Science* 295, 1278–1280.

Institute of Medicine., 2002. *Dietary Reference Intakes. Energy, Carbohydrate, Fiber, Fat, Fatty Acids, Cholesterol, Protein, and Amino Acids.* National Academy Press, Washington, DC, *passim.*

Ioannidis, J.P.A., 2005. Contradicted and initially stronger effects in highly cited clinical research. J. Am. Med. Assoc. 294, 218–228.

Jenike, M.R., 2001. Nutritional ecology: diet, physical activity and body size. In: Panter-Brick, C., Layton, R.H., and Rowley-Conwy, P. (Eds.), *Hunter-Gatherers. An Interdisciplinary Perspective.* Cambridge Univ. Press, Cambridge. pp. 205–238.

Kuhn, T.S., 1996. *The Structure of Scientific Revolutions*, 3rd ed. Univ. Chicago Press, Chicago, pp. 10–15.

Layman, D.K., Boileau, R.A., Erickson, D.J., et al. 2003. A reduced ratio of dietary carbohydrate to protein improves body composition and blood lipid profiles during weight loss in adult women. *J. Nutr.* 133, 411–417.

Mannino, D.M., Homa, D.M., Akinbami, L.J., Moorman, J.E., Gwynn, C., and Redd, S.C., 2002. Surveillance for asthma—United States, 1980–1999. *MMWR* 51, 1–13.

McAuley, K.A., Hopkins, C.M., Smith, K.J., McLay, R.T., Williams, S.M., Taylor, R.W., and Mann, J.I., 2005. Comparison of high-fat and high-protein diets with a high-carbohydrate diet in insulin-resistant obese women. *Dialectology* 48, 8–16.

Mokyr, J., 2002. *The Gifts of Athena.* Princeton University Press, Princeton, N.J., pp. 193–195.

New York Times, 1998. Sunday, March 22, p. WK 4.

Sebastian, A, Frassetto, L.A., Sellmeyer, D.E., Merriam, R.L., and Morris, R.C., 2002. Estimation of the net acid load of the diet of ancestral preagricultural *Homo sapiens* and their hominid ancestors. *Am. J. Clin. Nutr.* 76, 1308–1316.

Sinclair, A., 1992. Was the hunter-gatherer diet prothrombotic? In: Sinclair, A., and Gibson, R., (Eds.), *Essential Fatty Acids and Eicosanoids.* American Oil Chemical Society, Champaign, IL, pp. 318–324.

United States Department of Health and Human Services, 2004. *Bone Health and Osteoporosis: A Report of the Surgeon General.* U.S. Dept. Health and Human Services, Washington DC, pp. 69–87.

21

Limits to Knowledge on the Evolution of Hominin Diet

PETER S. UNGAR

We tend to fill conferences, not to mention magazines and airwaves, with what we know. We much less often explore and disclose the limits to our knowledge. Few experts like or bother to write terra incognita on their maps. Yet, disclosing the limits to our knowledge is often among the most useful of acts. It helps people choose where to explore, and it helps people to hedge their bets.
—Jesse Ausubel, Director of the Alfred P. Sloan Foundation's Known, Unknown, and Unknowable Program, April 22, 2005

What did early hominins eat? What are the limits to our knowledge of the diets of early hominins? Philosophers call these first- and second-order questions, respectively (Rosenberg, 2001). The chapters in this volume address both. The basic idea is that by reconstructing the diets of early hominins, we can understand something of the adaptations of our distant ancestors and how we came to be the way we are today. Further, by assessing the limits to our knowledge, and by taking stock in our approaches to reconstructing diet, we should be able to make better progress toward answering questions about the adaptations and behaviors of the early hominins.

In this chapter, I summarize and synthesize some of the issues raised by the contributors to this volume. First, I consider what we actually know (or, at least, think we know) about early hominin diets. Second, I consider second-order questions raised by contributions in this volume. Why should we study early hominin diets? What are the limits to our knowledge? What directions might our research take in the future?

The Known

There is no clear consensus on what is "known" about early hominin diets. Indeed, there are as many differing opinions as there are researchers to express them. Nevertheless, because the editor gets the last word (one of the few rewards for this task), some of my own thoughts are presented here. Because there are simply too many good ideas in this volume for a comprehensive summary, I hope individual authors will forgive my inadvertent omissions.

Evidences of early hominin diets can be divided into evidence that comes from the fossils themselves and that derived from context and models. Evidence from the fossils includes tooth chemistry and dental microwear analyses, which bear telltale signs of what specific individuals ate during their lifetimes. It also includes functional morphology of teeth and jaws, reflecting properties of the foods species are adapted to consume.

Contextual evidence includes archaeological remains; the durable remnants of foods eaten by hominins and the tools used to prepare them. Other lines of evidence include paleoenvironmental reconstructions and the paleobotanical, ethnographic, ethological, physiological, and nutritional data used to establish models that set logical limits on the ranges of possible hominin diets. Taken together, these lines of evidence can provide us with important clues.

Fossil Evidence

Epigenetic Signals

Epigenetic, or nongenetic, signals provide information about the lives of individuals and so are more independent of phylogenetic history than is adaptive morphology. The two main epigenetic signals considered here are dental microwear (Teaford, chapters 5 and 7) and tooth chemistry (Sponheimer, Lee-Thorp, and de Ruiter, chapter 8; Schoeninger, chapter 9). What do these tell us about early hominin diets?

As Teaford (chapter 7) noted, while "the shapes of unworn teeth can tell us a great deal about what a tooth is *capable* of processing, tooth *wear* can give us insights into how a tooth was actually used" (emphasis added). Dental microwear shows significant differences between early hominins, such that, for example, *Paranthropus robustus* consumed harder, more brittle foods than did *Australopithecus africanus* (Grine, 1986). More recent work has demonstrated considerable overlap in microwear patterns between the species, suggesting that diet differences were likely not in preferred foods but in occasional ones with challenging fracture properties (Scott et al., 2005). This pattern also seems to hold when comparing microwear of early *Homo* species (Ungar, Grine, and Teaford, 2006). In this case, a perceived limit of microwear analysis, the fact that microwear reflects only a few days diet, actually allows us to look at short-term variations when reasonably large sample sizes are available. Microwear patterns are consistent with a diet dominated by soft fruits or other foods with less-challenging fracture properties but still occasional fall back on items that were more difficult to fracture.

Isotope ratios, however, can reflect several months worth of diet, as elements are laid down during tooth formation. Sponheimer and colleagues (chapter 8) note that, for example, early hominins from South Africa tended to have ^{13}C enriched diets. This suggests that these hominins probably had fairly broad diets, including both C_3 foods, such as forest fruits, and C_4 plant resources, such as grasses, sedges, or perhaps animals that eat these.

Genetic or Adaptive Signals

Morphological adaptations provide genetic signals for reconstructing diet. Because the teeth and jaws function principally to fracture food, the sizes (Lucas, chapter 3), shapes, (Ungar, chapter 4; Daegling and Grine, chapter 6), and underlying structures of these elements (Teaford, chapters 5 and 7; Daegling and Grine, chapter 6) have all been examined for evidence of feeding adaptations. These lines of evidence are typically examined in the context of biomechanical models for what early hominins were capable of eating.

Functional studies of tooth size in early hominins date back to Robinson's (1954) pioneering work suggesting that differences among species in relative tooth sizes relate to differences in their diets. He noted that *Paranthropus* had relatively larger molar teeth than did *Australopithecus*, implying to him that the former chewed great quantities of low-quality herbaceous foods. whereas the latter was more likely to have been an omnivore. Indeed, many of the chapters in this book also invoke tooth size as a dietary signal. While the role of dental allometry in the diets of the earliest hominins remains unclear (see Teaford and Ungar, 2000), reduced tooth size in more recent *Homo* species may well be related to changes in food fracture properties associated with cooking or other methods of preparation (Lucas, chapter 3).

Occlusal morphology in early hominins has also offered some clues. For example, early *Homo* teeth evince more occlusal relief than do those of *Australopithecus afarensis* (Ungar, 2004), and *Paranthropus robustus* molars are less sloping than are those of *Australopithecus africanus* (Ungar, in press). These data suggest that early *Homo* would have been capable of shearing and slicing tough items such as young leaves, some fruits, and meat more efficiently than could their australopith predecessors. Further, *P. robustus* could have crushed hard, brittle foods such as some nuts or seeds more efficiently than could *A. africanus*. The variation in occlusal morphology among these hominins, however, is not as dramatic as some introductory texts might have us believe; it is, at most, on the order of magnitude of that between living sympatric chimpanzees and gorillas, species that tend to differ mostly in fallback foods.

Contextual and Other Lines of Evidence

Archaeological Evidence

Early hominins began eating things in a manner that leaves durable traces about 2.5 Mya (Blumenschine and Pobiner, chapter 10; Bunn, chapter 11; Shea, chapter 12). Bones preserve cutmarks that, by their frequencies and locations, tell us something about how hominins processed animals for food. Bone fracture patterns and elements found at archaeological sites can also provide clues to the diets and subsistence practices of early hominins. Despite debates about methods of acquisition, processing, and transport of carcasses, it is clear that by 2.5 Mya, hominins at least occasionally consumed meat, marrow, and perhaps other tissues from a variety of animals (Bunn, chapter 11). We can even begin to put together a species list, or "menu," of animals eaten by early hominins, which included mammals ranging from 10 kg to 2,500 kg (Blumenschine and Pobiner, chapter 10). Larger concentrations of archaeological remains are known, beginning around the Plio-Pleistocene boundary, perhaps suggesting an increased reliance on such foods by some hominins.

The association of Oldowan stone tools with large mammal bones bearing cutmarks and percussion fractures confirm the acquisition and almost certain consumption of animal products—our first tangible evidence of hominins entering the carnivore guild (Blumenschine and Pobiner, chapter 10; Bunn, chapter 11; Shea, chapter 12). Limited microwear polishes found on Oldowan tools hint at their use to process a range of materials (Keeley and Toth, 1981), but further study has been frustrated by a lack of lithics suitable for analysis (Walker, chapter 1). Still, these simple flakes and cores may well have been what Shea (chapter 12) refers to as general-purpose instant technology, providing access to a broad range of potential foods.

Models

Models are useful for studies of early hominin diet both because they provide a source of hypotheses for testing by direct evidence and because they offer logical limits to the constellation of possible subsistence practices of these species. Models can be based on many sources of information, from paleoenvironmental reconstructions, to ethnographic and ethological analogy, to studies of nutrition and energetics.

Paleoenvironmental Reconstructions Paleoenvironmental reconstructions can give us an idea of resources available to the early hominins and foods they would likely have eaten. As Peters (chapter 13) notes, "we need to be able to picture the land systems and vegetation ecostructure to postulate patterns of selection for common and divergent hominin forms of locomotion and food acquisition." Reed and Rector (chapter 14) argue, for example, that ecomorphology and other lines of evidence indicate environmental changes during the Plio-Pleistocene that would have reduced fruit availability compared with earlier times. This implies possible selective pressures to increase leaf consumption. Environmental instability (e.g., see Potts, 1998) may also have made it difficult to maintain a diet of soft-fruit frugivory given fluctuations in availability.

Peters (chapter 13) elaborates on the potential plant food niche, suggesting that environmental reconstructions indicate a hominin demographic sink, with many resources, including fruits, leaves and shoots, seeds and pods, and roots all readily available, depending on rainfall and other variables. He ranks potential foods by nutritional quality and suggests that hominin generalists would have done well on the East African plateau.

Indeed, there are many variables that can affect food availabilities, and as Sept (chapter 15) points out, our models must balance realism and precision. She suggests that agent-based models can help us cut through some of the complexity by focusing on effects of foraging decisions under different habitat types and other conditions. She posits that, for example, that if we consider meat to have been an important fall-back resource for early hominins, there would have been less competition for dwindling plant resources during crunch times.

Ethnographic and Ethological Analogs Studies of living peoples and nonhuman primates can also offer analogies for early hominin diets. Ethnographic evidence shows that meat, tubers, and other resources have been important food sources for many recent foraging peoples (Wrangham, chapter 16; Cordain, chapter 19; O'Connell, Hawkes, and Jones, 1999), though researchers have yet to reach a consensus on ratios of meat to plant matter consumed (Milton, 2000). Ethnographic studies also demonstrate that cooking is a cultural universal, and so must have its origin in the distant past. As with other tools, fire offers unprecedented access to nutrients in foods that would otherwise be inaccessible to hominins (Lucas, chapter 3; Wrangham, chapter 16). While there remains debate as to when cooking was established, this tool would certainly have been available to hominins by the middle Pleistocene.

Studies of nonhuman primate diets likewise provide potential insights into the diets of our ancestors (Stanford, 2001; Rodman, 2002). As Wrangham (chapter 16), and I (Ungar, chapter 4) point out, for example, African apes today have a penchant for easy to digest, sugar-rich fruits. It is mostly at times of resource stress that gorillas "fall back" on tougher, lower-quality food resources. Early hominins, too, then may have preferred higher-quality food sources such as fruits except at "crunch times." Lambert (chapter 17) develops this idea further distinguishing different types of fallback foods. Her comparisons of chimpanzees and bonobos suggest possible links between biome variability, diet, and tool use by hominins.

Nutrient Composition and Energetics Models Studies of nutrient composition give us ideas of food quality given availability. Tubers offer little in the way of nutritional value, whereas other resources, such as fruits, honey, nuts, and animal products available throughout the year would provide energy and protein (Schoeninger, chapter 9; Bunn, chapter 11). Such studies allow us to establish a list of potential foods by quality (Peters, chapter 13).

Some researchers have focused on energy as a currency for measuring fitness (Aiello and Wells, 2002; Leonard et al., chapter 18). Their models consider locomotor efficiency, energetic requirements of enlarged brains and larger bodies, and other factors to infer minimal nutritional need of hominins, with their implications for dietary requirements.

Second-Order Questions

One thing that should be obvious from this discussion is that we really do not know much at all about early hominin diets. As Peters (chapter 13) noted, "the fact is we remain, at this stage in the development of our science, largely ignorant of what life was like in the past." This begs two questions, both of which were raised and considered in the chapters of this volume: (1) What are the limits to our knowledge of early hominin diets? and (2) Is it worth the effort to learn more? While the first question is difficult to answer, the second is easier to address.

Why Study the Evolution of Hominin Diet?

Some have argued that science operates according to the law of diminishing returns. The more we learn, it is argued, the closer we come to the limits of knowledge and the more expensive and difficult it becomes to make new discoveries (Horgan, 1996). We may then ask whether it makes sense to continue to focus so much effort on reconstructing early hominin diets. The answer is, in my opinion, "absolutely"! First, we *have* made progress over the past quarter century (Walker, chapter 1), and most authors in this volume make compelling arguments that there is much left to learn. Moreover, our motivations to learn more are both academic and practical.

Diet is "the single most important parameter underlying behavioral and ecological differences among living primates" (Fleagle, 1999). The same was certainly true of our early hominin forbearers. From an academic perspective then, inferences of diet promise better insights into hominin paleoecology and the evolution of our lineage.

More practical reasons for understanding hominin diets are articulated by Eaton (chapter 20) and Cordain (chapter 19). These and other proponents of evolutionary medicine suggest that many of the chronic degenerative diseases facing our society result from a discordance between what we eat today and what our ancestors have evolved to eat. In a sense, nature may be selecting against those of us ill-adapted to our diets. The choices are evidently to change what we eat, or to face hypertension, myocardial infarction, diabetes mellitus, various cancers, and other diseases of affluence.

This "Paleolithic Prescription" has received a great deal of attention in the media and has led to many popular diet books (e.g., Eaton, Shostak, and Konner, 1989; Audette and Gilchrist, 1996; Somer, 2001; Cordain, 2002; among others). This approach has been criticized on both practical and theoretical grounds. First, as is clear from the chapters in this volume, we have little knowledge of specific nutrient contents of our hominin ancestors. Second, given the wide range of environments and broad spectrum of resources consumed by past peoples, it is clear that there was no single ancestral paleodiet to which we are adapted (Ungar and Teaford, 2002). Even if there were, because selection is so dependent on niche, we cannot expect to use early hominin models to infer an optimal or ideal diet for us.

Nevertheless, there is practical value in drawing attention to early hominin diets. Some vegans and raw-food enthusiasts claim that eating animal products and cooking are somehow "unnatural" (M. Plavcan, personal communication). Modified animal bones from the late Pliocene and evidence of fire from the Middle Pleistocene show that these claims are simply untrue (Lucas, chapter 3; Blumenschine and Pobiner, chapter 10; Bunn, chapter 11; Wrangham, chapter 16). Further, while we do not know much about the menus of our Plio-Pleistocene forbearers, we can be pretty certain that they did not include the "ubiquitous processed foods" that have come to dominate the typical American diet (Cordain, chapter 19). We *are* eating foods other than those our ancestors evolved to eat. An evolutionary medicine paradigm reminds us of this and points out that disease prevention can be more cost-effective than treatment (Eaton, chapter 20).

Limits to Our Knowledge of Early Hominin Diets

Gomory (1995) wrote, "we grow up thinking more is known than actually is." The most popular notion of our state of knowledge concerning early hominin diets is succinctly summarized in *Robert Atkins' New Diet Revolution*. Atkins writes in this 10 million plus bestseller, "the human animal was able, for millions of years, to remain strong and healthy in conditions of often savage deprivation by eating the fish and animals that scampered and swam around him, and the fruits and vegetables and berries that grew nearby" (Atkins, 2002). Such statements clearly go way beyond the limits of our knowledge. As Peters (chapter 13) notes, "limitations in our paleoanthropological knowledge sometimes invite overly rich interpretations and speculations uninformed in ways that are critical."

These limitations come in two forms: temporary and intrinsic (Hut et al., 1998). Temporary limits are often practical in nature and may be breachable. These present challenges at the frontier of science. Intrinsic limits, however, are permanent and intractable.

Practical Limits to Knowledge of Early Hominin Diets

Many practical, but perhaps not intractable limits to our knowledge of early hominin diets are discussed in the chapters of this volume. These limits are related to both our data, and to our methods for interpreting these data.

Data-Related Limits The meagerness of our fossil record imposes a significant barrier to our knowledge of early hominin diets. Indeed, limited data are a classic liability in historical sciences in general (Hut et al., 1998). Small samples can make it difficult to identify subtle but real differences between species. How can we hope, for example, to compare ingestive behaviors in species of early *Homo* when the combined published sample of *H. habilis, H. rudolfensis,* and *H. erectus* includes only five maxillary central incisors (Ungar et al., 2006)? Continued fieldwork may to help remedy this problem though, as might new approaches to increasing the useable fraction of those specimens we already have access to (Ungar and M'Kirera, 2003).

Taphonomic alteration of specimens further limits our samples, particularly for studies of epigenetic phenomena, such as dental microwear and stable isotope ratios. Surface etching and erosion typically reduce dental microwear samples from Plio-Pleistocene fossil sites by 75% or more (Teaford, 1988, chapter 7; El Zaatari et al., 2006; Ungar et al., 2006), and diagenetic effects of fossilization can affect both trace element composition and stable isotope ratios (Schoeninger, chapter 9). Still, such effects may be understood and at least in part controlled for through taphonomic studies. It may even be possible, in the spirit of turning lemons into lemonade, for us to use diagenetic "noise" to reconstruct aspects of a site's ecological history!

Even if we can assemble large numbers of specimens, however, it may be difficult to divide them into meaningful samples for statistical analyses. The degree of speciosity one is willing to accept, for example, determines both the sizes and compositions of our groups. As Henry and Wood (chapter 2) note, "even if researchers can use several lines of evidence to recover reliable information about the diet of a single fossil hominin specimen, that specimen and therefore those data, still have to be attributed to a taxon in order to enable those data to inform us about the paleobiology of early hominins."

Systematics issues are even more problematic, as increasing numbers of researchers are becoming pessimistic about our abilities to ascribe ancestor-descendant relationships to those hominin taxa we have identified. Without the ability to assign species to lineages, it will be very difficult to reconstruct the evolution of hominin diet.

The ultimate sample-related limitation rests with our efforts to infer dietary behaviors that leave no traces and therefore no samples for us to analyze. As Blumenschine and Pobiner (chapter 10) note, for example, "assigning aspects of hominin carnivory to 'the known, the unknown, and the unknowable' depends on how well these can be linked to the production of dietary trace fossils."

Methodological Limits Most authors use the present to interpret the past. Uniformitarian assumptions limit our retrojections to the way things work today (Leroi et al., 1994; Doughty, 1996). As Walker (chapter 1) writes, "paleontologists and historians both know of the danger of the pull of the present (presentism for historians) whereby the past can be distorted by our knowledge and immersion in present day circumstances." In the worst-case scenario, this would lead us to reconstruct "a world of the past that is an exact replica of the present" (Kelley, 1993). It may well be, however, that as L. P. Hartley wrote in *The Go-Betweens*, "the past is a foreign country; they do things differently there." This is a problem because, without the comparative method, paleobiological interpretation may be left in "undecipherable chaos" (Martin, 1991). The best approach may be to focus on underlying processes rather than outcomes (Teaford et al., 1993). It then becomes possible to use the comparative method to document unique conditions in the past (e.g., Reed and Rector, chapter 14) and to recognize evolutionary novelties, such as grazing giraffes (Solounias et al., 1988), or hard-object feeding apes (Ungar, 1996).

We can certainly benefit in such cases from a better understanding of general relationships between form and function today. Lucas (chapter 3) writes, for example, that "little is known about dental-dietary relationships in mammals, but I contend that diet can be 'read' from the dentition of many of them if enough theory is developed to understand the mechanical interactions involved in dental function." Teaford (chapter 7) adds "clearly, we need a better understanding of primate feeding and the properties of primate foods." In the same vein, Daegling and Grine (chapter 6) opine with reference to mandibular biomechanics that "we are limited more by our lack of understanding of the relationship between mechanical stress environment and bone form than by the paucity of the fossil record."

Theoretical or Intrinsic Limits to Our Knowledge of Hominin Diets

Although new data and better theories will probably increase our knowledge about early hominin diets, we cannot expect to ever know all the details. Kant (1783) wrote "every answer given on principle of experience begets a fresh question, which likewise requires its answer." Kant's principle of question propagation is clearly seen in studies of hominin diet. Teaford (chapter 7) notes that our studies "often raise more questions then they answer, largely because they give new and different glimpses of the intricacies of previous behavior." According to Rescher (1999), we will never know everything not because we cannot answer specific questions but because we will not run out of questions to answer.

Another Sisyphusian frustration accompanying increasing knowledge is the perception that things become ever more complex as we learn more about them. As Rescher (1999) notes, "in a complex world, the natural dynamics of rational inquiry will inevitably exhibit a tropism toward increasing complexity." The more we learn, the more it seems there is to know. Every time we reach the crest of a hill, we find a higher peak to climb off in the distance. This presents a considerable challenge because, just as our ability to predict an event depends on the number of factors leading to it (Gomory, 1995), several chapters in this volume suggest that our ability to infer diet from morphology, microwear, stable isotope ratios, and other clues depends on the number of factors that affect these.

What then are the ultimate limits to our knowledge? Comprehensive scientism argues that science has no boundaries and that we can ultimately answer all questions (see Stenmark, 2001). This view sits in stark contrast to postmodern irrationalism, to the insistence of Popper that nothing can be proved, and to the claims of Kuhn that science draws us no closer to truth. Both sets of views, it seems to me, are unreasonable and counterproductive. On the one hand, it is overly optimistic to expect accurate micronutrient composition estimates for a typical *Homo erectus*. On the other hand, to say that gross categorizations like frugivore are "as good as it gets" is overly pessimistic.

In the end, we probably cannot know the ultimate limits to our knowledge, at least not until science can reflexively predict its own future performance. As Rescher (1999) wrote, "we cannot discern the substance of future science with sufficient clarity to say just what it can and cannot accomplish." We can, however, identify the current boundaries to our knowledge, and in doing so, begin to nudge them.

Directions for the Future

While we cannot know what the future will bring, progress on early hominin diet research shows no abatement. If anything, as new fossils and new technologies to study them become available, advancements seem to be accelerating. As Sponheimer and coauthors (chapter 8) aptly write, "there can be little doubt that some 50 years hence, many will look back at what we 'know' today and shake their heads with wry amusement." Advances will likely come with better data questions, and better ways of addressing those questions.

Better Questions

Theory

The questions we ask depend on the body of theory, or principles of inquiry, on which they are based. Theory guides our questions and gives meaning to our data. According to Jacob (1994), it "allows us to imagine and explore a wider range of worlds than ours, giving new perceptions and questions about how our world came to be as it is." Without evolutionary theory, for example, biology becomes basic descriptive natural history.

Peters (chapter 13) identifies five roles for theory in paleoanthropology: (1) "clarifying thinking about method; (2) motivating alternative working hypotheses; (3) introducing conceptualizations and model building relevant to the state of development of a discipline; (4) creating new interpretations while evaluating old ones; (5) helping to push back the boundaries of the unknown and the apparently unknowable." While theory in paleoanthropology, according to Peters, is still "relatively unrefined and imprecise," some basic concepts, such as fundamental niche and ecosystem, and the interplay between hominin biology, use of technology and environmental dynamics, are being developed.

This development is an iterative process. As we learn more about early hominin diets, we can create more mature and sophisticated theories on all levels. These theories, in turn, would inevitably lead to new questions, new data, and new theoretical perspectives. Theory building guides the accumulation and interpretation of facts to create knowledge and so will be an important key to the future of studies on the evolution of hominin diets.

The Role of Modeling

Several of the chapters in this book also tout modeling as a route to asking more informed questions of our data. Sept (chapter 15) writes that models "can provide a critical framework for evaluating evidence and refining hypotheses about diet." Her metaphor of the modeler as puppeteer is apt; we control variables and assess outcomes, much like pulling the strings of a puppet. These offer a virtual world that allows us to generate ideas about how complex systems might work. The greatest challenge to this virtual world is establishing a balance between precision, realism, and generality (Peters, chapter 13; Sept, chapter 15). As Leonard and coauthors (chapter 18) acknowledge, this can lead to "false quantification" and drawing inferences with unreasonably high levels of precision. They go on to note, however, that assigning numbers to variables in a model can still be valuable. This can allow us to explore how variables might interact with one another to produce specific outcomes. Cast in this light, the precise nutrient composition and energy expenditure estimates reported in this volume (Leonard, chapter 18; Cordain, chapter 19; Eaton, chapter 20) can be of great heuristic value.

Better Data

Perhaps the most progress in studies of early hominin diet will come as we improve our approaches to data collection and analysis. According to Marchetti (1998), knowability is limited by the resolution with which we sense the world. Several examples of how new technologies can lead us in new directions are presented in this volume. Teaford (chapter 5) suggests that noninvasive imaging techniques should allow us to map distribution of enamel prisms, and perhaps even crystallites, to better understand relationships between dental structure and function. Further, Sponheimer and coauthors (chapter 8) note that laser ablation should soon give us sufficient microsamples to look at changing isotope ratios during dental development, and new advances may even allow us to tap enamelin and osteocalcin for traces of nitrogen to assess its stable isotope ratios. Finally, Walker (chapter 1) portends that advances in comparative genomics will soon lead us to explore the genetics of digestion to make inferences about the evolution of diet in our lineage.

New approaches often lead to a feedback loop between research and technological innovation. Improved analytical tools for microwear, for example, have led to flurries of activity, new insights, and the need for even better tools (Ungar et al., 2003). Most recently, as Walker (chapter 1) and Teaford (chapter 7) both point out, the application of confocal microscopy and scale-sensitive fractal analysis to the interpretation of dental microwear offers a precision that for the first time allows us to address within species variation (Scott et al., 2005).

Consilience

One additional approach to improving our understanding of hominin diets has come up repeatedly in the chapters of this volume. Peters (chapter 13) argues that our theories may best be developed and evaluated using Whewell's principle of consilience. Whewell (1858) wrote, "the evidence in favour of our induction is of a much higher and more forcible character when it enables us to explain and determine cases of a kind different from those which were contemplated in the formation of our hypothesis. The instances in which this have occurred, indeed, impress us with a conviction that the truth of our hypothesis is certain." Hypotheses can be shown to be robust by colligating disparate lines of evidence. The various approaches described in this volume form a whole greater than the sum of its parts. The inferences drawn from the different methods can be used to evaluate and inform one another (e.g., Reed and Rector, chapter 14; Sponheimer et al., chapter 8). For example, microwear, isotope, and morphology analyses all sample diet over different time intervals, and together, may help us better understand the relative roles of preferred foods and fallback resources in early hominin paleoecology. As Teaford (chapter 7) writes, "only by considering the total range of evidence can we appreciate what is known, unknown, and unknowable about this complex topic."

Acknowledgments I thank Matt Cartmill for his helpful and insightful comments and criticisms—he is very good at seeing the forest from the trees. I also thank Mike Plavcan and Rob Scott for their willingness to engage in inane philosophical chatter.

References

Aiello, L.C., and Wells, J.C.K., 2002. Energetics and the evolution of the genus *Homo*. *Annu. Rev. Anthropol.* 31, 323–338.
Atkins, R. C., 2002. *Dr. Atkins' New Diet Revolution*. HarperCollins, New York.
Audette, R., and Gilchrist, T., 1996. *Neanderthin: A Caveman's Guide to Nutrition*. Paleolithic Press, Dallas, TX.
Cordain, L., 2002. *The Paleo Diet: Lose Weight and Get Healthy by Eating the Food You Were Designed to Eat*. Wiley, New York.
Doughty, P., 1996. Statistical analysis of natural experiments in evolutionary biology: Comments on recent criticisms of the use of comparative methods to study adaptation. *Am. Nat.* 148, 943–956.
Eaton, S. B., Shostak, M., and Konner, M., 1989. *The Paleolithic Prescription: A Program of Diet and Exercise and a Design for Living*. HarperCollins, New York.
El Zaatari, S., Grine, F.E., Teaford, M.F., and Smith, H.F., 2006. Molar microwear and dietary reconstruction of fossil Cercopithecoidea from the Plio-Pleistocene deposits of South Africa. *J. Hum. Evol.* in press.
Fleagle, J. G., 1999. *Primate Adaptations and Evolution*. 2nd ed. Academic Press, New York.
Gomory, R.E., 1995. The known, the unknown and the unknowable. *Sci. Am.* 272, 120.
Grine, F.E., 1986. Dental evidence for dietary differences in *Australopithecus* and *Paranthropus*: A quantitative analysis of permanent molar microwear. J. Hum. Evol. 15, 783–822.
Horgan, J., 1996. *The End of Science*. Addison-Wesley, Reading, MA.

Hut, P., Ruelle, D., and Traub, J., 1998. Varieties of limits to scientific knowledge. *Complexity* 3, 33–38.

Jacob, F., 1994. *The Possible and the Actual.* University of Washington Press, Seattle.

Kant, I., 1783. Prolegomena to any future metaphysic, sect. 57. *Akad.* 352.

Keeley, L.H., and Toth, N., 1981. Microwear Polishes on Early Stone Tools from Koobi Fora, Kenya. *Nature* 293, 464–465.

Kelley, J., 1993. Taxonomic implications of sexual dimorphism in *Lufengpithecus.* In: Kimbel, W. H., and Martin, L.B. (Eds.), *Species, Species Concepts, and Primate Evolution.* Plenum Press, New York, pp. 429–458.

Leroi, A.M., Rose, M.R., and Lauder, G.V., 1994. What does the comparative method reveal about adaptation? *Am. Nat.* 143, 381–402.

Marchetti, C., 1998. Notes on the limits to knowledge explored with Darwinian Logic. *Complexity* 3, 22–35.

Martin, L., 1991. Paleoanthropology: Teeth, sex and species. *Nature* 352, 111–112.

Milton, K., 2000. Reply to plant-animal subsistence ratios and macronutrient energy estimations in worldwide hunter-gather diets. *Am. J. Clin. Nutr.* 72, 1590–1592.

O'Connell, J.F., Hawkes, K., and Jones, N.G.B., 1999. Grandmothering and the evolution of *Homo erectus. J. Hum. Evol.* 36, 461–485.

Potts, R., 1998. Variability selection in hominid evolution. *Evol. Anthropol.* 7, 81–96.

Rescher, N., 1999. *The Limits of Science.* University of Pittsburgh Press, Pittsburgh, PA.

Robinson, J.T., 1954. Prehominid dentition and hominid evolution. *Evolution* 8, 324–334.

Rodman, P. S., 2002. Plants of the apes: Is there a hominoid model for the origins of the hominid diet? In: Ungar, P. S., and Teaford, M.F. (Eds.), *Human Diet: Its Origin and Evolution.* Bergin & Garvey, Westport, CT, pp. 77–109.

Rosenberg, A., 2001. *Philosophy of Science: A Contemporary Introduction.* Routledge, London.

Scott, R.S., Ungar, P.S., Bergstrom, T.S., Brown, C.A., Grine, F.E., Teaford, M.F., and Walker, A., 2005. Dental microwear texture analysis shows within-species diet variability in fossil hominins. *Nature* 436, 693–695.

Solounias, N., Teaford, M., and Walker, A., 1988. Interpreting the diet of extinct ruminants: The case of a non-browsing giraffid. Paleobiology 14, 287–300.

Somer, E., 2001. *The Origin Diet: How Eating Like Our Stone Age Ancestors Will Maximize Your Health.* Holt, New York.

Stanford, C. B., 2001. A comparison of social meat-foraging by chimpanzees and human foragers. In: Stanford, C. B., and Bunn, H. T. (Eds.), *Meat-Eating and Human Evolution.* Oxford University Press, New York, pp. 122–140.

Stenmark, M., 2001. *Scientism: Science, Ethics and Religion.* Ashgate, Aldershot, Hampshire.

Teaford, M.F., 1988. Scanning electron microscope diagnosis of wear patterns versus artifacts on fossil teeth. *Scanning Microsc.* 2, 1167–1175.

Teaford, M.F., and Ungar, P.S., 2000. Diet and the evolution of the earliest human ancestors. *Proc. Natl. Acad. Sci.* 97, 13506–13511.

Teaford, M.F., Walker, A., and Mugaisi, G.S., 1993. Species discrimination in *Proconsul* from Rusinga and Mfangano Islands, Kenya. In: Kimbel, W.H., and Martin, L.B. (Eds.), *Species, Species Concepts, and Primate Evolution.* Plenum Press, New York, pp. 373–392.

Ungar, P.S., 1996. Dental microwear of European Miocene catarrhines: Evidence for diets and tooth use. *J. Hum. Evol.* 31, 335–366.

Ungar, P.S., 2004. Dental topography and diets of *Australopithecus afarensis* and early *Homo. J. Hum. Evol.* 46, 605–622.

Ungar, P.S., 2006. Dental topography and human evolution: With comments on the diets of *Australopithecus africanus* and *Paranthropus robustus.* In: Bailey, S., and Hublin, J.J. (Eds.), *Dental Perspectives on Human Evolution: State of the Art Research in Dental Anthropology.* Springer-Verlag, New York, in press.

Ungar, P.S., Brown, C.A., Bergstrom, T.S., and Walkers, A., 2003. Quantification of dental microwear by tandem scanning confocal microscopy and scale-sensitive fractal analyses. *Scanning* 25, 185–193.

Ungar, P.S., Grine, F.E., and Teaford, M.F., 2006. Diet in early *Homo*: A review of the evidence and a new model of adaptive versatility. *Annu. Rev. Anthropol.*, in press.

Ungar, P.S., and M'Kirera, F., 2003. A solution to the worn tooth conundrum in primate functional anatomy. *Proc. Natl. Acad. Sci.* 100, 3874–3877.

Ungar, P.S., Teaford, M.F., 2002. Perspectives on the evolution of human diet. In: Ungar, P.S., and Teaford, M.F. (Eds.), *Human Diet: Its Origin and Evolution*. Bergin & Garvey, Westport, CT, pp. 1–6.

Whewell, W., 1858. *Novum Organon Renovatum*. J. W. Parker, London.

Index